Springer Climate

Springer Climate is an interdisciplinary book series dedicated to climate research. This includes climatology, climate change impacts, climate change management, climate change policy, regional climate studies, climate monitoring and modeling, palaeoclimatology etc. The series publishes high quality research for scientists, researchers, students and policy makers. An author/editor questionnaire, instructions for authors and a book proposal form can be obtained from the Publishing Editor.

Now indexed in Scopus® !

Christopher Ndehedehe

Hydro-Climatic Extremes in the Anthropocene

Springer

Christopher Ndehedehe
School of Environment and Science
Griffith University
Brisbane, QLD, Australia

ISSN 2352-0698 ISSN 2352-0701 (electronic)
Springer Climate
ISBN 978-3-031-37726-6 ISBN 978-3-031-37727-3 (eBook)
https://doi.org/10.1007/978-3-031-37727-3

This Springer imprint is published by the registered company Springer Nature Switzerland AG
The registered company address is: Gewerbestrasse 11, 6330 Cham, Switzerland

Foreword

Water is increasingly becoming the hot topic of the Anthropocene—water resources are under pressure due to expanding population growth as well as human-induced global climate change. Humans have changed the entire global water cycle. This book brings together solid background information on the Anthropocene as an epoch and why understanding how humans impact the natural environment on a global scale is important. The book then focuses on how humans in the Anthropocene have changed the global climate and thereby impacted the hydro-climate, changing the frequency and duration of floods and droughts. An important aspect of understanding these hydro-climatic changes resides in the human need for water security. How do we monitor, understand, and plan for future changes in water resources? The book explores some of the cutting-edge methods for monitoring (remote sensing) and analyzing (machine learning, deep belief networks, and neural networks) hydro-climatic data at the global scale. The book finishes by exploring the impacts of the Anthropocene, through changes in hydro-climatic variability, on extreme weather events, such as droughts, floods, and wildfires.

The increasing use of remote sensing and satellite imagery to capture data at large spatial scales has rapidly increased our understanding of hydrological processes across a range of systems. This rapid increase in data also presents challenges; the volume and complexity of the data can challenge our analytical capabilities. This book not only provides a detailed background context of the Anthropocene and hydro-climatic variability, but it also uses several case studies to demonstrate how different machine learning analytical techniques can be used. This makes this book a wonderful resource for anyone wishing to understand and further explore hydrological processes in a range of systems.

Dr. Christopher Ndehedehe is a Senior Lecturer in the School of Environment and Science at Griffith University where he leads the GIS and Remote Sensing teaching. Dr. Ndehedehe is also a research member of the Australian Rivers Institute, and a recipient of an Australian Research Council (ARC) Discovery Early Career Researcher Award (DECRA) for a project "Assessing the impacts of droughts and water extraction on groundwater". Despite being an early career researcher, Dr. Ndehedehe is a global leader in hydrological remote sensing and environmental geoscience with more than 70 peer-reviewed articles in leading high-impact journals. His recent research has included the development of new geospatial tools to support the assessment of climate change impacts on surface water and groundwater as well as explore the complex analytical approaches required to understand such large datasets. The book provides all the tools required to successfully apply remote sensing in terrestrial hydrological assessments and should be essential reading for anyone with an interest in this subject.

April 2023

Prof. Fran Sheldon
Head of School, School
of Environment and Science
Griffith University
Brisbane, QLD, Australia

Preface

Prior to receiving the prestigious Australian Research Council Fellowship for the Discovery Early Career Researcher Award in 2022, I conducted preliminary assessments of extreme events, such as droughts, and their impacts on hydrological systems, including groundwater. Through leading research and collaborating with experts in multidisciplinary geoscience, including climate science, machine learning, and remote sensing, I developed methods and frameworks to improve understanding of hydro-climatic extremes in the Anthropocene. A key motivation for this came from my strong desire and curiosity to understand how groundwater aquifers and global terrestrial hydrology are affected by extreme climatic conditions and water extraction. Due to the complexities in hydrological systems, some of the important outcomes from my research collaborations included discussion around the need to employ machine learning techniques to help track the footprints of climate modes and large-scale climate variability indices (e.g., ENSO, PDO) on terrestrial stored water. Why is this important in the Anthropocene and how does it help with mitigation and planning? This book provides more context on this, including addressing other important questions in the Anthropocene dispensation. These include the following: Is climate change real? Are we running out of freshwater? What if the rains do not come? How can we maintain balance and stability in the Earth system in an era of unprecedented human-driven changes that coalesce with climate change to amplify the instability of the Earth system in the Anthropocene? How can operational monitoring through increased collaborative investments in satellite programmes facilitate freshwater reporting on local and water management scales? The book provides contemporary insight into changes in the Earth system in the Anthropocene and how human civilization and rising population have played a crucial role. It further elaborates on efficient monitoring that relies on remote sensing data and innovative geospatial tools that underpin sustainable management of water resources.

Collaborative investments in satellite radar programmes are increasing because of the limitations of passive remote sensing systems. These radar satellites are now emerging as excellent substitutes and in other cases are very complementary to passive remote sensing systems. The first results from the recently launched satellite radar mission, the Surface Water Ocean Topography are discussed to illustrate the significant advancement of the mission to provide a fundamental understanding of surface water hydrology. This book also discusses the uniqueness of the NISAR (an acronym for NASA-ISRO Synthetic Aperture Radar) radar mission, which will help consolidate the gains of ongoing satellite radar in tracking environmental changes in the Anthropocene. The novel NISAR mission is discussed in detail, including its objectives, scope, and broad capabilities in environmental assessment. Among the capabilities discussed, the potential of its combination with satellite gravity missions like the Gravity Recovery and Climate Experiment to improve the assessment of groundwater status is stressed. Given the multidisciplinary capabilities and applications of the NISAR mission detailed in this book, it is expected to be a blessing to the Anthropocene. While policy dimensions and benefits of various satellite hydrology missions like Gravity Recovery and Climate Experiment (GRACE) are also discussed, there is a quick mention of the planned European Space Agency (ESA) satellite hydrology mission, called Mass change and Geosciences International Constellation (MAGIC). MAGIC is expected to build on the capability of GRACE and other gravity missions in tracking the composite influence of climate change and human water management on freshwater resources. MAGIC will be the latest of new generation gravity missions as well as ESA's major investment in satellite geodetic missions that are expected to underpin our understanding of Earth's geophysical processes including the water cycles.

The Anthropocene is a new era characterized by rapid changes in the environment driven largely by human activities and global climate change. Unlike the relative stability of the Earth system under Holocene conditions, in the Anthropocene, human activities (e.g., dam constructions, reservoirs, groundwater extraction, water transfer, and harvesting) are now the predominant drivers of these changes, surpassing those caused by natural processes and climate change. An integrated data-driven or assimilated deep machine learning approach therefore promises improved predictions and efficiency to reconstruct fluxes and groundwater in non-linear systems to aid the tracking of both human and climatic influence on hydrology in the Anthropocene. Novel machine learning frameworks (e.g., Convolution Neural Networks, Support Vector Machine, and Deep Belief Network) demonstrated here used climate modes and other satellite-based information as inputs to predict groundwater level fluctuations. The robustness and capabilities of the deep learning frameworks in helping to better understand groundwater dynamics is thus critical to enhance our ability to protect groundwater-dependent ecosystems and avoid consequences resulting from permanent loss of water supply. This book demonstrates these capabilities, showcasing the potential of these novel machine learning techniques as new tools for the Anthropocene.

In addition to discussing the benefits of these machine learning experiments, this book provides valuable insights into hydrological extremes and the Earth system in the Anthropocene. It identifies factors contributing to alterations of the Earth system, including the upsurge in man-made lakes and reservoirs, rise in sea level, rapidly increasing human population, land use and land cover change, groundwater depletion, and deforestation. Case studies, such as the Lake Chad predicament and the anthropogenic recklessness in Lake Urmia, emphasize the need for satellite-based monitoring frameworks and strong institutions to ensure natural resources are protected and sustainably managed. The book further identifies threats to water security in the twenty-first century. These threats are expected to rise, emanating mostly from population and economic growth, climate change, and poor management of water resources.

The need to pursue water security and build resilience is challenging because (i) water demand can vary in time and space due to the varying degree of consumption for domestic and industrial purposes, and (ii) investment in water infrastructures creates multiple dimensions of impacts that lead to loss of ecosystem services and disruption of freshwater ecosystem functioning, among others. Even though some regions are able to cope with water service delivery and functions during periods of disruptions caused by extreme events (e.g., floods and droughts), climate change will continue to pose a danger to water availability in the Anthropocene. This book therefore detailed a methodology to help assess drought characteristics and impacts in the Anthropocene, looking into the future by leveraging a suite of multi-scaled satellite earth observations and global climate models. These models incorporate the Shared Socioeconomic Pathways that assume considerable socio-economic changes. In assessing hydro-climatic extremes using these data, some important questions and hypotheses were addressed in this book. These include (i) can global climate models be trusted? (ii) will groundwater be affected by future changes in drought characteristics (e.g., duration, intensity, and extent)? and (iii) will global land areas under drought increase toward the end of the twenty-first century? The book therefore fills a critical gap, making it an important contribution to climate science and multidisciplinary aspect of planetary science. The book will set the stage for the next generation to provide a better understanding of the current challenges and issues in the Anthropocene and how to address them, leveraging the broad range of machine learning tools and emerging radar observations.

Brisbane, QLD, Australia
April 2023

Christopher Ndehedehe

Acknowledgements

Dr. Christopher Ndehedehe is grateful to the Australian Research Council (ARC) for his Discovery Early Career Researcher Award (DECRA) Fellowship. He is grateful to the ARC for funding the DECRA project (DE230101327), *Assessing the Impacts of Droughts and Water Extraction on Groundwater*. This book consists of preliminary machine learning experiments and testing on terrestrial hydrology, including groundwater conceptualized by Dr. Ndehedehe following the award of his ARC DECRA Fellowship in September 2022. These experiments as well as principles, and conceptual framework advocated in this book were inspired by this DECRA project. This project will enable unprecedented ways of harnessing big data from satellites to assess groundwater status, improving the capability of water resource management agencies to monitor water availability and sustainably manage these critical resources. Additionally, the DECRA project will further operationalize the use of novel satellite data and some of the machine learning technologies discussed in this book for environmental monitoring that could readily be incorporated into the workflows of existing online water monitoring platforms, generating publicly accessible groundwater information and contributing to expanding global markets for Australian satellite-based monitoring. He is thankful to Griffith University's Australian Rivers Institute where this project is hosted and for providing a conducive and sophisticated first-class strategic research environment in freshwater science.

Contents

Part I The Anthropocene

1 Understanding the Anthropocene 3
1.1 Introductory Remarks 3
1.2 The Anthropocene 5
1.2.1 What Is the Anthropocene? 5
1.2.2 The Earth System in the Anthropocene 6
1.3 The Twenty-First Century Environment 7
1.3.1 Man-Made Lakes and Reservoirs 7
1.3.2 Climate-Induced Changes in Terrestrial Stored Water and Contributions to Sea-Level Rise 10
1.3.3 Rise in Human Population 10
1.3.4 Land Use and Land Cover Change 12
1.3.5 Groundwater Depletion 13
1.3.6 Deforestation 14
1.3.7 An Altered Global Water Cycle: What Are the Consequences? 16
1.4 About This Book 18
References 19

Part II Hydro-Climatic Extremes

2 Hydro-Climatic Extremes: Climate Change and Human Influence 25
2.1 Introductory Remarks 25
2.2 Hydrological Extremes in the Twenty-First Century 26
2.2.1 Will Climate Change Affect Groundwater Recharge in the Future? 27
2.3 Hydrological Systems: Human Actions Versus Climate Change 30
2.3.1 Hydrological Processes in Semi-arid Domains 31
2.3.2 Droughts in the Murray Darling Basin 31

2.3.3 The Sahelian Drought ... 34
2.3.4 Lessons from the Lake Chad Predicament ... 35
2.3.5 A Tale of Two Extreme and Prolonged Drought Events ... 35
2.3.6 Lake Urmia: A Case of Anthropogenic Recklessness ... 36
2.3.7 Socio-economic Implications of Hydro-Climatic Extremes ... 37
2.4 The Earth System ... 44
2.4.1 Socio-hydrological Systems ... 44
2.4.2 Understanding Drought and Flood in the Anthropocene ... 45
2.5 Operational Monitoring ... 46
2.5.1 Earth Resources Satellites ... 46
2.5.2 Hydrological and Climate Models ... 48
References ... 51

Part III Water Resources

3 How Much Freshwater Is Available? ... 59
3.1 Introductory Remarks ... 59
3.2 Impacts of Climate Change on Freshwater Resources ... 61
3.3 Are We Running Out of Freshwater? ... 62
3.3.1 The Water Cycle ... 62
3.3.2 Are There Threats to Freshwater Availability? ... 63
3.3.3 Water Scarcity, Security, and Crises ... 67
3.3.4 Water Demand and Climate ... 69
3.4 Security, Sustainability, and Resilience ... 71
3.4.1 Pursuing Water Security and Sustainability ... 72
3.4.2 Building Resilience ... 74
3.5 Optimizing Monitoring Framework ... 75
3.5.1 Challenges with Observational Networks for Hydrology ... 75
3.5.2 Existing Gauge Observations ... 76
References ... 77

4 Satellite Hydrology Programmes: Capabilities and Benefits ... 81
4.1 Introductory Remarks ... 81
4.2 Multi-mission Satellite Hydrology Systems ... 82
4.2.1 Satellite Radar Systems ... 85
4.2.2 Satellite Geodesy Techniques ... 88
4.3 Quantifying Land Water Storage ... 90
4.3.1 Terrestrial Water Storage ... 92
4.3.2 Satellite Observations of Changes in Groundwater ... 94
4.3.3 Surface Water Hydrology ... 98
4.3.4 Reconstruction from Models and Observations ... 105

4.3.5 Reconstruction with Machine Learning Models 106
4.3.6 Downscaling Using Machine Learning Models 108
4.4 Benefits of Satellite Hydrology Systems 110
4.4.1 Tracking Climate Change 110
4.4.2 Hydrological Processes 115
4.4.3 Understanding Anthropogenic Influence 118
4.4.4 Filling Data Gaps to Improve Environmental Monitoring 120
4.4.5 Management and Policy Frameworks 124
References 126

5 Can the Wells Run Dry? 135
5.1 Introductory Remarks 135
5.2 Groundwater Resources 136
5.2.1 Global Groundwater Status 140
5.2.2 Challenges and Constraints 143
5.3 Groundwater Drought: A New Type of Drought? 146
5.3.1 Measuring Groundwater Drought 149
5.3.2 Impacts of Extraction and Drought 154
5.4 Groundwater Processes 157
5.4.1 Groundwater Recharge 159
5.4.2 Discharge 164
5.5 Sustainable Limits 166
5.5.1 Addressing Scarcity Through Policies 166
5.5.2 Human Population and Risk 167
5.5.3 Adaptation and Resilience to Climate Change 168
References 171

Part IV Environmental Sensing

6 Remote Sensing of the Environment 181
6.1 Introductory Remarks 181
6.2 New Capabilities in Satellite Remote Sensing Systems 182
6.2.1 Optical Satellite Systems 183
6.2.2 Radar Technology 185
6.3 The NISAR Mission: A Blessing to the Anthropocene? 185
6.3.1 NISAR Mission Science Overview and Scope 186
6.3.2 Environmental Applications of NISAR 188
6.4 Multi-satellite Systems to Improve Understanding of Water Storage Components 193
6.4.1 Remote Sensing and Geospatial Methods in Groundwater Assessment 193
6.4.2 Groundwater from GRACE Satellite 195
6.4.3 Satellite Data Fusion for Groundwater Monitoring 196
6.4.4 Integrating SAR with GRACE Follow-on to Assess Groundwater Status: The NISAR Prospect 199

6.4.5 Remote Sensing of Soil Moisture 201
6.4.6 Satellite-Derived Changes in Surface Water Storage 205
6.5 Monitoring to Improve Decision and Policies 207
6.5.1 Freshwater Management 207
6.5.2 Disaster and Risk Management 208
6.5.3 Sustainable Development Goals 213
References 215

Part V Machine Learning in Hydrology

7 Assimilated Deep Learning to Assess Terrestrial Hydrology 223
7.1 Introductory Remarks 223
7.1.1 Climatic Impacts on Freshwater Availability 225
7.1.2 Terrestrial Water Storage 228
7.1.3 Deep Learning in Hydrological Modeling 232
7.2 South America: Hydrology and Climatology 236
7.2.1 Impact of Climate Modes on South American Weather Systems 236
7.2.2 Evidence of Climatic Footprint on South American Terrestrial Hydrology 237
7.3 Data Types and Processing Methods 238
7.3.1 Independent Data Types 239
7.3.2 Dependent Data Types 242
7.3.3 Deep Learning Framework 243
7.3.4 Weight and Bias Initialization 244
7.3.5 Number of Hidden Layers 245
7.3.6 Number of Hidden Neurons 246
7.4 Methodology 247
7.4.1 Statistical Decomposition of Predictors 247
7.4.2 Back Propagation Neural Network 251
7.4.3 Implementation of Assimilated Deep Learning 254
7.4.4 Deseasonalization of the Dependent Data Type 257
7.5 Results and Discussion 258
7.5.1 Impacts of Global Climate Modes on GRACE Hydrological Signals 259
7.5.2 Identifying Dominant Independent Patterns 261
7.5.3 Predicting Total Water Storage 263
7.5.4 Training and Testing Phase 264
7.5.5 Prediction Step and Performance Assessment 267
7.6 Concluding Remarks 269
References 271

8 Deep Belief Network for Groundwater Modeling 279
8.1 Introductory Remarks 279
8.1.1 Understanding Global Change 282
8.1.2 Growing Concerns on the Impacts of Climate Change on Groundwater Dynamics 283
8.1.3 Transboundary Water Resources 284
8.1.4 Where Deep Learning Comes In 284
8.2 Exemplifying Deep Learning in the Great Artesian Basin 287
8.2.1 Hydro-Climatology and Groundwater 287
8.2.2 Groundwater Challenges and Water Use 288
8.3 Hydrogeology of the Subsurface 288
8.3.1 Surface Water, Evapotranspiration, and Precipitation 289
8.3.2 Saturated Groundwater 290
8.4 Deep Learning Datasets 295
8.4.1 Groundwater Level Data 295
8.4.2 Terrestrial Water Storage 295
8.4.3 Precipitation 296
8.4.4 Temperature 296
8.4.5 Climate Indices 296
8.5 Methodology 299
8.5.1 Deep Learning Framework 299
8.6 Deep Belief Network 301
8.6.1 Data Assimilation 305
8.6.2 Quantile Regression: An Uncertainty Evaluation Approach 312
8.7 Results and Discussion 312
8.7.1 Model Performance Evaluation 312
8.7.2 Parameter Selection 313
8.7.3 Implementation of the DBN Model 314
8.7.4 Analysis of the Representative Stations 315
8.7.5 Validation of the Change in Groundwater Level Estimates 316
8.8 Concluding Remarks 318
References 320

9 Integrated Machine Learning in Satellite Hydrology 325
9.1 Introductory Remark 325
9.1.1 Traditional Machine Learning Techniques 326
9.1.2 The Emergence of Integrated Machine Learning Approach 326
9.1.3 Satellite-Derived Terrestrial Water Storage Changes 329
9.1.4 GRACE Data Gaps and Implications 330
9.1.5 Reconstruction of GRACE Satellite Data 331

9.1.6 Some Gap Filling Techniques in Satellite Hydrology ... 333
9.1.7 Singular Spectrum Analysis ... 333
9.1.8 The Water Balance Approach ... 334
9.1.9 Data Assimilation ... 335
9.1.10 Statistical and Data Mining Techniques ... 335
9.2 The Murray–Darling Basin ... 337
9.3 Data and Methodology ... 339
9.3.1 Independent Parameters ... 339
9.3.2 Overview of the Convolution Neural Networks and Support Vector Machines ... 341
9.3.3 Integrating CNN with SVM—a Convolution Based Support Vector Machine Framework ... 343
9.3.4 One Filter One Variable Greedy Update Mechanism ... 345
9.4 Results and Discussions ... 346
9.4.1 Fitting the Proposed Algorithm to the Three Level-2 GRACE Products ... 346
9.4.2 Estimating the Impacts of HF's, CI's, and SSTs in Regulating TWS ... 348
9.4.3 Performance of the Different Kernel Variants Applied in the Proposed System ... 351
9.4.4 Comparison Between the C-SVM and Other Conventional Models in TWS Regressions ... 353
9.5 Concluding Remarks ... 355
References ... 356

10 Regularized Neural Network for Tide Modeling ... 361
10.1 Introductory Remarks ... 361
10.1.1 Machine Learning and Statistical Models for Tide Forecast ... 362
10.1.2 Why Tide Forecast? ... 363
10.2 Data and Methods ... 364
10.3 Results ... 370
10.4 Discussion ... 377
10.5 Concluding Remarks ... 379
References ... 379

Part VI Impacts of Extreme Events

11 Drought Characteristics and Impacts in the Anthropocene ... 385
11.1 Introductory Remarks ... 385
11.1.1 Drought Impact ... 386
11.1.2 Drought Duration ... 387
11.2 Changes in Global Drought Characteristics ... 388
11.2.1 Future Droughts and Hydrological Systems ... 389

11.2.2 Understanding Future Drought Evolutions is Important ... 390
11.3 Framework to Assess Global Changes in Drought Characteristics ... 391
11.3.1 Standardized Precipitation Evapotranspiration Index ... 391
11.3.2 Standardized Precipitation Index ... 392
11.3.3 Other Drought Indicators ... 394
11.3.4 Ensemble Climate Models ... 394
11.3.5 Global Groundwater From Space ... 395
11.3.6 Drought Extent ... 395
11.3.7 Statistical Analyses ... 396
11.4 Understanding Changes in Global Drought Characteristics ... 396
11.4.1 Drought Events: Extents and Duration ... 397
11.4.2 Drought Impacts on Groundwater ... 400
11.4.3 Spatial and Temporal Drought Variability ... 401
11.4.4 Climatic Hotspots of Drought Variability and Impact ... 401
11.5 Implications on Global Hydrology ... 404
11.5.1 Twenty-First Century Global Land Areas Under Drought will Decrease but Their Durations will not ... 404
11.5.2 Can Global Climate Models be Trusted? ... 405
11.5.3 Clime Model Prediction of Impacts on Water Availability ... 406
11.5.4 Will Groundwater be Affected by Future Changes in Drought Characteristics? ... 407
11.5.5 Global Climate Teleconnection Patterns are Important Predictors of Drought Impact ... 408
11.5.6 How can we Maintain Balance, Justice, and Stability? ... 409
References ... 410

12 Floods and Wildfire Events ... 415
12.1 Introductory Remarks ... 415
12.2 Global Flood Events ... 419
12.2.1 Flood Risk and Impacts in the Anthropocene ... 419
12.2.2 Assessing Flood Risk ... 421
12.2.3 Changes in the Magnitude of Flood Events ... 423
12.3 Flood Prediction ... 425
12.3.1 Climate Models ... 425
12.3.2 Statistical and Machine Learning Methods ... 426
12.4 Wildfires ... 427
12.4.1 Long-Term Changes in Global Fire Occurrence ... 430
12.4.2 Improved Techniques and Systems for Enhanced Wildfire Assessment ... 433

12.4.3 Cloud-Computing for Wildfire Mapping 440
12.4.4 Impacts of Wildfire on Freshwater (Groundwater and Surface Water) 441
References 444

Index 449

About the Author

Dr. Christopher Ndehedehe is a Senior Lecturer at the School of Environment & Science, Griffith University, where he concurrently serves as an Australian Research Council (ARC) DECRA (Discovery Early Career Researcher Award) Fellow (2023–2026). He joined Griffith University in June 2018 under its Areas of Strategic Investment as a Research Fellow in Griffiths Australia's Rivers Institute. This was after the completion of an outstanding Ph.D. thesis, which received the Vice Chancellors commendation in early 2018 at Curtin University, Western Australia. Christopher is a leading expert in satellite hydrology and environmental geo-informatics with a strong international reputation and experience. This is exemplified in his pioneering cutting-edge cross-disciplinary research in environmental remote sensing and satellite hydrology that connects climate, freshwater resources, and the environment. Christopher has published widely in several inter-disciplinary areas that cut across machine learning, computational geoscience, natural hazards, geomatic engineering, and earth system science, including a leading and widely downloaded Springer book (https://link.springer.com/book/10.1007/978-3-030-99577-5) on satellite hydrology, and more than 80 peer-reviewed high-impact factor (IF) journal articles and book chapters. Some of these journals include *Nature* (IF = 69.500), *Remote Sensing of the Environment* (IF = 13.85), *Journal of Cleaner Production* (IF = 11.07), *Science of The Total Environment* (IF = 10.75), *Earth's Future* (IF = 9.27), *Pattern Recognition* (IF = 8.51), *Scientific Data* (IF = 8.501), *Journal of Advances in Modeling Earth Systems* (IF = 8.47), *Journal of Hydrology* (IF = 6.71), *Ecological Indicators* (IF = 6.26), *Atmospheric Research* (IF = 5.96), *Climatic Change* (IF = 5.17), etc. He is a recipient of prestigious international awards, including the 2018 D. B. Johnston Award for Excellence in the Spatial Sciences area. He won the 2023 Griffith Sciences Pro Vice Chancellor's Excellence of an Early Career Researcher Award at Griffith University. Christopher is a member of the American Geophysical Union (AGU) and has been funded as a keynote speaker at the prestigious AGU Chapman conference. He leads the remote sensing research group within Australian Rivers Institute at Griffith University and several strategic research in remote sensing of the environment. He is at the forefront of remote sensing hydrology research, providing leadership on cutting-edge cross-disciplinary research in big satellite data analysis for environmental applications. This includes *assessing the impacts of climate change and water extraction on groundwater resources* supported by an ARC grant (DE230101327). Christopher has provided leadership on several consultancies and government projects, including those funded by the Australian Government. He has been a chief investigator and co-PI in research projects valued at more than 2.5 million dollars, and has engaged and collaborated with international experts and agencies, including the Earth Commission (https://earthcommission.org/) and Asian Development Bank on environmental water-related issues. Some of these engagements have received media attention and publicity (https://www.canberratimes.com.au/story/8018614/callfor-federal-action-over-nt-water-plan/). He has also served as a water expert informing on water planning and regulation (e.g., https://www.ecnt.org.au/water_experts_letter_cm_fyles) through non-governmental agencies (https://www.ecnt.org.au/currell_report) working with local communities to bring environmental justice and hold government to account for key decisions and governance on environmental issues. Christopher is an Associate Editor in Water and Climate within Frontiers in Water and sits on the editorial boards of two Elsevier journals (*Resources, Environment and Sustainability* and *Sustainable Horizons*). He has also Guest Edited several special issues on remote sensing and climate-related subjects and continues to serve as a key grant assessor for several international and government agencies across the world, including the ARC.

Part I
The Anthropocene

Chapter 1
Understanding the Anthropocene

1.1 Introductory Remarks

The rise in global mean sea level, acceleration of the water cycle, loss of forest ecosystems, rapid declines in groundwater, desiccation of large freshwater bodies, and rapid urbanization are clear indications that humans have transited from the Holocene conditions into a different era (Rockström et al. 2023; Ndehedehe 2023). The global average sea-level rise for the 1880–2009 period is estimated at 210 mm with an increasing linear rates of 1.7 ± 0.2 mm/year between 1900 and 2009 (Church and White 2011). Since 2000, glacier mass loss every year was estimated at 267 ± 16 gigatonnes, contributing approximately $21 \pm 3\%$ to observed sea-level rise (Hugonnet et al. 2021; WCRP Group 2018). Ocean thermal expansion is the highest contributor (42%) to global mean sea level in the twenty-first century, followed by glaciers, Greenland (15%), and Antarctica (8%) (WCRP Group 2018). Between 1992 and 2017, the Antarctic ice sheet lost $2{,}720 \pm 1{,}390$ billion tonnes of ice, which correspond to a mean sea-level rise of 7.6 ± 3.9 millimetres (IMBIE 2018). Recent evidence from satellite gravimetry and altimetry observations suggests that the loss of Antarctic ice sheet will continue into the future (Kim et al. 2022). For instance, due to rising temperatures, projected glacier mass loss by 2100 will range from $26 \pm 6\%$ (+1.5 °C) to $41 \pm 11\%$ (+4 °C) (Rounce et al. 2023). Not only are the glaciers retreating and groundwater stocks declining, global climate teleconnection patterns such as the El Niño–Southern Oscillation, and Pacific Decadal Oscillation are increasing the duration, frequency, and severity of extreme events (drought and floods) across several regions of the world (e.g., Ndehedehe et al. 2021; Brito et al. 2018; de Marsily and Abarca-del Rio 2016; Spinoni et al. 2014; Sheffield and Wood 2008). Climate is indeed changing!

The rising trajectory of global mean sea level (Fig. 1.1) apparently suggests the human planetary system is at risk and its climatic elements are being disturbed largely due to anthropogenic factors. Direct anthropogenic contributions to sea-level rise generally emanate from water impoundment by dams, groundwater depletion, deforestation, changes in land cover and land use, wetland degradation, and large

C. Ndehedehe, *Hydro-Climatic Extremes in the Anthropocene*, Springer Climate,
https://doi.org/10.1007/978-3-031-37727-3_1

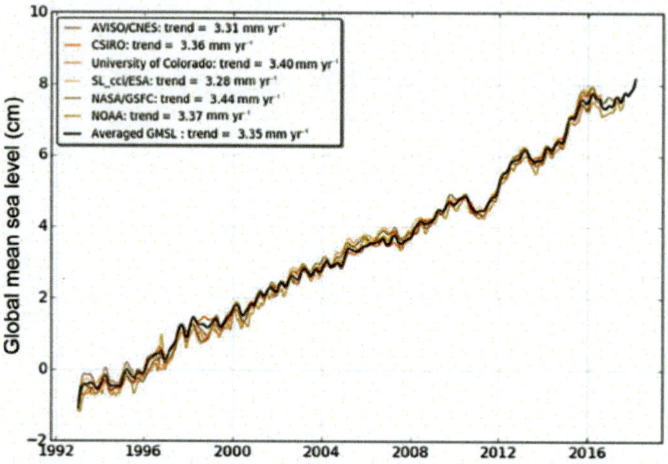

Fig. 1.1 The rise in global mean sea level since 1993. *Source* WCRP Group (2018)

water storage changes in man-made lakes and water reservoirs (Bierkens and Wada 2019; WCRP Group 2018; Reager et al. 2016). These anthropogenic actions in turn regulate net contributions to sea-level rise, though some of these (e.g., groundwater depletion and evaporation from water reservoirs) can be directly driven by climate as well. Unfortunately, humans are the direct recipients of the dangers associated with sea-level rise. For instance, by the end of the twentieth century, about 200 million people are expected to be impacted by rising sea levels and more than one billion people will face water shortage and food insecurity (Hugonnet et al. 2021).

The environmental challenges in contemporary society are alarming. From water scarcity in the West African Sahel and Southern Africa, terrific groundwater depletion in, e.g., India, California Central Valley, Middle East, and North Africa, permanent surface water storage losses in Lake Chad and Aral Sea, to frequent wildfire in Australia, the Earth system is experiencing a seemingly chain of complexities and transformations that will continue to affect human well-being as well as natural ecosystems. The impacts of climate change on water availability have triggered concerns on water scarcity even though mankind consumes only 10% of maximum renewable freshwater resources (Oki and Kanae 2006). These concerns are coming from the considerable spatial and temporal variability of freshwater availability, that is, it is unevenly distributed in time and space (Oki and Kanae 2006). Climate change (e.g., changes in the duration and intensity of droughts) increases this variability, which in turn drives the quantity of freshwater available to mankind. Water from precipitation transitions into freshwater in lakes, rivers, water holes, floodplains, soil moisture, and groundwater, which occur through infiltration and recharge. Evaporation, transpiration, and sublimation are natural channels and processes that transport or convert this freshwater from the land surface back to the atmosphere. However, humans intercept some of this available water for several uses. But the quantity being intercepted by humans in today's society is increasingly changing due to increased water demand caused by rising population. When the effects of greenhouse gases on the human planet are superimposed on this alteration of the natural process and the amount of fluxes transported, increase in the variability of freshwater availability at a given time and location are further intensified.

The rising demand for food and water has been linked to population growth, economic development, and dietary changes, the latter being one of the important drivers of expansion of irrigated agriculture, especially in typically dry areas where rainfall and surface water are limited (Bierkens and Wada 2019). With rising urban settlements, the increased reliance on non-renewable groundwater resources will be high and further decline in groundwater aquifers due to an inevitable groundwater over-exploitation. Excessive groundwater pumping draws down the water table and under prolonged drought conditions and higher evaporation, groundwater depletion and low well yields will result in groundwater drought (when the wells run dry). Groundwater wells run dry because infiltration or recharge is unable to replace groundwater as quickly as pumping removes it. Groundwater depletion as well as dewatering of groundwater-dependent springs and wetlands causes loss of unique biodiversity and unprecedented levels of mortality in trees, resulting in the collapse of forest ecosystems (Rossini et al. 2018). This depletion also triggers sea-level rise and land subsidence, affecting buildings and public infrastructures like roads (e.g., Bierkens and Wada 2019). When the wells run dry, farmers are unable to irrigate and access to reliable and convenient drinking water is limited. This is the reality of the new age—the Anthropocene!

1.2 The Anthropocene

1.2.1 What Is the Anthropocene?

The Anthropocene is a new era characterized by rapid changes in the environment driven largely by human activities and global climate change. As opposed to the relative stability of the Earth system under Holocene conditions, in the Anthropocene, human activities (e.g., dam constructions, reservoirs, groundwater extraction, water transfer, and harvesting) are now the predominant drivers of these changes, surpassing those caused by natural processes and climate change (e.g., Ndehedehe et al. 2023; Steffen et al. 2011). These human activities continue to be important indicators of the Anthropocene and have negative feedback on hydrological processes, e.g., increasing drought termination duration (Margariti et al. 2019). Furthermore, the phrase *new geologic epoch* has been used by scientists to describe the Anthropocene, indicating that it is an era that marks a fundamental change in the way mankind has interacted with the Earth system (Lewis and Maslin 2015). It is a period with stupendous increase in global prosperity driven by technological innovation and population growth, especially in urban areas around the world. The scale of human activity driving these changes in the Anthropocene is global and also the leading predictor of a substantially large proportion of environmental changes in contemporary society. A clear distinction of the Anthropocene era with the Holocene epoch is the magnitude of changes caused by human population, including the transformation of the land surface, which have contributed to changing the composition of the atmosphere (Lewis and Maslin 2015).

With the global human population expected to range from 9 to 11 billion toward the end of the twenty-first century, humans are seen as important drivers and agents of planetary change in the Anthropocene (Ndehedehe 2023; Folke et al. 2021). Unlike the stable environment-friendly Holocene dispensation where complex human societies developed, conditions in the Anthropocene are relatively more hostile and unfriendly to contemporary society (e.g., Steffen et al. 2011). This is particularly so, given the rise in sea level due to rapid losses in glacier mass and groundwater aquifers, as well as increased water demand and frequency in water scarcity, wildfire, and devastating drought and flood events (Hugonnet et al. 2021; Bierkens and Wada 2019; Di Baldassarre et al. 2017). The disturbance and instability of the earth system in the Anthropocene will continue to increase, leading to more deleterious impacts on both human well-being and natural ecosystems. The disturbance of natural systems (hydrological cycle) through the diversion of surface water can have both positive feedback like slowing down groundwater depletion rates and decreasing the duration of drought termination (Long et al. 2020; Margariti et al. 2019). However, there are other potential consequences on freshwater ecosystems and the ecological services they support.

1.2.2 The Earth System in the Anthropocene

The combination of socio-economic prosperity and population growth will imply unsustainable energy use and increased resource extraction that will further enhance changes in landscape. With planned increase in cultivated areas in semi-arid regions, expected increase in water demand for agriculture would be the aftermath of rising temperature and declines in precipitation and runoff (Barría et al. 2019; Steffen et al. 2018). These changes will potentially amplify the impacts of climate change on humanity, including drought characteristics such as frequency, intensity, duration, and termination, among others (e.g., Naumann et al. 2018). The collective effort to change the trajectory of the Earth system in the Anthropocene from potentially damaging conditions to a more stable habitable state has been advocated (Steffen et al. 2018). The resilience of the Earth system is being challenged in the Anthropocene and changing this trajectory is thus crucial. This is because of the interconnection between the biosphere and the broader Earth system, which also includes the atmosphere and the climate system and the implications of such connectivity on human well-being and prosperity (Folke et al. 2021). The pioneering work of Steffen et al. (2011) on the changes in the Anthropocene identified environmental degradation, human-induced biodiversity loss, changes in atmosphere and ocean, as well as urbanization as some challenges in contemporary society. They advocated the need for effective planetary stewardship and also argued on the potential risk of triggering more irreversible hostile environment in the Earth System as mankind moves further into the Anthropocene. Consequently, the assessment of Earth system boundaries to ensure a safe and just planet therefore is now receiving attention and from a perspective that acknowledges human well-being as a function of Earth system stability and resilience. The Earth Commission (https://earthcommission.org/) is pioneering this assessment underpinned by several working groups as well as partnerships with advocacy groups (e.g., science, business, and philanthropy) (Rockström et al. 2023). Consistent with key strategic objectives that complement ongoing assessments by the Intergovernmental Panel on Climate Change and other agencies, the Earth Commission has documented evidence indicating that some of the boundaries or thresholds for a healthy stable planet and people, e.g., for climate, water, aerosol, and biosphere, have already been crossed (Rockström et al. 2023). The Commission argued that achieving resilience and stability of the Earth system where humans are substantially less vulnerable to risk will require that the safe and just targets recently quantified be not transgressed further (Rockström et al. 2023).

The alterations of the Earth system are increasingly being driven by a plethora of anthropogenic activities. These human actions include construction of dams, reservoirs, water transfer, deforestation, intensive groundwater extraction, and deforestation, among others. Together with climate change, they coalesce to regulate net contributions to sea-level rise (Hugonnet et al. 2021; Bierkens and Wada 2019; WCRP Group 2018). Their (e.g., dams and water harvesting) negative impacts on drought characteristics are also being studied. For instance, a rise in drought termination rates has been reported across six catchments in Europe where reservoir operations are in place (Margariti et al. 2019). Surface water extraction impacts on drought intensity at regional and national scales and anthropogenic forcings now contribute to changes in the characteristics of short-term rainfall deficits across the globe (Chiang et al. 2020; Wan et al. 2017). However, the Anthropocene is also rapidly changing and evolving due to climatic changes. The impacts (e.g., extreme floods and droughts) of these changes on human societies contributed to the collapse of ancient civilization, migrations, and shifts in cultural values during the relatively stable Holocene era (Steffen et al. 2018; Di Baldassarre et al. 2017). However, the impacts of climate change through increase in the frequency and duration of extreme events have intensified, thus making the Anthropocene a geologic epoch dominated by the composite influence of climate change and human activities. For instance, much of the groundwater depletion across the globe will end up in the ocean, through runoff and evaporation thus contributing to considerable rise in global sea level, which in turn causes several devastations (Wada et al. 2010).

1.3 The Twenty-First Century Environment

1.3.1 Man-Made Lakes and Reservoirs

Anthropogenic footprints through human-managed reservoirs and water infrastructures are notable drivers of global surface water variability (Ndehedehe 2022b; Cooley et al. 2021). Efforts to manage and mitigate drought risk and its impacts led to increased investments in water infrastructures. Moreover, large flood events have been managed by constructing large water infrastructures/reservoirs and ensuring these reservoirs are empty to accommodate excess water from extreme rainfall and flood conditions (Di Baldassarre et al. 2017). However, surface water storage is insufficient and often unavailable in many regions of the world, especially as a result of geographic isolation, resulting in the development of considerably large water infrastructures. Some gigantic human water resource projects include the Three Gorges Dam in China and Lake Volta in Ghana, which is the largest man-made lake in the world. As of 2005, more than 45,000 dams with average height greater than 15 m existed and 300 of these dams globally fall into the gigantic category, characterized in terms of height (greater than 150 m), volume (more than 15 million m^3), or storage capacity (greater than 25 km^3) (Nilsson et al. 2005). These projects, e.g., the Akosombo dam construction, led to large floods, which affected local communities, resulting in massive resettlements of local populations (Ndehedehe 2022a).

The impacts of dam construction on large river systems have been identified and discussed and include higher irrigation pressure and increased economic activity in catchments that could potentially pose ecological risks (Nilsson et al. 2005). Satellite measurements for 227,386 water bodies between October 2018 and July 2020 revealed that about 57% seasonal surface water storage variability occurred in human-managed reservoirs, with this variability averaging 0.86 metres, unlike natural systems (e.g., natural water bodies), which varied by only 0.22 metres (Cooley et al. 2021). As illustrated in Fig. 1.2, the assessment of these human-managed reservoirs using ICESat-2 underpinned by the Global Reservoirs and Dams showed that about 1,151 (12.8%) of these reservoirs indicated variations on a seasonal scale that were greater than 5 m, while 405 (4.5%) showed seasonal variations greater than 10 m. The increased developments of these water resources infrastructure can be linked to economic development and population growth. However, the impact of climate change, e.g., changes in the magnitude and frequency of extreme drought, can continue to increase the pressures on global water resources.

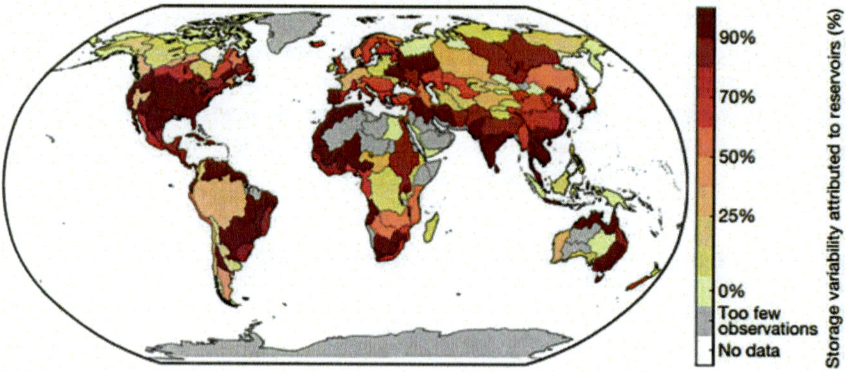

Fig. 1.2 Ratio of seasonal freshwater storage variability correlated with reservoirs by hydrologic basins from 2018 to 2020. Greater effect on human-managed reservoirs on freshwater storages is depicted using darker colors, while less effect is depicted using lighter colors. Map was originally adapted from the work of Cooley et al. (2021) for use in Ndehedehe (2022b)

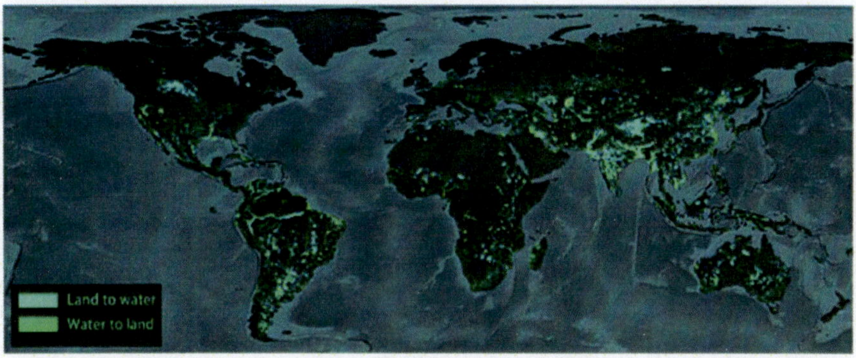

Fig. 1.3 Heat map of global surface water and land variations. Map was originally adapted from the work of Donchyts et al. (2016) for use in Ndehedehe (2022d)

The study undertaken by Donchyts et al. (2016) summarized the changes in global surface water, indicating that approximately 115,000 km^2 of surface water extent was gained over a 30-year period. Figure 1.3 illustrates this further and shows a heat map representing global surface water and land variations during the 1985–2015 period using Landsat images. The blue lighting depicts where land was translated to water, whereas the green lighting depicts where water was translated to land during the study period (Fig. 1.3). The spatial magnitude of variations shown in Fig. 1.3 is illustrated in the color intensity. These changes in the presence of water on global landscapes were further characterized as *"disruptive versus gradual'*. For example, changes in the Aral Sea were labeled *disruptive* because of the unregulated practice of water extraction for irrigation resulting in about 27,650 km^2 loss of surface water extent while the desiccation and disappearance of Lake Mead, which provides the largest freshwater supply in the United States, was characterized as gradual (Donchyts et al. 2016).

Surface water is an important component of the water cycle. Its changes in space and time reflect this but more importantly, it emphasize how human civilization, urbanization, socio-economic prosperity, or human response to unfavorable hydrological conditions like drought can drive such changes. Analyzing satellite data between 1984 and 2015, the work undertaken by Pekel et al. (2016) showed that about 90,000 square kilometres of surface water extent were lost over this period. However, during the same period, they observed that new permanent surface water covering a total land area of 184,000 square kilometres emerged in other locations. Much of the increase in surface water extent are likely to emanate from man-made reservoirs, dams, and lakes as a result of expansion of agricultural lands to meet dietary changes of a growing population and/or response to devastating sever droughts. For instance, Brazil is one of such regions where the proliferation of water infrastructures like dams was a direct response to several precarious drought events. As of 2016, the National Water Agency in Brazil reported a total of 19,361 man-made reservoirs used for irrigation and hydro-power, among several other important uses (ANA 2017). Apart from their socio-economic importance, these water infrastructures were also in response to aridity and to cushion impacts of droughts in the region (Ndehedehe et al. 2020). The rise in water infrastructures is because they are perceived as pathways to water security. The construction of more than 5000 large dams in India is based on this perception, making the country the third largest in dam constructions (Pradhan and Srinivasan 2022). Even though dams may have negative impacts on water security as argued by Pradhan and Srinivasan (2022), they have articulated an integrative approach that considers the design and implementation of physical water infrastructures as socio-technical systems.

1.3.2 Climate-Induced Changes in Terrestrial Stored Water and Contributions to Sea-Level Rise

The rise in sea level is another obvious characteristics of the Anthropocene. As indicated in Fig. 1.1, the rising trajectory of global mean sea level will continue into the future. Again, as detailed by WCRP Group (2018), human activities resulting from water impoundment, groundwater depletion, changes in land cover and land use, and large water storage changes in man-made lakes and water reservoirs, among others are important agents and drivers of this increase. Additionally, the rapid loss of ice sheets through climate change are inducing changes in global hydrology, contributing to rising global sea level. The work undertaken by Hugonnet et al. (2021) provides glaring statistics of such contributions. Their analyses revealed that global glaciers showed a mass loss that is 47% higher than those of the Greenland Ice Sheet and the Antarctic Ice Sheet between 2000 and 2019. This glacier loss would translate to $21 \pm 3\%$ contribution in observed sea-level rise. Groundwater depletion caused by over-extraction leads to the redistribution of land water storage, which in turn contributes to rising sea-level rise. The work of Bierkens and Wada (2019) summarized several studies, which demonstrated water storage contribution to sea-level rise due to increasing groundwater depletion as well as discussing the mechanisms behind such process. The interplay between land water storage and changes in sea level was earlier captured by Reager et al. (2016). They found that climate-induced changes (2002–2014), which increased land water storage to 3200 ± 900 gigatons of water slowed down sea-level rise. Changes in important water cycle indicators, e.g., rainfall, evaporation, and runoff, are channels through which such changes occur. In other words, even though humans intercept some proportion of freshwater within the global water budget for several uses through groundwater extraction, this freshwater is never completely lost as it ends up in the sea and subsequently the water cycle. However, the contributions to sea-level rise are problematic for mankind.

1.3.3 Rise in Human Population

The Anthropocene is characterized by rapidly increasing human population, which is a major factor in the Earth's alteration. Because human–landscape interactions will change the Earth's morphology in the Anthropocene (Sofia et al. 2014), climate, topography, landscapes, climate, hydrology, and ecosystems will considerably change due to growing human population. Deforestation, land reclamation, and drainage networks are major landscape transformations caused by urbanization and pressures from increasing human population (Sofia et al. 2014). In the last 50 years, urban sprawl in several big cities around the world has skyrocketed due to rising global human population. From the start of the nineteenth century to the current age, a fivefold increase in global population puts the value at approximately 8 billion people and is expected to reach 9.7 and 11.2 billion people by 2050 and 2100, respectively (Laurance and Engert 2022). From an estimated 2.5 billion people in 1950, global human population according United Nations is expected to reach over 10 billion by 2059, doubling population count from the 5 billion people reported in 1987 (UN 2022). The most populous regions of the world as of 2022 were those in Asia, accounting for a total of 55% (approximately 4.4 billion people) of global population (Fig. 1.4). Essentially, China and India are home to the largest populations in Asia, accounting for more than 1.4 billion each in 2022 (UN 2022). Most of the world's continents including Asia, Africa, Europe, North and South America are projected to increase in the future (Fig. 1.4). As further explained by United Nations (UN 2022), more than 50% of the projected increase in global population during 2020–2050 is expected to be concentrated in two continents—Africa (Democratic Republic of the Congo, Egypt, United Republic of Tanzania, Ethiopia, and Nigeria) and Asia (India, Pakistan, and the Philippines). The increasing trends in global population implies a rise in urbanization. This is usually caused by the search for greener pastures, including jobs, education, and healthcare services. To accommodate the large chunk of migrants, urban centers are expanding through increased infrastructure developments, housing, etc.

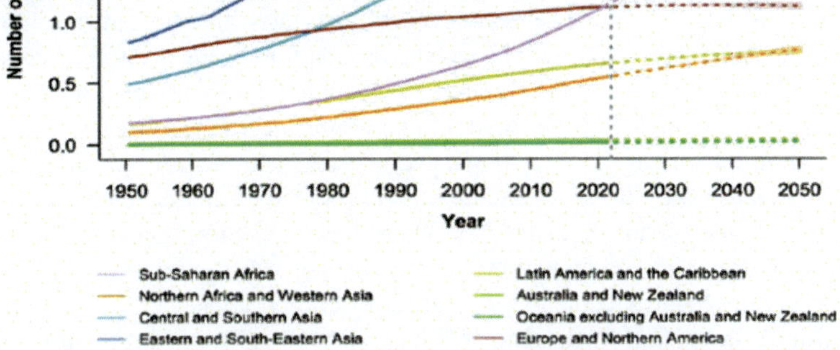

Fig. 1.4 Global population estimates (1950 and 2022) and projections for the 2022–2050 period for each region. *Source* UN (2022)

1.3.4 Land Use and Land Cover Change

Land is the main source of food, energy, and human development. The rise in human population and economic development means increased competition for space (land) for shelter and agriculture and other purposes. One of the most obvious indicators of living in the twenty-first century is changes in land cover and without doubt is the result of growing population. The change in global land cover between 1982 and 2016 reported by Song et al. (2018) provides a vivid example of how humanity is rapidly driving changes in the environment that is contributing to alteration in the Earth's energy balance and biogeochemical cycles. They showed that available open space (bareground cover) decreased by 1.16 million km^2 because they were converted to agricultural. They also reported that direct human activities accounted for 60% of all land changes. Apart from climate change-induced changes in land cover, which accounted for 40% of the land change reported in Song et al. (2018), historically, humans are the biggest drivers of changes in land, be it agricultural or urban landscapes. Laurance and Engert (2022) described this further, indicating that between 1990 and 2000, urban areas rose by 1.9 million ha per year and is expected to increase to about 2.3–4.4 million ha per year between 2015 and 2050. The aftermath of rising human population is urban expansion and impacts biodiversity, forest ecosystems, land cover changes, coastal resources, and the environment generally. The work undertaken by Simkin et al. (2022) detailed these impacts including, loss of habitats for 30,393 species of terrestrial vertebrates moving into the future (2015–2050). With economic growth and technological innovations, such impacts are expected to continue into the future. For instance, there is a link between land-use change and financial credit in South Florida, thus linking real estate developers to financial markets (Sealey et al. 2018). The rise in human population in the twenty-first century could translate to some challenges in agriculture, including increasing food production under a changing climate.

Changes in global land cover are expected to result in other challenges, including thermal stress, which in turn impacts energy demands. The work undertaken by Adeyeri et al. (2023) revealed these challenges, and showed that there is an increasing loss of global forest as they are being converted to rangeland and croplands. They showed that the falling trends in forest (also see Sect. 1.3.6) increased the severity of thermal stress, and that future climate projections suggest thermal stress is likely to be more intense. Figure 1.5 shows the spatial distribution of annual thermal stress and land cover land-use change for the year 1959, 2014, and future projections (2014 and 2100) under two Coupled Model Intercomparison Project Phase 6 (CMIP6) Shared Socioeconomic Pathways (SSP 370 and 585). In all of the epochs considered (e.g., 1959, 2014, and 2100), there is generally a wide spread of moderate heat stress that coincides with the expansion of secondary non-forest, pasture, annual cropland, and rangeland pastureland and annual cropland (Fig. 1.5). However, some regions like Russia experienced strong cold to moderate cold stress because of the distribution of forested and primary non-forested areas (Fig. 1.5). In other European countries (Spain and France), the evolution of rangeland led to slightly cold stress, while in places like Turkey, the transition of other land covers to rangeland and annual cropland modified the distribution of slight cold stress (Fig. 1.5). The clarification on the perceived

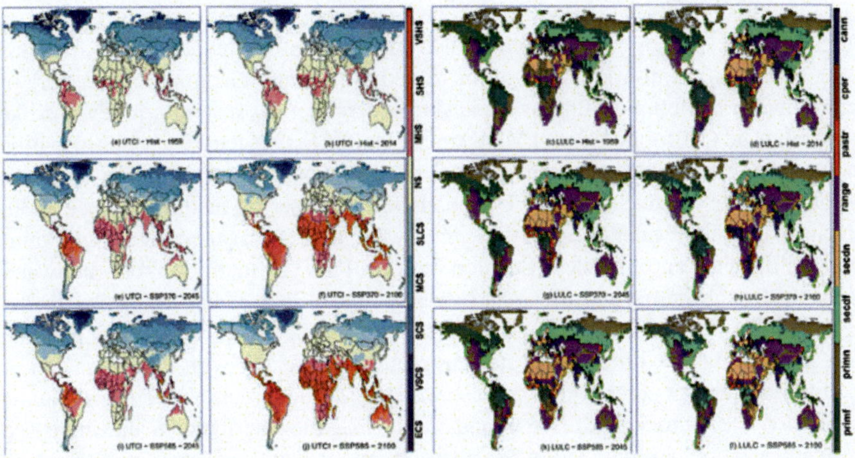

Fig. 1.5 Spatial distribution of annual thermal stress and land cover land-use change for the years 1959, 2014, and future projections (2014 and 2100) under SSPs 370 and 585. The colorbar description for the thermal stress on the left panel is *VSHS*—very strong heat stress; *SHS*—strong heat stress; *MHS*—moderate heat stress; *NS*—no thermal stress; *SLCS*—slight cold stress; *MCS*—moderate cold stress; *SCS*—strong cold stress; *VSCS*—very strong cold stress; *ECS*—extreme cold stress. The colorbar description for the land cover on the right panel is as follows: *primf*—Primary Forest; *secdf*—Secondary Forest; *secdn*—Secondary Non-Forest; *cann*—Annual Cropland; *past*—Pasture; *cper*—Permanent Cropland; *range*—Rangeland; *primn*—Primary Non-Forest. UTCI on the map is universal thermal comfort index. *Source* Adeyeri et al. (2023)

role of changes in land use land cover in shaping the future trajectory of thermal stress (Fig. 1.5) is another pointer to mankind (through human population increase, urban expansion, etc.) being the dominant driver of changes in the Anthropocene. For instance, future projections (SSP585 2100) suggest very strong heat stress will be experienced in much of Africa and these coincide with considerable transitions of land cover, including large losses of primary forest around West and Central Africa. Beyond this (land cover changes driving thermal stress), and the corresponding impacts on the water cycle and biodiversity, the established connection between transitions in land cover units (e.g., forest; also see Sect. 1.3.6) and thermal stress (Adeyeri et al. 2023) underpins science-based solutions for adaptation strategies.

1.3.5 Groundwater Depletion

Groundwater depletion is another fundamental problem in many regions around the world and an important indicator of human-driven changes in the twenty-first century environment. More than 50% of the world's largest aquifers are showing rapid declines in groundwater storage (Solander et al. 2017), which can jeopardize global food and water security and the sustainability of ecosystems, constraining human adaptation to climate change (Jasechko and Perrone 2021). Groundwater depletion is caused by long-term water extraction and prolonged extreme drought (Ojha et al. 2020). The review by Bierkens and Wada (2019) identified expansion of irrigated agriculture into semi-arid areas and increased urbanization as important drivers of groundwater depletion. With this expansion observed in most semi-arid regions like Australia and California Central Valley in the United States, the reliance of irrigation on groundwater is expected to increase. Presently, about 60% of irrigation water use is taken from groundwater. In the US, about 50% of observed historical (since 1900) declines in groundwater occur in these highly irrigated landscapes, e.g., High Plains and California Central Valley (Scanlon et al. 2012). The total global groundwater depletion in sub-humid and arid areas increased from 126 ± 32 km^3/yr in 1960 to 283 $\pm$ 40 km^3/yr in 2000 (Wada et al. 2010). In the Anthropocene, rising water demand directly correlates with groundwater abstraction, especially with the expansion of cultivated areas in semi-arid domains, which leads to groundwater depletion. In summary, the factors that drive expansion in irrigation agriculture and increased groundwater use are population-driven economic growth and have also triggered sudden rise in water demand.

However, increased water extraction will disrupt recharge, contributing to groundwater loss in time and space during prolonged drought events. The latter increase dependence on groundwater and decreased surface water flow and connection with aquifers. This is because change in recharge is sensitive to and affected by changes in rainfall intensity, flow, and temperature (Taylor et al. 2013). These are the factors that regulate water balance and induce changes in the amount or timing of groundwater recharge. The influence of global climate teleconnection patterns like El Nino–Southern Oscillation, Pacific Decadal Oscillation, and Indian Ocean Dipole regulate changes in the severity and intensity of droughts in some regions. If these changes occur around large aquifer systems, they can impact the amount of recharge and this will be evident in the loss of terrestrial water storage and declines in surface water availability. The loss in groundwater storage will be due to the rates of human water withdrawal exceeding how quickly water is being replaced by recharge. This shows vividly how the interactions between climate and humans exacerbate groundwater depletion. However, excessive groundwater withdrawal is a major driver of groundwater depletion. Since the advent of GRACE satellite, both climatic and human footprints on groundwater depletion have been studied. A detailed discussion of the impacts of groundwater withdrawal and loss has been provided by Bierkens and Wada (2019), and this includes land subsidence, contribution to sea-level change, groundwater salinization, effects on groundwater-dependent ecosystems, and enhancing hydrological drought.

1.3.6 Deforestation

Vegetation (including forest) is no doubt an important source of food and energy and also provides critical services like regulating the climate system, storing of carbon, and hosting biodiversity (e.g., Trumbore et al. 2015; Ndehedehe 2022c).

However, with increasing human population and expansion of urban centers in the Anthropocene, the delivery of these services can be affected due to deforestation, which is increasingly the major factor driving the destruction of vegetation and forest ecosystems (Ndehedehe 2022c). Based on the analysis of a suite of Landsat images, the work of Hansen et al. (2013) provided an insight to the evolutionary patterns of forest. They characterized global forest extent and change during the 2000–2012 period, reporting a global forest loss of 2.3 million Sq Km with rising losses in forest of about 2101 Sq km each year in the tropics. Wildfire (See Chap. 12) and deforestation were major triggers of these observed losses in forest cover around the world. In the Central African rainforest, for example, forest loss of 721.0 ha per year to deforestation was reported for the period between 2000 and 2010 and a rising trend of 0.3–0.5% per year has been projected for the 2020–2030 period (Nogherotto et al. 2013). Forest loss in Brazil is the second highest globally, and several other countries in Asia (Malaysia and Cambodia), Africa (Cote d'Ivoire and Tanzania), and South America (Argentina and Paraguay) are also showing increasing loss in forest cover (Hansen et al. 2013). As most large urban areas are closer to the coast, the removal of mangrove forests is also common. For instance, the spatial distribution of mangrove forest along the West African coast in the Niger Delta region of Nigeria showed a loss of 21,340 hectares between mid-1980s and 2003 (James et al. 2007). Additionally, there is substantial evidence in the literature to confirm that forest and vegetation systems along the coast and in inland areas are increasingly being replaced by urbanization and rising trajectory in urban settlements (e.g., Okwuashi and Ndehedehe 2021).

Moreover, the partnership between Global Forest Watch and Global Land Analysis and Discovery laboratory at the University of Maryland[1] led to the continuous tracking of global forest changes (loss and gain), thus providing consistent annual changes in global-scale forest data, including its extents. The illustration in Fig. 1.6 shows the changes (gains and losses) and spatial distribution in forest cover per year between 2000 and 2021. In this updated data (Fig. 1.6), Forest Cover Loss is defined *as a stand-replacement disturbance, or a change from a forest to non-forest state*, between 2000 and 2021. Forest Cover Gain is defined *as the inverse of loss, or a non-forest to forest change entirely within the period*, that is, 2000–2012.

Apparently, the Amazon and Congo basins in South America and Africa, respectively, have the largest or highest forest and tree cover distribution in the world (Fig. 1.6). The updated change detection of global-scale forest for the last 21 years shows a significant proportion of the Earth and continents across are losing their forest cover (Fig. 1.6). However, these data also show areas where both forest gain and forest loss are occurring but the spread of forest losses is far more higher (Fig. 1.6). Deforestation is a major problem globally and as global population increases, some of the hotspots shown in Fig. 1.6 may continue to experience the loss of forest ecosystems. These continued losses will lead to reduced precipitation and drying conditions. Using a regional climate model, this has been demonstrated in the Congo basin where decline in rainfall was observed over deforested areas with decreased evaporation

[1] https://storage.googleapis.com/earthenginepartners-hansen/GFC-2021-v1.9/download.html.

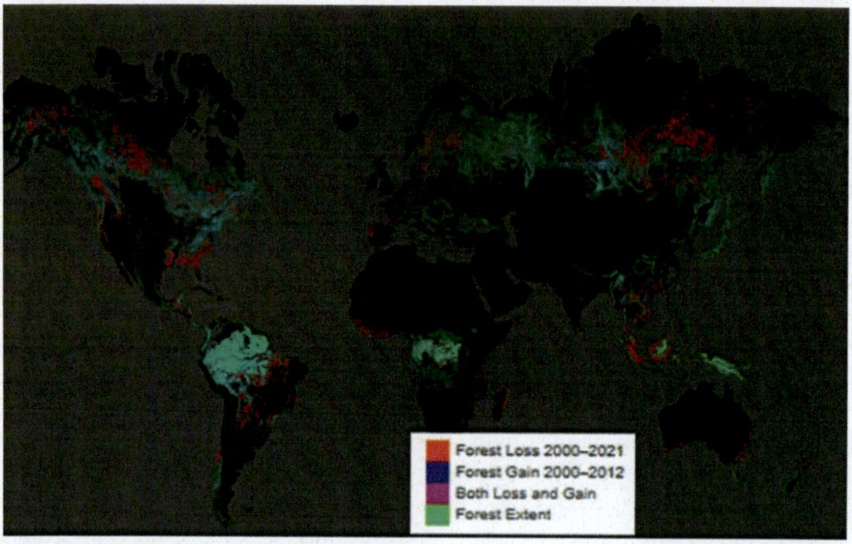

Fig. 1.6 Global forest extent and change during the 2000–2021 period. This data was prepared from the publicly available Hansen et al. (2013) data via the University of Maryland data portal (https://storage.googleapis.com/earthenginepartners-hansen/GFC-2021-v1.9/download.html)

(Nogherotto et al. 2013). Apart from human removal of forests and vegetation, the loss of forest ecosystems due to increasing frequency of wildfires is another driver of vegetation cover loss. In addition to destroying human life, wildfires affect ecology (e.g., vegetation communities), resulting in damages that disrupt several natural and ecological processes on a large scale (Ndehedehe 2022c). More details on wildfire being one of the key events in the Anthropocene are provided in Chap. 12.

1.3.7 An Altered Global Water Cycle: What Are the Consequences?

Indeed, the Anthropocene era is characterized by several impacts caused by increasing human activities as well as climate change. In the foregoing sections, rising human-managed reservoirs, and rapidly increasing human population, human water extraction, water harvesting, and deforestation were identified as typical human actions that contribute to the alterations of the Earth system in the Anthropocene. Furthermore, alterations in the global nitrogen cycle caused by conversion of atmospheric nitrogen to ammonia for use as fertilizer, changes in land use for the production of food, fuel, fiber, as well as other sundry evolutionary pressures like the development of novel genetically engineered organisms, etc. are other impacts on the Earth system in the Anthropocene (Lewis and Maslin 2015). Ultimately, several of these human actions, if not all, have resulted in a new Anthropocene-type of water cycle that has been altered and with negative impacts on humanity

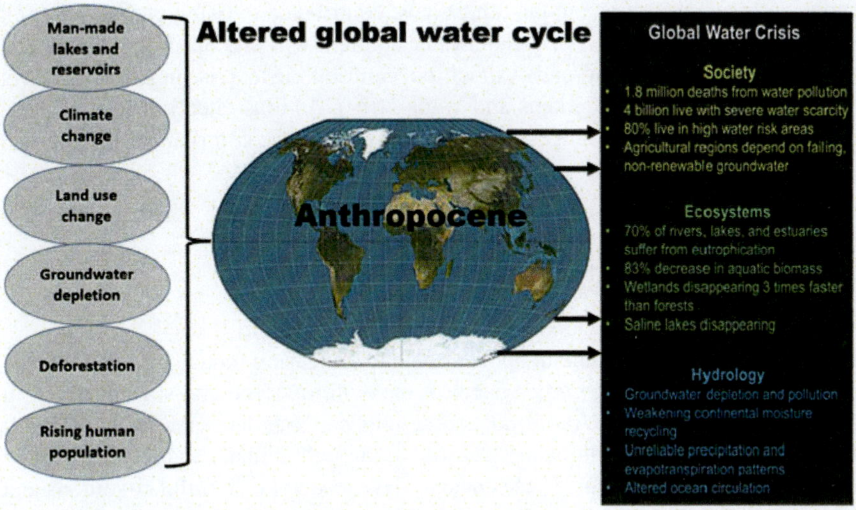

Fig. 1.7 Illustrating the consequences of an altered global water cycle in the Anthropocene due to the composite influence of climate change and multiple strings of human activities. *Source* modified from Abbott et al. (2019b)

(Abbott et al. 2019b). Figure. 1.7 summarizes some of these impacts on human society (e.g., water scarcity, death due to water pollution, and large human population residing in water risk regions), ecosystems (e.g., eutrophication in surface water, wetland loss, and disappearance of saline lakes), and hydrology (e.g., groundwater depletion and strong variability in precipitation and evapotranspiration rates), among others. The water cycle mechanism and the aftermath of humans intercepting freshwater from the water cycle have been discussed further in Chap. 3 of this book.

As freshwater use is now half of global river discharge, the work undertaken by Abbott et al. (2019a) offers some insight on human interference with the water cycle. Beyond water consumption being a critical aspect of human interference, they argued that changes in land use also played an important role in the alteration of water cycle through impacts on land–atmosphere interactions and movement of land water storage (includes groundwater, soil moisture, and surface water in lake and reservoirs). Moreover, they captured total human water use (approximately 24000 $km^3\ yr^{-1}$), identifying the contributions of soil moisture for crops and rangelands (green water), consumptive water for agriculture, industry and domestic activity (blue water), and water necessary to dilute human pollutants (gray water) to the global water cycle. Comparing the contemporary appropriation of these various freshwater components or types (blue, green, and gray) with previous periods, they revealed increasing water usage from man-made reservoir storages, non-renewable groundwater, and groundwater recharge in recent times. This was in contrast to the last decade when usage from global water pools (e.g., renewable groundwater and lakes, rivers) was significantly lower (Abbott et al. 2019a). Water impoundment in reservoirs, groundwater over-exploitation, large water storage changes in man-made lakes, and waterbreakreservoirs, among others have contributed to rising global sea level. For instance, the depletion of groundwater aquifers will end up in the ocean, and this happens through runoff and evaporation, contributing to a rise in global sea level (e.g., Ndehedehe 2023; Bierkens and Wada 2019). As illustrated in Fig. 1.7, these human interferences or changes in the Earth system have contributed to increased acceleration and alteration of the water cycle in several ways.

1.4 About This Book

This book explores how human civilization and increasing human population have influenced changes in the Earth system in the Anthropocene era, specifically with regards to water resources. Anthropogenic activities such as dams, reservoirs, groundwater over-exploitation, and water transfer, along with climate change, are driving changes in the Earth system. These changes have resulted in harmful conditions and an unfriendly environment in the Anthropocene era, where climate change and human activities are the primary causes of global environmental change. The book takes a trans-disciplinary approach to the discussion of these issues, highlighting the various impacts of human development and global climate change on hydrological processes, drawing from several case studies. It also discusses the multidisciplinary framework that relies on satellite technology to support the latest knowledge on changes in environmental systems. The book aims to generate new insights into socio-hydrological systems to enhance our understanding of current challenges and issues in the Anthropocene era and to identify ways to address them. By leveraging the broad range of machine learning tools, the book also elaborates on the efficient monitoring of water resources, demonstrating the feasibility of remote sensing data and innovative geospatial tools. Within this context, the book covers various operational monitoring capabilities, including optical satellite systems, data assimilation, satellite geodetic platforms, radar systems, and hydrological and climate models to improve our understanding of climate change impacts on the environment, especially freshwater.

Investments in innovative satellite programs that could optimize policy decisions and aid sustainability are yet to be fully utilized. Demonstrating the practical utility of big earth observation data will facilitate environmental monitoring and understanding of the impact of climate change in the Anthropocene. This, in turn, could create innovative opportunities to address several socio-economic and environmental challenges facing society. This book delves into a plethora of important issues, including how freshwater resources (e.g., groundwater and surface water) respond to human and climatic impacts. It also examines the changes in extreme events such as droughts, floods, and wildfires, and their implications on environmental assets, land-use change, water resources development, and the twenty-first century in the context of climate change, groundwater depletion, and a rising human population. Additionally, it explores building resilience to climate change and new methods to improve our understanding of climate processes in the Anthropocene. The book also provides case-specific regional examples that illustrate the combined use of novel satellite programs, like GRACE, with other optical sensors to further explore capabilities in freshwater accounting. This information serves as a foundation for water resources planning. Policy issues have also been raised as an important aspect that strengthen the management of water resources, particularly in emerging economies where observational data is often lacking, limited, or difficult to access.

Moreover, this book discusses the implications of the Anthropocene era on freshwater resources and their management. It highlights the lessons learned from freshwater hotspots, such as Lake Chad and Lake Urmia, where prolonged droughts and human activities have led to a permanent loss of surface water. The book provides a detailed overview of the strategic applications of novel remote sensing tools for water management policy and climate change adaptation in these hydrological hotspots. Moreover, the book delves into observational data from new multi-satellite missions, methods to measure and characterize extreme events, and their links to global climate. The aim is to provide tools and a scientific framework to aid our understanding of environmental and hydrological systems, and the complexities of socio-hydrological systems in the Anthropocene. Apart from the widely used Earth resources satellites mentioned earlier, this book discusses the capability of recently launched and planned novel satellite radar missions that support environmental assessment in a changing climate. The goal is to generate new knowledge to help humans adapt to climate change and increase resilience based on our capability to measure and predict changes. The book fills a critical gap in the multidisciplinary aspect of planetary science, particularly in understanding the impacts of climate change and human actions on freshwater resources, as well as the stability of the Earth system. Quantitative and empirical region-specific case studies have been used to explore the complexities of socio-hydrological systems in the Anthropocene and also demonstrate the utility of big earth observation data to improve the understanding of these complexities.

References

Abbott B, Bishop K, Zarnetske J, Minaudo C et al (2019a) Human domination of the global water cycle absent from depictions and perceptions. Nature Geosci

Abbott BW, Bishop K, Zarnetske JP, Hannah DM, Frei RJ, Minaudo C, Chapin III F, Krause S, Conner L, Ellison D, Godsey SE, Plont S, Marçais J, Kolbe T, Huebner A, Hampton T, Gu S, Buhman M, Sayedi SS, Ursache O, Chapin M, Henderson KD, Pinay G (2019b) A water cycle for the anthropocene. Hydrol Process

Adeyeri OE, Zhou W, Laux P, Wang X, Dieng D, Widana LA, Usman M (2023) Land use and land cover dynamics: implications for thermal stress and energy demands. Renew Sustain Energy Rev 179:113274. https://doi.org/10.1016/j.rser.2023.113274

ANA (2017) Brazilian water resources report 2017. National Water Agency, Brazil

Barría P, Rojas M, Moraga P, Muñoz A, Bozkurt D, Alvarez-Garreton C (2019) Anthropocene and streamflow: Long-term perspective of streamflow variability and water rights. Elementa: Sci Anthropocene 7(2). https://doi.org/10.1525/elementa.340

Bierkens MFP, Wada Y (2019) Non-renewable groundwater use and groundwater depletion: a review. Environ Res Lett 14(6):063002. https://doi.org/10.1088/1748-9326/ab1a5f

Brito SSB, Cunha APMA, Cunningham CC, Alvalá RC, Marengo JA, Carvalho MA (2018) Frequency, duration and severity of drought in the semiarid Northeast Brazil region. Int J Climatol 38(2):517–529. https://doi.org/10.1002/joc.5225
Chiang F, Mazdiyasni O, AghaKouchak A (2020) Evidence of anthropogenic impacts on global drought frequency, duration, and intensity. Nature Commun 12:2754. https://doi.org/10.1038/s41467-021-22314-w
Church J, White N (2011) Sea-level rise from the late 19th to the early 21st century. Surv Geophys 35:585–602. https://doi.org/10.1007/s10712-011-9119-1
Cooley SW, Ryan JC, Smith LC (2021) Human alteration of global surface water storage variability. Nature 591(7848):78–81. https://doi.org/10.1038/s41586-021-03262-3
de Marsily G, Abarca-del Rio R (2016) Water and food in the twenty-first century. Surv Geophys 37:503–527. https://doi.org/10.1007/s10712-015-9335-1
Di Baldassarre G, Martinez F, Kalantari Z, Viglione A (2017) Drought and flood in the Anthropocene: feedback mechanisms in reservoir operation. Earth Syst Dyn 8(1):225–233. https://doi.org/10.5194/esd-8-225-2017
Donchyts G, Baart F, Winsemius H et al (2016) Earth's surface water change over the past 30 years. Nature Climate Change 6:810–813. https://doi.org/10.1038/nclimate3111
Folke C, Polasky S, Rockström J et al (2021) Our future in the anthropocene biosphere. Ambio 50:834–869. https://doi.org/10.1007/s13280-021-01544-8
Hansen MC, Potapov PV, Moore R et al (2013) High-resolution global maps of 21st-century forest cover change. Science 342(6160):850–853
Hugonnet R, McNabb R, Berthier E et al (2021) Accelerated global glacier mass loss in the early twenty-first century. Nature 592:726–731. https://doi.org/10.1038/s41586-021-03436-z
IMBIE (2018) Mass balance of the Antarctic Ice Sheet from 1992 to 2017. Nature 558:219–222. https://doi.org/10.1038/s41586-018-0179-y
James GK, Adegoke JO, Saba E, Nwilo P, Akinyede J (2007) Satellite-based assessment of the extent and changes in the mangrove ecosystem of the Niger Delta. Marine Geodesy 30(3):249–267. https://doi.org/10.1080/01490410701438224
Jasechko S, Perrone D (2021) Global groundwater wells at risk of running dry. Science 372(6540):418–421. https://doi.org/10.1126/science.abc2755
Kim B-H, Seo K-W, Lee C-K, Kim J-S, Chen J, Wilson CR (2022) Antarctic ice mass change (2003–2016) jointly estimated by satellite gravimetry and altimetry. J Geophys Res Solid Earth 127(9):e2021JB023297. https://doi.org/10.1029/2021JB023297
Laurance WF, Engert J (2022) Sprawling cities are rapidly encroaching on earth's biodiversity. Proc Natl Acad Sci 119(16):e2202244119. https://doi.org/10.1073/pnas.2202244119
Lewis S, Maslin M (2015) Defining the anthropocene. Nature 519:171–180. https://doi.org/10.1038/nature14258
Long D, Yang W, Scanlon B et al (2020) South-to-north water diversion stabilizing Beijing's groundwater levels. Nature Commun 11:3665. https://doi.org/10.1038/s41467-020-17428-6
Margariti J, Rangecroft S, Parry S, Wendt DE, Loon AFV (2019) Anthropogenic activities alter drought termination. Elementa Sci Anthropocene 7. https://doi.org/10.1525/elementa.365
Naumann G, Alfieri L, Wyser K et al (2018) Global changes in drought conditions under different levels of warming. Geophys Res Lett
Ndehedehe C (2022a) Anthropogenic influence on terrestrial hydrology, pp 283–298. Springer, Cham. https://doi.org/10.1007/978-3-030-99577-5_12
Ndehedehe C (2022b) Impacts of water resources development on hydrology, pp 389–437. Springer, Cham. https://doi.org/10.1007/978-3-030-99577-5_15
Ndehedehe C (2022c) Remote sensing of surface vegetation, pp 131–176. Springer, Cham. https://doi.org/10.1007/978-3-030-99577-5_7
Ndehedehe CE (2022d) Floodplain productivity in wet-dry tropics, pp 557–626. https://doi.org/10.1007/978-3-030-99577-5_18
Ndehedehe CE (2023) Understanding the Earth system in the Anthropocene. Resour Environ Sustain 12:100113. https://doi.org/10.1016/j.resenv.2023.100113

Ndehedehe CE, Ferreira VG, Adeyeri OE, Correa FM, Usman M, Oussou FE, Kalu I, Okwuashi O, Onojeghuo AO, Getirana A, Dewan A (2023) Global assessment of drought characteristics in the Anthropocene. Resour Environ Sustain 12:100105. https://doi.org/10.1016/j.resenv.2022.100105

Ndehedehe CE, Ferreira VG, Agutu NO, Onojeghuo AO, Okwuashi O, Kassahun HT, Dewan A (2021) What if the rains do not come? J Hydrol 595:126040. https://doi.org/10.1016/j.jhydrol.2021.126040

Ndehedehe CE, Haile GG, Agutu NO, Ferreira VG, Getirana A, Okwuashi O (2020) Hydrological hotspots of climatic influence in Brazil: a two-step regularization approach. Atmos Res 246:105116. https://doi.org/10.1016/j.atmosres.2020.105116

Nilsson C, Reidy CA, Dynesius M, Revenga C (2005) Fragmentation and flow regulation of the world's large river systems. Science 308(5720):405–408. https://doi.org/10.1126/science.1107887

Nogherotto R, Coppola E, Giorgi F, Mariotti L (2013) Impact of Congo basin deforestation on the African monsoon. Atmos Sci Lett 14(1):45–51. https://doi.org/10.1002/asl2.416

Ojha C, Werth S, Shirzaei M (2020) Recovery of aquifer-systems in Southwest US following 2012–2015 drought: evidence from InSAR, GRACE and groundwater level data. J Hydrol 587. https://doi.org/10.1016/j.jhydrol.2020.124943

Oki T, Kanae S (2006) Global hydrological cycles and world water resources. Science 313(5790):1068–1072. https://doi.org/10.1126/science.1128845

Okwuashi O, Ndehedehe CE (2021) Integrating machine learning with Markov chain and cellular automata models for modelling urban land use change. Remote Sens Appl Soc Environ 21:100461. https://doi.org/10.1016/j.rsase.2020.100461

Pekel JF, Cottam A, Gorelick N, Belward AS (2016) High-resolution mapping of global surface water and its long-term changes. Nature 540(7633):418–422. https://doi.org/10.1038/nature20584

Pradhan A, Srinivasan V (2022) Do dams improve water security in India? a review of post facto assessments. Water Secur 15:100112. https://doi.org/10.1016/j.wasec.2022.100112

Reager JT, Gardner AS, Famiglietti JS et al (2016) A decade of sea level rise slowed by climate-driven hydrology. Science 351(6274):699–703. https://doi.org/10.1126/science.aad8386

Rockström Johan GJ, Qin D, Lade SJ, Abrams JF et al (2023) Safe and just earth system boundaries. Manuscript in Press, Nature

Rossini RA, Fensham RJ, Stewart-Koster B, Gotch T, Kennard MJ (2018) Biogeographical patterns of endemic diversity and its conservation in Australia's artesian desert springs. Divers Distrib 24(9):1199–1216. https://doi.org/10.1111/ddi.12757

Rounce DR, Hock R, Maussion F, Hugonnet R, Kochtitzky W, Huss M, Berthier E, Brinkerhoff D, Compagno L, Copland L, Farinotti D, Menounos B, McNabb RW (2023) Global glacier change in the 21st century: every increase in temperature matters. Science 379(6627):78–83. https://doi.org/10.1126/science.abo1324

Scanlon BR, Faunt CC, Longuevergne L, Reedy RC et al (2012) Groundwater depletion and sustainability of irrigation in the us high plains and central valley. Proc Natl Acad Sci 109(24):9320–9325. https://doi.org/10.1073/pnas.1200311109

Sealey KS, Binder P-M, Burch RK (2018) Financial credit drives urban land-use change in the United States. Anthropocene 21:42–51. https://doi.org/10.1016/j.ancene.2018.01.002

Sheffield J, Wood EF (2008) Global trends and variability in soil moisture and drought characteristics, 1950–2000, from observation-driven simulations of the terrestrial hydrologic cycle. J Climate 21(3):432–458. https://doi.org/10.1175/2007JCLI1822.1

Simkin RD, Seto KC, McDonald RI, Jetz W (2022) Biodiversity impacts and conservation implications of urban land expansion projected to 2050. Proc Natl Acad Sci 119(12):e2117297119. https://doi.org/10.1073/pnas.2117297119

Sofia G, Prosdocimi M, Dalla Fontana G, Tarolli P (2014) Modification of artificial drainage networks during the past half-century: evidence and effects in a reclamation area in the veneto floodplain (Italy). Anthropocene 6:48–62. https://doi.org/10.1016/j.ancene.2014.06.005

Solander K, Reager J, Wada Y et al (2017) Grace satellite observations reveal the severity of recent water over-consumption in the united states. Sci Rep 7:8723. https://doi.org/10.1038/s41598-017-07450-y

Song XP, Hansen MS, Stehman SV et al (2018) Global land change from 1982 to 2016. Nature 560:639–643. https://doi.org/10.1038/s41586-018-0411-9

Spinoni J, Naumann G, Carrao H, Barbosa P, Vogt J (2014) World drought frequency, duration, and severity for 1951–2010. Int J Climatol 34(8):2792–2804. https://doi.org/10.1002/joc.3875

Steffen W, Persson A, Deutsch L, Zalasiewicz J, Williams M, Richardson K, Crumley C, Crutzen P, Folke C, Gordon L, Molina M, Ramanathan V, Rockström J, Scheffer M, Schellnhuber H, Svedin U (2011) The Anthropocene: from global change to planetary stewardship. Ambio 40:739–761. https://doi.org/10.1007/s13280-011-0185-x

Steffen W, Rockström J, Richardson K et al (2018) Trajectories of the earth system in the Anthropocene. Proc Natl Acad Sci 115(33):8252–8259. https://doi.org/10.1073/pnas.1810141115

Taylor R, Scanlon B, Döll P et al (2013) Ground water and climate change. Nature Climate Change 3:322–329. https://doi.org/10.1038/nclimate1744

Trumbore S, Brando P, Hartmann H (2015) Forest health and global change. Science 349(6250):814–818. https://doi.org/10.1126/science.aac6759

UN (2022) World population prospects 2022: summary of results. United Nations Department of Economic and Social Affairs, Population Division. UN DESA/POP/2022/TR/NO.3

Wada Y, van Beek LPH, van Kempen CM, Reckman JWTM, Vasak S, Bierkens MFP (2010) Global depletion of groundwater resources. Geophys Res Lett 37(20). https://doi.org/10.1029/2010GL044571

Wan W, Zhao J, Li H-Y, Mishra A, Ruby Leung L et al (2017) Hydrological drought in the Anthropocene: impacts of local water extraction and reservoir regulation in the U.S. J Geophys Res Atmos 122(21):11,313–11,328. https://doi.org/10.1002/2017JD026899

WCRP Group GSLB (2018) Global sea-level budget 1993-present. Earth Syst Sci Data 10(3):1551–1590. https://doi.org/10.5194/essd-10-1551-2018

Part II
Hydro-Climatic Extremes

Chapter 2
Hydro-Climatic Extremes: Climate Change and Human Influence

2.1 Introductory Remarks

Extreme events like floods and droughts are important parts of human history that date back to ancient civilization. These events have caused major environmental and societal changes for more than 2000 years and contributed to the collapse of some ancient civilizations (Cook et al. 2022; Di Baldassarre et al. 2017). In this day and age, industrialization and the emergence of big cities have contributed to driving changes in the characteristics (e.g., extent, frequency, magnitude, and intensity) of hydro-climatic extremes or events. The duration, frequency, and magnitude of extreme events like droughts have increased since the last half of the twentieth century (Ndehedehe et al. 2023; Spinoni et al. 2014). Every year, drought affects approximately 55 million people globally and its frequency, duration, and magnitude are expected to increase in several regions of the world (Schumacher et al. 2022). Additionally, approximately 650,000 people have died between 1970 and 2019 due to droughts, thus making them one of the world's deadliest natural disasters.[1] The predictions from climate models under different scenarios indicate human societies, be it on a regional or global scales will continue to experience different intensities of hydro-climatic extremes and impacts associated with these events.

Changes in drought intensities and characteristics, for instance, are beginning to challenge the wet-gets-wetter and dry-gets-dryer paradigm. This is because of the anticipated increase in future water consumption or requirements, which currently exceed availability in several geopolitical boundaries. The pressure on water resources is triggering increased upstream consumption thus reducing the availability of water on seasonal time scales at the downstream areas (Smolenaars et al. 2022). The construction of large water infrastructure on transboundary river systems is driven by the need to fix the water challenges in highly populated cities around the world (Godinez-Madrigal et al. 2020). This is regardless of the financial, socio-political,

[1] https://www.weforum.org/agenda/2022/08/nasa-lake-mead-water-drought/.

C. Ndehedehe, *Hydro-Climatic Extremes in the Anthropocene*, Springer Climate,
https://doi.org/10.1007/978-3-031-37727-3_2

and environmental problems or other hidden costs associated with developing such infrastructures, including water transfer projects.

Environmental hazards like floods and droughts are indeed a major cause of concern to modern society. This is because the perception of these risks can be damaging as it can contribute to several unpredicted outcomes and impacts, including loss of lives and property as well as shaping economic investments. The loss of about USD59 billion due to the 2011 flood in Thailand (Hall et al. 2014) is an example of an unpredictable outcome of changes in global climate that impacts socio-economic systems. To date, the impacts of droughts and floods on human societies have continued unabated. Living with these extreme events (droughts or floods) in the twenty-first century is not just a way of life that we are now accustomed to but also a concept that must be understood.

This understanding will underpin an essential framework and blueprint to enhance plans for adaptation, risk-based decisions, mitigation strategies, and drought resilience, looking forward. Furthermore, in the Anthropocene, the causes and drivers of droughts and floods are different. These have been explored in past studies (Van Loon et al. 2016), but the need to adequately understand how humans have contributed to changes in the characteristics of hydro-climatic extremes is critical to help manage its impacts going forward. This chapter will discuss issues around the twenty-first century environment, focusing on contemporary understanding of socio-hydrological systems. It will also explore operational monitoring frameworks and systems, which are critical to improving knowledge of the earth system in the Anthropocene.

2.2 Hydrological Extremes in the Twenty-First Century

Reducing the devastating impacts of extreme events is now the major focus in several global climate change meetings. From the Paris Climate Change conference in 2015 to Glasgow Climate Change conference of 2021, the discourse on climate change takes the lead in the agenda of these meetings. Key decisions revolve around strengthening efforts to build resilience to climate change and its impacts as well as reducing greenhouse gas emissions, among others. From these meetings, there is a concerted effort among countries to achieve these key objectives, working in close collaboration. Additionally, the agreement by participating countries is mutual, particularly in limiting emissions to 1.5 degrees. However, there is still a critical gap in contemporary understanding of the various characteristics and physical mechanisms behind complex hydro-climatic processes in the twenty-first century, an era that has come to be known as the Anthropocene (see Chap. 1). The composite influence of climate and human activities on the environment are fundamental characteristics of the Anthropocene. In other words, the interplay between climate and various human actions is changing the characteristics, structure, and patterns of extreme events. For the most parts, the global water cycle and the Earth systems in general have been affected by these changes. The rise in urban settlements and changes in dietary conditions led to the expansion of cultivated areas around the globe, thus increasing human water use and demand, especially in arid and typically dry regions of the world (Bierkens and Wada 2019). Even some freshwater-rich and highly humid river basins of the world like the Amazon and Congo basins are experiencing water crises due to the impact of prolonged extreme droughts (e.g., Getirana et al. 2021; Ndehedehe et al. 2019, 2020c).

The water crises in Brazil (Fig. 2.1) confirms that even the most humid regions are not exempted from the deleterious effects of climate change (e.g., drought). Despite having one of the largest freshwater stores in the world, one of the biggest energy crisis in Brazil was caused by drought (Getirana 2016). About two-thirds of surface flows in the Amazon River (domiciled in Brazil) can arguably meet the world's freshwater demand, yet the ongoing water crises in Brazil caused by drought is the worst in 20 years. Getirana et al. (2021) highlighted the current water crises in Brazil and identified some of its impacts. This included several important reservoirs across the country (Fig. 2.1) being at less than 20% capacity, more than a 100% rise in electricity bills, and about 67% increase in prices of major exportable agricultural products like the Soya beans. Increasing water demand is not just a function of climate change, but rising human population and industrialization are now important predictors. This increasing influence of several human activities under a changing climate complicates our understanding of climatic feedback processes and restricts efforts to pursue water and food security. These issues are further discussed in what follows, and illustrations from region-specific case studies are highlighted.

2.2.1 *Will Climate Change Affect Groundwater Recharge in the Future?*

Groundwater is an important source of global freshwater resource for more than two billion people on the planet. It provides much of the world's drinking water supply and supporting food production on a large scale through irrigation water supply. Rapid declines in groundwater in the last 15 years have been reported in India, Australia, California Central Valley, Middle East, and North China Plain (Chen et al. 2016). Given the need to improve decision-making, groundwater assessments on local and region scales are deemed more efficient to support the development of water policy (Meixner et al. 2016). While several research programmes have explored the impacts of climate change on surface water hydrology (e.g., Ndehedehe et al. 2020b, 2021a; Pekel et al. 2016), similar assessments for groundwater are limited. Climate change has substantial impacts on groundwater recharge, and this can be exacerbated with intensive human water abstraction (Ndehedehe et al. 2021a). But what will groundwater recharge in the future look like under a changing climate? To improve the understanding of this impact on recharge for the US, Meixner et al. (2016) addressed important hypotheses including "how projected climate changes interact with individual recharge mechanisms". A conceptualization of different recharge

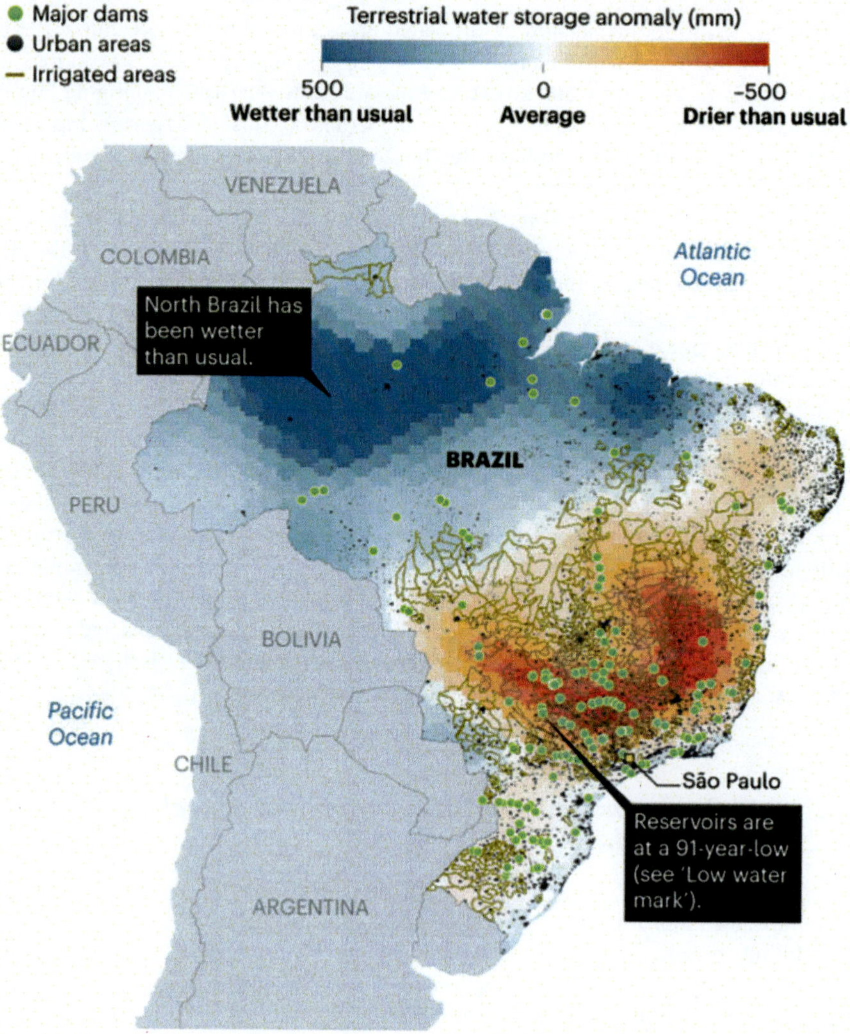

Fig. 2.1 Water crises in Brazil caused by drought. *Source* Getirana et al. (2021)

mechanisms for the twentieth century climate is illustrated in Fig. 2.2. Based on a detailed case-specific review for the US, four categories of recharge mechanisms were identified, and they include diffuse, focused, mountain system recharge (MSR), and irrigation (Fig. 2.2a, b).

As was described by Meixner et al. (2016), diffuse recharge originates from precipitation and then percolates to the water table. Focused recharge comes from perennial surface water interactions and occurs when precipitation concentrates on the Earth's surface through the process of runoff, which then infiltrates and percolates, driving

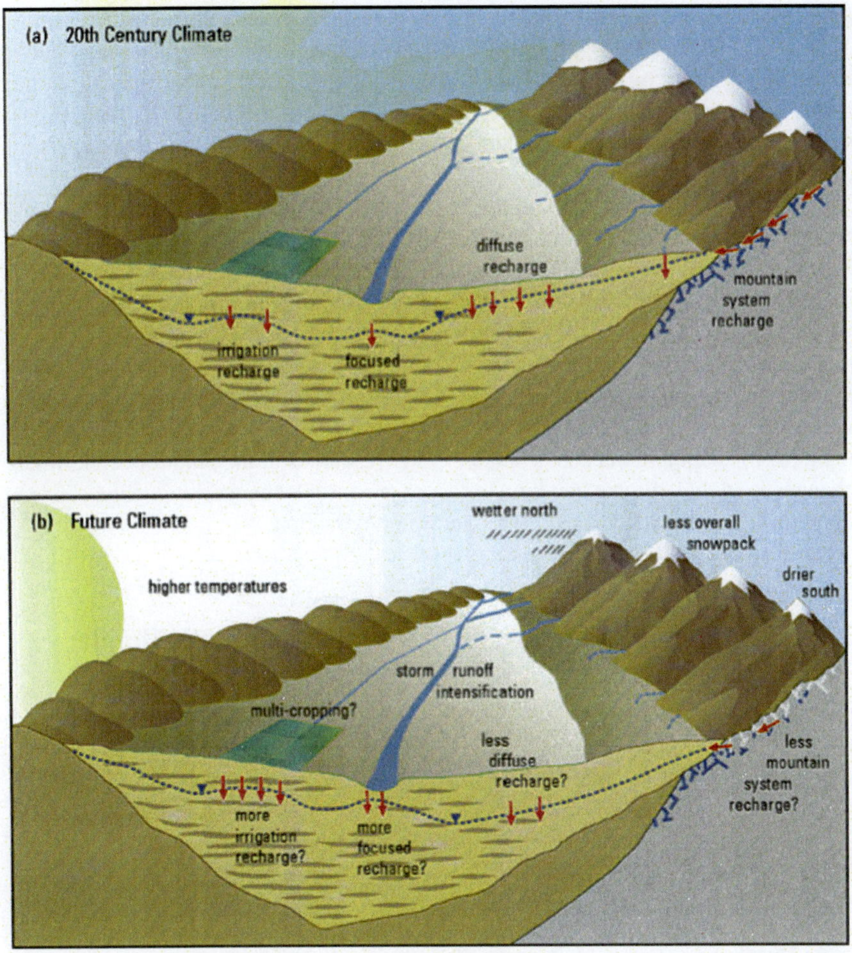

Fig. 2.2 An illustration of groundwater recharge in the future and expected impacts of climate change. *Source* Meixner et al. (2016)

recharge of runoff on the landscape. But MSR is related to recharge from stream loss at mountain fronts. Notably, the MSR also includes transfer of groundwater stocks from the mountain block to the surrounding adjacent alluvial aquifer (Meixner et al. 2016). Another form of recharge identified in the literature is irrigation recharge. This is defined as excess irrigation water that moves to the water table and can emanate from both surface water and groundwater (Meixner et al. 2016; Scanlon et al. 2002). Overall, recharge can be sourced from precipitation, groundwater, irrigation, and surface water, be it natural or man-made. As precipitation and surface waters are expected to be impacted by climate change, these various forms of recharge will ultimately be negatively affected by future climate. As with recharge emanating from rainfall and surface water, decline in mountain system recharge is also expected under future climate change scenarios (Fig. 2.2b). For the West Coast of United States, this decline in mountain groundwater recharge is associated with decreased snowpack, but this impact reduces with higher elevation and latitude (Meixner et al. 2016). The expansion of irrigated agriculture in the future could also imply a rise in irrigation recharge (Fig. 2.2b).

Projected changes in the climate system are ultimately expected to drive substantial changes in groundwater systems and all forms of subsurface water storage and hydrological stores, including unconfined and confined aquifer storage. This is because under several climate change scenarios, anticipated rise in the variability of water budget variables like precipitation, temperature, and evapotranspiration will result in changes in groundwater systems (Green et al. 2011). Therefore, the broad interest on how changes in these water budget indicators will impact groundwater will continue to be the focus in the coming years. This is because of the increasing need to adapt and manage available water resources in the Anthropocene where changes in global climate and human responses to such changes, arguably, can lead to increased and a rather reckless groundwater extraction.

2.3 Hydrological Systems: Human Actions Versus Climate Change

Human activities have interfered with the propagation process of drought characteristics and the transition from meteorological to hydrological (Van Loon et al. 2016). Non-climatic factors like groundwater withdrawals and reservoir storage have the potential to influence hydrological regime. As was reported by Kubiak-Wójcicka and Bąk (2018), interactions between non-climatic factors and extreme climatic conditions can influence the onset or termination of a hydrological drought despite an ongoing meteorological drought event. This means both groundwater extraction and drought affect groundwater systems. So, to understand drought impacts on groundwater changes in agricultural fields or cultivated areas, implications of groundwater abstraction for agriculture on the water cycle of the catchment must first be understood. This is because across catchments, such impacts can be different. For instance, Park et al. (2021) found the impact of drought on groundwater level in cultivated catchments in Korea to be significantly different compared to forest catchments. This highlights the inter-relationships of catchment characteristics with climate and human actions.

2.3.1 Hydrological Processes in Semi-arid Domains

The response of flows to rainfall conditions is expected to be linear for most catchments or regions. This will imply that water supply for agriculture will be limited during prolonged drought events. However, cases of rising groundwater or upsurge in water table during prolonged drought events have been reported in some areas of the West African Sahel. Pioneering studies (e.g., Favreau et al. 2009; Leduc et al. 2001) that reported this hydrological paradox have argued in favor of land clearing and land use change as principal triggers of this phenomena. The aforementioned studies identified land clearing as an important driver of long-term changes in surface water hydrology (surface runoff) of the dry Sahelian region. When water table rises, it is usually considered as the aftermath of increased recharge, obviously caused by improved rainfall. But a fall in water table hides increase in recharge, when groundwater withdrawal under this condition (increased recharge rates) is intensive. The relationship between land clearing for cultivation or agriculture and changes in groundwater is now established, following the numerous examples detailed in past literature (e.g., Scanlon et al. 2005; Favreau et al. 2009). With human activities like land use and land cover change driving increase in groundwater recharge in some areas of Australia and the US (Scanlon et al. 2005; Allison et al. 1990), the relationship between rainfall conditions and water availability may be somewhat complex in some hydrological settings. The combination of observational climate data (1960–2017) and models to assess changes in drought and flood conditions in continental Australia revealed that changes in rainfall conditions alone are inadequate to express changing drought conditions (Wasko et al. 2021).

2.3.2 Droughts in the Murray Darling Basin

The prolonged droughts in Australia popularly referred to as Millennium Drought left the southeastern part of the country with an altered watershed streamflow regime. The recent work of Peterson et al. (2021) provides more perspective on this. Their study found that runoff is yet to recover (that is, returning to pre-drought levels) in 37% of watersheds (approximately one-third), 7 years after the notorious Millennium Drought ended. This was despite substantial improvement in rainfall conditions after this drought. They argued that watersheds are resilient to hydrological disturbances but the impacts of hydrological droughts could be infinite much after the termination of meteorological droughts. The Millennium Drought was the longest drought experienced in Australia since the European settlement in the late eighteenth century. Much of the Millennium Drought period between 2001 and 2009 showed drought extent of more than 50% (top, Fig. 2.3). Moreover, averaged drought duration based on the Australian Bureau of Meteorology rainfall product shows hotspots along the east coast and northern territories where droughts have persisted for at least 12 months (bottom, Fig. 2.3). Even though this drought was immediately followed

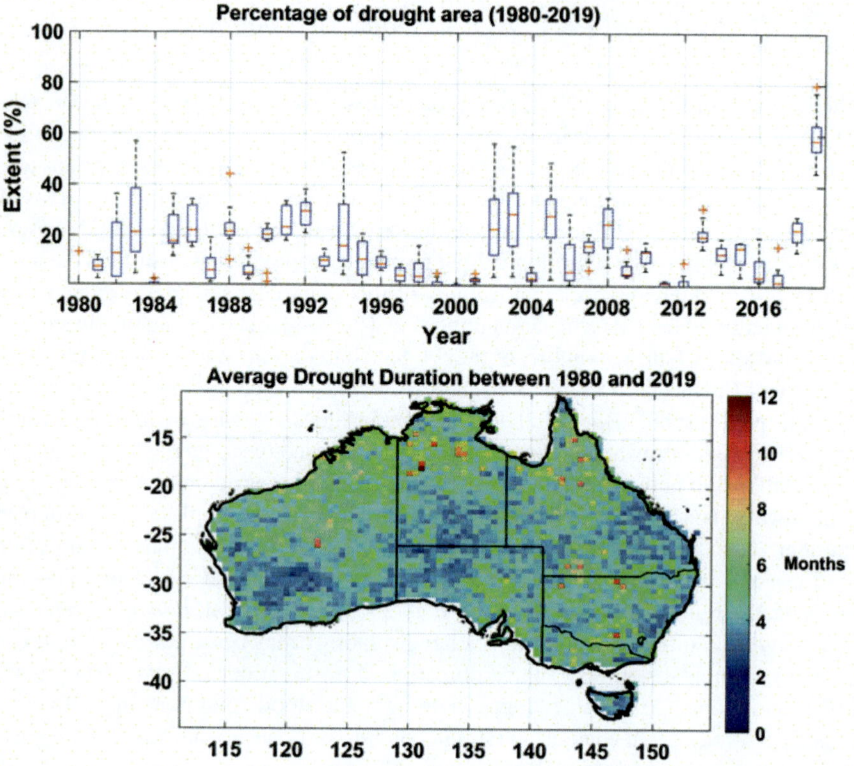

Fig. 2.3 Long-term changes in drought impacts (1980–2019) based on the Standardized Precipitation Index (SPI). (Top) temporal variations of drought extents and (bottom) the spatial distribution of drought duration. The SPI is based on precipitation data source from historical observations of the Australian Bureau of Meteorology

by a strong La Niña event that led to extreme big wet periods in the country (Fig. 2.4), it seems extreme and prolonged drought can permanently damage hydrological systems, especially in semi-arid domains. California's Central Valley is an ideal example that illustrates this. The groundwater loss in California's Central Valley during the 2007–2010 extreme drought triggered a compaction in the aquifer system that could not be reversed, resulting in a permanent loss of about 2% of aquifer-system storage (Ojha et al. 2018).

Furthermore, the impacts of climate oscillations or large-scale atmospheric circulation patterns on hydrological conditions (wetness in this case) in Australia appear to be changing. The linearity of the Australian climate makes it predictable in springtime when indices of oceanic variability like the El Niño–Southern Oscillation (ENSO) typically shows large amplitudes. For example, the strong La Niña event in 2010/2011 made Australia very wet, nourishing much or all of its hydro-ecological systems (floodplains, wetlands, rivers, and lakes). Reservoirs, lakes, and river channels were filled and the increase in flows were clearly evident in the productivity of

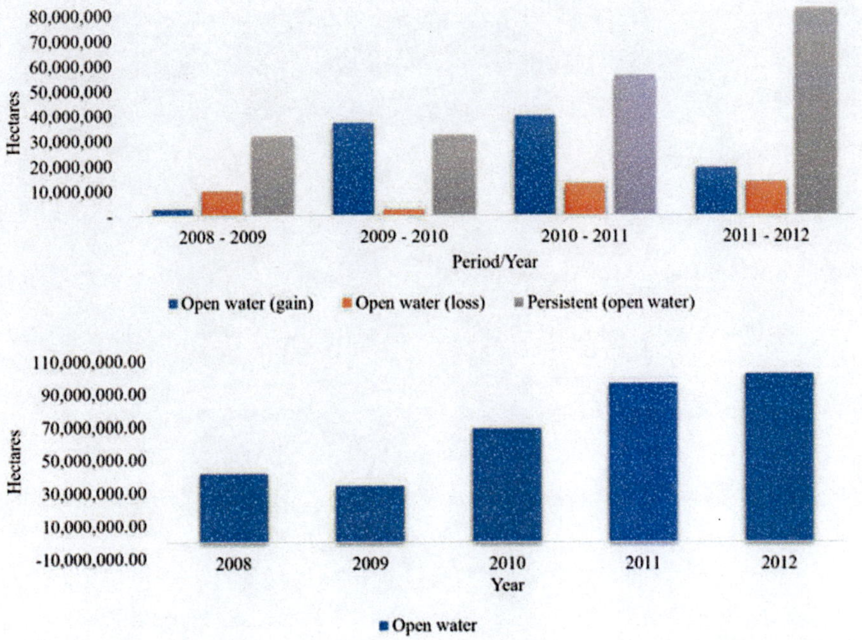

Fig. 2.4 Open surface water distribution in the Murray Darling basin between 2008 and 2012. **a** Change detection of surface water extent and **b** the temporal variations of surface water extent. *Source* Ndehedehe et al. (2021a)

floodplain wetlands in the wet–dry tropical areas of the country (e.g., Ndehedehe et al. 2021b). As with 2000/2001 wet period in Australia (Fig. 2.4), the temporal variations of drought extents shown in the top panel of Fig. 2.3 obviously allude to no drought event in 2011. There was more surface water between 2010/2011 (when the wet conditions started) and 2011/2012 (when the wet conditions were highest) (Fig. 2.4a). Changes in surface water extent in between years during the 2008–2012 period and the substantial rise in surface water extent after 2010 (Fig. 2.4a) were linked to climate teleconnection patterns (e.g., Ndehedehe et al. 2021a; Kiem et al. 2016; van Dijk et al. 2013). However, despite the La Niña, which occurred during the spring of 2020, Australia was not wet due to changes in global climate system (Lim et al. 2021). Changes in the interactions of atmospheric patterns like the negative phase of the Indian Ocean Dipole ending prematurely can trigger a relatively "drier" La Niña event in Australia during Spring (Lim et al. 2021). Therefore, changes in the characteristics of climate system regulate the occurrence of drought and flood events, and the response of hydrological systems to either of these extreme events may differ depending on their severity, duration, and magnitude as well as human activities, for example, change in agricultural practice, deforestation, construction of water infrastructures for agriculture, and water harvesting.

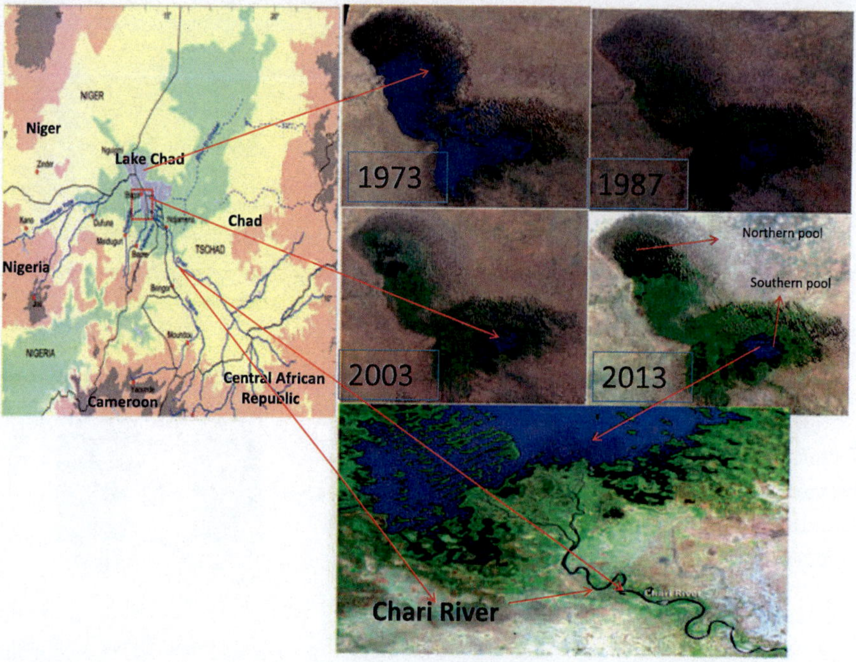

Fig. 2.5 Historical changes in the Lake Chad's surface water extent during the 1973–2013 period. *Source* Ndehedehe et al. (2016)

2.3.3 *The Sahelian Drought*

Prior to the prolonged drought episodes of the 1980s, Lake Chad was a full lake with a surface area of about 25,000 km^2 as of 1950s. However, after the droughts of the early 1980s, the lake lost a substantial proportion (more than 95%) of it aerial extent, declining to approximately 1700 km^2 as of 2000 and 2001 period (Ndehedehe 2019; Wald 1990). To date, Lake Chad remains largely desiccated and now segmented into smaller northern and southern pools that infrequently get inundated (Fig. 2.5). It permanently lost much of its surface water, and only water transfer projects as proposed (Ndehedehe et al. 2016; Lemoalle et al. 2012) can revert the lake to its original extents. The construction of smaller dams and water infrastructures in the upstream regions, the Chari/Logon river system (Fig. 2.5), that provide about 90% of the water that feeds into Lake Chad exacerbated the lake's desiccation (Ndehedehe et al. 2020b). Outgoing water flux like evapotranspiration is one of the highest in the region and the surrounding lake, and can also deprive the basin of freshwater. As we now know, Lake Chad has been in deficits for more than 30 years (Fig. 2.5), lacking sufficient freshwater to support agriculture and the functioning of the surrounding ecosystems. Changes in land use increased water abstraction, and deforestation interacts in non-complex ways to drive ocean-atmospheric processes, which in turn regulate the severity and magnitude of these extreme events (e.g., Ndehedehe et al. 2023, 2020a; Kiem et al. 2016).

2.3.4 Lessons from the Lake Chad Predicament

There are some key lessons from the Lake Chad predicament. First, the influence of large-scale climate teleconnection patterns like the ENSO and Indian ocean induced drought conditions. As argued by Giannini et al. (2003), the synergy between abnormally warm sea surface temperature of the eastern Atlantic and Indian ocean played a critical role in the prolonged drought of the early 1980s. Second, human water withdrawal for agriculture was preceded by insufficient rains and climate variability. After the prolonged drought, agricultural activities increased around the Lake Chad basin and the need for freshwater also doubled during the period. For a river basin that is overtly reliant on rainfed agriculture and surface water, during extreme droughts, especially when it becomes prolonged, the expected response will be increased demand for more water. Unfortunately, the Lake Chad is the only available and easily accessible freshwater in the region, further exacerbating drought impact on surface water. Finally, this led to what is now called the current Lake Chad, which is less than 153,000 Ha as of 2019. Change detection analysis of open water features undertaken by Ndehedehe et al. (2020b) revealed this new surface extent of the lake and also showed that between 2015 and 2019, on average, only 28.4% of inundated areas persisted in the southern pool.

2.3.5 A Tale of Two Extreme and Prolonged Drought Events

In terms of impacts, intensity and duration, the Sahelian drought of the 1980s appears to take after the similitude of the Millennium Drought in Australia. The Millennium Drought affected the availability in surface water (Fig. 2.4a) and its distribution in freshwater habitats and isolated water holes and river channels (e.g., Ndehedehe et al. 2021a; van Dijk et al. 2013). In major river basins like the Murray Darling basin, limited surface water due to drought led to substantial groundwater use. However, despite the severity of this drought, Australia came out of the Millennium Drought, having developed a robust water resources plan. However, the hydrology of Lake Chad and its corresponding basin (Lake Chad basin) has still not recovered from the impacts of the Sahelian drought. Lake Chad (Fig. 2.5) is the fourth largest lake in Africa and with more than 40 million people from riparian countries relying on the water resources of the lake (Riebe and Dressel 2021). An Important question is will the Lake Chad ever recover from the impact of past drought? From a hydro-climatology viewpoint, the vanishing Lake Chad is still very much a hydrological puzzle that still requires more scientific investigation to fully understand the role of drought and human water mismanagement in its desiccation. The ugliness and peril of climate-induced water scarcity by itself is considerably damaging to the economy and community life. In the early 1960s, the Lake Chad basin area, for instance, was a beehive of economic activities, agriculture, social integration, and trade, apparently enhanced by the presence of a nearly full Lake Chad. After the demise of a substantial proportion of the lake's surface water storage, these activities ceased or reduced considerably, leading to migration, poverty, and social conflicts. The aftermath of this loss in the basin's freshwater led to violent conflicts and tensions through competition by pastoralists and herdsmen. However, poor governance, and weak and corrupt institutions also played major roles in the Lake Chad predicament.

2.3.6 *Lake Urmia: A Case of Anthropogenic Recklessness*

Lake Urmia in Iran is another case study of altered surface water hydrology. The debates on key drivers of water loss in the lake generally allude to the composite influence of human activities and droughts, the latter being a secondary driver. This section looks into the main drivers of desiccation in Lake Urmia and also highlights the need for a monitoring framework that can alert stakeholders and the public sector to the dangers of mismanagement to help facilitate water governance and management.

2.3.6.1 What Caused the Desiccation in Lake Urmia?

Measurements using multi-sensor satellite observations revealed about 70% (220 km^2/yr) loss in the aerial extent of the Lake between 2002 and 2014 (Tourian et al. 2015). The Lake Urmia region is typically dry and arid, and, along with climate change, for example, increased evaporation, the construction of a causeway in the lake for transportation altered the flow patterns of the lake. This construction divided the lake into two, northern and southern sections with only 1500 m gap available for the movement of water (Eimanifar and Mohebbi 2007). Additionally, the developments of four dams accounted for approximately 65% and 25% of the observed desiccation in Lake Urmia, respectively (Hassanzadeh et al. 2012). As with other Middle Eastern countries and the West African Sahel, Iran's aridity is very normal as well as the limited rainfall it receives. Despite this, overwhelming evidence clearly suggests anthropogenic recklessness is the principal reason for the loss of freshwater in Lake Urmia. The construction of a causeway in the lake as well as dams that take substantial proportion of flows away from the lake can at best be described as reckless. Earlier on, AghaKouchak et al. (2015) described the construction of dams to support hydro-economic projects as *"aggressive"* and *"ill-conceived"*. The report by Hassanzadeh et al. (2012) mentioned that more than 200 new dams to support irrigation projects were either completed or at the approval stage or under construction. With no significant change in drought pattern observed in the region (AghaKouchak et al. 2015), the peril of human actions to surface water hydrology cannot be overemphasized. Several decades of unsustainable water use through the construction of water infrastructures desiccated the lake, reducing it to approximately 10% of its former size. This decline became very obvious after water was diverted from the lake's major source of inflow to support irrigation (AghaKouchak et al. 2015). In retrospect, this lake was once the largest lake in the Middle Eastern region and today much of its surface extent exists as extensive salt marshes.

2.3.6.2 The Dangers of Mismanagement and Need to Facilitate Water Governance

The anthropogenic recklessness in the Lake Urmia case is a clarion call to building strong institutions and implementing monitoring frameworks that will ensure natural resources and environmental asset of this nature are protected and sustainably managed. The impacts of human footprints as seen in Lake Urmia can be subtle and sometimes go undetected until the situation becomes worse. The satellite-derived temporal changes in the lakes' surface extent since 1972 suggest this (Fig. 2.6). The lake appears to have been full during the 1970–1990 period with no noticeable change in spatial extent (Fig. 2.6). But a gradual shrinking after 1998 became obvious and by 2014 more than 80% of the surface area was lost (Fig. 2.6). A monitoring framework therefore can alert stakeholders and the public sector to the dangers of mismanagement and help facilitate water governance approaches. As a final thought on this, Lakes Urmia and Chad are, arguably, closely related in some ways. They are both inland freshwater bodies in semi-arid domains as well as obvious prototypes that demonstrate the composite influence of climate change (drought) and anthropogenic factors as key drivers of water loss. Additionally, the receding Lake Urmia has also contributed to declines in groundwater resources, but this is an interesting concept yet to be explored or investigated in the Lake Chad area. With increasing human population, the pressure on the environment, freshwater, and ecological resources is expected to rise. This will continue to amplify the variability of hydro-climatic conditions and their impacts on a broad range of ecosystem services.

2.3.7 Socio-economic Implications of Hydro-Climatic Extremes

Notable impacts of droughts include decline of primary production, land degradation, loss of national income, famine, food insecurity, hydro-power failure, and disturbed riparian habitats among others (e.g., Ndehedehe et al. 2019; Schroth et al. 2016; Brown et al. 2011). Much of these impacts are those resulting from the lack of or reduced freshwater and hydrological stores (e.g., surface water, reservoirs, groundwater, and soil moisture). The losses of historical surface water in some regions and territories as revealed in the global-scale satellite analysis of changes are impacting social behaviors in some of these regions (Pekel et al. 2016). This has contributed to

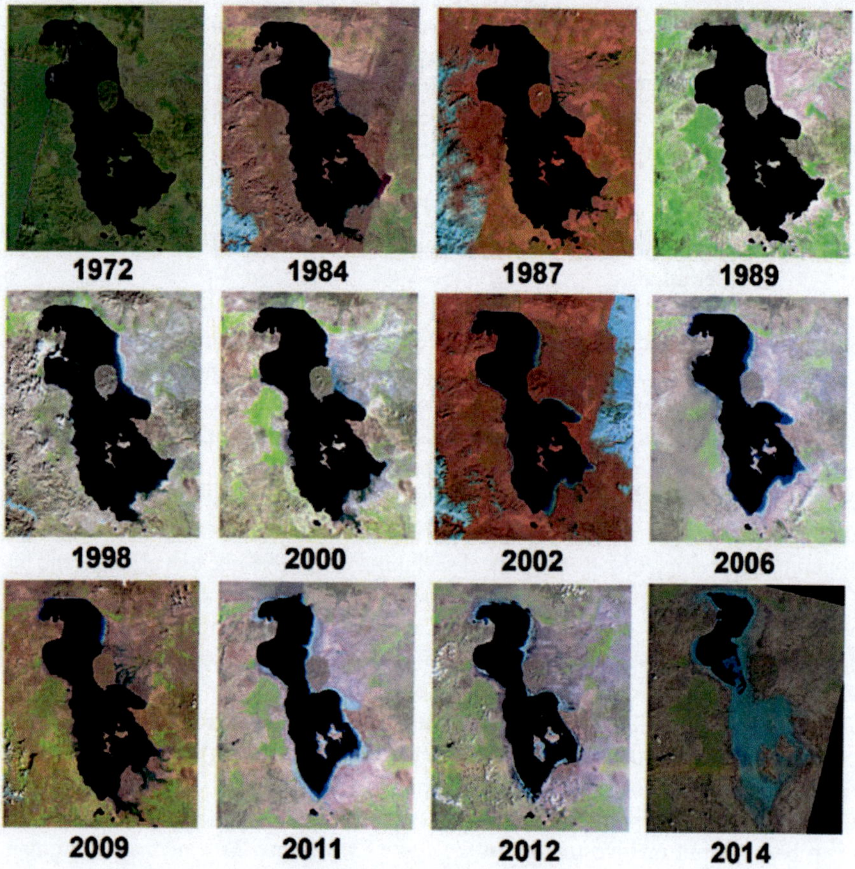

Fig. 2.6 Satellite observations of historical changes in Lake Urmia. Areas with the presence of water are shown in black. *Source* AghaKouchak et al. (2015)

some changes in local water management policy as well as discussions on directions for agricultural production. For instance, some tropical wet areas of the Gulf catchment in northern Australia have been considered and prioritized for future surface water developments to support agricultural expansion.[2] The competition for surface water of Lake Chad and the resulting communal conflicts and agitations in the Lake Chad basin as will be discussed later in this chapter emphasize the importance of surface water as important predictors of social behaviors. In this section, some specific examples of socio-economic impacts of droughts on human populations are discussed further.

2 https://www.industry.gov.au/sites/default/files/June%202018/document/pdf/nawp-fullreport.pdf?acsf_files_redirect.

2.3.7.1 Famine and Food Insecurity

The impacts of droughts can be far-reaching and intensely pervasive, impacting distant populations through price volatility and trade disruption (FAO 2016). This obviously broadens the impacts of droughts, subjecting low-income, rural agro-communities to poverty and other risks associated with increased prices of agricultural goods caused by drought. As an example, China experienced an unusually dry winter between 2010 and 2011, which reduced the production of wheat. The impacts of these shortages increased the prices of bread in Egypt, the world's largest importer of wheat (Papaioannou 2016). Generally, in economically less developed regions, if there is any food scarcity caused by drought or extreme event, the low-income households are usually the worst hit and more likely to be subjected to hunger.

In 2020, famine caused by droughts killed more than 10 people in Kenya (East Africa) and more than a million people were devastated by hunger and lack of water (Ndehedehe et al. 2020a). Other predictors of famine in East Africa, including constraints on asset and labor, and inability to enforce policy and adopt risk-averse production systems due to poverty were identified among agro-pastoral households in Kenya (Ifejika Speranza et al. 2008). While famine can also be linked to complex interactions involving politics and socio-economic factors, water deficits caused by extreme drought overwhelmingly account for famine in much of Africa's sub-region where agriculture is largely reliant on rainfall (Agutu et al. 2017; Schroth et al. 2016; Shiferaw et al. 2014; Brown et al. 2011; Ifejika Speranza et al. 2008). The impacts of droughts on crop production or agricultural systems have been measured and for East Africa and semi-arid regions like Australia, the results confirmed that climatic factors like drought will continue to affect crop production, triggering food insecurity.

For instance, this relationship was assessed in Greater Horn of Africa by measuring the response of annual crop production with several drought indicators. A case study was undertaken by Agutu et al. (2017) over Kenya, Tanzania, and Uganda in Greater Horn of Africa where drought indicators (meteorological, agricultural, and hydrological) were used as input to predict variability in annual crop (maize and wheat) production. In their study (Agutu et al. 2017), nine drought indicators were constructed from different precipitation (GPCC and CHIRPS); soil moisture (MERRA-2, ERA, GLDAS, FLDAS, and CPC); and terrestrial water storage—TWS (GRACE and MERRA-2) products as well as VCI—vegetation condition index (Fig. 2.7a–f). Precipitation (CHIRPS, GPCC), soil moisture (FLDAS), and vegetation-based drought indicators were better predictors of annual crop production, having accounted for much of the variabilities, be it in Kenya, Uganda, or Tanzania (Fig. 2.7a–f). There are regions, for example, in Kenya and Uganda where drought indices based on one and three months' aggregation showed considerably strong association with crop production (Fig. 2.7d, f). Similar outcomes have been reported in Australia where a multivariate approach to measuring the response of crop production to drought events confirmed that meteorological and agricultural drought indicators are better predictors of crop production (Ndehedehe et al. 2021a). Ultimately, the suitability of these drought metrics for agricultural drought (when lack of rainfall starts affecting available water in the soil resulting in poor yield) assessment in much of Greater Horn

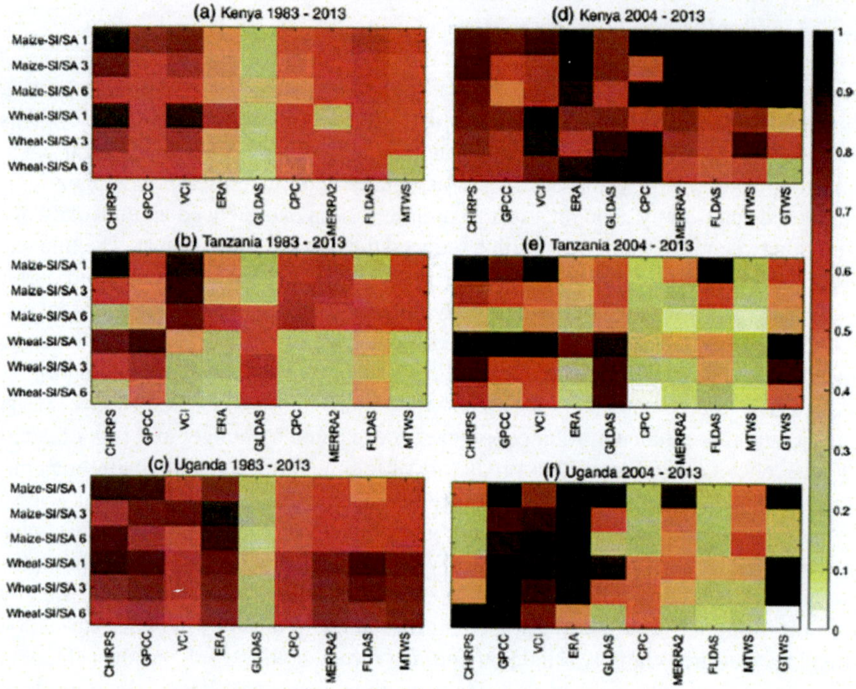

Fig. 2.7 Relationship of annual crop (maize and wheat) production (using coefficient of determination-R2) with different drought indicators aggregated on different time scales (3, 6, and 12 months). Focused region is the Greater Horn of Africa (Kenya, Tanzania, and Uganda). The crops (maize/wheat) are listed on the x-axis where SI and SA are standardized indices and standardized anomalies, respectively. Drought indicators are shown on the x-axis of the plot. *Source* Agutu et al. (2017)

of Africa (Fig. 2.7a–f) provides a monitoring capability to assess drought impacts on agro-based income and vulnerability to famine in the region.

Drought affects water availability. The latter is the basis for food security and environmental functions, and has been identified as the key driver of global patterns of agricultural production and trade (Odorico et al. 2020). When drought impacts water supply or when its availability is prolonged or even more frequent, increases in prices of groceries and food supplies are inevitable (Ndehedehe et al. 2021a). This is because the quantity of exportable agricultural goods produced in drought-affected regions will be limited. When the exports of agricultural products are limited, the regions, which rely on these goods and services, will be affected. In other words, the indirect consumption of water resources from drought-affected regions by nations not experiencing drought could alter trade patterns and exports of agricultural products. With rising prices in goods and services, local communities in economically less developed countries are the most impacted, increasing poverty and nutrition problems. Even agricultural activities at the subsistence level in most African sub-regions are also affected. On a much larger scale, agricultural losses of $40 and $30 billion was reported in the US due to the 1988 and 2012 extreme drought events, respectively (Rippey 2015). These are some examples of drought impacts on national income. Our ability to predict drought impacts is therefore important to guide future investments in drought resilience as well as engage in the prudent management and allocation of available water resources.

2.3.7.2 Migration and Pastoralism

Climate shocks like strong changes in annual temperature and rainfall can lead to migration because of people seeking for greener pastures. Climate variability is an important predictor of human migration. In Southern American countries, human migration responses to climate variability was quantified and monthly temperature and rainfall shocks were found to induce migration (Thiede et al. 2016). The movement of pastoralists and herders in the West African Sahel in search of water for their livestock can be exacerbated when water availability is limited due to prolonged drought. The hypothesis surrounding human migration patterns and its association with climate change has not been disputed and such connections have been well established in the literature (e.g., Tubi and Feitelson 2016; Thiede et al. 2016; Reuveny 2007). As was reported in the case of Bedouin herders and Jewish farmers in Israel's northern Negev, migration associated with drought can lead to conflict (Tubi and Feitelson 2016). Similarly, the conflict in the Lake Chad basin area among herdsmen and pastoralists is the aftermath of migration, resulting in increased competition of available water resources (Ndehedehe et al. 2020a; Okewu et al. 2019). Drought-induced migration could cause agitations and harmful disputes in receiving areas, especially when it involves people from economically less developed regions (Reuveny 2007).

2.3.7.3 Wars and Communal Conflicts

Historically, water has been the basis for wars among nations as well as local agitations and conflicts. Nations have besieged one another, using water as a weapon in battle and destroying the water infrastructures of enemy nations (Gleick 1993). For instance, in the 2022 war between Russia and Ukraine, a major aggressive military operation by the Russian forces in an attempt to invade Ukraine was destroying the dam on the North Crimean Canal. Crimea being an ally of Russia benefited from this as water flowed back to Crimea. In addition to several political reasons that initiated the war, water became a very handy weapon that Russians leveraged in their military operation against Ukrainians. Other water-related disputes in some geopolitical zones with shared water resources also exist. This can either be in the form of cutting off or limiting access to sufficient supply or deliberate destruction of existing water infrastructure or systems as evident in the Russian–Ukraine conflict. However, water-induced conflicts can also be triggered when its availability is limited due to climate change or when its demand exceeds availability.

Climate change will induce considerable evaporative losses and the rise in water demands caused by increasing temperatures exacerbates these losses. And when rainfall deficit is prolonged, reservoir storage, surface runoff, and recharge decline. As a direct consequence, there will be failure in hydro-power, reduced agricultural production, famines, and declines in groundwater, among others. As climate change may affect human development and people's livelihoods, Barnett and Adger (2007) implied earlier on that climate change can contribute to human security problems, arguing that such impacts may increase the risk of violent conflict in some circumstances. Given the vulnerability of freshwater to droughts, increased freshwater demand, and the influence of geopolitics, regions like countries in the Middle East are prone to water-related conflict as well as some countries in the African Sahel. The growing competition for the limited freshwater resources in much of these regions, including Asia and East Africa, makes the likelihood of water-related disputes nearly inevitable (Gleick 1993). This is the reality in today's Lake Chad basin where limited freshwater in Lake Chad has created violent competition and interstate tensions among riparian communities (Ndehedehe et al. 2020b).

As a case study, Lake Chad provides an example of how climate change can trigger not only economic challenges but also social problems. The development of water resources infrastructures were part of strategic solutions or approaches to resolve drought impacts on the riparian communities of the basin. But this further restricted inflows to the lake, amplifying water shortage or water scarcity already created by prolonged extreme droughts and poor rains (Ndehedehe et al. 2020b). Human interventions of this nature culminated in social problems, increasing local agitations, and armed conflicts as well as migration in the region because of the limited shared water resources of Lake Chad. In riparian countries of the Lake Chad basin like Nigeria, a significant relationship between rainfall deviations and intensity in conflict was found, especially in subsistence agricultural settings, and appears stronger in agro-ecological zones which are very vulnerable to climate variability (Papaioannou 2016). The connection between climate and conflict appears to be well established in the literature. From the studies (see, e.g., Tubi and Feitelson 2016; Papaioannou 2016) that showed this connection, drought appears to be a common denominator used in illustrating how agricultural activities in local communities can be disturbed by lack of water. This disturbance created tensions that potentially led to major violent disputes. Furthermore, transboundary river basins, which cover approximately 47.1% of global land area (excludes Antarctica) account for about 60% of global freshwater discharge (Guo et al. 2022). And much of these conflicts are common in transboundary river basins, where water resources are shared among riparian member nations, for example, major river basins in Africa—Lake Chad, Nile, and Victoria. Because transboundary water resources conflicts are inevitable, the use of water diplomacy to foster cooperation and management of water-related conflicts emerged as an important management initiative that underpins regional water security and stability (e.g., Ndehedehe et al. 2020a). As a step further to build upon these initiatives, a framework to enable the tracking of conflict on transboundary rivers was proposed by Guo et al. (2022). The outcome of their work shows the data generated from this framework is critical to strengthening and managing transboundary water conflict .

2.3.7.4 Floods and Socio-economic Impacts

The impacts of floods on nations and big cities are rapidly escalating, costing billions of dollars. The disruption of public services and physical infrastructures, including properties as well as human lives makes it economically catastrophic to humans (Allaire 2018; Hall et al. 2014). Globally, be it in arid or humid regions, flood events are increasing and nations continue to strive, working to improve current efforts on mitigation and adaptation measures, including flood control infrastructure, and optimizing predictive models for early warning. Away from physical impacts and damages, floods do have socio-economic implications. A comprehensive review of socio-economic costs or impacts of floods has been detailed by Allaire (2018). Even though quantifying these costs can be broadly complex in some ways, four areas of socio-economic impacts were identified and these can either be direct or indirect (Allaire 2018). The first is business interruption, which is mostly impacts associated with delayed and forgone production of goods and services. The second is the cost connected to damages of building and its contents as well as other infrastructures like roads. Even though insurance cover minimizes the burden of this cost in developed economies, rural communities in less developed countries can be made homeless for sometime as the cost of renovation could be unaffordable. The cost of cleaning roads and debris or even revamping broken bridges, etc. can be tremendous.

The extent of damage caused by floods depends on flood intensity and magnitudes. The third is impacts on the well-being of households, or welfare. Floods can lead to changes in (i) prices of goods and services, (ii) consumption, and (iii) income and employment, which have negative impacts on people (e.g., Tellman et al. 2021; Allaire 2018; Hall et al. 2014). Such negative effects can be facilitated by the disruption of public services, thus intensifying impacts on the quality of life, more generally. As with drought impacts on food production (e.g., Ndehedehe et al. 2021a; FAO 2016; Ifejika Speranza et al. 2008), floods can also limit the quantity of food items produced or the amount imported, leading to higher prices for low-income earners. The loss of jobs during such disasters can be daunting, affecting people's ability to acquire goods and services. Finally, flood has impacts on health and can be even direct and devastating including illness, death from drowning, and electrical injuries, among several others. Extreme floods are obvious manifestations of a changing climate and the risks associated with its frequency and intensity could be devastating to society and the economy (Hall et al. 2014).

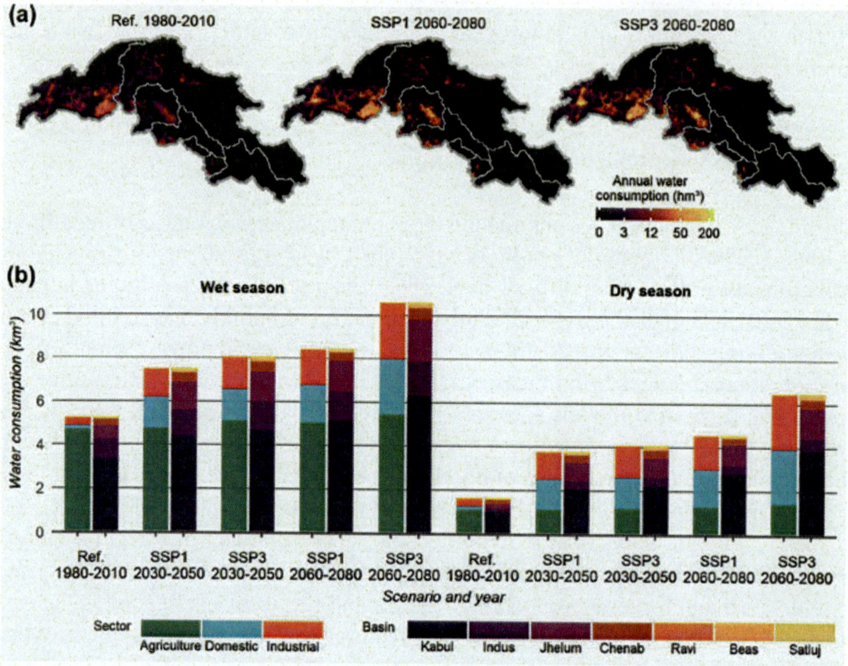

Fig. 2.8 Future water consumption in the sub-basins of the upper Indus Basin scenarios. *Source* Smolenaars et al. (2022)

2.4 The Earth System

2.4.1 Socio-hydrological Systems

In the next three decades, global urban population is projected to increase by 2.5 billion, and changes in land cover are expected to be an important predictor of biodiversity and habitat loss (Simkin et al. 2022). The contemporary society is therefore expected to experience several environmental challenges, including pressure on water resources and hydrological systems. Urbanization will result not only in the expansion of agricultural areas, but urban centers will also expand as a direct consequence of growing human population. Challenges of water security, water allocation, and water governance may arise largely due to higher demand and increased consumptive water use, especially for agriculture, which accounts for approximately 70% of water use. Regarding increased water consumption, an important observation under the Anthropocene is the impact of upstream water consumption on the downstream areas (Smolenaars et al. 2022). The illustration in Fig. 2.8 shows an estimated impact of water use in upstream areas on water availability in the downstream region based on water accounting framework (Fig. 2.8a, b). The annual water consumption in Fig. 2.8a suggests an increasing water consumption scenario for the future (2060–2080) in the upstream catchment (upper Indus Basin). Much of this consumption is expected to be used for industrial applications (Fig. 2.8b).

2.4.2 Understanding Drought and Flood in the Anthropocene

Climate projections under global temperature increase of 1.5 °C and 2.0 °C indicate drought intensity and areal coverage across China might double and are expected to be accompanied by drought losses (Su et al. 2018). In several other regions, rising precipitation and evapotranspiration patterns will lead to alternating extremes of wet and dry conditions. However, the impacts associated with the emergence of these events are expected to be different because of these intensities. Limited water availability and loss of biodiversity and important ecological resources in the future are very common problems that have been identified (Usman et al. 2021a). As would be expected based on the current understanding of climate change impacts on the environment, these hydro-climatic extremes are also expected to have varying magnitudes of impacts on agriculture and the natural ecosystem in the future.

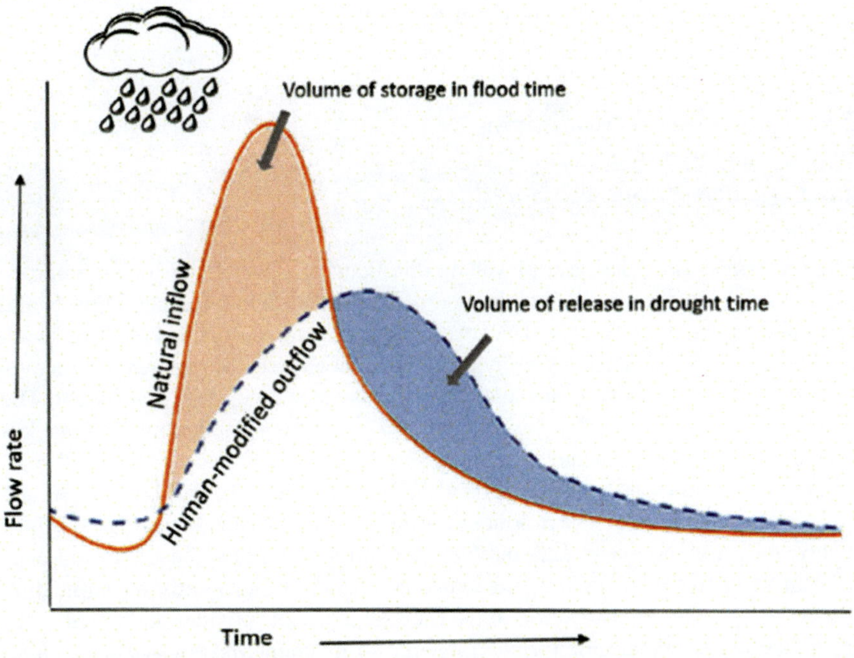

Fig. 2.9 Illustration of human impact on hydrological extremes. *Source* Di Baldassarre et al. (2017)

In the Anthropocene era, there's a relationship between human activities and hydrological extremes (Fig. 2.9). Human developments and population growth, in the modern day society, shape these extremes as was discussed earlier in this chapter. For example, the increasing rate of deforestation in the Congo basin is contributing to drought and freshwater loss in the region. The proliferation of dams and water infrastructures are strongly correlated with rising water demand (Ndehedehe et al. 2021a). However, these extremes in turn shape human societies through several processes, early warning systems, and strategic decisions to ensure adaptation and resilience to future impacts (Di Baldassarre et al. 2017). To ensure resilience and mitigation of impacts from extreme events, mankind has responded to the occurrence of drought or flood events in many ways. This includes the construction of reservoirs and dams, and water harvesting during big floods as well as developing policies and water accounting framework to inform the management of urban water systems. The construction of dams and reservoirs is very popular and are quick fixes that have helped in some ways, particularly in addressing both impacts of floods and droughts. But these risk reduction measures have turned out to be important predictors of change in the characteristics (frequency, magnitude, and extents) of hydrological extremes (Di Baldassarre et al. 2017). The change in hydrological behavior is illustrated in Fig. 2.9 and shows that water infrastructure developments like dams limit the effects of hydrological extremes, be it flood or drought. That is, in much of these dams and water infrastructures, outflows are reduced when inflows are high while higher outflows are maintained when inflow is low (Fig. 2.9).

2.5 Operational Monitoring

2.5.1 Earth Resources Satellites

Earth resources satellites have played an important role in helping society improve understanding of the environment, which has changed due to climate impacts and sustained human development. The development of new sensors and platforms with capabilities to assess changes in hydrology has emerged, complementing innovative optical systems in baseline studies and freshwater ecosystem assessment. The Landsat programme (Fig. 2.10) is a major Earth resources satellite mission that has supported this assessment and monitoring for more than four decades now. Substantial progress in the sensors onboard Landsat systems (Fig. 2.10) has been made, and the continuity and open-access policies of the Landsat mission have made a robust archive of historical images and satellite observations possible. The launch of Landsat 9 in September 2021 as was scheduled (Fig. 2.10) is apparently a continuation of that legacy. The sophistication in the Landsat 9 sensor provides additional capability in environmental monitoring compared to its counterparts Landsat 8. Other investments in satellite programmes include the novel Sentinel (1–6) missions of the European Space Agency. These missions have been broadly used in several

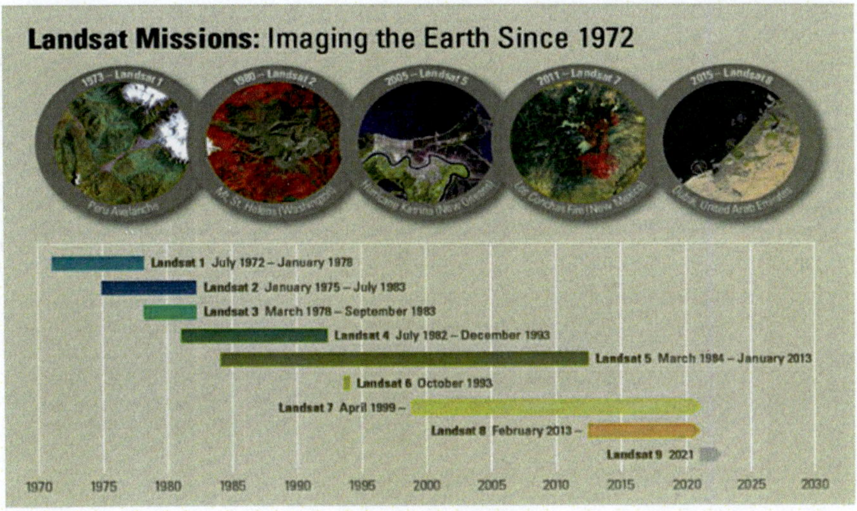

Landsat Mission Timeline.

Fig. 2.10 The Landsat mission. *Source* NASA/USGS

areas including measuring sea-surface topography, monitoring of changes in surface water, land surface temperature, land cover, and weather monitoring, among others. Additionally, several commercial satellite systems also exist. This includes, for example, Ikonos, Quick Bird, and WorldView-3 and, together, these systems provide robust applications in assessing environmental systems and can facilitate our understanding of global changes in the Anthropocene.

Its use rapidly increased over the years, and applications include environmental monitoring, change detection to assess key drivers and processes surrounding human civilization. Today, along with the European Space Agency multi-satellite systems like the suite of Sentinel data, the application of satellite earth observation is skyrocketing. Increasing applications of these satellite observations in several disciplines, e.g., water resources, geomorphology, geography, climate change, and disaster management, among others are well known. These applications have contributed to new directions in policy, creating several new opportunities that underpin better approaches to address socio-economic and environmental issues in today's society. New satellites such as the Sustainable Development Goals Science Satellite 1 (SDGSAT-1) is a dedicated mission to enable the achievement of specific policies and goals like the United Nation 2030 Agenda for Sustainable Development (http://en.people.cn/n3/2021/1112/c90000-9918780.html). The planned *Landsat Next* (discussed in Chap. 6) in late 2030 is another follow-up and future mission to complement the suite of existing Landsat missions in Earth resources monitoring as it is expected to come with improved temporal frequency, and spectral (26 spectral channels) and spatial resolution (10 and 20 m).[3]

[3] https://landsat.gsfc.nasa.gov/satellites/landsat-next/.

The potential of these optical systems as well as new radar methods like the NASA-ISRO SAR (see Chap. 6) and Surface Water and Ocean Topography (SWOT) (https://swot.jpl.nasa.gov/) missions and the opportunities that exist through the use of innovative satellite gravity methods like the Gravity Recovery and Climate Experiment (GRACE Tapley et al. 2004) to improve environmental monitoring in the Anthropocene have been discussed in Chaps. 5 and 6 of this book. In the Anthropocene, targeted water service delivery and policies that strengthen water resources management could benefit from these satellite technologies, especially the combination of state-of-the-art satellite hydrology missions such as GRACE and SWOT with innovative optical satellite systems. For instance, the launch of Landsat 9 in late 2021 resulted in new capabilities that will advance the monitoring of environmental systems.

2.5.2 *Hydrological and Climate Models*

Apart from satellite data obtained from remote sensing systems, global hydrological models forced by global climate models (GCMs), global reanalysis data, and model data have been used to study hydrological systems around the world and assess water resources and droughts. Common among these data is the Global Land Data Assimilation System (GLDAS, Rodell et al. 2004). GLDAS simulates more than twenty five land surface states and fluxes through an assimilation technique that integrates satellites with ground-based observational data. These fluxes and state variables include soil moisture, rainfall, canopy water storage, surface flow, snow depth, baseflow-groundwater runoff, wind speed, evapotranspiration, and net long-wave radiation flux, among several other land surface variables (Rodell et al. 2004). The WaterGAP Global Hydrology Model is another important model data, which provides estimates of groundwater recharge from surface water bodies in semi-arid and arid regions (Döll and Müller Schmied 2014). The WaterGAP model is based on a simple water balance technique and captures changes in groundwater across the world. Together with satellite data like GRACE, these models (WaterGAP and GLDAS) have enabled the assessment of hydrological systems and their response to climate change. For instance, the suitability of aquifers in East Africa to support groundwater irrigated agriculture using WaterGAP and GRACE groundwater was investigated Agutu et al. (2019). Their results showed these data (WaterGAP and GRACE) have considerable potential to provide an overview of changes in groundwater in the region. This provides a monitoring capability that underpins better understanding of long-term sustainability of groundwater aquifers to support water resources planning and impact assessment of drought.

The global reanalysis data called the Modern-Era Retrospective Analysis for Research and Applications is useful in land surface hydrological studies (Döll and Müller Schmied 2014; Rienecker et al. 2011). These model data have been combined with satellites to generate comprehensive metrics that provide a balance for the characterization prediction of land surface conditions. However, some

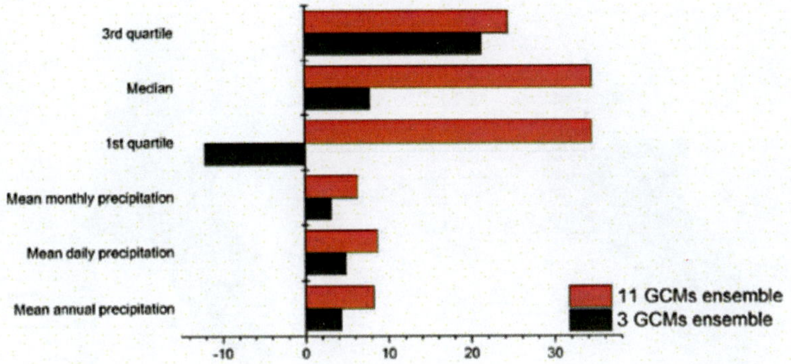

Fig. 2.11 Performance of global climate models ensembles in capturing different characteristics of observed precipitation (mm). *Source* Usman et al. (2021a)

limitations associated with these data have been discussed further (Ndehedehe 2019) and include

- poor representation of surface water balance
- bias and conceptual model uncertainties
- dependence on computational estimates
- lack of feedback processes involving anthropogenic impacts
- model physics and choice of parameterizations.

Their continued evaluation and/or calibration in data-deficient regions using observational data was earlier advocated and has been the focus of ongoing research (Ndehedehe 2019; Agutu et al. 2017). The capabilities of models in operational assessment through improved monitoring and forecasting have been exemplified in the development of several hydrological models, such as GLDAS, LISFlood model (i.e., the European Flood Alert System), and the European Drought Observatory, among several other warning systems relevant to water resources and prediction of extreme events like droughts and floods (van Dijk and Renzullo 2011). In the context of water resource monitoring systems on large spatial scales, the Australian Water Resources Assessment is also exemplary, demonstrating the importance of a coupled landscape model assimilation systems in supporting accurate and improved water balance assessment for freshwater reporting and management (van Dijk and Renzullo 2011). Similar water resources architecture or a framework that operationally integrates emerging satellite observations with other Earth observation data is required in other regions for improved water resource assessment systems.

The evaluation of climate and hydrological models is gradually emerging with much of it being undertaken on catchment and basin scales. For instance, Fig. 2.11 shows how several ensemble GCMs performed in capturing different characteristics of observed precipitation in the Cooper Creek-Bulloo River basin located in the Australian east coast (Fig. 2.12). Several of these climate projection data were obtained from general circulation models. They include CNRM-CM5,

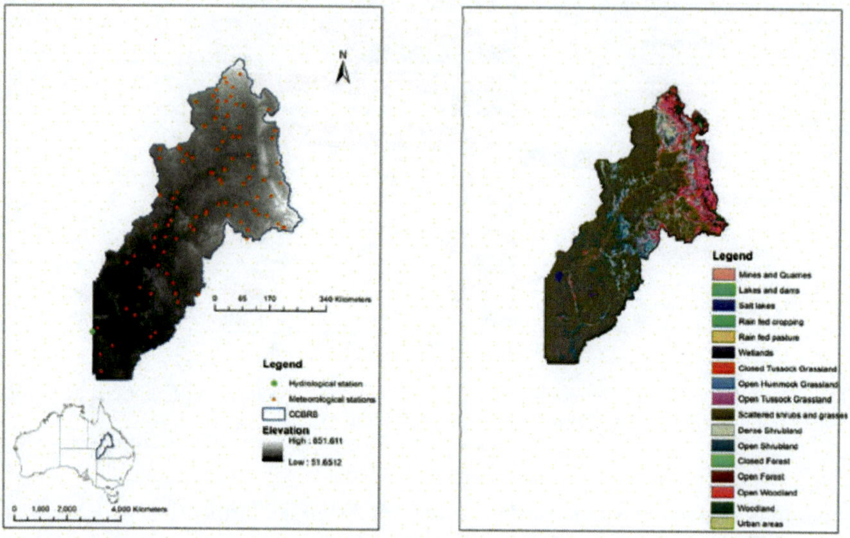

Fig. 2.12 The Cooper Creek-Bulloo River basin in the Australian east coast showing hydro-meteorological gauged stations and landcover. *Source* Usman et al. (2021a)

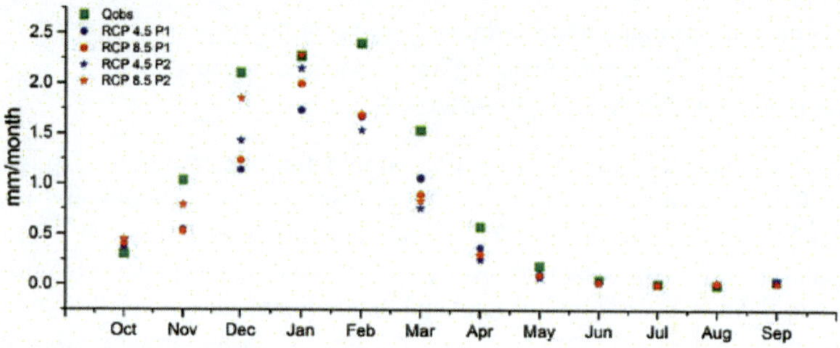

Fig. 2.13 Projected streamflow under the RCP 4.5 and RCP 8.5 emission scenarios for 2040–2069 and 2070–2099 periods. The 2050 and 2080 periods are referred to as P1 (period 1) and P2 (period2), respectively. This was compared to the observed values for the 1986–2015 period. *Source* Usman et al. (2021a)

GFDL-ESM2M, HadGEM2-CC, MIROC5, NorESM1-M, ACCESS1-0, CSIRO-Mk3.6, GFDL-CM3, GFDL-ESM2M, HadGEM2, ACCESS1.0, etc. In the Cooper Creek-Bulloo River basin, three of these GCM ensembles showed smaller difference with respect to observed precipitation compared to the total ensemble of eleven GCMs indicated in Fig. 2.11. To assess the impacts of climate change on streamflow in this basin, Usman et al. (2021a), optimally selected climate model outputs forced with RCP 4.5 and RCP 8.5 emission scenarios were used to generate streamflow for the 2040–2069 and 2070–2099 periods. Their simulated streamflow (Fig. 2.13) showed future water availability will be restricted. Decrease in annual precipitation in both periods is expected under the RCP 4.5 and RCP 8.5 emission scenarios. Consequently, averaged annual streamflow was also projected to decrease (Fig. 2.13). Even though streamflow was also projected to decrease in period 2 under the RCP 8.5 emission scenario, precipitation is expected to increase. Increased temperature may lead to rising evapotranspiration, which in turn can impact negatively on streamflow and net precipitation (i.e., water availability). Furthermore, in the dry and semi-arid Chirah and Dhoke Pathan sub-catchments of Pakistan, climate models and the hydrological model HBV-light have been used to assess the impacts of climate change on the hydrological regime (Usman et al. 2021b). Again, streamflow projections under average (RCP 4.5) and extreme (RCP 8.5) emission scenarios for the 2040–2064 and 2074–2098 periods as reported by Usman et al. (2021b) are expected to decrease.

The lack of observational soil moisture data to help improve the understanding of interactions between land surface and atmosphere has increased reliance on model outputs and reanalysis data. For data-deficient regions like Africa, the evaluation of these products to quantify their uncertainties is particularly critical to assess suitability in agricultural drought-related studies. Pioneering studies like the work undertaken by Agutu et al. (2021) in East Africa evaluated land surface models from several models and reanalysis outputs, including MERRA-2, ERA5, GLDAS components (NOAH, CLM, and VIC), FLDAS NOAH, and CPC. Even though this will be discussed in detail in subsequent chapters, there is evidence that the reliability of these models is region-dependent due to the influence of topography and other land surface conditions (Agutu et al. 2021).

References

AghaKouchak A, Norouzi H, Madani K et al (2015) Aral Sea syndrome desiccates Lake Urmia: call for action. J Great Lakes Res 41(1):307–311. https://doi.org/10.1016/j.jglr.2014.12.007

Agutu N, Awange J, Zerihun A, Ndehedehe C, Kuhn M, Fukuda Y (2017) Assessing multi-satellite remote sensing, reanalysis, and land surface models' products in characterizing agricultural drought in East Africa. Remote Sens Environ 194:287–302. https://doi.org/10.1016/j.rse.2017.03.041

Agutu N, Awange J, Ndehedehe C, Kirimi F, Kuhn M (2019) GRACE-derived groundwater changes over Greater Horn of Africa: temporal variability and the potential for irrigated agriculture. Sci Total Environ 693:133467. https://doi.org/10.1016/j.scitotenv.2019.07.273

Agutu N, Ndehedehe C, Awange J, Kirimi F, Mwaniki M (2021) Understanding uncertainty of model-reanalysis soil moisture within Greater Horn of Africa (1982–2014). J Hydrol 603:127169. https://doi.org/10.1016/j.jhydrol.2021.127169

Allaire M (2018) Socio-economic impacts of flooding: a review of the empirical literature. Water Secur 3:18–26. https://doi.org/10.1016/j.wasec.2018.09.002

Allison G, Cook P, Barnett S, Walker G, Jolly I, Hughes M (1990) Land clearance and river salinisation in the western Murray Basin, Australia. J Hydrol 119(1):1–20. https://doi.org/10.1016/0022-1694(90)90030-2

Barnett J, Adger WN (2007) Climate change, human security and violent conflict. Polit Geogr 26(6):639–655. https://doi.org/10.1016/j.polgeo.2007.03.003

Bierkens MFP, Wada Y (2019) Non-renewable groundwater use and groundwater depletion: a review. Environ Res Lett 14(6):063002. https://doi.org/10.1088/1748-9326/ab1a5f

Brown C, Meeks R, Hunu K, Yu W (2011) Hydroclimate risk to economic growth in sub-Saharan Africa. Climatic Change 106(4):621–647. https://doi.org/10.1007/s10584-010-9956-9

Chen J, Famigliett JS, Scanlon BR, Rodell M (2016) Groundwater storage changes: present status from GRACE observations. Surv Geophys 37(2):397–417. https://doi.org/10.1007/s10712-015-9332-4

Cook B, Smerdon J, Cook E et al (2022) Megadroughts in the common era and the Anthropocene. Nature Rev Earth Environ. https://doi.org/10.1038/s43017-022-00329-1

Di Baldassarre G, Martinez F, Kalantari Z, Viglione A (2017) Drought and flood in the Anthropocene: feedback mechanisms in reservoir operation. Earth Syst Dyn 8(1):225–233

Döll P, Müller Schmied H et al (2014) Global-scale assessment of groundwater depletion and related groundwater abstractions: combining hydrological modeling with information from well observations and GRACE satellites. Water Resour Res 50(7):5698–5720. https://doi.org/10.1002/2014WR015595

Eimanifar A, Mohebbi F (2007) Lake (Northwest Iran): a brief review. 3(5). https://doi.org/10.1186/1746-1448-3-5

FAO (2016) Climate change and food security: risks and responses. Food and Agricultural Organisation of the United Nations. Retrieved from http://www.fao.org/3/a-i5188e.pdf 20 Sept 2016

Favreau G, Cappelaere B, Massuel S, Leblanc M, Boucher M, Boulain N, Leduc C (2009) Land clearing, climate variability, and water resources increase in semiarid southwest Niger: a review. Water Resour Res 45:W00A16. https://doi.org/10.1029/2007WR006785

Getirana A (2016) Extreme water deficit in brazil detected from space. J Hydrometeorol 17(2):591–599. https://doi.org/10.1175/JHM-D-15-0096.1

Getirana A, Libonati R, Cataldi M (2021) Brazil is in water crisis – it needs a drought plan. Nature 600:218–220. https://doi.org/10.1038/d41586-021-03625-w

Giannini A, Saravanan R, Chang P (2003) Oceanic forcing of Sahel rainfall on interannual to decadal time scales. Science 302(5647):1027–1030. https://doi.org/10.1126/science.1089357

Gleick PH (1993) Water and conflict: fresh water resources and international security. Int Secur 18(1):79–112. https://doi.org/10.2307/2539033

Godinez-Madrigal J, Van Cauwenbergh N, van der Zaag P (2020) Unraveling intractable water conflicts: the entanglement of science and politics in decision-making on large hydraulic infrastructure. Hydrol Earth Syst Sci 24(10):4903–4921. https://doi.org/10.5194/hess-24-4903-2020

Green TR, Taniguchi M et al (2011) Beneath the surface of global change: impacts of climate change on groundwater. J Hydrol 405(3):532–560. https://doi.org/10.1016/j.jhydrol.2011.05.002

Guo L, Wei J, Zhang K, Wang J, Tian F (2022) Building a methodological framework and toolkit for news media dataset tracking of conflict and cooperation dynamics on transboundary rivers. Hydrol Earth Syst Sci 26(4):1165–1185. https://doi.org/10.5194/hess-26-1165-2022

Hall JW, Grey D, Garrick D, Fung F, Brown C, Dadson SJ, Sadoff CW (2014) Coping with the curse of freshwater variability. Science 346(6208):429–430. https://doi.org/10.1126/science.1257890

Hassanzadeh E, Zarghami M, Hassanzadeh Y (2012) Determining the main factors in declining the Urmia Lake level by using system dynamics modeling. Water Resour Manag 26:129–145. https://doi.org/10.1007/s11269-011-9909-8

Ifejika Speranza C, Kiteme B, Wiesmann U (2008) Droughts and famines: the underlying factors and the causal links among agro-pastoral households in semi-arid Makueni district, Kenya. Global Environ Change 18(1):220–233. https://doi.org/10.1016/j.gloenvcha.2007.05.001

Kiem AS, Johnson F, Westra S, van Dijk A et al (2016) Natural hazards in Australia: droughts. Climatic Change 139(1):37–54. https://doi.org/10.1007/s10584-016-1798-7

Kubiak-Wójcicka K, Bąk B (2018) Monitoring of meteorological and hydrological droughts in the Vistula basin (Poland). Environ Monit Assess 190(11):691. https://doi.org/10.1007/s10661-018-7058-8

Leduc C, Favreau G, Schroeter P (2001) Long-term rise in a Sahelian water-table: the continental terminal in south-west Niger. J Hydrol 243(1–2):43–54. https://doi.org/10.1016/S0022-1694(00)00403-0

Lemoalle J, Bader J-C, Leblanc M, Sedick A (2012) Recent changes in Lake Chad: observations, simulations and management options (1973–2011). Global Planet Change 80–81:247–254. https://doi.org/10.1016/j.gloplacha.2011.07.004

Lim E, Hudson D, Wheeler M et al (2021) Why Australia was not wet during spring 2020 despite La Niña. Sci Rep 11:18423. https://doi.org/10.1038/s41598-021-97690-w

Meixner T, Manning AH, Stonestrom DA, Allen DM, Ajami H, Blasch KW, Brookfield AE, Castro CL, Clark JF, Gochis DJ, Flint AL, Neff KL, Niraula R, Rodell M, Scanlon BR, Singha K, Walvoord MA (2016) Implications of projected climate change for groundwater recharge in the western United States. J Hydrol 534:124–138. https://doi.org/10.1016/j.jhydrol.2015.12.027

Ndehedehe CE, Ferreira VG, Adeyeri OE, Correa FM, Usman M, Oussou FE, Kalu I, Okwuashi O, Onojeghuo AO, Getirana A, Dewan A (2023) Global assessment of drought characteristics in the Anthropocene. Resour Environ Sustain 100105. https://doi.org/10.1016/j.resenv.2022.100105

Ndehedehe CE, Agutu N, Ferreira VG, Getirana A (2020a) Evolutionary drought patterns over the Sahel and their teleconnections with low frequency climate oscillations. Atmos Res 233:104700. https://doi.org/10.1016/j.atmosres.2019.104700

Ndehedehe CE, Ferreira VG, Onojeghuo AO, Agutu NO et al (2020b) Influence of global climate on freshwater changes in Africa's largest endorheic basin using multi-scaled indicators. Sci Total Environ 737:139643. https://doi.org/10.1016/j.scitotenv.2020.139643

Ndehedehe CE, Haile GG, Agutu NO, Ferreira VG et al (2020c) Hydrological hotspots of climatic influence in Brazil: a two-step regularization approach. Atmos Res 246:105116. https://doi.org/10.1016/j.atmosres.2020.105116

Ndehedehe CE, Ferreira VG, Agutu NO et al (2021a) What if the rains do not come? J Hydrol 595:126040. https://doi.org/10.1016/j.jhydrol.2021.126040

Ndehedehe CE, Onojeghuo AO, Stewart-Koster B, Bunn SE, Ferreira VG (2021b) Upstream flows drive the productivity of floodplain ecosystems in tropical Queensland. Ecol Ind 125:107546. https://doi.org/10.1016/j.ecolind.2021.107546

Ndehedehe CE (2019) The water resources of tropical West Africa: problems, progress and prospect. Acta Geophysica 67(2):621–649. https://doi.org/10.1007/s11600-019-00260-y

Ndehedehe CE, Agutu NO, Okwuashi OH, Ferreira VG (2016) Spatio-temporal variability of droughts and terrestrial water storage over Lake Chad Basin using independent component analysis. J Hydrol 540:106–128. https://doi.org/10.1016/j.jhydrol.2016.05.068

Ndehedehe CE, Anyah RO, Alsdorf D, Agutu NO, Ferreira VG (2019) Modelling the impacts of global multi-scale climatic drivers on hydro-climatic extremes (1901–2014) over the Congo basin. Sci Total Environ 651:1569–1587. https://doi.org/10.1016/j.scitotenv.2018.09.203

Odorico P, Chiarelli DD, Rosa L et al (2020) The global value of water in agriculture. Proc Natl Acad Sci 117(36):21985–21993. https://doi.org/10.1073/pnas.2005835117

Ojha C, Shirzaei M, Werth S, Argus DF, Farr TG (2018) Sustained groundwater loss in California's Central Valley exacerbated by intense drought periods. Water Resour Res 54(7):4449–4460. https://doi.org/10.1029/2017WR022250

Okewu E, Misra S, Sanz LF, Ayeni F, Mbarika V, Damaševiius R (2019) Deep Neural Networks for curbing climate change-induced farmers-herdsmen clashes in a sustainable social inclusion initiative. Problems Sustain Dev 14(2):143–155

Papaioannou KJ (2016) Climate shocks and conflict: evidence from colonial Nigeria. Polit Geogr 50:33–47. https://doi.org/10.1016/j.polgeo.2015.07.001

Park S, Kim H, Jang C (2021) Impact of groundwater abstraction on hydrological responses during extreme drought periods in the Boryeong Dam catchment, Korea. Water 13(15):2132. https://doi.org/10.3390/w13152132

Pekel J-F, Cottam A, Gorelick N, Belward AS (2016) High-resolution mapping of global surface water and its long-term changes. Nature 540:418–422. https://doi.org/10.1038/nature20584

Peterson TJ, Saft M, Peel MC, John A (2021) Watersheds may not recover from drought. Science 372(6543):745–749. https://doi.org/10.1126/science.abd5085

Reuveny R (2007) Climate change-induced migration and violent conflict. Polit Geogr 26(6):656–673. https://doi.org/10.1016/j.polgeo.2007.05.001

Riebe K, Dressel A (2021) The impact on food security of a shrinking lake chad. J Arid Environ 189:104486. https://doi.org/10.1016/j.jaridenv.2021.104486

Rienecker MM, Suarez MJ, Gelaro R et al (2011) MERRA: NASA's modern-era retrospective analysis for research and applications. J Climate 24(14):3624–3648. https://doi.org/10.1175/JCLI-D-11-00015.1

Rippey BR (2015) The U.S. drought of 2012. Weather Climate Extremes 10(Part A):57–64. https://doi.org/10.1016/j.wace.2015.10.004

Rodell M, Houser PR, Jambor U et al (2004) The global land data assimilation system. Bull Am Meteorol Soc 85(3):381–394. https://doi.org/10.1175/BAMS-85-3-381.R

Scanlon B, Healy R, Cook P (2002) Choosing appropriate techniques for quantifying groundwater recharge. Hydrogeol J 10:18–39. https://doi.org/10.1007/s10040-001-0176-2

Scanlon BR, Reedy RC, Stonestrom DA, Prudic DE, Dennehy KF (2005) Impact of land use and land cover change on groundwater recharge and quality in the southwestern US. Global Change Biol 11(10):1577–1593. https://doi.org/10.1111/j.1365-2486.2005.01026.x

Schroth G, Läderach P, Martinez-Valle AI et al (2016) Vulnerability to climate change of cocoa in West Africa: patterns, opportunities and limits to adaptation. Sci Total Environ 556:231–241. https://doi.org/10.1016/j.scitotenv.2016.03.024

Schumacher D, Keune J, Dirmeyer P et al (2022) Drought self-propagation in drylands due to land-atmosphere feedbacks. Nature Geosci. https://doi.org/10.1038/s41561-022-00912-7

Shiferaw B, Tesfaye K, Kassie M, Abate T, Prasanna B, Menkir A (2014) Managing vulnerability to drought and enhancing livelihood resilience in sub-Saharan Africa: technological, institutional and policy options. Weather Clim Extrem 3:67–79. https://doi.org/10.1016/j.wace.2014.04.004

Simkin RD, Seto KC, McDonald RI, Jetz W (2022) Biodiversity impacts and conservation implications of urban land expansion projected to 2050. Proc Natl Acad Sci 119(12):e2117297119. https://doi.org/10.1073/pnas.2117297119

Smolenaars WJ, Dhaubanjar S, Jamil MK et al (2022) Future upstream water consumption and its impact on downstream water availability in the transboundary indus basin. Hydrol Earth Syst Sci 26(4):861–883. https://doi.org/10.5194/hess-26-861-2022

Spinoni J, Naumann G, Carrao H, Barbosa P, Vogt J (2014) World drought frequency, duration, and severity for 1951–2010. Int J Climatol 34(8):2792–2804. https://doi.org/10.1002/joc.3875

Su B, Huang J, Fischer T et al (2018) Drought losses in china might double between the 1.5c and 2.0c warming. Proc Natl Acad Sci 115(42):10600–10605. https://doi.org/10.1073/pnas.1802129115

Tapley B, Bettadpur S, Watkins M, Reigber C (2004) The gravity recovery and climate experiment: mission overview and early results. Geophys Res Lett 31:1–4. https://doi.org/10.1029/2004GL019920

Tellman B, Sullivan J, Kuhn C, Kettner A, Doyle C, Brakenridge G, Erickson T, Slayback D (2021) Satellite imaging reveals increased proportion of population exposed to floods. Nature 596:80–86. https://doi.org/10.1038/s41586-021-03695-w

Thiede B, Gray C, Mueller V (2016) Climate variability and inter-provincial migration in South America, 1970–2011. Global Environ Change 41:228–240. https://doi.org/10.1016/j.gloenvcha.2016.10.005

Tourian M, Elmi O, Chen Q, Devaraju B, Roohi S, Sneeuw N (2015) A spaceborne multisensor approach to monitor the desiccation of Lake Urmia in Iran. Remote Sens Environ 156:349–360. https://doi.org/10.1016/j.rse.2014.10.006

Tubi A, Feitelson E (2016) Drought and cooperation in a conflict prone area: Bedouin herders and Jewish farmers in Israel's northern Negev, 1957–1963. Polit Geogr 51:30–42. https://doi.org/10.1016/j.polgeo.2015.11.009

Usman M, Ndehedehe CE, Farah H, Manzanas R (2021a) Impacts of climate change on the streamflow of a large river basin in the Australian tropics using optimally selected climate model outputs. J Cleaner Prod 315:128091. https://doi.org/10.1016/j.jclepro.2021.128091

Usman M, Ndehedehe CE, Manzanas R et al (2021b) Impacts of climate change on the hydrometeorological characteristics of the Soan River Basin, Pakistan. Atmosphere 12(6). https://doi.org/10.3390/atmos12060792

van Dijk AIJM, Renzullo LJ (2011) Water resource monitoring systems and the role of satellite observations. Hydrol Earth Syst Sci 15(1):39–55. https://doi.org/10.5194/hess-15-39-2011

van Dijk AIJM, Beck HE, Crosbie RS, de Jeu RAM, Liu YY, Podger GM, Timbal B, Viney NR (2013) The millennium drought in southeast Australia (2001–2009): natural and human causes and implications for water resources, ecosystems, economy, and society. Water Resour Res 49(2):1040–1057. https://doi.org/10.1002/wrcr.20123

Van Loon A, Stahl K, Di Baldassarre G, Clark J, Rangecroft S et al (2016) Drought in a human-modified world: reframing drought definitions, understanding, and analysis approaches. 20:3631–3650. https://doi.org/10.5194/hess-20-3631-2016

Wald L (1990) Monitoring the decrease of Lake Chad from space. Geocarto Int 5(3):31–36. https://doi.org/10.1080/10106049009354266

Wasko C, Shao Y, Vogel E et al (2021) Understanding trends in hydrologic extremes across Australia. J Hydrol 593:125877. https://doi.org/10.1016/j.jhydrol.2020.125877

Part III
Water Resources

Chapter 3
How Much Freshwater Is Available?

3.1 Introductory Remarks

It is true that water availability is essential to every aspect of human life and the functioning of ecosystems. For humans, it is particularly so as it drives global patterns of agricultural production and trade and the peaceful coexistence of some regions where shared water resources could be problematic due to competition and increased demand (Ndehedehe 2022b; Odorico et al. 2020). Yet, about 2.3 billion of the world's population (about one-third) are currently facing water stress and/or living in areas with extreme water shortages (United Nations-Water).[1] However, this number is expected to double by 2050 due to frequent extreme droughts, signaling a more challenging water deficit in the Anthropocene. The Earth's freshwater is limited (2.5% of the world's total available water) as most (68.7% of freshwater) of it is locked up in ice and glaciers, leaving only 29.9 and 0.26% as groundwater resources and surface waters, respectively (Shiklomanov 2000). The available renewable freshwater resource on Earth sourced from rainfall (840 mm $year^{-1}$) is pegged at 113,000 km^3/year, and only less than 10% of this amount is directly consumed by mankind while the remaining proportion is appropriated in the functioning of natural ecosystems, and biodiversity (de Marsily and Abarca-del Rio 2016). Of the 36,000 km^3/year or 32% of the renewable freshwater resources (113,000 km^3/year) that represent flows within continents (popularly known as *Blue Water*), only about 10,000 km^3/year (or about 9% of 113,000 km^3/year) is available to recharge aquifers and sustain rivers during drought, yet increasing global human water consumption through irrigation continues to deplete groundwater stocks across the globe (de Marsily and Abarca-del Rio 2016).

Available freshwater in the form of surface waters is present in lakes, reservoirs, and river systems, though some lakes and reservoirs are man-made (Ndehedehe 2022b). While this means freshwater is limited, the argument that the amount of annual water

[1] https://www.unwater.org/water-facts/water-scarcity/.

C. Ndehedehe, *Hydro-Climatic Extremes in the Anthropocene*, Springer Climate,
https://doi.org/10.1007/978-3-031-37727-3_3

withdrawal (3800 km^3/yr) is greater than what is stored in rivers (2000 km^3) implies it can be competitive (Oki and Kanae 2006) as well as vulnerable to intensive human use. The vulnerability of freshwater stored in lakes, reservoirs, and groundwater aquifers to extreme droughts and human use raises concerns on its availability to continuously provide essential ecosystem functions and livelihood. The 2021 water crises in south central Brazil resulted in a substantial loss of freshwater stored in rivers, lakes, soil, and aquifers estimated at 267 km^3 (Getirana et al. 2021). An earlier account of drought impact in Brazil indicated freshwater storage losses in more than ten reservoirs were the aftermath of a direct consequence of prolonged extreme drought (Getirana 2016), thus reaffirming the vulnerability of freshwater resources to climate change (Fig. 3.1). If available water for consumption in a particular region is perceived as streamflow and precipitation amounts persisting through vegetation growth, or net precipitation (i.e., difference between rainfall and evapotranspiration) (Chen et al. 2023; Andam-Akorful et al. 2017), a changing climate is expected to increase water demand, which may result in the decline of available water in the said regions. An assessments of China's oases (productive and vulnerable ecosystems in drylands) by Chen et al. (2023) exemplified this as water demand showed a significant rising trend over a period of 30 years (1987–2017) in these oases. This increasing water demand accounted for the decline in water availability in about 60% of the oasis regions, despite the latter being compensated by substantial increases in rainfall and streamflow. The availability of freshwater in most arid and semi-arid ecosystems is thus a critical issue in the Anthropocene. As will be discussed in this chapter, there is evidence of strong changes in water budget at various spatial scales. Such changes are obvious indicators of climate change impacts inducing rising water demand.

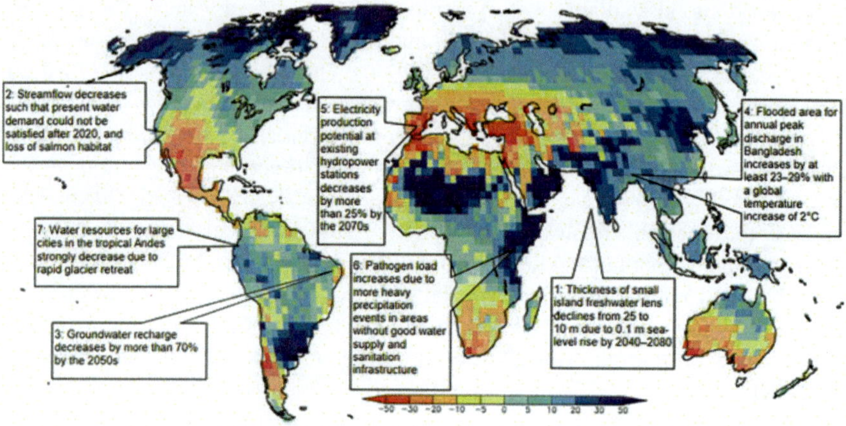

Fig. 3.1 Projected impacts of climate change on global freshwater resources. The map shows ensemble mean change of annual runoff (in percentage) for the 1981–2000 and 2081–2100 periods based on A1B emissions scenario. *Source* Kundzewicz et al. (2008)

3.2 Impacts of Climate Change on Freshwater Resources

Freshwater stored in the Earth's water column is never lost as it is recycled within the inner Earth, the oceans and rivers, and the atmosphere, thus making it available to Earth's surface. However, with increased human extraction due to a rise in water demand, which could also be exacerbated by climate change (Fig. 3.1), the amount available, when, and where are some reasons warranting the concerns about humans running out of freshwater. In other words, the high spatial and temporal variability in freshwater may lead to critical water problems and crises in some regions where these occur (Oki and Kanae 2006). These water problems have been reported in several rapidly urbanizing regions of the world, including the Middle East, South Africa, South and North America, and Asia, among others. As discussed later in this book (e.g., Chaps. 5 and 6), groundwater over-exploitation has led to critical impacts on public infrastructure (e.g., land subsidence) and ecosystems. de Marsily and Abarca-del Rio (2016) argued that groundwater stocks are at the risk of running out as about 10% of the irrigation water emanates from groundwater, especially in India and China, where such usage is relatively higher and water crises are eminent. In some regions of the world, future projections of climate change suggest the amount of groundwater recharge will decline by more than 70%.

The impacts of climate change on freshwater resources have long been detailed in the literature and are evident in several freshwater hotspots like Lakes Chad and Urmia as previously discussed in Chap. 2 of this book. Future implications of these impacts were also highlighted by Kundzewicz et al. (2008) in their rapid communication, highlighting climate change impacts on freshwater resources and the threats they pose to sustainable management. As illustrated in Fig. 3.1, a decrease of 10–30% annual average river runoff and water availability are expected in mid-latitudes and dry tropics by mid-century. Climate change will also affect the production of electricity, resulting in a 25% reduction by 2070 (Fig. 3.1). This kind of impact is common in West Africa where there is heavy reliance on hydro-power (Ndehedehe 2022c). For instance, during hydrological drought years, Ghana suffered power rationing, which affected the production of goods as well as reduced gross domestic products (Ndehedehe et al. 2017). The Akosombo and Kainji dams as well as other reservoirs in Western Africa are major water infrastructures that are frequently affected by climate change (drought), resulting in hydro-power failure in the region (Ndehedehe 2019).

Furthermore, Fig. 3.1 summarizes the nature of future impacts on freshwater based on model simulations. Some of these include decrease in streamflow leading to loss of habitat for Salmon, glacier retreat affecting water resources in large cities, decrease in groundwater recharge by 2050, more than 23% increases in flooded areas, rising pathogen loads resulting from high annual rainfall amounts in regions with poor sanitation, etc. (Fig. 3.1). There is a considerable amount of uncertainties that come with these impacts on the forecast of freshwater resources (Kundzewicz et al. 2008). Developing early warning systems will underpin adaptation and mitigation measures and are critical as these events are expected to increase in magnitude and severity with implications on water management operations and the way water is allocated.

In today's world, evidence of critical freshwater problems in the Middle East (e.g., Iran), West African Sahel, South Africa, Central California Valley, and East Africa exists. Managing freshwater resources is therefore now an important aspect of governance, and global interest focused on groundwater sustainability is evidenced in the number of international forums and meetings, (e.g., International Panel Climate Change, World Climate Conference-3, and United Nations Climate Conference) that discuss these groundwater problems (Green et al. 2011). This chapter articulates some of these issues, providing new perspectives on the state of the world's water resources as well as discussing important water concepts (e.g., water security, water demand, and water crises), sustainability of freshwater and monitoring framework that underpin freshwater reporting.

3.3 Are We Running Out of Freshwater?

3.3.1 The Water Cycle

The water cycle is a mechanism by which water moves over, above, and below the Earth's surface. The world's water exists in various forms (solid, liquid, and gas—Fig. 3.2) and includes water present in oceans, seas, lakes, rivers, ice caps, in the atmosphere, and as groundwater. The transition or movement of water can occur through components of the water cycle, including precipitation, infiltration, runoff, groundwater, discharge, and evapotranspiration. The various processes used to describe these transitions and transformations include evaporation, transpiration, sublimation, etc. For instance, evaporation describes the transition of some amount of water in oceans, seas, lakes, rivers, and streams to water in the form of molecules (gas). Moisture transition to the atmosphere emanating from plants occurs through transpiration. When ice or snow melts and changes to gas, the process is called sublimation. These processes (evaporation, transpiration, and sublimation) are the channels for moisture transports (Fig. 3.2). But the availability of freshwater resources (in lakes, rivers, water holes, floodplains, soil moisture, and groundwater) is limited and will depend on the amount of precipitation, infiltration, and recharge and the rate at which humans intercept some of the available water for several uses. Humans intercept freshwater in various ways (Fig. 3.2). This can occur through over-abstraction of groundwater for domestic water supply and irrigation, construction of large dams to increase surface water storage for agricultural purposes, water transfer through redirecting of rivers, water harvesting, and even draining water from wetlands through reclamation programmes for urban expansion (Ndehedehe 2022b). In developing economies, hydro-power is the main source of electricity, and this requires large in-land surface water storage facilities. These various uses of freshwater make it highly

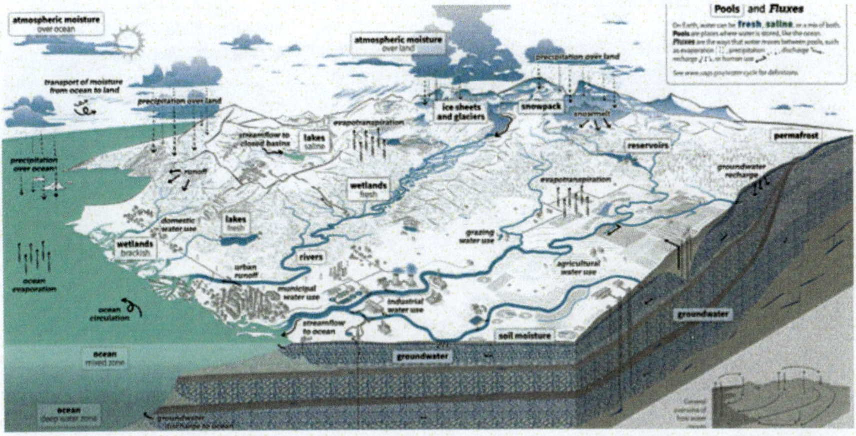

Fig. 3.2 The global water cycle. Image and further definitions of key words and explanation of concepts available at https://www.usgs.gov/media/images/water-cycle-png. *Source* United States Geological Surveys

variable, especially when its restriction by climate variability increases the amount being used or needed.

Apparently rainfall is the primary input to any hydrological system, regulating the amount or recharge and changes in groundwater. The balance between input and output fluxes may not always be maintained, especially when abstraction is not being rapidly replaced by rainfall or subsurface flow, thus resulting in water shortages. For instance, despite a significant rise in annual rainfall (4.9 ± 0.02 km^3/yr), Iran lost approximately 211 ± 34 km^3 of its terrestrial water storage in the last twenty years with groundwater depletion of 28 ± 1.4 cm/yr (Saemian et al. 2022). The spatial variability of the observed rates in terrestrial water storage and rainfall during the common period in Iran (Fig. 3.3) confirms that human interception of freshwater for use can impact its availability. This is particularly so when rainfall is not quickly replenishing what was taken or its intensity is dissimilar to what was used or taken from the aquifer. The threat to freshwater and acceleration of the water cycle is indeed real and for this reason, some regions (e.g., India and South Africa) are showing the possibility of running out of water.

3.3.2 *Are There Threats to Freshwater Availability?*

As previously mentioned, available freshwater in lakes, reservoirs, groundwater, and river systems are limited and yet critical for terrestrial life and ecosystem functioning. Freshwater is a valuable resource to billions of people in the world, and plays key roles in agriculture, socio-economic development, and food security. For instance,

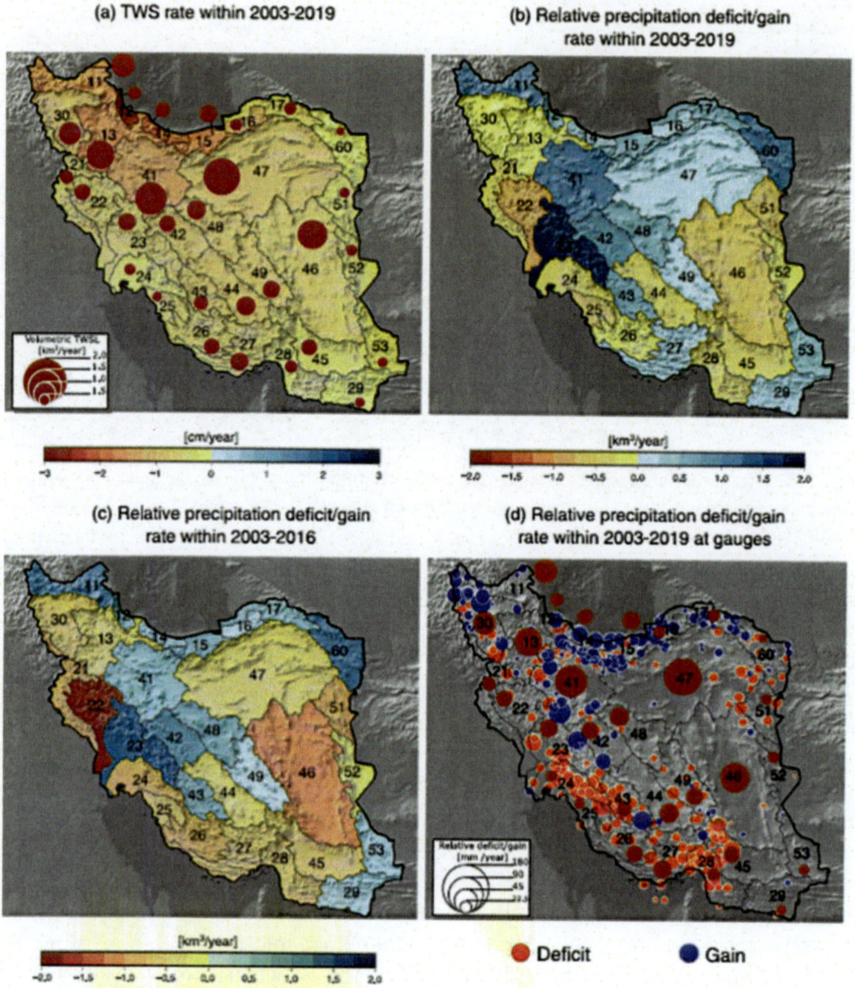

Fig. 3.3 Trends in precipitation and terrestrial water storage distribution illustrating threats to freshwater availability. *Source* Saemian et al. (2022)

as illustrated in Fig. 3.4, irrigated agriculture is the biggest consumer of global water resources. While this is true, the proportion of water used for livestock is substantially different across several regions of the world with African sub-regions indicating the highest consumption of water for livestock (Fig. 3.4). With projected availabilities of freshwater being less than 1000 m^3/capita/yr in the world's densely populated arid areas, water scarcity will be a major restriction to agricultural development in the coming decades (Rijsberman 2006). Average water demand is now expected to increase from about 8200 km^3/year in 2000 to around 13,000 km^3/year in 2050, and with a projected increase of about 9.5 billion people by 2050 (de Marsily and

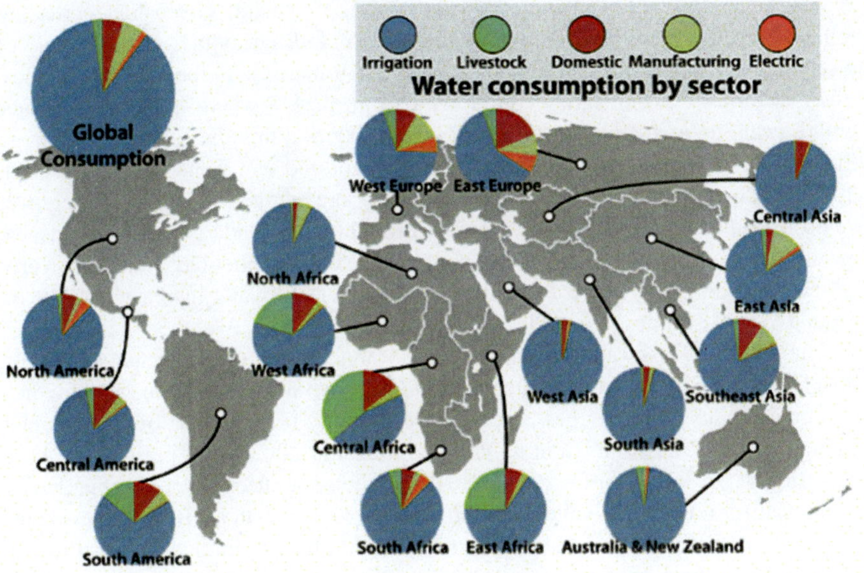

Fig. 3.4 Global water consumption by sector. Irrigation is the leading water consumption sector for all sub-regions of the world. *Source* Brauman et al. (2016)

Abarca-del Rio 2016), the threats to freshwater can only get worse. Given the rise in global population, rapid urbanization, and industrialization, the demand for freshwater is inevitably expected to rise. For example, a large number of river basins that make up 12% of the total global area of basins with limited annual available water is already under pressure because of increased water withdrawals (Alcamo et al. 1997). Water availability is more likely to restrict economic development in these basins, and an estimated 1.7 billion people live in regions where groundwater resources and groundwater-dependent ecosystems are under threat (Gleeson et al. 2012). Globally, more than 1.5 billion people rely on groundwater as a major source of drinking water and is the only source of water in most arid and desert regions. Given the importance of groundwater, in some arid regions, for example, in Israel, groundwater management is integrated as part of the operation of the national system. Groundwater abstractions have tripled and depletion rates have doubled between 1960 and 2000 in Asia and North America (Gleick 2000).

The study undertaken by Schewe et al. (2013) has shown that there is about 40% increase in the number of people living under absolute water scarcity, up from 15% of the global population with a severe decrease in water resources. As in the year 2000, total water withdrawals amounted to 8.4% of global water resources, and this amount is expected to increase up to 12.2% (i.e., between 4600–5800 km^3/yr) by 2025 (Shiklomanov 2000). Vörösmarty et al. (2000) projected the estimate of global water use by 2025 is conservative and puts the value at 4700 km^3/yr. The rise in urban population and changing climate will lead to more withdrawals of groundwater as surface water could be limited due to extreme drought. Given the increasing reliance on groundwater for urban water supply and agriculture during times of droughts and water shortage, global water use will increase in the coming decades. The California Central Valley aquifer system in the US where large groundwater withdrawals for irrigation have been reported exhibits this pattern, showing large fluctuations in aquifer volume in between drought and wet years (e.g., Ojha et al. 2020; Vasco et al. 2019). This means aquifer storage recovers during extremely wet years after substantial declines in previous drought years, underpinned by excessive groundwater pumping. Arguably, this could be the case in other large aquifers in arid or semi-arid domains where drought and limited surface water increase reliance on groundwater resources (e.g., Ndehedehe et al. 2021).

The threat to freshwater availability is real, making it a critical issue, especially in such regions where groundwater pumping is much higher than the rate at which it is being replaced (recharge). For instance, satellite observations of groundwater changes in the California Central Valley during the 2003–2010 period showed a depletion of 3.1 km^3/year (Vasco et al. 2019). The continued losses of groundwater in the California Central Valley and other regions like northern India are evidence of threats to freshwater availability and also highlight how human actions through excessive pumping drive the water cycle. Satellite observations of terrestrial water storage in the last two decades across different regions of the world show significant decline (Scanlon et al. 2023; Rodell et al. 2018). Recently, the loss of about 310 to 260 km^3 in terrestrial water storage observed by the Gravity Recovery and Climate Experiment mission (Chap. 4) has been reported, and the composite influence of climate and human water management accounts for substantial loss of surface water area (90,000 km^2) between 1984 and 2015 (Scanlon et al. 2023). Decreasing trends in terrestrial water storage or hydrological stores such as groundwater and aquifers and surface water are some indications of threats to freshwater availability.

A good illustration to the underlying threats to freshwater availability is the satellite-based analysis of fluxes, including rainfall, groundwater, and terrestrial water storage in Iran over a twenty-year period (Fig. 3.3). Despite the increase in precipitation during the 2003–2019 and 2003–2016 periods in several areas, terrestrial water storage declined in several basins and sub-regions in Iran (Fig. 3.3). The significant terrestrial water storage loss during these periods coincided with significant groundwater depletion in the region (Saemian et al. 2022). Yes, freshwater availability is under threat, and the aggressive exploitation of non-renewable freshwater or groundwater resources in several semi-arid regions poses a lot of danger to its future availability. Moreover, as seen in the Iran case study (Saemian et al. 2022), long-term changes in climatic conditions could exacerbate water use, thus contributing to changes in groundwater rates. In the next section, some issues associated with freshwater changes and availability in the Anthropocene are highlighted. Some of the challenges in assessing terrestrial water systems are also mentioned, and the opportunities that exist through the use of new satellite remote sensing methods and frameworks are also discussed.

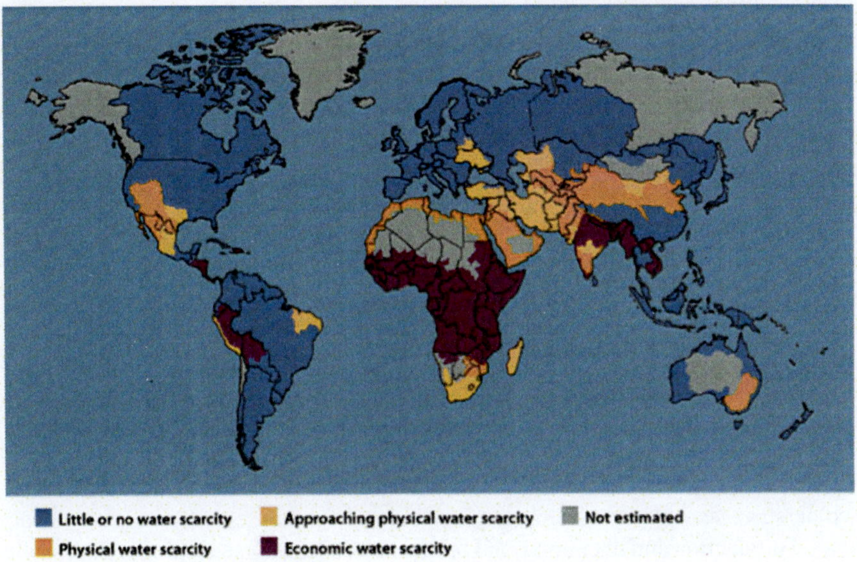

Fig. 3.5 Global patterns of water scarcity. *Source* World Resources Institute. https://www.wri.org/data/physical-and-economic-water-scarcity

3.3.3 Water Scarcity, Security, and Crises

Many regions across the globe are now facing threats to food and water security because of increased acceleration of the water cycle. In regions that will be impacted by extreme drought, there is the likelihood of running out of existing freshwater, culminating in water scarcity problems. Globally, water scarcity is becoming a major challenge and concern around the world (Fig. 3.5). Water scarcity occurs when water demand exceeds the available water supply. As illustrated further in Fig. 3.5, the nature of water scarcity differs across several regions and can be physical or economical. As was highlighted in de Marsily and Abarca-del Rio (2016) review of current status of water and food, physical water-scarce regions are those regions where more than 75% of river flow have been extracted by humans for various purposes. Notably, these areas are largely arid and semi-arid regions (Fig. 3.5) where large irrigation systems are in place to support large-scale agriculture. As opposed to regions experiencing physical water scarcity or those approaching physical water scarcity (Fig. 3.5), de Marsily and Abarca-del Rio (2016) described economical water scarcity as regions where there is surface water or river flow "*but not exploited for lack of resources to finance the equipment to do so (dams, canals, irrigation systems, etc.)*". Much of these areas or zones of economical water scarcity are located in Africa (Fig. 3.5) and also explain the famine and malnourishment in some hotspots despite having numerous water resources (Ndehedehe 2019).

One indicator of water scarcity is extensive declines in groundwater levels. For example, India's widespread groundwater depletion is alarming due to extensive annual withdrawal for irrigating agricultural lands. Groundwater is an important source of freshwater, especially in arid and semi-arid regions of the world where surface water is considerably limited (e.g., Australia and the Middle East). Regions that are heavily dependent on groundwater such as India, the world's largest user of groundwater (Chindarkar and Grafton 2019), are at risk of water scarcity or could be at risk of acute water shortage. The International Water Management Institute estimates that nearly 20% of the world's population lives in areas of physical water scarcity, where water use for various purposes exceeds 75% of river flows (Gleick 2000). This underscores the risk of water scarcity in these regions. Water scarcity projections based on other metrics estimate that about 57 and 69% of the world's population will be living in countries facing high water stress by 2025 and 2075, respectively (Alcamo et al. 1997). These values may be somewhat too high as argued by Gleick (2000); they, however, suggest the urgent conversations needed to raise awareness on the susceptibility of global water resources. The numerical projections and perspectives on water scarcity and anticipated future use underscore the threats to food production and livelihood, and for the non-industrialized regions of the world (e.g., South Africa), stormwater harvesting has been suggested as a possible alternative water resource that could supplement traditional urban water supplies, cushion the impacts of climate change, and improve water security (Fisher-Jeffes et al. 2017). Water scarcity by itself represents one of the characteristics (e.g., degree of scarcity) that could make water a potential source of strategic rivalry (Gleick 1993). This is true for Texas where there are some indications that the water requirements for hydraulic fracturing are already creating conflicts with other uses and could restrict future natural gas production in some areas (see Gleick 2000).

Water crises as defined by the World Economic Forum is *a significant decline in the available quality and quantity of freshwater resulting in harmful effects on human health and/or economic activity*. With a focus on the quantitative aspect of freshwater, water crises is arguably the result of water scarcity. The latter is expressed as the ratio between water demand and water availability. Following this ratio, a threshold of 0.4 is usually adopted as severe water scarcity. The Lake Chad is a good example of water crises. On the one hand, a failing climate as was observed in the Lake Chad basin limited freshwater availability. On the other hand, access to water resources could become difficult due to rising population growth and pressure from increased water needs. Apparently, about 90% decline of the Lake's surface area declined due to prolonged below normal rainfall Ndehedehe (2022b). Inflows from major river systems (Chari–Logone river) that provide about 95% freshwater of the lake were restricted due to this change in rainfall. Due to climate change and human influence, the present Lake Chad is segmented to two pools with the northern pool completely dried up during drought periods Ndehedehe (2022b). This led to water crises and armed conflicts due to competition for the limited and shared water resources in the region. However, water withdrawals from the Lake for irrigation purposes during this water deficit continued unabated. With increasing population accompanied by a rise in water demand, the northern portion of the lake dried up with little water left in the southern pool. This condition created water crises that have escalated to agitations, tensions, and even armed conflicts among local communities from the riparian countries sharing the lake's water resources. Several other freshwater hotspots such as lakes and rivers are at the risk of desiccation while others occasionally run dry. As reported by Ndehedehe (2022b), the Aral Sea experienced a gradual decline in the last decade and may be completely desiccated with time. The Colorado river in the US, Indu river in Pakistan and the Hai and Huai rivers in China are other examples of highly vulnerable freshwater bodies that became dry and diminished due to the combined effects of climate and human actions such as water diversion. With increasing human population and water demand, all these scenarios could create potential water crises and scarcity (Fig. 3.5), especially during prolonged drought periods.

3.3.4 Water Demand and Climate

Rising human population is triggering increased reservoir and water resources development. More dams are expected to be constructed to support irrigation activities that underpin food security. However, following the analyses of about 6695 reservoirs across the world (during the 1984–2015 period) by Hou et al. (2022), evidence have emerged of the impacts of climate variability on water availability in reservoirs. For instance, they reported a decline of 23% of reservoirs while 21% of other reservoirs showed increase. Much of these declines according to their study were found in southeastern Australia, southwestern USA (10%), and eastern Brazil (9%), while water storage increases were observed in the Nile and Mediterranean basins and southern Africa. Lake Mead in the US, for instance, is a major reservoir that has been supplying water for drinking and irrigation to about 25 million people since 1937. Despite being a critical source of water in the region, satellite images show the lake is experiencing one of the lowest water levels since its inception in 1937 (Fig. 3.6). The lake whose full capacity is approximately 9.3 trillion gallons of water was last full in the summers of 1983 and 1999, but was only 27% full as of July 2022,[2] having been subjected to varying levels of drought intensities (e.g., extreme and exceptional drought) and impacts. The impacts of drought affecting more than 60% land areas in continental United States since 2020 is not only increasing the risk of heat-related medical conditions and wildfires but also impacting the availability of freshwater.[3] Observed changes in spatial patterns of Lake Mead (Fig. 3.6) relative to the 2000 baseline period coincide with long-term drought in the western region of US, one of the worst in more than a thousand years. Notably, this loss in aerial extent of the lake resulted in historic fall in water levels in 2022, a period that coincides with the disappearance of a substantial proportion (greater than 70%) of the lake and other tributaries (Fig. 3.6).

---

[2] https://www.weforum.org/agenda/2022/08/nasa-lake-mead-water-drought/.

[3] https://www.weforum.org/agenda/2022/05/drought-united-states-climate-change.

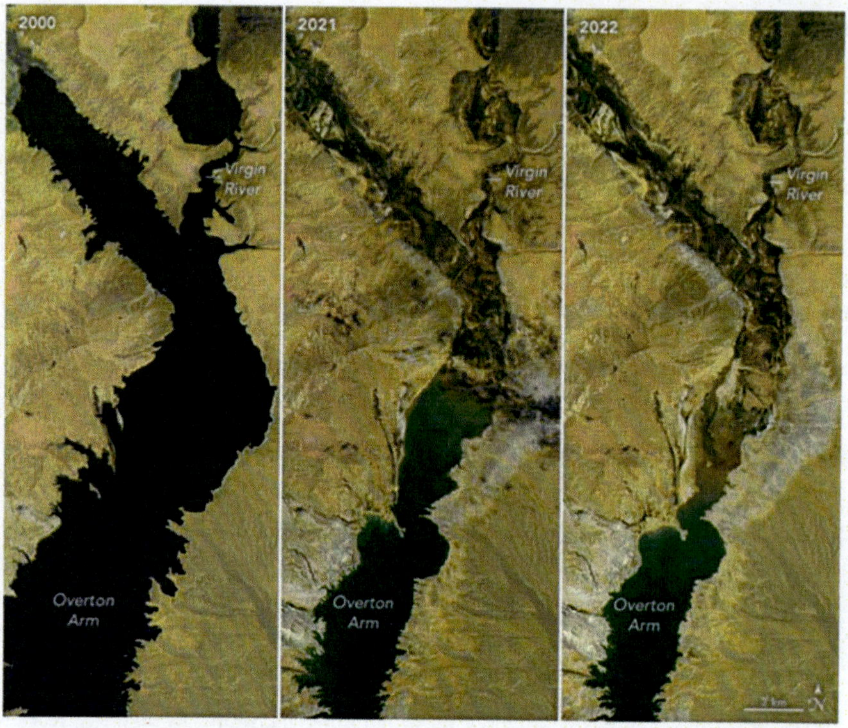

Fig. 3.6 Water storage changes in Lake Mead, US, between 2000 and 2022. *Source* NASA

Climate risk continues to amplify in regions with critical water challenges. Arid regions, e.g., Australia, Israel, the African Sahel, and entire Middle East face these challenges due to aridity, which limits the availability of surface water. Groundwater abstraction in these countries results in drastic declines in aquifer levels. With rapidly expanding agricultural activities in most water deficit regions, their strategic water management policies have favored the development of several water infrastructures. These were targeted efforts aimed at addressing the susceptibility of these regions to dangerous climatic thresholds and aridity. The investments in hydraulic and water infrastructures serve as risk-based decisions, mitigation strategies, and science-based adaptive approaches to boost water availability. These investments are some traditional approaches to addressing the problem of water insecurity and they include dams, reservoirs, pipes, canals, drains, and cement-lined streams for the movement of water. Notably, the economic basis to invest in water infrastructures by developed nations has improved their capacity to cushion the impacts of extreme drought events and threats to freshwater resources (Vorosmarty et al. 2010). However, regions with no infrastructure and economic instruments to manage and develop mechanisms to share risks associated with the cascading impacts of extreme droughts on water resources continue to be vulnerable to the impacts of climate variability (e.g., Vorosmarty et al. 2010).

In geopolitical zones with shared transboundary water resources, there have been recent reports and recommendations on the use of diplomacy and institutional frameworks to foster cooperation and management of water-related conflicts (e.g., Akhmadiyeva and Abdullaev 2019; Atef et al. 2019). Water diplomacy is now perceived as a relevant tool for the cooperative management of shared water resources, multi-stakeholder dialogues, policy and political engagement to support productive discussions related to the legal nature of water sharing, and increase the effectiveness of transboundary water cooperation, among others (see, e.g., Honkonen and Lipponen 2018; Grech-Madin et al. 2018; Schmeier and Shubber 2018). As a template that can be extended to support regional cooperation on transboundary water sharing in other African regions, Al-Saidi and Hefny (2018) highlights the institutional arrangements and the merits of regional cooperation in addressing the underlying issues of water diplomacy in the Eastern Nile Basin. Water diplomacy and negotiation have been advocated as a vehicle to foster the cooperation over transboundary freshwater resources. Inter-basin water transfers, the movement of water from one watershed to another, is now emerging as part of a risk-base solution to cushion the effects of climate change and water crises.

3.4 Security, Sustainability, and Resilience

The impacts of drought on water resources systems can create a plethora of risks and challenges in different sectors, including hydro-power, agriculture, etc. These risks can be amplified in regions that are heavily reliant on rainfall for agriculture and food production (e.g., Getirana et al. 2021; Ndehedehe et al. 2020a; Agutu et al. 2017). The West African Sahel and Eastern Africa are examples of such regions where drought impacts could be catastrophic. From multiple deaths, displacements of rural communities to social (e.g., armed conflict and migration) and economic (e.g., loss of national income) problems, drought impacts in these regions may continue to escalate into the future. Historically, droughts in Ethiopia and Sudan claimed the lives of 450,000 people in 1985.[4] In recent times, dry conditions and droughts continue unabated in these sub-regions (i.e., East Africa) and as critical water resources dry up, about 21.1 million people are expected to be affected by acute food insecurity and other devastating impacts and social problems, including migration. However, since drought is now a way of life in the Anthropocene, some regions of the world are beginning to adapt, increase sustainability, and build resilience to ensure water security. Australia, for instance, is one such region where despite prolonged and frequent episodes of extreme droughts, innovative technologies, including solar-powered desalination, drip-irrigation, and water holding, have been employed to

[4] https://www.ifad.org/en/web/latest/-/east-africa-is-experiencing-its-worst-drought-in-decades-it-s-time-to-invest-in-climate-adaptation.

increase water security and cushion drought impacts. Other regions have pursued water security through investments in satellite-based monitoring to inform on water management practices. Additionally, through stake-holder engagements and participation, a public policy framework has been designed to allocate funds and water distribution to small-scale agro-industries with the goal to cushion drought impacts (e.g., Ndehedehe et al. 2020b; Brito et al. 2018). However, others are increasing drought resilience or "drought-proofing" through water transfer projects and promoting drought-tolerant crops, e.g., millet, sorghum, cowpeas, and cassava. These are some lessons for emerging economies in Asia and Africa to cushion drought impacts similar to what has been adopted in regions like Australia. In this section, some progress and challenges along these lines are identified, focusing on water security, sustainability, and resilience.

3.4.1 Pursuing Water Security and Sustainability

The threats to water security in the twenty-first century are expected to rise and will emanate mostly from population and economic growth, climate change, and poor management of water resources (Vörösmarty et al. 2021). Other causes of water insecurity may include changes in increasing demands, poor water quality and governance, as well as the composite influence of climate and land-use cover changes (Jain 2021). These threats as well as challenges to water resources availability are increasing in several hotspots around the world, including regions that are typically humid. The 2021 water crises in Brazil (Getirana et al. 2021), for instance, is a clarion call to respond to all forms of water security threats, be it human or climate induced. In the pursuit of water security, water management agencies and stake holders are beginning to increase their investments in developing water infrastructures to boost water security underpinned by an integrated water assessment. Teixeira et al. (2021) who tracked the evolutionary patterns of water security for different periods in the São Francisco River Basin in Brazil identified some of these efforts. This included the establishment of water security index in 2019 to guide the investments in water security and planning as well as prioritize investment pathway using a set of United Nations-derived indicators, e.g., Human, Economic, Resilience, and Ecosystem. Additionally, a 5-billion dollar investment to construct water storage infrastructure (reservoirs and dams) running till 2035 are other efforts in place in the pursuit of water security in Brazil Teixeira et al. (2021). Other forms of targeted investments to ensure water security include inter-basin water transfer projects. The water diversion project in the South-to-North Water Diversion project in North China Plain is one of the largest of such projects (e.g., Ndehedehe 2022a; Zhang et al. 2020). In the Sahel, water transfer from the Congo basin to sustain the dying Lake Chad is targeted to help increase water supply for the riparian countries of the Lake Chad basin.

Water demand or water availability can vary in time and space due to the varying degrees of consumption for domestic and industrial purposes. Areas with

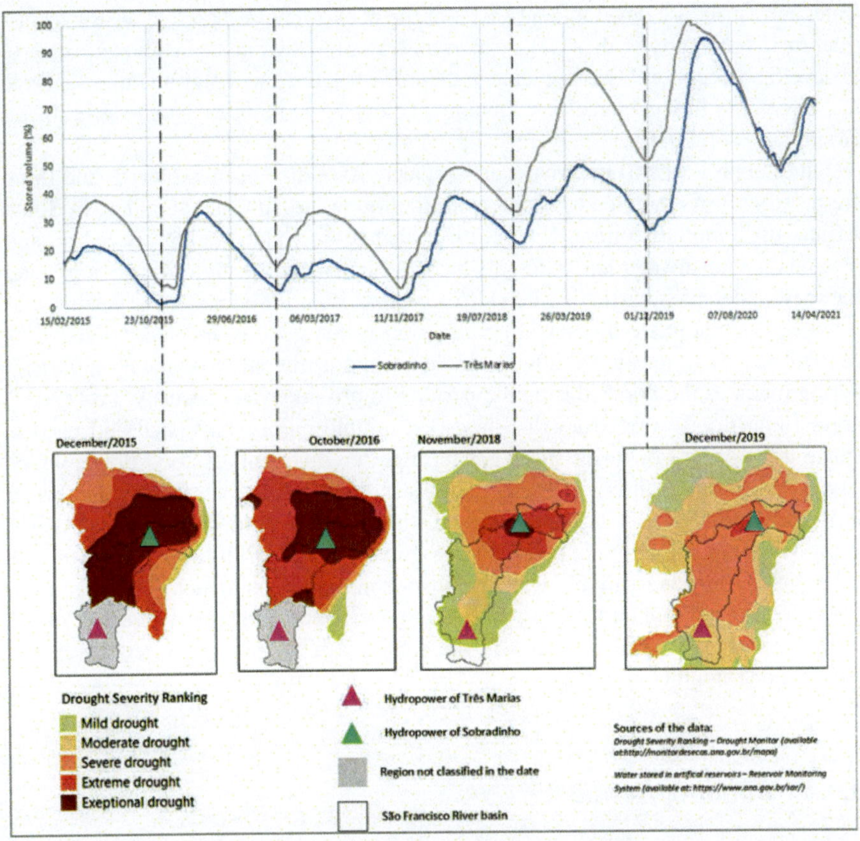

Fig. 3.7 Impacts of droughts on water storage in reservoirs and dams. A case study for the Sobradinho and Três Marias dams here illustrates the vulnerability of stored water in dams to strong changes in climatic conditions. *Source* Teixeira et al. (2021)

considerably higher demands may result in the presence of more reservoirs given the need to increase storage and boost water supply for different applications. However, as illustrated in Fig. 3.7, extreme droughts affect water storage in dams and reservoirs. During the periods of extreme droughts, water storage in the Sobradinho and Três Marias dams were significantly low between 2015 and 2017 as opposed to the 2019–2020 period when these dams had the highest water storage (Fig. 3.7). The challenge of climate extremes and rising water demand underpinned by rising human populations makes it critical to pursue water security. In São Francisco River Basin, Teixeira et al. (2021) identified the development of water infrastructure is playing an important role to regulate threats to water security in some areas though some upstream regions continue to face continuous water security threats that also impact the downstream areas. The need to maintain water security in emerging economies is expected to culminate in the rapid expansion of several investments in water infrastructure developments even at the cost of nature-based assets or infrastructure (Vörösmarty et al. 2021). As part of strategic investments to alleviate poverty in developing economies, these large water infrastructures will play an important role in 21st century in several regions, including emerging economies (Ndehedehe 2022c; Grantham et al. 2019).

Other approaches to improve water security have also been explored, and these are mostly efforts targeted to supplement traditional urban water supplies. In South Africa, the cumulative impacts of climate change on water availability led to suggestions on stormwater harvesting as a cost-effective approach to improving water security (Ndehedehe 2019; Fisher-Jeffes et al. 2017). Flood water harvesting was proposed as a suitable measure to addressing water security in India (Jain 2021). More than 75% of the annual flows in India occur during four monsoon months, triggering much of the floods during the period. To provide water security, Jain (2021) argued that flood water from this monsoon rainfall can be harvested and used to cushion the effects of water shortages through the diversion of flood flows, which can also recharge aquifers. In the Middle East, the extensive development of water infrastructure to exploit available water resources in the Jordan River Basin is evidenced in the construction of more than 45 dams (FAO and Delft 2020). Even though water governance is relatively well established in the region, to achieve water security, a framework that underpins good water governance principles in Jordan, which is categorized as one of countries in the world with the highest water-scarcity, is advocated (Daoud et al. 2022).

3.4.2 Building Resilience

Changes in climatic conditions and the characteristics of key hydrologic metrics (magnitude, frequency, etc.) are raising new concerns about the vulnerability of freshwater resources, including those stored in dams and reservoirs. This is because the frequency and magnitude of extreme events like floods, storms, and droughts continue to change, regularly falling outside observed historical records of climate change for which the water infrastructure were designed. This reason was highlighted by Grantham et al. (2019) who also argued that the uncertainties in climatic conditions will translate to a mismatch in water infrastructure projects, meaning the latter can either be over-constructed or under-built. While such possibility (uncertainty in climate) is making water resources engineers seek novel ways to improve infrastructure performance (Stakhiv 2011), building resilience under these changing climatic conditions for sustainable human development is imperative. Building the resilience of human and natural systems is urgently critical given the severe effects of changes in the Earth system on the human–social–ecological systems (Boltz et al. 2019). That is, these investments in water infrastructures create multiple dimensions of impacts that lead to the loss of ecosystem services, disruption of freshwater ecosystem functioning, and large-scale declines in biodiversity, among others. In promoting ecological resilience for freshwater ecosystems due to these investments in water resources, four important principles and levels of management have been advocated by Grantham et al. (2019). They include managing for

- flow variability,
- hydrologic connectivity,
- spatial heterogeneity, and
- catchment or basin-scale.

These principles are templates to help guide or improve how water management decisions are made and implemented (Grantham et al. 2019). Moreover, these levels of management are important because changes in natural flows and hydrologic connectivity in freshwater systems emanate from water infrastructure development. On the other hand, groundwater abstraction from aquifers can directly affect hydrologic connectivity, resulting in declining surface water flow (Ndehedehe et al. 2021; Grantham et al. 2019).

The awareness on the deleterious impacts of climate change on the environment and the threats posed by water to mankind are increasing the interest of stakeholders and water managers in resilience. Exploring the concept of resilience in terms of persistence, adaptability, and transformability, Hall et al. (2019) acknowledged the capability of developed economies like England to assess the persistence of water supply system during periods of climate shocks and their adaptability over different periods. Such capability (i.e., ability to persist) implies such regions are able to cope with water service delivery and functions during periods of disruptions caused by extreme events (e.g., floods and droughts). Emergency water supplies to customers during water scarcity or periods of disruption caused by floods or droughts exemplify resilience for water security (Hall et al. 2019). Changes in climate and water availability are likely to impact more broadly on water supply systems in emerging economies where agriculture is rainfed (Ndehedehe 2019; Agutu et al. 2017). The development of risk-based decision-making and modeling frameworks is therefore important to help guide water resource management in these regions. Exploring other management strategies that increase resilience and adaption as well as mitigation are also crucial. Some of these strategies have recently been re-echoed, and they include increasing water supplies through desalination, wastewater reuse and enhancing artificial surface water storage in reservoirs and depleted aquifers, as well as water transfers, among others (Scanlon et al. 2023).

3.5 Optimizing Monitoring Framework

3.5.1 Challenges with Observational Networks for Hydrology

Changes in terrestrial water storage, be they groundwater, soil moisture, or surface waters (i.e., lakes, wetlands, and rivers), remain one of the most critical components of the hydrological cycle. However, estimating these changes in stored water remains a major challenge due to sparsity of in situ monitoring stations and lack of large-scale hydrological data, and unreliable field measurements, among others. In some instances, degraded hydrological infrastructures contribute to creating gaps in routine measurements of relevant hydrological variables (e.g., river discharge, groundwater). The poor or lack of investments in observational data in several regions has necessitated the need to investigate how decision makers can leverage geospatial information from space-borne measurements to improve knowledge on surface water hydrology, and support intervention measures and risk assessments from extreme droughts and their impacts. Knowledge gaps in large-scale terrestrial water storage changes and the impacts of climate on the hydrology exist primarily because of limited or lack of observational networks to provide the current state of hydrologic information. Gauge stations for rainfall and river discharge measurements are in decline globally (e.g., Alsdorf and Lettenmaier 2003; Lettenmaier 2005), and it is even worse in non-industrialized nations where gauge networks are not only extremely sparse, discontinuous, and lacking but their density falls far below the World Meteorological Organization guidelines (e.g., Vörösmarty et al. 2001).

3.5.2 *Existing Gauge Observations*

Some investments in gauge measurements for meteorological and hydrological applications exist in non-industrialized nations, but some regions have little or no history of hydrological measurements due to incessant political instabilities, lack of civilization, and the costs implications for the installation of gauges to characterize flow dynamics. Sparse observational records in some gauged river basins have contributed to the lack of capacity to face future challenges of climate change impact on water resources. This problem is further intensified by political concerns, and legal and institutional constraints that have hindered the acquisition of existing data for scientific purposes. Research institutions in these regions find it difficult to engage in regional water management and transboundary water sharing conversations because hydro-climatic data from national archives and government institutions are frequently withheld mostly for political and security reasons. This could only be more damaging for countries where water sharing across political boundaries is vastly enormous. Apparently, the paucity of high-quality base line data remains a major constraint to understanding the climatic influence on changes in the hydrological cycle of several regions (e.g., Ndehedehe et al. 2016; Todd et al. 2011; Farnsworth et al. 2011). The consistent monitoring of the environment to quantify these impacts is required for risk-management solutions to support adaptation plans and increase resilience to climate change. Ultimately, it is also anticipated that such monitoring will lead to a better understanding of climate change processes and the cascading effects of extreme climate events such as droughts and floods on agricultural and socio-economic systems.

References

Agutu N, Awange J, Zerihun A, Ndehedehe C, Kuhn M, Fukuda Y (2017) Assessing multi-satellite remote sensing, reanalysis, and land surface models' products in characterizing agricultural drought in East Africa. Remote Sens Environ 194:287–302. https://doi.org/10.1016/j.rse.2017.03.041

Akhmadiyeva Z, Abdullaev I (2019) Water management paradigm shifts in the Caspian Sea region: review and outlook. J Hydrol 568:997–1006. https://doi.org/10.1016/j.jhydrol.2018.11.009

Al-Saidi M, Hefny A (2018) Institutional arrangements for beneficial regional cooperation on water, energy and food priority issues in the Eastern Nile Basin. J Hydrol 562:821–831. https://doi.org/10.1016/j.jhydrol.2018.05.009

Alcamo J, Döll P, Kaspar F, Siebert S (1997) Global change and global scenarios of water use and availability: an application of WaterGAP1.0. Center for Environmental Systems Research (CESR), University of Kassel, Germany. Retrieved from: https://pdfs.semanticscholar.org/a788/80a9f9952f0e04bc88c7c5699ea9c9e0fa30.pdf on 18th Aug 2016

Alsdorf DE, Lettenmaier DP (2003) Tracking fresh water from space. Science 301(5639):1491–1494. https://doi.org/10.1126/science.1089802

Andam-Akorful S, Ferreira V, Ndehedehe CE, Quaye-Ballard J (2017) An investigation into the freshwater variability in West Africa during 1979–2010. Int J Climatol 37(S1):333–349. https://doi.org/10.1002/joc.5006

Atef SS, Sadeqinazhad F, Farjaad F, Amatya DM (2019) Water conflict management and cooperation between Afghanistan and Pakistan. J Hydrol 570:875–892. https://doi.org/10.1016/j.jhydrol.2018.12.075

Boltz F, LeRoy Poff N, Folke C, Kete N, Brown CM, St. George Freeman S, Matthews JH, Martinez A, Rockström J (2019) Water is a master variable: solving for resilience in the modern era. Water Secur 8:100048. https://doi.org/10.1016/j.wasec.2019.100048

Brauman KA, Richter BD, Postel S, Malsy M, Flörke M (2016) Water depletion: an improved metric for incorporating seasonal and dry-year water scarcity into water risk assessments. Elementa 4(83). https://doi.org/10.12952/journal.elementa.000083

Brito SSB, Cunha APMA, Cunningham CC, Alvalá RC, Marengo JA, Carvalho MA (2018) Frequency, duration and severity of drought in the semiarid Northeast Brazil region. Int J Climatol 38(2):517–529. https://doi.org/10.1002/joc.5225

Chen P, Wang S, Liu Y, Wang Y, Wang Y, Zhang T, Zhang H, Yao Y, Song J (2023) Water availability in China's Oases decreased between 1987 and 2017. Earth's Future 11(4):e2022EF003340. https://doi.org/10.1029/2022EF003340

Chindarkar N, Grafton RQ (2019) India's depleting groundwater: when science meets policy. Asia Pac Policy Stud 6(1):108–124. https://doi.org/10.1002/app5.269

Daoud I, Dehnavi S, Ribbe L (2022) Towards good water governance: an analysis of Jordan's National Water Strategy. Environ Manag 69:847–860. https://doi.org/10.1007/s00267-022-01606-x

de Marsily G, Abarca-del Rio R (2016) Water and food in the twenty-first century. Surv Geophys 37:503–527. https://doi.org/10.1007/s10712-015-9335-1

FAO and IHE Delft (2020). Water accounting in the jordan river basin. FAO WaPOR water accounting reports. https://doi.org/10.4060/ca9181en

Farnsworth A, White E, Williams CJ, Black E, Kniveton DR (2011) Understanding the large scale driving mechanisms of rainfall variability over central africa, in african climate and climate change: physical, social and political perspectives. African climate and climate change: physical, social and political perspectives, pp 101–122. https://doi.org/10.1007/978-90-481-3842-5_5.

Fisher-Jeffes L, Carden K, Armitage NP, Winter K (2017) Stormwater harvesting: improving water security in South Africa's urban areas. South Afr J Sci 113(1/2):4. https://doi.org/10.17159/sajs.2017/20160153

Getirana A (2016) Extreme water deficit in brazil detected from space. J Hydrometeorol 17(2):591–599. https://doi.org/10.1175/JHM-D-15-0096.1

Getirana A, Libonati R, Cataldi M (2021) Brazil is in water crisis – it needs a drought plan. Nature 600:218–220. https://doi.org/10.1038/d41586-021-03625-w

Gleeson T, Wada Y, Bierkens MFP, van Beek LPH (2012) Water balance of global aquifers revealed by groundwater footprint. Nature 488:197–200. https://doi.org/10.1038/nature11295

Gleick PH (1993) Water and conflict: fresh water resources and international security. Int Secur 18(1):79–112. https://doi.org/10.2307/2539033

Gleick PH (2000) The world's water: Biennial report on freshwater resources 2000–2001. Island Press, Washington, DC

Grantham TE, Matthews JH, Bledsoe BP (2019) Shifting currents: managing freshwater systems for ecological resilience in a changing climate. Water Secur 8:100049. https://doi.org/10.1016/j.wasec.2019.100049

Grech-Madin C, Döring S, Kim K, Swain A (2018) Negotiating water across levels: a peace and conflict "toolbox" for water diplomacy. J Hydrol 559:100–109. https://doi.org/10.1016/j.jhydrol.2018.02.008

Green TR, Taniguchi M et al (2011) Beneath the surface of global change: impacts of climate change on groundwater. J Hydrol 405(3):532–560. https://doi.org/10.1016/j.jhydrol.2011.05.002

Hall JW, Borgomeo E, Bruce A, Di Mauro M, Mortazavi-Naeini M (2019) Resilience of water resource systems: lessons from England. Water Secur 8:100052. https://doi.org/10.1016/j.wasec.2019.100052

Honkonen T, Lipponen A (2018) Finland's cooperation in managing transboundary waters and the UNECE Principles for Effective Joint Bodies: value for water diplomacy? J Hydrol 567:320–331. https://doi.org/10.1016/j.jhydrol.2018.09.062

Hou J, van Dijk AIJM, Beck HE, Renzullo LJ, Wada Y (2022) Remotely sensed reservoir water storage dynamics (1984–2015) and the influence of climate variability and management at a global scale. Hydrol Earth Syst Sci 26(14):3785–3803. https://doi.org/10.5194/hess-26-3785-2022

Jain SK (2021) Providing water security in India by conserving and utilizing flood flows. Water Secur 14:100105. https://doi.org/10.1016/j.wasec.2021.100105

Kundzewicz ZW, Mata LJ, Arnell NW, Doll P, Jimenez B, Miller K, Oki T, Şen Z, Shiklomanov I (2008) The implications of projected climate change for freshwater resources and their management. Hydrol Sci J 53(1):3–10. https://doi.org/10.1623/hysj.53.1.3

Lettenmaier DP (2005) Observations of the global water cycle – global monitoring networks. In: Anderson MG (ed) Encyclopedia of hydrological sciences, vol 5, pp 2719–2732. Wiley

Ndehedehe C (2022a) Anthropogenic influence on terrestrial hydrology, pp 283–298. Springer, Cham

Ndehedehe C (2022b) Global freshwater systems, pp 19–32. Springer, Cham. https://doi.org/10.1007/978-3-030-99577-5_2

Ndehedehe C (2022c) Impacts of water resources development on hydrology, pp 389–437. Springer, Cham. https://doi.org/10.1007/978-3-030-99577-5_15

Ndehedehe C, Awange J, Agutu N, Kuhn M, Heck B (2016) Understanding changes in terrestrial water storage over West Africa between 2002 and 2014. Adv Water Resour 88:211–230. https://doi.org/10.1016/j.advwatres.2015.12.009

Ndehedehe CE (2019) The water resources of tropical West Africa: problems, progress and prospect. Acta Geophysica 67(2):621–649. https://doi.org/10.1007/s11600-019-00260-y

Ndehedehe CE, Agutu N, Ferreira VG, Getirana A (2020) Evolutionary drought patterns over the Sahel and their teleconnections with low frequency climate oscillations. Atmos Res 233:104700. https://doi.org/10.1016/j.atmosres.2019.104700

Ndehedehe CE, Awange J, Kuhn M, Agutu N, Fukuda Y (2017) Analysis of hydrological variability over the Volta river basin using in-situ data and satellite observations. J Hydrol Reg Stud 12:88–110. https://doi.org/10.1016/j.ejrh.2017.04.005

Ndehedehe CE, Ferreira VG, Agutu NO et al (2021) What if the rains do not come? J Hydrol 595:126040. https://doi.org/10.1016/j.jhydrol.2021.126040

Ndehedehe CE, Haile GG, Agutu NO, Ferreira VG, Getirana A, Okwuashi O (2020) Hydrological hotspots of climatic influence in Brazil: a two-step regularization approach. Atmos Res 246:105116. https://doi.org/10.1016/j.atmosres.2020.105116

Odorico P, Chiarelli DD, Rosa L et al (2020) The global value of water in agriculture. Proc Natl Acad Sci 117(36):21985–21993. https://doi.org/10.1073/pnas.2005835117

Ojha C, Werth S, Shirzaei M (2020) Recovery of aquifer-systems in Southwest US following 2012–2015 drought: evidence from InSAR, GRACE and groundwater level data. J Hydrol 587. https://doi.org/10.1016/j.jhydrol.2020.124943

Oki T, Kanae S (2006) Global hydrological cycles and world water resources. Science 313(5790):1068–1072. https://doi.org/10.1126/science.1128845

Rijsberman FR (2006) Water scarcity: fact or fiction? Agric Water Manag 80(1–3):5–22. https://doi.org/10.1016/j.agwat.2005.07.001

Rodell M, Famiglietti JS, Wiese DN, Reager JT, Beaudoing HK, Landerer FW, Lo MH (2018) Emerging trends in global freshwater availability. Nature 557(7707):651–659. https://doi.org/10.1038/s41586-018-0123-1

Saemian P, Tourian MJ, AghaKouchak A, Madani K, Sneeuw N (2022) How much water did Iran lose over the last two decades? J Hydrol Reg Stud 41:101095. https://doi.org/10.1016/j.ejrh.2022.101095

Scanlon B, Fakhreddine S, Rateb A, de Graaf I, Famiglietti J, Gleeson T et al (2023) Global water resources and the role of groundwater in a resilient water future. Nature Rev Earth Environ 4:87–101. https://doi.org/10.1038/s43017-022-00378-6

Schewe J, Heinke J, Gerten D, Haddeland I, Arnell NW et al (2013) Multimodel assessment of water scarcity under climate change. PNAS 111(9):3245–3250. https://doi.org/10.1073/pnas.1222460110

Schmeier S, Shubber Z (2018) Anchoring water diplomacy – the legal nature of international river basin organizations. J Hydrol 567:114–120. https://doi.org/10.1016/j.jhydrol.2018.09.054

Shiklomanov IA (2000) Appraisal and assessment of world water resources. Water Int 25(1):11–32. https://doi.org/10.1080/02508060008686794

Stakhiv EZ (2011) Pragmatic approaches for water management under climate change uncertainty. JAWRA J Am Water Resour Assoc 47(6):1183–1196

Teixeira AL, Bhaduri A, Bunn SE, Ayrimoraes SR (2021) Operationalizing water security concept in water investment planning: case study of são francisco river basin. Water 13(24):3658. https://doi.org/10.3390/w13243658

Todd M, Andersson L, Ambrosino C, Hughes D, Kniveton DR, Mileham L, Murray-Hudson M, Raghavan S, Taylor R, Wolski P (2011) Climate change impacts on hydrology in Africa: case studies of river basin water resources, in African climate and climate change: physical, social and political perspectives. African climate and climate change: physical, social and political perspectives, pp 123–153. https://doi.org/10.1007/978-90-481-3842-5_5

Vasco D, Farr T, Jeanne P et al (2019) Satellite-based monitoring of groundwater depletion in California's Central Valley. Sci Rep 9:16053. https://doi.org/10.1038/s41598-019-52371-7

Vörösmarty C, Askew A, Grabs W et al (2001) Global water data: a newly endangered species. EOS, Trans Am Geophys Union 82(5):54–58. https://doi.org/10.1029/01EO00031

Vörösmarty CJ, Green P, Salisbury J, Lammers RB (2000) Global water resources: vulnerability from climate change and population growth. Science 289(5477):284–288. https://doi.org/10.1126/science.289.5477.284

Vorosmarty CJ, McIntyre PB, Gessner MO et al (2010) Global threats to human water security and river biodiversity. Nature 467:555–555–561. https://doi.org/10.1038/nature09440

Vörösmarty CJ, Stewart-Koster B, Green PA et al (2021) A green-gray path to global water security and sustainable infrastructure. Global Environ Change 70:102344. https://doi.org/10.1016/j.gloenvcha.2021.102344

Zhang C, Duan Q, Yeh P, Pan Y et al (2020) The effectiveness of the South-to-North water diversion middle route project on water delivery and groundwater recovery in North China Plain. Water Resour Res 56(10):e2019WR026759. https://doi.org/10.1029/2019WR026759

Chapter 4
Satellite Hydrology Programmes: Capabilities and Benefits

4.1 Introductory Remarks

The capacity to monitor the hydrosphere from space has reformed not only our perspective of the Earth as an interconnected system, but also our understanding of the dynamic processes that takes place above, on, and beneath the earth's surface. The year 2022 marks the 65th year since the launch of Sputnik 1, which was the first man-made object to be placed into orbit. The successful launch of Sputnik-1 in the late 1950s by the National Aeronautics and Space Administration (NASA) marked the definition of a new era, the space age. Sputnik 1 broadcasted a dual-frequency radio transmission spanning a period of 21 days, until its batteries ran out. However, this was an important leap and advancement in science, as it made notable contributions to the spatial sciences domain, thus triggering the "space race". Notably, the earliest satellite systems were used for advancing military reconnaissance operations and surveillance. But scientists at the time quickly recognized the value of space-based sensors in monitoring weather and climatic conditions on large scales (Nordberg 1965). In the years following the launch of Sputnik, several satellite systems (Fig. 4.1) were launched. This included the Television and Infrared Observation Satellite (TIROS), Nimbus 1, and the Environmental Science Services Administration (ESSA-1) satellites in 1960, 1964, and 1966, respectively. However, the first civilian focused multispectral sensors were launched fifteen years after Sputnik, i.e., the Landsat 1 mission in 1972. Since then, successive Landsat programmes continue to provide unrivaled record of changes in landscapes for over fifty years without interruption, and with Landsat 9 being the latest of the Landsat family (Ndehedehe 2022b; Hemati et al. 2021; Wulder et al. 2008).

After the launch of these early satellite missions, there have been notable advancements in remote sensing platforms and sensors. Some of the advances include the gradual development from early panchromatic to red–green–blue imagery, the inclusion of multi- and hyperspectral visible to near infrared bands to sensor technology, multi-channel microwave emissions, etc. These have all contributed to redefining man's knowledge of the earth systems and its dynamics. However, the emergence

C. Ndehedehe, *Hydro-Climatic Extremes in the Anthropocene*, Springer Climate,
https://doi.org/10.1007/978-3-031-37727-3_4

of new sensor (Fig. 4.1) capabilities revolutionized the utility of satellite remote sensing products, thus increasing their multidisciplinary capabilities in earth system monitoring. Some of these missions have redefined the remote sensing discipline to include novel concepts such as *Remote Sensing Hydrology* (Ndehedehe 2022b). This is largely because of new investments in satellite geodetic programmes like the Gravity Recovery and Climate Experiment (Tapley et al. 2004) and the Surface Water and Ocean Topography[1] missions, which are satellite hydrology-specific missions and dedicated technologies that provide level 1 information on the quantitative aspects of freshwater resources (e.g., lakes, rivers, and groundwater).

Remote sensing techniques like Gravity Recovery and Climate Experiment, Surface Water and Ocean Topography, and satellite altimetry missions are examples that illustrate the *Remote Sensing Hydrology* concept (see Ndehedehe 2022b). This is because they are different from other earth resources satellites such as the suite of European Space Agency-based Sentinel products (e.g., Sentinels 1-2) and Landsat systems, which can be used to measure changes in land surface (e.g., surface water, vegetation, and temperature). Satellite hydrology missions (e.g., Gravity Recovery and Climate Experiment and Surface Water and Ocean Topography) are dedicated remote sensing systems that focus on improving our understanding of freshwater (terrestrial stored water) and large-scale hydrological processes and patterns by measuring the amount of water in surface water bodies (e.g., river or lake) and changes in land water storage (e.g., soil moisture, snow and ice cover, canopy water storage, and groundwater) (Ndehedehe 2022b). This chapter will discuss some of these new capabilities and methods to assess global terrestrial hydrology underpinned by the recent advances in multi-mission satellite technology. Some of these satellite programmes including the Surface Water and Ocean Topography mission have radar capabilities that will help track the pace of climate change. These satellite geodetic missions including the Gravity Recovery and Climate Experiment and satellite altimetry have contributed to improving the understanding of impacts of climate change and human influence on the water cycle, thus helping with policy and management decisions related to water resources.

4.2 Multi-mission Satellite Hydrology Systems

The European, American, Japanese, Indian, Chinese, Canadian, and other national space agencies manage a vast number of satellite systems that deliver a wide range of remote sensing products with different spatial and temporal resolutions that benefit the scientific community. Some of these satellite systems are jointly owned and operated by two or more countries. For example, the International Space Station is jointly owned by the US, Russia, Europe, Canada, and Japan. NASA alone has eighteen primary and active earth science missions in orbit, whereas the European Space Agency (ESA) has eleven earth observation missions in operation. There are

[1] https://swot.jpl.nasa.gov/.

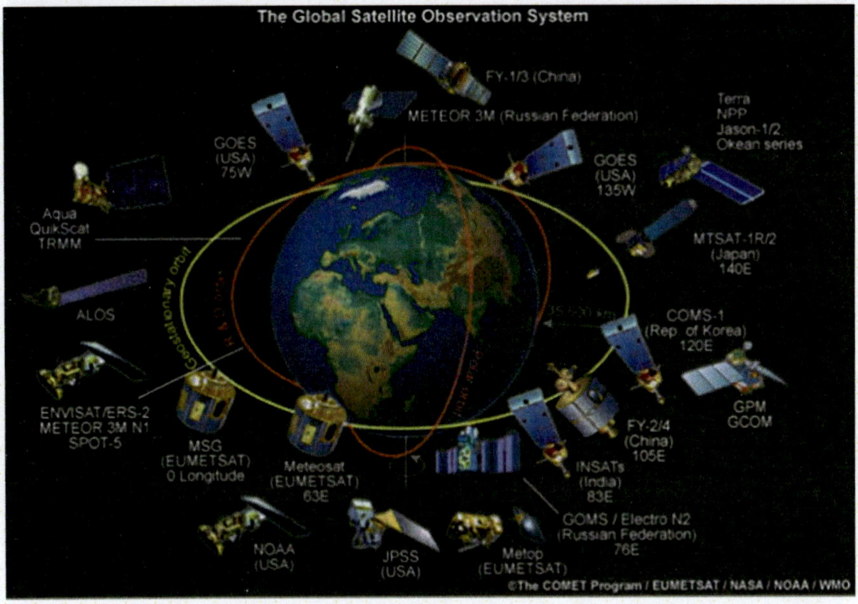

Fig. 4.1 Earth resources satellite systems. *Source* NAS/NOAA

also a range of future satellites which are in their advanced stages of planning and their launch is underway. Various international agencies also have their own different earth observation instruments that operate onboard small satellites and Cubesats, whereas others are mounted within the International Space Station. A large proportion of these satellite missions (Fig. 4.1) are focused on monitoring earth resources, including climate and hydrological processes, and are notably useful in providing key information on the physical properties and extents of land surface conditions.

The use of optical satellite systems has aided the detection of surface water around the globe, including mapping various flood intensities. These earth observation satellites are now important data sources for hydrologists, environmental scientists, and water managers to improve understanding environmental changes caused by climate and human activities. The petabytes of big data from these missions make this possible. Additionally, the potential for better monitoring, analysis, and forecasts of water balance and assessing land surface processes and atmospheric interactions using these data is increasing (Ndehedehe 2022c). However, with the emergence of dedicated satellite hydrology missions, which provide quantitative assessments of water resources, and snow and ice from local to global scales, satellite applications in water resources have improved further.

Due to the impact of climate change, the need to pay close attention to satellite hydrology systems has soared over the years. The launch of key missions like Gravity Recovery and Climate Experiment (GRACE) has really helped in understanding freshwater changes through a global-scale assessment, making such effort better

Fig. 4.2 Illustration of the Gravity Recovery and Climate Experiment satellite mission. The two identical twin satellites popularly known as *Tom* and *Jerry* are linked with inter-satellite ranging system and equipped with global positioning system receivers and attitude sensors and high precision accelerometers. The ranging system is super sensitive and can detect separation changes about one-tenth the width of a human hair over a distance of 220 km (NASA 2002). The GRACE satellites are designed to map the global gravity field every thirty days. Image adapted from http://spaceinfo.com.au/2010/07/23/gravity-mission-down-under/

than what was achieved before now. The GRACE satellite mission was launched in March 2002 from Plesetsk, Russia, and was the result of a joint collaboration between NASA and Germany. The two identical satellites (Fig. 4.2) are linked with inter-satellite ranging system and equipped with global positioning system receivers and attitude sensors and high precision accelerometers (NASA 2002), and by using their distance apart (220 km), they provide continuous high-resolution changes of gravity in relative motion. After the end of the GRACE mission in October 2017, the GRACE-Follow On satellite became operational, consolidating on the gains of the GRACE mission in improving understanding of freshwater changes around the globe. Some of these gains include broad applications in assessing extreme events (droughts, floods), terrestrial water budget, as well as providing capabilities in monitoring aquifer-system processes, and evaluating groundwater resources (Ndehedehe 2019).

Several imaging sensors in the optical and microwave spectral range are available, much of which can be used for snow and ice observations. Aiming to overcome some of the challenges associated with spatial heterogeneity and retrieval performance, some studies have proposed the use of active microwave instruments, such as Synthetic Aperture Radar (SAR) in the retrieval of surface water estimates (e.g., Yueh and Chaubell 2012). The application of polarimetric and radar interferometry techniques also has the potential of yielding precise hydrological measurements, including snow water equivalent and snow height estimates. However, space-borne sensors with the needed SAR frequencies for backscattering estimates or the required coverage and revisit time for interferometric observations are currently not available. Notwithstanding, it is important to note that satellite missions with the capability to monitor the water cycle are currently being studied, and their planned dual-frequency polarized microwave scatterometer promises to achieve high spatio-temporal resolutions for hydrological assessments, including snow water equivalent observations. For now, GRACE is the only satellite geodetic mission providing the capability to track the acceleration of global water cycle (Ndehedehe 2022c).

However, the recently launched (December 15, 2022) Surface Water And Ocean Topography mission (SWOT) pioneers a new dimension in hydrological sciences. The SWOT mission will, for the first time, improve the capability to track changes in freshwater availability in lakes, rivers, reservoirs, etc. Such capabilities will also include measuring surface inundation extents and surface water elevations concurrently and over contiguous river sections that have the potential of extending to 100 km or even more. The first results (Sect. 4.3.3.2) from this mission are underway and scientists and researchers have developed algorithms and a framework that can infer channel parameters including cross-sectional area, roughness, and ultimately discharge with the use of time series observations of long river profiles. The work undertaken by Durand et al. (2023) is one of the pioneering efforts on a framework for global river discharge from the SWOT mission. They detailed how discharge from SWOT will be estimated, highlighting the uncertainty (less than 30%) and expected accuracy of such estimates for global reaches. In this section, the focus will be on discussing important satellite hydrology missions like SWOT and GRACE that underpin our better understanding of terrestrial hydrology.

4.2.1 Satellite Radar Systems

As one of the most critical breakthroughs of Second World War, radar (**R**adio **D**etection **A**nd **R**anging) technology has evolved into a more valuable tool in environmental monitoring, including atmospheric phenomena and helping with navigation among others. These modern-day applications of radar are different from its original military applications after the technology was publicly made available post World War II. Today, satellite radar observations like the Synthetic Aperture Radar (SAR) have been used to generate stacks of interferograms to monitor surface deformation associated with natural (earth movement caused by earthquakes), or those caused by human activities such as intensive groundwater extraction.

Changes in groundwater levels in aquifers lead to ground displacements that are discernible at the Earth's surface. Surface displacements are interconnected to the geo-mechanical characteristics of the aquifer, and land surface displacement observations can be implemented as a proxy to constrain variations in the level of groundwater within an artesian monitoring bore. The relationship between groundwater extraction and land subsidence has been established in major cities and agricultural landscapes where irrigation water use is substantially higher. The association between long-term irrecoverable subsidence and the inelastic compaction of aquifers, which is largely driven by the overexploitation of groundwater has also been documented (Galloway and Burbey 2011). In Australia, for instance, long-term subsidence (approximately 3–6mm/year) since the 70's and 90's has been measured from continuously operating Global Position System (GPS) instruments, repeated differential leveling, and by using the Interferometric Synthetic Aperture Radar (InSAR) technique.

Multi-temporal satellite observations of interferometry or InSAR provides large-scale, high spatial resolution maps of surface displacements. Interferograms are usually generated by weighing against the phase of successive SAR images. Assuming the phase change impact from the satellite orbit and topography is removed, the remaining interferogram phase changes observed in multiple slices of the satellite wavelength are basically the result of displacement of the Earth's surface, including the dynamics in the refractivity of the atmosphere (Bürgmann et al. 2000). InSAR can serve as a tool to obtain head change observations that are spatially denser and cover a larger spatial extent than discrete borehole records, and for delineating hydro-geological structures and features within a groundwater basin (Chaussard et al. 2014).

Some recent applications of InSAR to monitor groundwater-induced land surface displacements suggest radar missions have satellite hydrology capability. For instance, Sneed and Brandt (2020) used InSAR (C-band from ERS1/2, Envisat and Sentinel-1) from 1995 to 2017 over the Coachella Valley, California, USA, to examine increases in aquifer-wide groundwater levels. This application was undertaken following artificial recharge from other groundwater sources, conservation, and multiple managed aquifer recharge facilities. The illustration in Fig. 4.3 re-echoes the importance of how the continued monitoring of water levels and subsidence can help inform on future mitigation measures underpinned by our capability to track regions with overdrafted aquifers and map subsidence-prone locations. In the main deformation map, elevation differences of up to 400 m can be seen in the south of La Quinta, and the corresponding InSAR-derived time series (Fig. 4.3b) of this location (LQ, 6S/7E-16A2) on the map shows falling water surface elevation since 1992. InSAR-based time series for other locations on the map also show large land subsidence rates (Fig. 4.3b). Additionally, Neely et al. (2021) use C-band Sentinel-1 to focus on the uplift from natural recharge and reduction in groundwater extraction, thereby relating this to a pattern of seasonal amplitude. In their study, the uplift was considered seasonal rather than directly attributable to managed aquifer recharge.

Over the last decade, satellite remote sensing of the temporal dynamics of groundwater levels and changes in gravity has advanced our understanding of groundwater storage and variations. The GRACE mission proves very useful in this regard as it measures mass changes caused by continental water redistribution (Fig. 4.4). InSAR techniques, on the other hand, are better suited for ground deformation detection

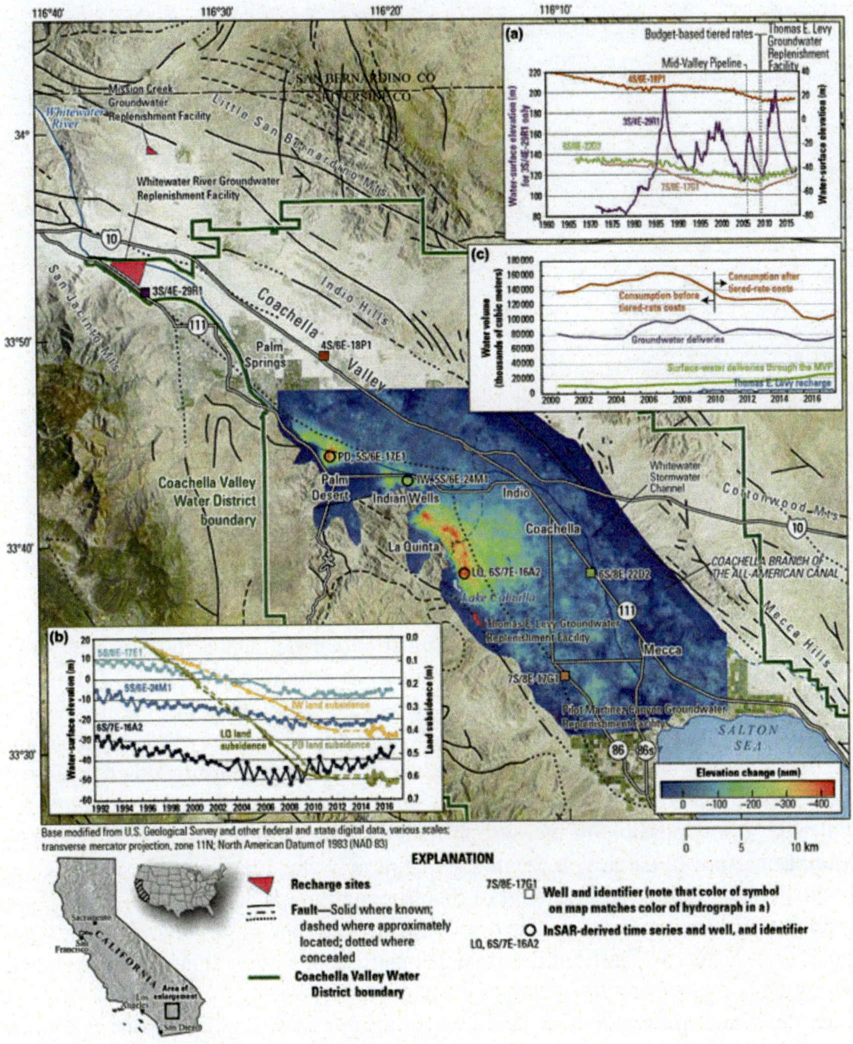

Fig. 4.3 An interferogram showing land surface deformation for 1995–2017 period. **a** Time series of groundwater levels (1960–2017), **b** InSAR-derived deformation in specific locations and groundwater levels (1992–2017) for some well locations; **c** groundwater volumes and surface water deliveries. *Source* Sneed and Brandt (2020)

through the processing and analysis of interferometry using the synthetic aperture radar satellite data. Both systems are very complementary, providing the unique capability to measure the effects of climate variability and water extraction on changes in groundwater storage. However, their sensitivities to aquifer systems could be different given the spatial scales at which GRACE detects gravity fluctuations associated with groundwater changes. The GRACE mission is sensitive to mass changes on a

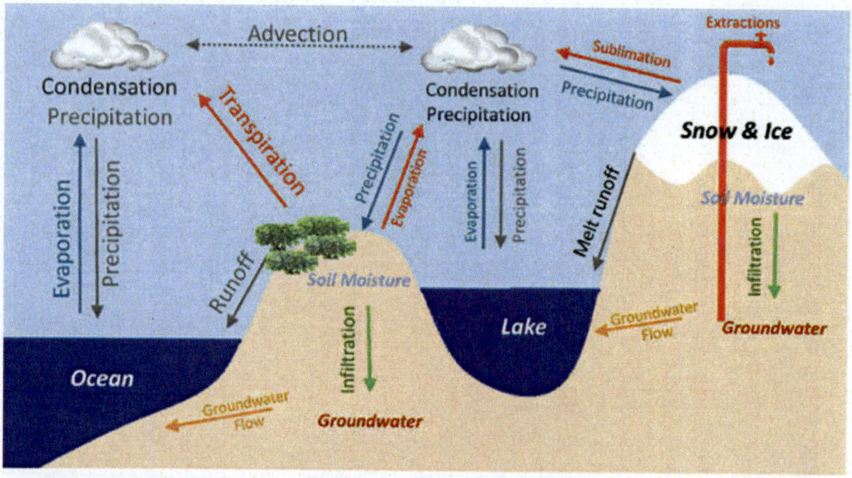

Fig. 4.4 Terrestrial water storage components and hydrological stores (e.g., surface water, soil moisture, snow/ice, and groundwater) observed by the Gravity Recovery and Climate Experiment satellite. *Source* Masood et al. (2022)

much larger scale (>10,000 km^2), observing continental changes in terrestrial water storage, including variations caused by groundwater pumping (Fig. 4.4). Therefore, it is not able to provide groundwater storage change estimates at local scales, which is the case in most aquifer systems globally, and at which most groundwater management schemes are implemented. Even though a machine learning downscaling of GRACE-derived groundwater has recently emerged to circumvent this problem (Ali et al. 2022), it is possible to measure ground displacement due to aquifer response to fluid pressure dynamics in localized spatial or water management scales using InSAR. Furthermore, the inversion of displacement maps derived from the InSAR techniques into volumes of depleted groundwater storage is possible through horizontal and vertical variability of sediment constriction. Both techniques have shown increasing relevance and interest over the past decade, especially in the hydrological community and are efficient in complementing few in-situ observations that have been in existence for some time now.

4.2.2 *Satellite Geodesy Techniques*

4.2.2.1 GRACE, GOCE, and CHAMP

There has been an increased investment in satellite geodetic programmes in the last decade. These programmes such as the GRACE and altimetry missions rely on key principles of geodesy to measure large geodetic changes within the earth's surface, including key components of the global water cycle (Ndehedehe 2022c). Generally, this principle entails the inter-relationship among surface mass variations in the atmosphere, oceans, land hydrology, and ice sheets. For several years, satellite laser ranging measurements to geodetic satellites have provided useful information of the low-degree coefficients of the earth's gravity field. With the advent of the GRACE mission (Fig. 4.2), it is now possible to obtain information on time-variable gravity fields at high spatial resolutions. Prior to the advent of the satellite missions dedicated to gravity field recovery, i.e., the CHAllenging Minisatellite Payload (CHAMP), GRACE, the Gravity field and steady-state Ocean Circulation Explorer (GOCE), and geodetic satellite laser ranging (SLR) satellites were mostly utilized in the determination of the earth's gravity field models (Sośnica et al. 2015). The high-degree coefficients of the SLR-derived gravity field proved to be of poor quality as a result of the sparse and inhomogeneous SLR network available and due to the correlations existing between coefficients of similar order.

The launch of the CHAMP mission in the year 2000 brought about the first satellite dedicated to the recovery of the earth's gravity field. It produced high-quality static gravity models which were obtained from continuous GPS observations. During its mission, the CHAMP generated, for the first time, simultaneously high precise gravity and magnetic field observations. These observations have aided scientists and hydrologists to detect the spatial dynamics of both fields (gravity and magnetic) including their variability with time. The CHAMP mission initiated a new era in geopotential research by

- mapping of the earth's global long to medium wavelength gravity field and temporal dynamics,
- mapping of the earth's global magnetic field and temporal dynamics, and atmosphere sounding.

With these advancements, CHAMP data has shown significant applications in global climate studies, weather forecasting, oceanography, disaster research, navigation, geophysics, solar terrestrial physics, and all branches of geodesy. CHAMP-derived datasets serve as an ideal basis for refinements of modern satellite surveying methods. They are also effective in cartography and remote sensing applications, which include the construction of digital terrain models that span large land and ice areas. The evaluation of all the three types of signals that CHAMP observes allows a comprehensive and integrated modeling of the structure and dynamics of the earth's core and mantle. These improvements strongly improve studies about the structure and composition of the earth's interior and will open new application areas in solid earth physics, geodesy, and hydrology.

After the launch of the GRACE mission in March 2002, the knowledge of the mass transport within the earth's system was substantially improved (Tapley et al. 2004). The twin satellites (Fig. 4.2) allowed defining the relationship between mass deviation in the oceans, atmosphere, land hydrology, and ice sheets with high temporal and spatial resolutions. The lowest degree coefficients of the gravity field remain better defined through the geodetic SLR satellites. This is as a result of the insensitivity of the K-band GRACE signals to the coefficients of degree 1 (geocenter), which is attributed to a differenced measurement type, while the coefficients C20 is degraded due to long period signals, largely because some C20 signals have the same period as the S2 and S1 tidal analysis with the GRACE orbits (Chen et al. 2009). Therefore, we still admit that the SLR satellites maintain a non-negligible potential to establish the low-degree parameters of the earth's gravity field.

4.2.2.2 MAGIC

The quest to improve the monitoring of freshwater changes to help track climate change impact on the global water cycle has led to a new proposal by the European Space Agency (ESA) for a more sophisticated satellite hydrology mission called **Mass change and Geosciences International Constellation (MAGIC)**. As with GRACE, MAGIC[2] will provide the capability to directly assess changes in terrestrial stored water by measuring changes in the Earth's gravity fields to help with understanding mass transport and impacts of climate change on surface water, ice sheets, glaciers, and groundwater as well as global sea level. It is also expected that MAGIC would aid the monitoring of tectonic motion and earthquakes around the globe. MAGIC will be the latest of new generation gravity missions and ESA's key investment in satellite geodetic missions that are expected to underpin our understanding of Earth's geophysical processes including the water cycle as well as the solid Earth. Similar to the role of GRACE in pioneering freshwater reporting, MAGIC will complement this, especially in water management applications as well as improving knowledge on sea-level change.

4.3 Quantifying Land Water Storage

The sustainable management of land water storage resources (Fig. 4.4) needs a complete understanding of the dynamics of discrete components of the water cycle as well as the integrated balance. The water cycle (Fig. 4.4) primarily involves hydrological fluxes including, precipitation (P), evapotranspiration (ET), and river discharge (R), which for a natural equilibrium maintains a steady state of $P - ET - R = 0$. Any divergence from the naturalized equilibrium such as a flux deficit in $P - ET - R$ caused by changing climate and environmental or anthropogenic factors suggests variations in land water storage, such that $P - ET - R = TWS$ changes, which is what GRACE measures (Fig. 4.4). The quantitative assessment of land water storage dynamics is critical to several applications in hydrological research. It helps to improve our understanding of large-scale hydrological processes, and track the acceleration of the water cycle and rising sea level, as well as assess impacts from extreme climate events. While the quantification of the various water cycle constituents can be achieved by precise point-based assessments, large-scale characterization of hydrological fluxes considering spatial and temporal changes can be quite difficult. Even with the adoption of robust monitoring programmes, full data access can be sometimes very restricted. Remote sensing techniques are now increasingly applied in

[2] https://www.esa.int/Applications/Observing_the_Earth/FutureEO/It_s_a_kind_of_MAGIC.

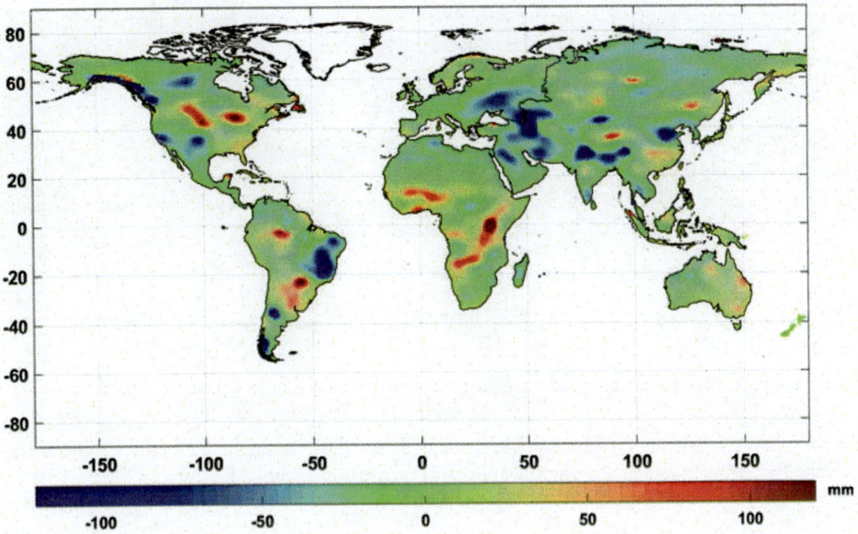

Fig. 4.5 Mean distribution of changes in terrestrial water storage (mm) observed by the Gravity Recovery and Climate Experiment (GRACE) satellite during the 2002–2021 period. Antarctica and Greenland GRACE hydrological signals have been masked out

hydrologic applications to address data limitations or spatial uncertainty of localized ground-based measurements. The capabilities of these techniques underpin insightful approaches that make it easier to accurately quantify and assess hydrological fluxes, measure their deficits, and provide detailed reports of their influences on the ecosystem.

Remote sensing-based gravimetric observations from the GRACE mission (Fig. 4.2) is capable of providing discrete spatial fields of terrestrial mass or surface mass variations (Fig. 4.5), making it feasible to directly and accurately estimate and assess variations in water storages across different regions of the world. Apparently, TWS vary in space and across different regions of the world (Fig. 4.5), and such variability depends on hydro-climatology and several human activities, including groundwater abstraction, water diversion, and construction of large water infrastructures. For instance, regions (e.g., Patagonia Ice Fields, Middle East, north east Brazil, and Asia) with large surface water bodies, permafrost, snow, and groundwater that are fast depleting show relatively stronger negative anomalies (less than -100 mm) and variabilities (Fig. 4.5). But humid areas (e.g., Amazon, Central and Western Africa) with substantial monsoon rainfall and presence of large surface water bodies (e.g., Lakes Victoria) as well as artificially constructed water bodies like Lake Volta indicate stronger positive anomalies (greater than 100 mm), suggesting the rise in land water storage or fluxes (Fig. 4.5). Furthermore, the monthly availability of the GRACE data on a global scale allows for integration with other hydrological fluxes or state variables like soil moisture, thus allowing continuous assessments of all forms of water fluxes present, near or below the Earth surface. For a more

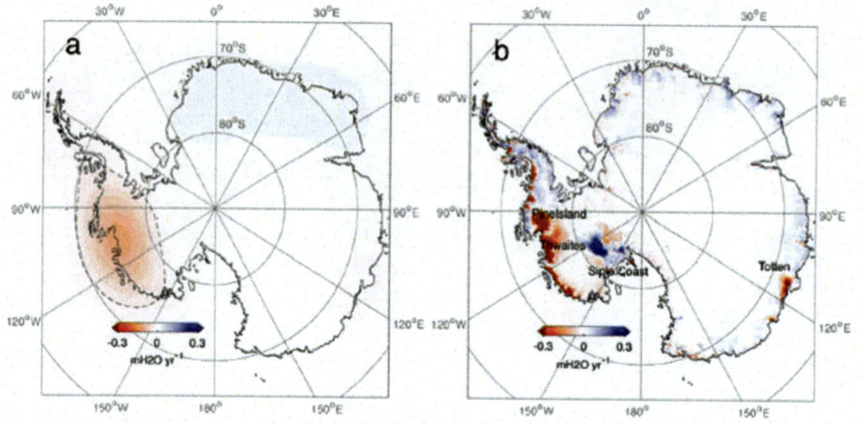

Fig. 4.6 Spatial distribution of mass loss in Antarctic ice sheets from the **a** Gravity Recovery and Climate Experiment data and **b** satellite altimetry observations. *Source* Kim et al. (2022)

accurate quantification of land water storage using the GRACE data it is important to eliminate background products, including oceanic and atmospheric mass variations, or even glacial isostatic corrections and earthquake parameters if present, in the region of interest (Ndehedehe 2022c).

4.3.1 Terrestrial Water Storage

Terrestrial water storage (TWS) is made up of five dynamic components, i.e., groundwater, surface water, soil moisture, canopy, and snow water. TWS as measured from GRACE (Fig. 4.5) and other satellite data has aided the tracking of the impacts of climate change by accurately measuring changes in Antarctic ice sheet, glacier mass balance, ocean bottom pressure, and the water cycle (Tapley et al. 2019; IMBIE 2018). This has contributed to improving knowledge of mass transport and how the global climate system influences and regulates key hydrological processes. However, combining satellite radar altimetry with GRACE data, the capability to monitor and quantify changes in the Antarctic ice sheet has substantially improved even at a better spatial resolution (Kim et al. 2022). Even though both data can independently assess ice mass change (Fig. 4.6), the recent work by Kim et al. (2022) provided improved estimates of ice sheet loss over Antarctic glaciers, consistent with past glacier mass change predictions. The linear trend in Antarctic ice mass variations indicated in Fig. 4.6a, b for the 2003–2016 period shows large amplitudes of mass loss in West Antarctica.

In the absence of severe anthropogenic pressure and substantial hydro-climatic shifts, the water storage components of TWS remain range-bound with variability occurring naturally (Kalu et al. 2021). With the advent of Earth resources satellite, remotely sensed altimetry data, extensive data collection, and measurements for hydrological assessments have been reported. The latter has only in the last few years been underpinned by satellite measurements from GRACE, thus enabling a universal evaluation across considerably large areas (Ndehedehe 2022c). The terrestrial water changes shown in Fig. 4.5 are the global mascon products, which provide accurate grids of terrestrial water storage and capture all hydrological changes around the globe. The use of GRACE in global assessments of terrestrial water storage changes is well established, and its use in hydro-geological assessments include

- predicting and mapping of changes in land water storage and tracking the pace of climate change (Dannouf et al. 2022; Ndehedehe and Ferreira 2020),
- monitoring the fluctuation changes is groundwater storage (Agutu et al. 2019; Abdelmohsen et al. 2019; Chen et al. 2016),
- observing lakes, reservoirs, and surface water resources, terrestrial water budget, and ecosystem assessments (Ndehedehe 2019; Becker et al. 2018),
- subsurface water storage, aquifer-system processes, and evaluating groundwater rates (Chen et al. 2016; Castellazzi et al. 2016; Scanlon et al. 2012; Tiwari et al. 2009),
- validating and improving outputs from hydrological and land surface models (Ahmed et al. 2016),
- assessing the relationship between TWS changes and climatic events (Ndehedehe et al. 2018),
- assessing flood and drought events (Ahmed and Wiese 2019; Ndehedehe et al. 2018; Thomas et al. 2014; Zhang et al. 2016),
- evaluating evapotranspiration (Burnett et al. 2020), and
- assessing hydrological controls on surface vegetation (Ndehedehe et al. 2019; Yang et al. 2014a).

Soil moisture exerts a meaningful effect in general circulation models through its capacity for storing and discharging heat and its control on evapotranspiration. Strong variations in terrestrial water storage can likely result in an imbalance in sea-level changes that contribute to changes in the gravity fields and rotation of the earth (Kalu 2022). However, the impacts of climate variability on global hydrology are still being explored because the assessment of terrestrial water storage as a singular variable was not possible until the inception of the GRACE mission. Land water storage (or TWS) is an important source of freshwater for reservoirs, agriculture, aquatic ecosystem, etc. It equally provides freshwater for drinking and other domestic uses, and surface water is also used for hydro-power generation (i.e., the Kainji dam in Nigeria). Plants and animals also depend on soil moisture and surface water for replenishment and habitation. Furthermore, groundwater nourishes streams between episodes of surface runoff, and snowmelt recharges other stocks of freshwater. Given these landscape changes and benefits of TWS, assessing changes in TWS is critical.

4.3.2 Satellite Observations of Changes in Groundwater

Groundwater is a critical source of freshwater for both the terrestrial and aquatic ecosystems. However, given the increasing human demand for groundwater to support industrial activities, agricultural practices, population growth, and economic development, this freshwater resource is constantly being depleted at an alarming rate. Ecological communities and natural systems/environments (e.g., streams, lakes, wetland, vegetation, springs, etc) that depend on groundwater for their needs and existence are basically referred to as Groundwater Dependent Ecosystems (GDEs). Groundwater serves as a reliable reservoir and source of steady water supply, thermoregulation, and unique chemical composition to GDEs by aiding in the support of unique habitat conditions for phreatophytes, animals or microorganisms within aquatic, coastal, and terrestrial ecosystems (Currell and Ndehedehe 2022). To balance human and ecosystem water needs globally, sustainable groundwater management policies and regulations are being designed and implemented in many regions. However, to achieve groundwater sustainability, it is important that the water needs of varied set of users are assimilated into water management decisions. Groundwater information, especially depth-to-groundwater data in shallow aquifers that are well-known supporters of GDEs are usually limited as a result of the gaps in groundwater monitoring and the precedence given to monitoring wells in deeper portions of river basins.

4.3.2.1 GRACE-Derived Groundwater

The advent of new remote sensing techniques has enabled the assessments of groundwater status on different spatial scales (Castellazzi et al. 2016; Chen et al. 2016; Tiwari et al. 2009), making it possible to now fill groundwater data gaps (Díaz-Alcaide and Martínez-Santos 2019). For instance, the use of the GRACE satellites in detecting groundwater storage losses across large sub-regions, including the high plains aquifer in the Central United States (Scanlon et al. 2021), California's Central Valley (Scanlon et al. 2012), Northwest India (Rodell et al. 2009), etc. has proved to be very useful. Quantifying groundwater from GRACE requires subtracting soil moisture snow water, canopy water storage, and surface water data from GRACE terrestrial water storage (based on Eq. 6.1 in Chap. 6) using the water budget approach (Ndehedehe 2022a). Using the water budget approach, auxiliary data from hydrological models (soil moisture and surface water) like Global Land Data Assimilation System and WaterGAP help with separating groundwater storage signal from terrestrial water storage (Fig. 4.7). This water budget method used to estimate groundwater from GRACE data is described further in Chap. 6. The agreement between GRACE-derived groundwater and in situ observations is generally good in some areas and strong in other regions with correlations ranging 0.5–0.95, thus increasing confidence in the use of satellite-based estimates of groundwater to support water management

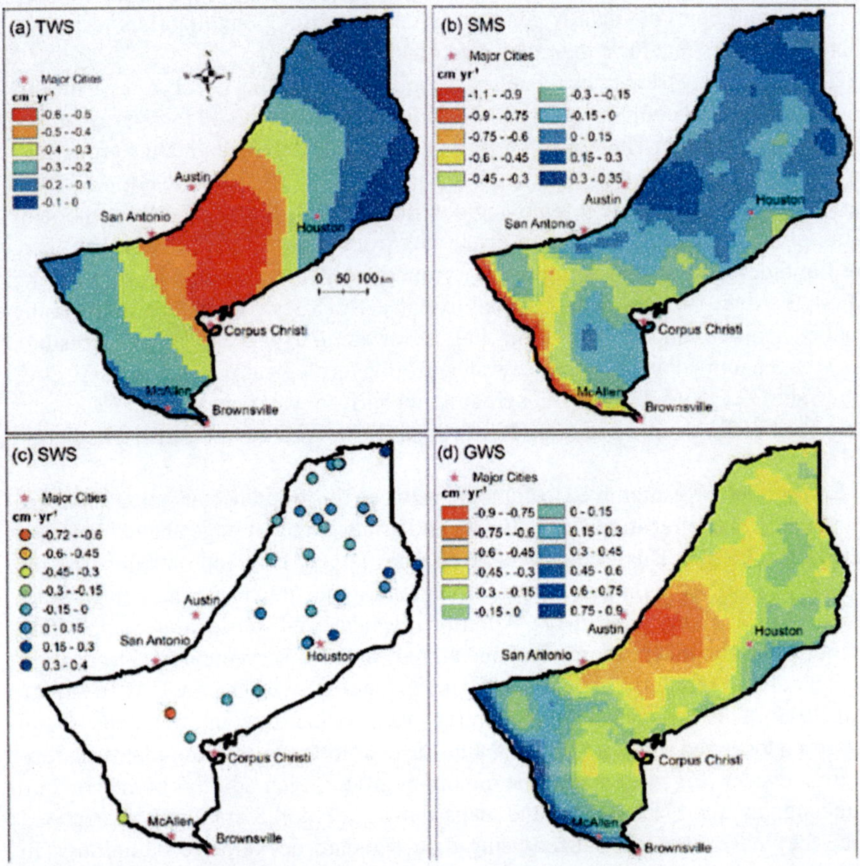

Fig. 4.7 Spatial distribution of trends in terrestrial water storage components, including groundwater derived from the Gravity Recovery and Climate Experiment data. *Source* Gyawali et al. (2022)

assessment and improve decision-making for sustainability (Gyawali et al. 2022; Rateb et al. 2020; Agutu et al. 2019).

Even though GRACE-derived groundwater storage variations correspond with groundwater level monitoring data in many major aquifers in the world, the spatial (>150,000 km^2) and temporal (monthly) resolution of GRACE data appear too coarse to replace sparse piezometric data over large sub-regions. This means that GRACE spatial resolution is insufficient to examine localized demand at a scale meaningful to water resources managers. This leads to a wide gap in the scales of monitoring, between point measurements from monitoring bore wells and basin-wide scale of GRACE datasets. Considering that groundwater level contour mapping with sparse borehole datasets can be enhanced using Bayesian data fusion approaches (Fasbender et al. 2008), substantial errors around groundwater-dependent ecosystems are likely if monitoring borehole data is penetrating deeper sub-regional aquifers instead of shallow aquifers that have their own water tables.

The GRACE mission tracks changes in continental TWS but not changes in the discrete hydrological components which includes surface water, soil moisture, canopy, ice, and groundwater. Therefore, the most popular approach for extracting groundwater products from these water storage components observed by GRACE (i.e., ΔTWS variable) is by isolating and removing contributions of all other TWS components using either auxiliary information or outputs from land surface models. But there are limitations associated with tracking groundwater resources and its status using GRACE. Apart from its coarse spatial extents, which restricts its use in assessing smaller aquifers, uncertainties from the use of auxiliary data (e.g., soil moisture, snow water equivalent, and surface water) from hydrological models (e.g., GLDAS and WaterGAP) and difficulty in synthesizing surface water are notable issues and have been well documented (e.g., Rateb et al. 2020; Agutu et al. 2019; Chen et al. 2016).

Past studies have attempted to compare between the groundwater storage changes obtained by disaggregating GRACE data and groundwater storage change obtained based on situ data. This evaluation exercise has largely centered on validating the accuracy of GRACE data at either sub-regional or continental scales (e.g., Ndehedehe 2022a; Henry et al. 2011). Additionally, advanced statistical decomposition methods like the independent component analysis have been employed to analyze the spatial and temporal variations of groundwater derived from GRACE (Agutu et al. 2019). This approach helps to identify spatio-temporal groundwater changes in important localized aquifers (Figs. 4.8 and 4.9). For instance, the temporal variations of groundwater (Fig. 4.9) in the nine important groundwater aquifers corresponding to the signals (the black box) in the maps shown in Fig. 4.8 are jointly interpreted since they have been normalized using their standard deviation to be unitless. In other words, the actual value (in mm) of groundwater change in each month/year (Fig. 4.8) can be obtained by multiplying it with the signal on the map in the black box (Fig. 4.8). If we take Fig. 4.9f (Kenya–Somalia area) as an illustration, the actual maximum value of groundwater change in that location in 2007 was 125 mm. Generally, the spatial groundwater patterns identified are mostly transboundary in nature (Fig. 4.8), and the general characteristics of these aquifers (names on top of each panel in Fig. 4.8a–i) and their potential to support the development of irrigated agriculture in the East Africa was detailed by Agutu et al. (2019). With limited ground monitoring bores in the region, satellite capability to assess groundwater thus provides an opportunity for large-scale monitoring of these aquifers.

4.3.2.2 InSAR-Based Groundwater

Geodetic techniques including ground deformation assessments provide a set of measurements that can be used to characterize water volume changes at intermediate scales between GRACE and well datasets. It has long been recognized that groundwater extractions may result in surface deformations including subsidence,

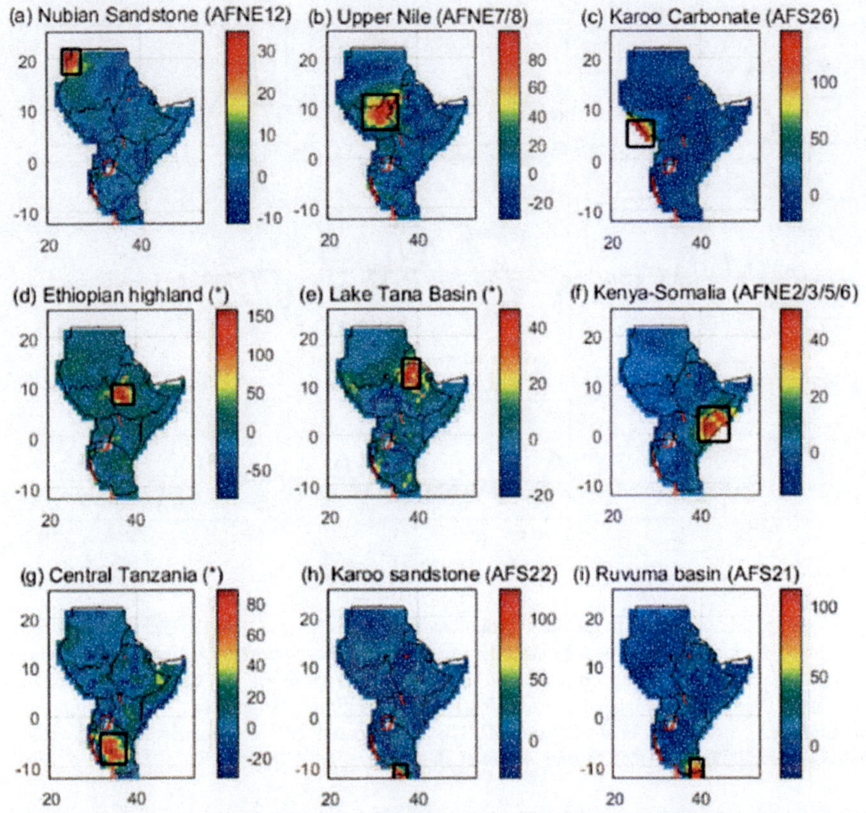

Fig. 4.8 Spatial patterns of localized groundwater changes (mm) derived from the Gravity Recovery and Climate Experiment (GRACE) data over Greater Horn of Africa. Nine important groundwater aquifers were identified using the Independent Component Analysis. These spatial patterns (black boxes) are jointly interpreted with the corresponding temporal patterns of groundwater changes in Fig. 4.9. The red polygons are surface water areas that were removed from the GRACE data to improve the accuracy of estimated groundwater. *Source* Agutu et al. (2019)

due to the changes in effective pressure within aquifers. More recently, satellite-based InSAR datasets have been applied for the assessment of surface deformations associated with groundwater extractions and recharge (e.g., Bell et al. 2008). These techniques (Fig. 4.3) are well established, having been used in California's Central Valley to assess and identify large-scale subsidence resulting from excessive groundwater extractions. Because radar satellites image or capture displacements that occur at the earth's surface, the procedure involving the application of the InSAR provides enhanced spatial resolutions in comparison to satellite-based gravity observations of the GRACE mission. However, it is important to note that ground deformation on its own is an indirect assessment of the internal variations and changes within an aquifer and additional analysis is needed to extract quantities needed by water agencies and

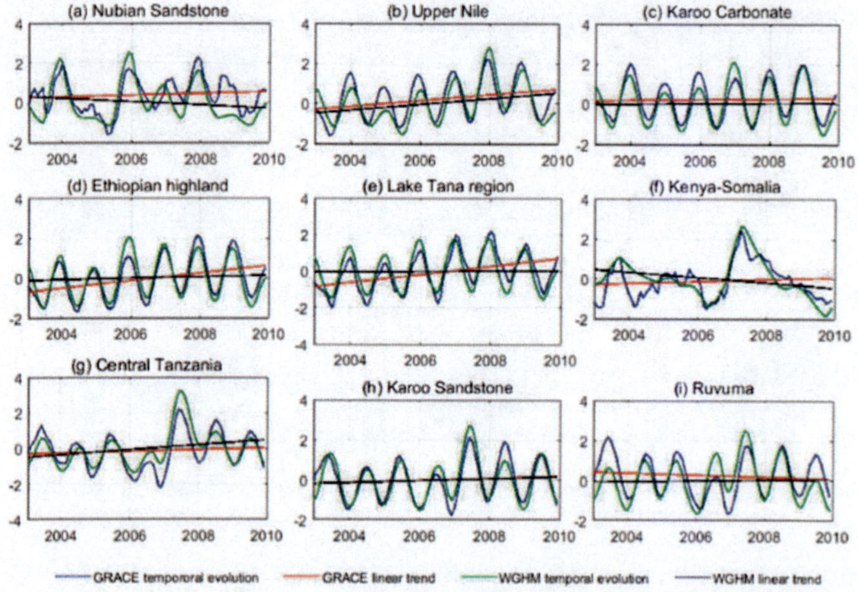

Fig. 4.9 Temporal variations of localized groundwater changes derived from the Gravity Recovery and Climate Experiment data over Greater Horn of Africa. These temporal patterns for the nine important groundwater aquifers correspond to the spatial patterns in Fig. 4.8 and are jointly interpreted since they were normalized using their standard deviation to be unitless. GRACE temporal evolution (i.e., groundwater changes from GRACE) is compared with model-derived groundwater (WaterGAP model or WGHM). *Source* Agutu et al. (2019)

managers to make informed decisions. Specifically, this requires solving the inverse problem to relate InSAR measurements to the aquifer volume changes (Vasco et al. 2016). Additional parameters describing the poroplastic and poroeslastic response of the medium are needed to link aquifer structural volume changes to the variations in the volume or the mass of the aquifer and its content (Hoffmann et al. 2003). The final step typically requires datasets that are difficult to get and are most likely to be unavailable, including the minimum effective pressure that the aquifer has ever been subjected to. These limitations make it very difficult to track groundwater level changes in refined scales from satellite observations, though there have been targeted efforts at reformulating these challenges to enable scientists make inferences directly from InSAR techniques, and geological and water level data (Smith et al. 2017).

4.3.3 *Surface Water Hydrology*

Surface water refers to water on the surface of the Earth including lakes, rivers, wetlands, and the ocean. Basically, the freshwater aspect of surface water hydrology does not include the ocean because it is too salty; however, smaller inland saline water bodies are normally included. Surface water bodies are very vital sources of freshwater resources for both terrestrial and aquatic ecosystems. They are of extreme importance and play significant roles in sustaining all forms of lives. Surface water helps in the preservation of biodiversity in wetland or riparian ecosystems by providing habitats to a plethora of flora and fauna. It does not only prove to be critical to ecosystem as a primary component of the hydrologic cycle but also affects other aspects of daily lives including electricity production, agriculture, safe drinking water, transportation, and industrial activities (Huang et al. 2018).

Surface water bodies are dynamic in nature (see Section Chap. 1) as they are susceptible to expansion, shrinkage, and can change their appearance and flow courses over time, as a result of natural and anthropogenic influences. Changes to surface water bodies can affect a range of other natural resources and human as well as environmental assets. Little changes in surface water volumes can lead to harsh consequences, and, in extreme cases, can result in floods, drought, fire outbreaks, heatwaves, etc. This is frequently the case in Europe, as several European countries including Germany, Spain, Portugal, France, the Netherlands, and the United Kingdom have had to endure droughts in recent summers. The impacts of these droughts have significantly affected farmers and food production levels, as well as relevant authorities in their various water uses and applications. It can be argued that these impacts are the aftermath of climate change in this region; however, it is important to point out that these are part of the consequences of decline in the volume of surface water, perceived as a response to severe drought events. For instance, Italy's worst droughts in decades have resulted in a reduction of the country's largest lake (Lake Garda) to one of its lowest levels ever recorded and warming the water temperatures that approach the average in the Caribbean Sea. Northern Italy has not seen significant rainfall for months, and snowfall in the year 2022 was down to 70%, drying up vital waterways such as the Po river, which flows across Italy's agricultural and industrial heartland. As a result of this drop in surface water levels, countries like France, Portugal, Germany, etc. are having difficulty with extinguishing extreme fire outbreaks. Several fire fighters in these regions have argued that the forest floors are considerably dry, making it rather difficult for fire fighters to extinguish these flames and they sometimes have to excavate the embers with tools (Aljazeera.com 2022). Therefore, it is important to efficiently determine available surface water, monitor its extraction rate and withdrawals to quantify its volume, and examine its dynamics for accurate assessment of sub-regional surface water hydrology and to mitigate some of the harsh impacts associated with mismanagement or climate change impacts.

4.3.3.1 Satellite Altimetry

Remote sensing datasets are vital for a thorough assessment of the Earth's surface water hydrology, especially since the amount of publicly available in situ datasets are on a steady decline. Satellite altimetry provides key details on the water levels and fluctuations of lakes, reservoirs, and rivers to aid in this assessment. Satellite radar altimetry provides sea surface height variations of surface water (Fig. 4.10) and is a well-known satellite mission over both ocean and land. In combination with GRACE

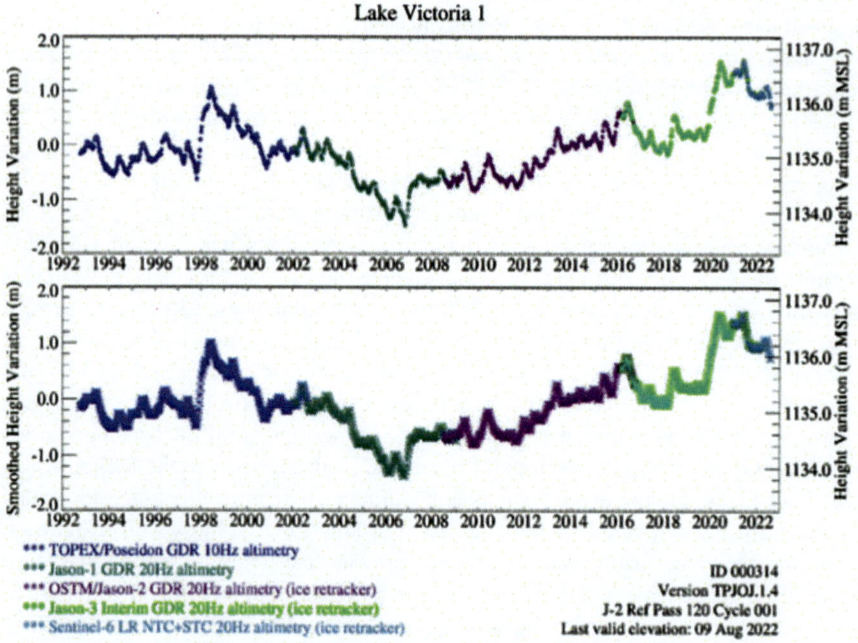

Fig. 4.10 Time series of lake height variations over Lake Victoria from TOPEX/POSEIDON, Jason series (Jasons1-3) and Sentinel-6 Altimetry. *Source* USDA/NASA G-REALM programme

observations, lake height variations from altimetry missions have been helpful in the separation of surface water storage changes to understand variability in other hydrological stores like groundwater and basin terrestrial water storage (Agutu et al. 2019; Ndehedehe et al. 2017). The monitoring of anthropogenic and climate change influence on surface water hydrology is underpinned by this capability. Altimetry products over land have successfully been applied in the assessment of surface variations of inland water bodies and rivers as well as improve outputs from hydrological models. As illustrated in Fig. 4.10, satellite altimetry has provided capabilities in monitoring water level changes in large lakes like Lake Victoria in East Africa where climate variability and human water management play important roles in water level fluctuations. However, as was mentioned earlier, the combination of altimetry with GRACE data can also facilitate the accurate assessment of changes in Antarctic ice sheets (Fig. 4.6).

Through near-real-time data assimilation, altimetry observations are also helping with forecast optimization of hydro-climatic processes. This includes the prediction of hurricane intensity and evaluation of floods to improve flood modeling and quantifying anthropogenic influence on changes in surface water hydrology (e.g., Ndehedehe et al. 2017; Moore and Williams 2014). The study undertaken by Getirana (2010) performed an assimilation of the altimetry data from Environmental Satellite (ENVISAT) into the MGB-IPH (Modelo de Grandes Bacias by Instituto de Pesquisas Hidraulicas) model with an optimization technique aimed at enhancing the parameters of the model over the Branco River basin. Furthermore, Michailovsky et al. (2013) also employed the ENVISAT water level datasets in improving the performance of a rainfall–runoff model. The study assimilated the altimetry datasets of over six virtual stations into the model over the Brahmaputra River to enhance its daily discharge outputs. Neal et al. (2009) on the other hand assimilated water level datasets from InSAR and digital terrain modeling (DTM) platforms to improve their discharge estimates. Biancamaria et al. (2011) integrated virtual Surface Water and Ocean Topography (Sect. 4.3.3.2) data to improve the modeling of the Arctic river basin, arguing that the approach minimized spatio-temporal channels of water depth (Emery et al. 2020).

4.3.3.2 First Results from the Surface Water Ocean Topography Mission

There are various developments of multi-mission sensors that could provide important and contemporary measurements of Earth's resources. However, there is still a major limitation in existing dedicated satellite missions that could provide continuous data on surface water storage changes and fluxes (Ndehedehe 2022b). This warranted the need for a space mission that could provide capabilities in monitoring changes in surface water hydrology, including rivers, reservoirs, wetlands, and lakes. The Surface Water Ocean Topography (SWOT) mission[3] provides these capabilities, including quantifying water storage changes and discharge variations in rivers across the globe to help hydrologists improve understanding of surface water availability and water resources managers increase preparedness for future impacts of floods and droughts. The SWOT mission is expected to provide a better avenue for the estimation and assessment of global oceans and its terrestrial surface waters. This capability will be facilitated by the SWOT system design such as its swaths (Fig. 4.11). For instance, SWOT is designed to have an overall width of approximately 120 km and is also equipped with an *interferometer that will produce two parallel tracks, with a Nadir track from a traditional altimeter in the gap between the swaths*[4] (Fig. 4.11). SWOT will collect data across the world's oceans and over freshwater areas at this wide swath (120-km) (Fig. 4.11) and will have a gap in the center for an altimetry track as indicated in Fig. 4.11. The 120-km swath width is for both the left and right interferometer swath (Fig. 4.11). Further, SWOT will have a significantly better spatial resolution compared to altimeters (Fig. 4.12). The illustration in Fig. 4.12 is the sea-level data in the Gulf Stream observed by SWOT's KaRIn instrument and has 10 times the spatial resolution of altimetry-derived data in the same location. This mission is jointly developed by NASA and the French's Centre National D'Etudes Spatiales (CNES) with contributions from the Canadian Space Agency (CSA) and the United Kingdom Agency.

---

[3] https://swot.jpl.nasa.gov/.

[4] https://swot.jpl.nasa.gov/resources/104/diagram-of-swot-data-collection/.

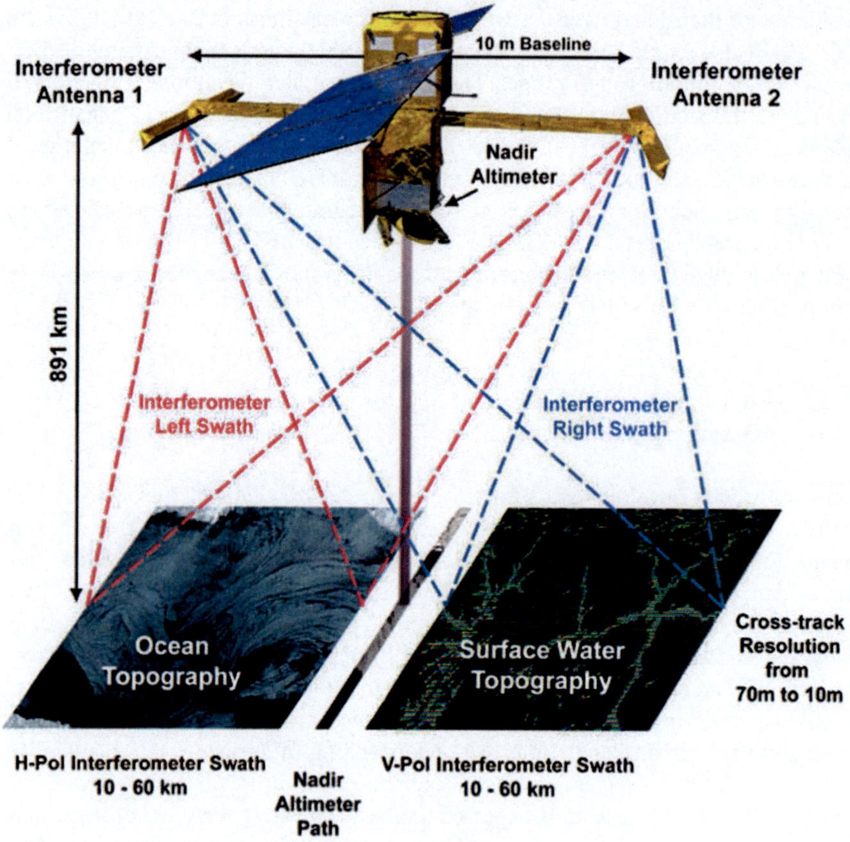

Fig. 4.11 The SWOT mission data collection architecture. *Source* https://swot.jpl.nasa.gov/resources/104/diagram-of-swot-data-collection/

The SWOT satellite was launched on December 15, 2022, atop a SpaceX rocket from Space Launch Complex 4E at Vandenberg Space Force Base in California, and is expected to be a 3-year mission, which will commence the collection of science data in about six months.[5] SWOT is the first of its kind and promises to provide the very first and detailed overview of the existing freshwater bodies of the earth from space and will also provide scientists with the opportunity of determining the change in volumes of freshwater across the globe at an unprecedented spatial and temporal resolution. SWOT has several benefits and applications as mentioned earlier and reported in Ndehedehe (2022b). But an important outcome is that the mission is expected to have a substantially large global coverage that will enable data collection on major lakes larger than 62,500 square meters and rivers that are

[5] https://swot.jpl.nasa.gov/news/81/nasa-launches-international-mission-to-survey-earths-water/.

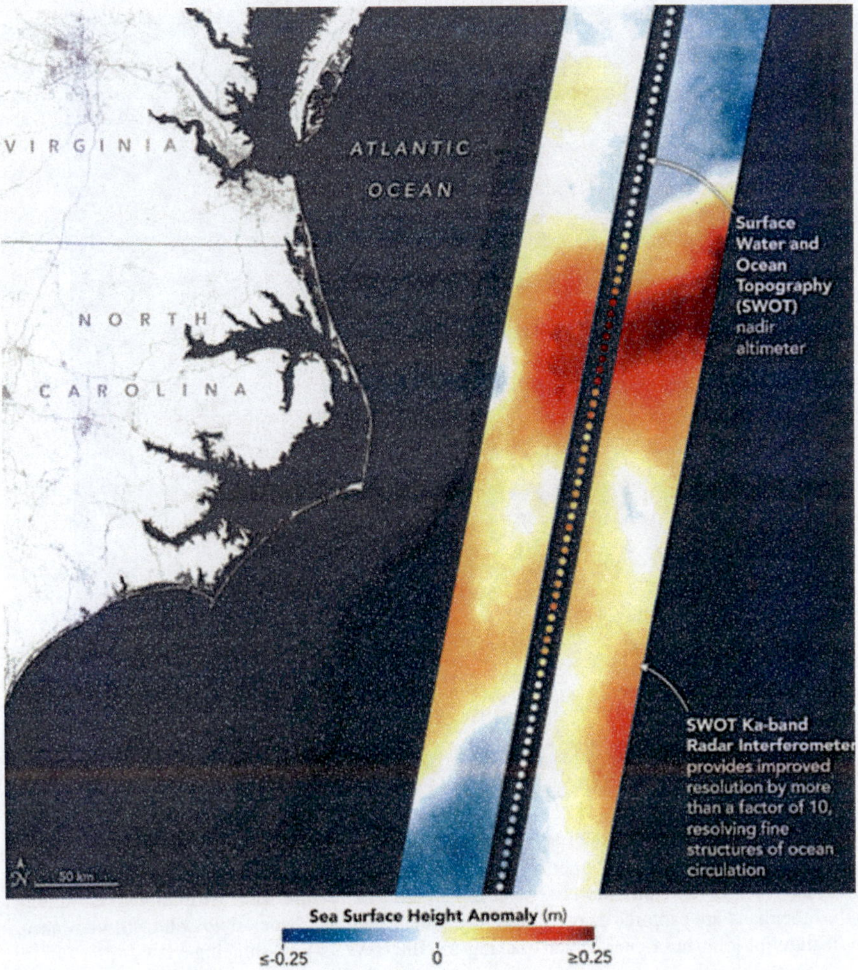

Fig. 4.12 Sea-level data from the Surface Water and Ocean Topography mission reported on January 21, 2023, for the Gulf Stream. *Source* https://www.nasa.gov/feature/jpl/joint-nasa-cnes-water-tracking-satellite-reveals-first-stunning-views

approximately 100 m wide (https://swot.jpl.nasa.gov/news/81/nasa-launches-international-mission-to-survey-earths-water/). With reliable measurements for millions of freshwater bodies to be made available by SWOT, hydrologists and water resource experts can track changes in water bodies, including estimating water gained or lost in lakes, wetlands, and reservoirs.

Given our fundamental need for freshwater resources, hydrological estimates of spatio-temporal variations in water volumes stored in lakes, rivers, and wetlands are critical. By estimating the water storage variations in lakes, rivers, and reservoirs more accurately using the SWOT mission, a significant advancement to the fundamental

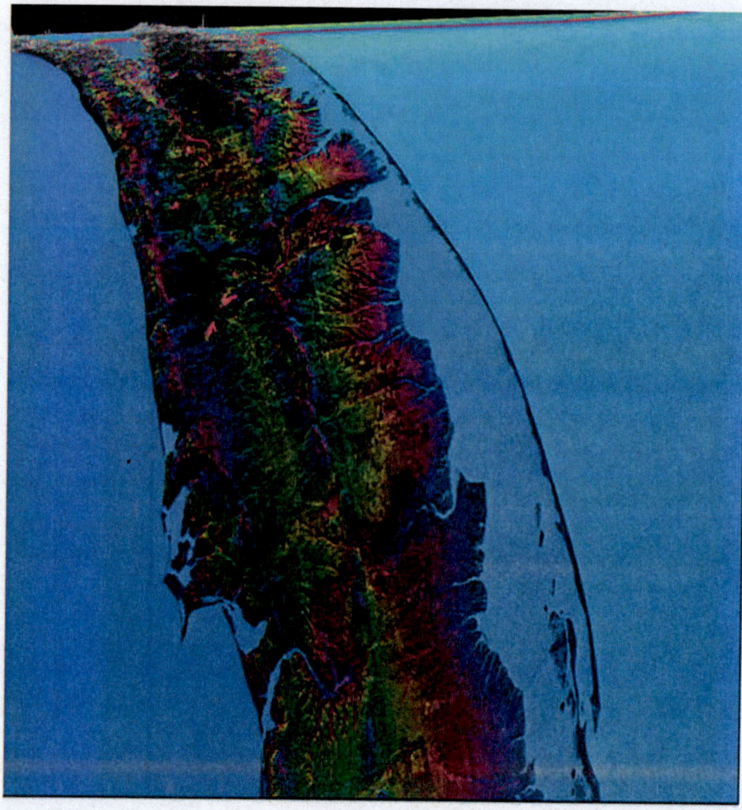

Fig. 4.13 First results from the Surface Water and Ocean Topography mission reported on January 21, 2023. The KaRIn instrument onboard the SWOT mission shows the distribution of elevations of water features *as bright pink splotches nestled within the landscape*. The distribution of elevations on the other hand are captured as purple, yellow, green, and blue shades. *Source* https://www.nasa.gov/feature/jpl/joint-nasa-cnes-water-tracking-satellite-reveals-first-stunning-views

understanding of the terrestrial component of global water cycle can be achieved. Additionally, by providing vital information on the earth's land-based water bodies twice every twenty-one days, the SWOT mission has the potential of revealing new features on the varying patterns of freshwater availability that could potentially impact large populations. Apart from assessing freshwater availability, the mission also has the potential of improving model simulations of fluxes and extreme events like floods. As an indication of future prospects, first results from the SWOT mission has already been published on NASA's website (https://www.nasa.gov/feature/jpl/joint-nasa-cnes-water-tracking-satellite-reveals-first-stunning-views). An example of this first result is indicated in Fig. 4.12 and shows SWOT-derived sea-level image in part of the Gulf Stream off the coast of North Carolina and Virginia. In this image (Fig. 4.12), sea levels with values higher than the global average are indicated in Red and orange while the shades of blue capture sea levels lower than global average. In Fig. 4.13, SWOT's KaRIn instrument shows water features on New York's Long

Island captured as bright pink splotches. The other colors (i.e., purple, yellow, green, and dark blue shades) in Fig. 4.13 indicate a broad range of land elevations, while the ocean is captured in lighter blue.

On land, a large percentage of freshwater accessible to humans is stored in rivers, lakes, soil, and groundwater. Through precipitation, water enters these reservoirs and flows from their surrounding watersheds. Water leaves through evapotranspiration and river discharge into the oceans. A simplified view of the water cycle depicts a balance between the water entering the land surface by precipitation and water leaving the land surface by evapotranspiration and runoff. This type of "balance equation" is the basis for computer algorithms to model hydrological conditions. One of the most convoluted and sophisticated computer models include data assimilation, due to its technique of taking into consideration and comparing the priori estimates with the newly received observations to update and improve the system model itself. The SWOT faces a challenge of generating robust hydrological models and assessing how the assimilation of its data will enhance the representation of the water cycle. For instance, the conventional understanding of the earth's water balance on land is poor. This is partially as a result of the limited global runoff data needed for accurate compute models. Today's model estimates can simulate very different patterns of runoff. Therefore, an important target of the SWOT mission is to provide runoff data at sufficiently finer scales to enhance and improve the knowledge of water balance on land.

4.3.4 Reconstruction from Models and Observations

Prior to the advent of the GRACE observations, traditional methods were the norm in attempting to quantify terrestrial water storage anomalies (TWSAs). This would include the combination of either global hydrological models or land surface models (Kumar et al. 2017) that integrate atmospheric drivers and surface features with some in situ observations to the modeling process. These processes are built on the physical approaches of water storage and dynamics based on the water balance model. They can be applied to estimate discrete hydrological parameters, such as groundwater, soil moisture, river discharge, evapotranspiration, and snow. However, they are inadequate in the provision of comprehensive hydrological information (e.g., terrestrial water storage and groundwater changes) derived from GRACE satellites, especially in areas where some aquifers and/or human water use are often overlooked, as well as regions characterized by cropland irrigation and deep groundwater storage. Therefore, some discrepancies may exist between what GRACE observes and model outputs, though models are based on sound physical principles of hydrological interactions and processes. The study by Scanlon et al. (2019), which evaluated trends in land water storage from two global hydrologic and water resources models (PC-GLOBWB and WGHM), five global land surface models (CLSM, CLM in the GLDAS, VIC, MOSAIC, and NOAH), and three GRACE solutions (i.e., Jet Propulsion Laboratory mascon, Center for Space Research mascon, and Center for Space

Research gridded spherical harmonics) over 186 global river basins, exemplified this. Results from their experiments showed that these models underestimated seasonal water storage amplitudes in some areas and while some models were consistent with those of GRACE, the performance of other hydrological models in terms of how well they compared with GRACE was restricted due to missing storage compartments. That is, these models lacked surface water and/or groundwater storage components or were poorly simulated. The possibility of these global hydrological models underestimating the trends in future climate change and human-induced water storage dynamics was also noted (Scanlon et al. 2019).

4.3.5 Reconstruction with Machine Learning Models

Recent developments in machine learning techniques and artificial intelligence have introduced novel methodologies in hydrology and related fields (e.g., Kalu et al. 2022). Machine learning methods are extremely useful in resolving problems that have overwhelmed the traditional techniques in the past and this includes data reconstruction. In assessing the impacts of climate change, GRACE data is limited in terms of availability of a longer time series, and reconstruction methods have been employed to predict past and future observations or at least fill existing data gaps as detailed in the chapters of this book. There has been a growing interest in the use and application of learning-based models to estimate various hydrological variables by learning the strength of correlation between the predictors and predictand (Kalu et al. 2022; Broxton et al. 2019). Sun et al. (2019) reconstructed GRACE-derived TWS by integrating hydrologic models with a deep convoluted neural network (CNN). They reported that the CNN model substantially enhanced the performance of the land surface models as it efficiently compensated for the groundwater variables missing in the land surface model. Adamowski et al. (2012) predicted the urban water demand of Montreal city in Canada by comparing autoregressive integrated moving average, artificial neural networks, multiple linear and non-linear regressions, and wavelet artificial neural network (ANN) models. The authors' projection that the models integrated wavelet with neural networks proved effective in predicting urban water demand better than traditional techniques. Pan et al. (2019) applied CNN in the forecast of daily precipitation rates and tested the model over 14 sites across the contiguous United States. Results from the analysis showed that with a comprehensive set of data, the precipitation estimates from the CNN could be prove to be more efficient than the statistical downscaling and reanalysis precipitation products.

Other techniques such as ANN, support vector machine, and random forest algorithm are also common methods that have emerged as efficient techniques in the reconstruction of GRACE-derived TWS. Be it the use of statistical approaches that establishes relationships of TWS anomalies (or changes in TWS) from GRACE with precipitation and temperature, or methods that rely on the relationship of in-situ river level records with TWS from GRACE (Humphrey et al. 2017; Becker et al. 2011), the techniques in reconstructing TWS are constantly evolving. For instance,

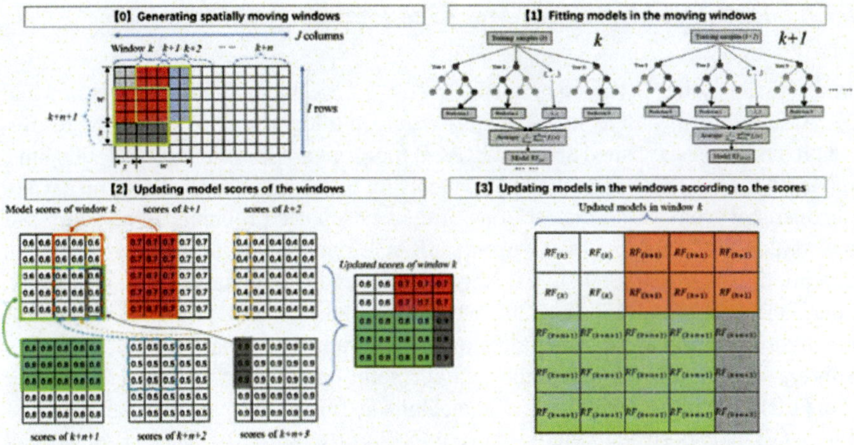

Fig. 4.14 A Random Forest regression framework to reconstruct changes in terrestrial water storage using spatially moving window structure. *Source* Jing et al. (2020)

a Random Forest regression-based reconstruction framework for GRACE TWS data based on model outputs from the Global Land Data Assimilation System has been developed (Jing et al. 2020). The illustration in Fig. 4.14 summarizes important steps on a spatially moving window structure employed to train a Random Forest model developed to reconstruct TWS in China. Given that GRACE-based TWS covers a relatively shorter period, backcasting or a backward reconstruction helps in filling data gaps for periods when the GRACE mission was unavailable. This is usually undertaken to aid the analysis of long-term climate change impacts on water resources. Within this context, the boosted regression trees (BRT) and the artificial neural network have recently been employed to reconstruct TWS during the 1982–2014 period (Dannouf et al. 2022). Comparatively, the BRT technique is insensitive to multicollinearity and outliers, making it suitable to predict hydrological conditions with complicated non-linear patterns, as well as deal with the impacts among predictors (Dannouf et al. 2022). The most recent GRACE gravity field solution is the Release Number 06 (RL06), which features the state-of-the-art processing of GRACE measurements with enhanced parameters, processing algorithms and background gravity models. The RL06 solution improved the estimation of mass redistribution and significantly minimized the noise in the GRACE gravity field solutions relative to previous releases (Göttl et al. 2018). Additionally, the RL06 reprocessing standards are the pioneer release of the GRACE-FO dataset. Therefore, filling data gaps between the GRACE and the GRACE-FO using the RL06 products as a reference will warrant the coherency between the GRACE and GRACE-FO time series. Studies in this area are, however, lacking, and more attention needs to be given to the use of novel machine learning frameworks in reconstructing GRACE and GRACE-FO datasets.

4.3.6 Downscaling Using Machine Learning Models

As discussed earlier in this chapter (see Sect. 4.3.2.1), one of the challenges of using GRACE to inform on the status of freshwater on a local scale is its spatial resolution. It has been established that the GRACE footprint is approximately 200,000 km^2, making it too coarse to meet the requirements for meaningful applications on a water management scale. Consequently, one of the key science problems being addressed now in this field in terms of the use of GRACE data is how to improve its spatial resolution. Using Artificial Neural Network framework and statistical downscaling, emerging studies (e.g., Yin et al. 2022; Vishwakarma et al. 2021) have integrated various hydrological data such as precipitation, soil moisture, evapotranspiration, temperature, and runoff to improve the spatial resolution of GRACE. There is evidence in the literature that the application of machine learning improved the spatial resolution of GRACE-derived groundwater (e.g., Seyoum et al. 2019). The earlier work of Yin et al. (2018) on the subject downscaled GRACE-derived groundwater anomalies from 110 to 2 km in the North China Plain, with in situ groundwater levels showing varying degrees of strong correlations (0.63–0.94 and 0.80) with the downscaled product. Such products are critical in supporting local water resources planning, but a downscaling method for GRACE-groundwater as proposed by Yin et al. (2018) is only feasible for regions where groundwater is strongly correlated with evapotranspiration. As significant as this progress is, novel machine learning approaches are still being investigated to improve the spatial resolution of GRACE-derived groundwater for hydrological applications. The workflow for statistical downscaling of terrestrial water storage illustrated in Fig. 4.15 used partial least regression and is an example of an ongoing effort to improve GRACE spatial resolution. As reported by Yin et al. (2022), the downscaled GRACE product was better improved showing a considerably strong correlation ($r = 0.98$) with original GRACE products (Fig. 4.15). Interestingly, deep learning machines are also emerging in the downscaling of GRACE data product for applications in drought assessment. The work undertaken by Foroumandi et al. (2023) is one of such that evaluated the feasibility of deep learning models like the Convolutional Long Short-Term Memory as well as shallow learning Feed Forward Neural Networks, and Random Forest in downscaling GRACE-derived terrestrial water storage changes. They reported better performance of the deep learning models compared to Random Forest and the Feed Forward Neural Networks techniques. The utility of these downscaled products in drought characteristics and impacts on water availability was also reported (Foroumandi et al. 2023). Be it traditional and statistical machine learning, or deep learning models, their use in downscaling GRACE product, including groundwater, has shown value and prospects in improving the spatial resolution of GRACE data for local-scale applications.

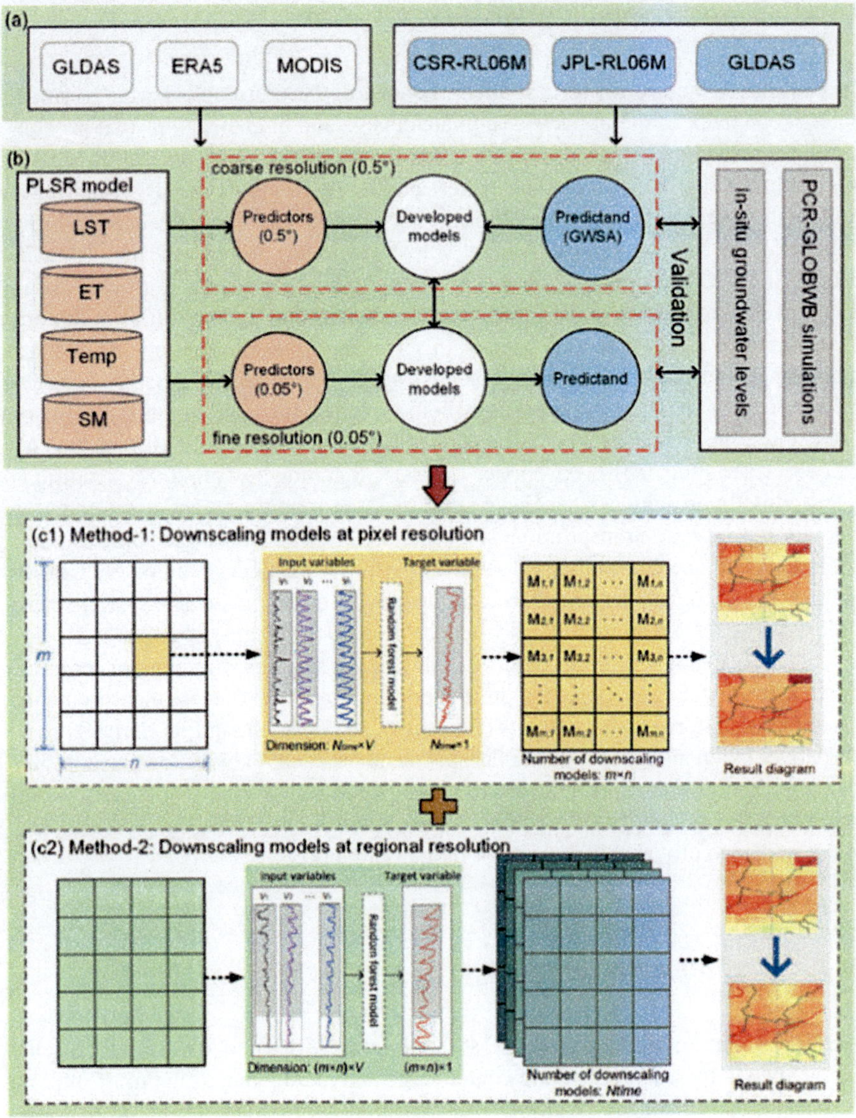

Fig. 4.15 Workflow for statistical downscaling of terrestrial water storage using multi-resolution data. *Source* Yin et al. (2022)

4.4 Benefits of Satellite Hydrology Systems

The continuous monitoring of various components of the hydrological cycle is critical to help keep track of the impacts of climate change and human actions that accelerate the water cycle. The past decade witnessed tremendous advancements in satellite programmes (Fig. 4.17), including novel satellite hydrology systems, which have resulted in several benefits to hydrologists and scientific community generally. From a hydrological science perspective, satellite hydrology programmes like GRACE and GRACE-FO missions and altimetry, among others, have driven process-based insights and provided novel and independent datasets that underpin the continuous assessments of the global water cycle. In retrospect, these developments and the advancements in satellite technology for hydrologic applications have provided several benefits (as detailed in subsections below), thus consolidating on the gains of past satellite missions (see, e.g., Ndehedehe 2022b; Becker et al. 2018; Smith et al. 2017; McCabe et al. 2017; Alsdorf et al. 2003; Schmugge et al. 2002). Additionally, satellite hydrology systems have simplified the monitoring and assessment of remotely sensed evapotranspiration, soil moisture content evaluations, precipitation estimations, surface water changes, as well as improve the understanding of changes in land water storage underpinned by satellite gravity-based methods (e.g., Kalu et al. 2021; Grace and Davenport 2021; Ndehedehe et al. 2016; Wagner et al. 2007). Leveraging spatial coverage of these satellite hydrological systems, a number of research efforts have also taken advantage of prolonged temporal series of these measurements to generate long-term global datasets (Beck et al. 2017). Such satellite-derived products deliver independent data that underpins the assessment of hydrological systems. The opportunities that exist in the use of these data to assess changes in global water cycle components could facilitate our understanding of hydrological systems and their response to climate change and or human influence.

4.4.1 *Tracking Climate Change*

In several scientific meetings (e.g., 21st Conference of Parties), the subject of climate change is increasingly taking center stage and is now one of the most important discussion. Given the changes in the Earth system (Chap. 1), the need for a reliable holistic approach is critical to help address threats to humanity. Monitoring frameworks that use satellite Earth observation to track climate change provide opportunity to assess such impacts on human lives and ecosystem functioning. Traditional techniques seem inadequate because of their limitations and inability to provide a continuous, thorough, and global-scale assessment of climatic variables to help with understanding such impacts on humanity. However, with the advent of satellite hydrology missions like GRACE and SWOT, it is now possible for hydrologists to collect continuous, reliable, and near-real-time data and information on changes in global freshwater around the world. This advancement in satellite technology (Figs. 4.16 and 4.17) provides us with large and continuous datasets that enhance future predictions as well as tracking climate change impacts in short lead times. Therefore, by assessing large-scale processes and spatio-temporal states of the atmosphere, land, and oceans, satellite remote sensing has provided major advances in understanding the climate systems and its changes.

With the use of satellites to track climate change, the opportunity to observe/measure and assess different states and processes of the atmosphere, ocean, and land at varying spatio-temporal scales is now feasible. For example, satellite remote sensing is one of the most efficient approaches of monitoring land cover and its changes over time on different spatial scales (e.g., Gong et al. 2013; Wulder and et al. 2008). Satellite datasets are usually integrated with climate models to simulate changes to the climate system and to enhance climate projections (Ghent et al. 2011). Satellite observations also contribute substantially to the enhancement of meteorological reanalysis products that are widely used for climate change research, for instance, the National Center for Environmental Prediction reanalysis. The Global Climate Observation System (GCOS) has listed 26 out of 50 important climate products as significantly dependent on satellite remote sensing observations (Yang et al. 2013). Additionally, satellite remote sensing datasets are substantially applied in the prevention, adaptation, and mitigation measures to cope with climate change impacts. Traditional land-based measurements are typically gathered at fixed intervals with limited spatial extent but satellite remote sensing techniques provide an opportunity for continual observation on a global scale. This has significantly advanced our understanding of the earth's climate system and its dynamics (Yang et al. 2013).

4.4.1.1 Global Warming

The warming trend in observed average surface temperature of the earth since the late nineteenth century confirms human and anthropogenic influences as significant drivers of global climate change. The trend was initially identified by studying anomalies in time series of near-surface air temperature over the land that were observed by available weather stations. However, the existence of this trend was constantly disputed because of the uncertainties prevalent in weather records, which are the aftermath of influence from land-use changes as well as poor siting of gauged facilities or instruments. With the advent of satellite data, a new, efficient, and independent avenue of investigating global temperature trends, especially at the ocean surfaces and in the atmosphere, is now possible.

4.4.1.2 Snow and Ice

The massive retreat of snow and ice cover is an important indicator of global warming and how it is affecting our planet (Chap. 1). Melting of the seasonal snow and ice cover can not only lead to a positive feedback by minimizing the albedo diffusion and intensity of the Earth's surface, but could also affect sea-level rise in tremendous ways. However, data from satellite remote sensing platforms like GRACE and Global Positioning Systems have played key roles in monitoring the dynamics and movements of snow extents and ice covers thus benefiting hydrologists over the globe (Khan et al. 2010). For instance, using advance multivariate methods, GRACE data captured the extensive and unabated mass loss in the Patagonia ice field, which was attributed to increased warming of the climate system (Ndehedehe and Ferreira 2020).

4.4.1.3 Sea-Level Change

A significant driver of sea level is climate change and or its variability. Data from monitoring sites with good-quality tide gauge records show that the global averaged sea-level rise was approximated to be 1.9 $\pm$ 0.4 mm/yr since 1961 (Church and White 2011). Historical estimates of sea-level variations from coastal and island sea-level measurements (1880–2009) capture the acceleration in the rate of sea-level rise. But satellite altimetry is now helping with tracking sea-level changes. The illustration in Fig. 4.16 shows an increase in sea-level rise during 1993–2009 based on a combination of different satellite altimetry missions (TOPEX, Jason-1, and OSTM/Jason-2). Satellite altimetry measurements using the TOPEX/Poseidon missions launched by NASA and the Center National d'Etudes Spatiales (CNES) focused on the mapping of ocean surface topography from 1992 to 2006, including its subsidiary and follow-up missions, which observed a global mean sea-level rise of approximately 3.2 $\pm$ 0.8 mm/yr between 1992 and 2010 (Leuliette et al. 2011).

Furthermore, satellite altimetry observations have been integrated with in-situ or tide gauge observations to reconstruct long-term sea-level time series having global coverage. With the use of satellite observations, it is now possible to uncover significant regional sea-level rise variations by monitoring their spatial and temporal trends. However, the possibility of these spatial trends getting influenced by short-term signals introduced by large-scale climatic processes such as El Nino/La Nina or even low frequency oscillations (e.g., Pacific Decadal Oscillation) may affect the distribution of these trends. The prevalence of mesoscale eddies in oceans has also been observed by satellite remote sensing techniques, improving our understanding of the existing inter-relationships between sea level and ocean circulation patterns (Fu et al. 2010).

4.4.1.4 Solar Radiation

Another driver of climate change is solar radiation. It is important to track the changes in the sun's luminosity (unit is Total Solar Irradiance, TSI) to determine if the natural changes in solar radiations have influenced climate change in any way. Satellite measurements of TSI began in 1978 with the use of active-cavity electrical-substitution radiometers. The climate simulation study undertaken by Hansen (2007) reported

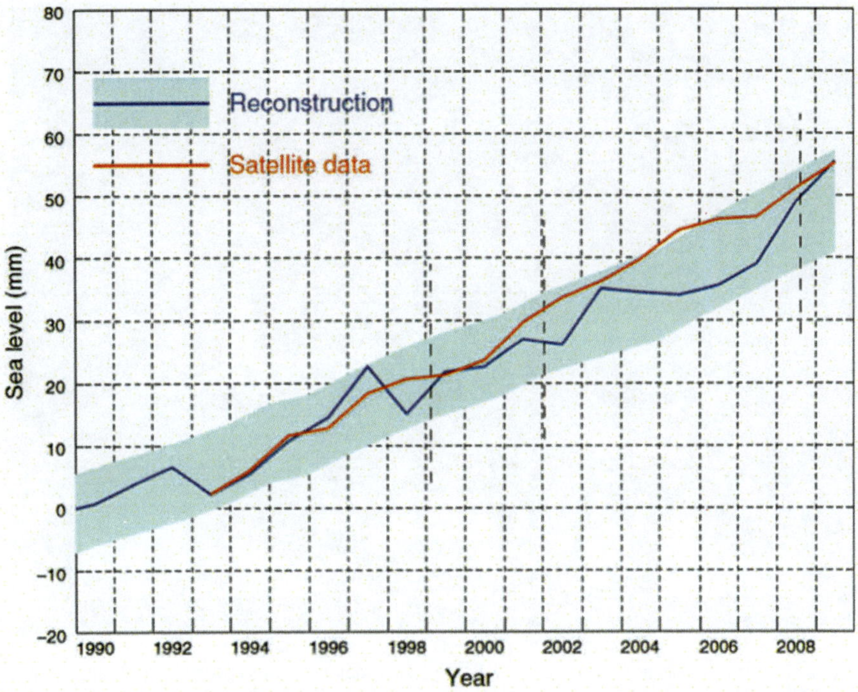

Fig. 4.16 Monitoring changes in global sea-level rise using satellite altimetry data (TOPEX, Jason-1, and OSTM/Jason-2) for the period between 1990 and 2009. Uncertainty estimates are indicated with light blue. *Source* Church and White (2011)

that variations in TSI have had minimal influence on global warming since 1880. A parallel study undertaken by Scafetta and West (2006) reported that the sun may have contributed an approximate 25–35% of the 1980 to 2000 global warming based on extracts from two alternative satellite TSI composites.

4.4.1.5 Aerosols

Particles in the atmosphere also known as aerosols can generate a cooling effect on the climate system, thus, counteracting the warming effects of anthropogenic greenhouse gases by influencing both cloud precipitation and atmospheric radiation processes. Using the aerosol optical depth (AOD) unit, recent changes in atmospheric aerosol concentrations have been recognized, and these estimates are derived from observations recorded by the visible and infrared optical sensors onboard different satellites (Fig. 4.17). With the help of these satellite observations, which provide direct and indirect climate forcing by aerosols, scientists and experts can make independent comparisons of these aerosol concentration measurements to other climate models, which have been in existence. The direct radiative forcing by anthropogenic aerosols

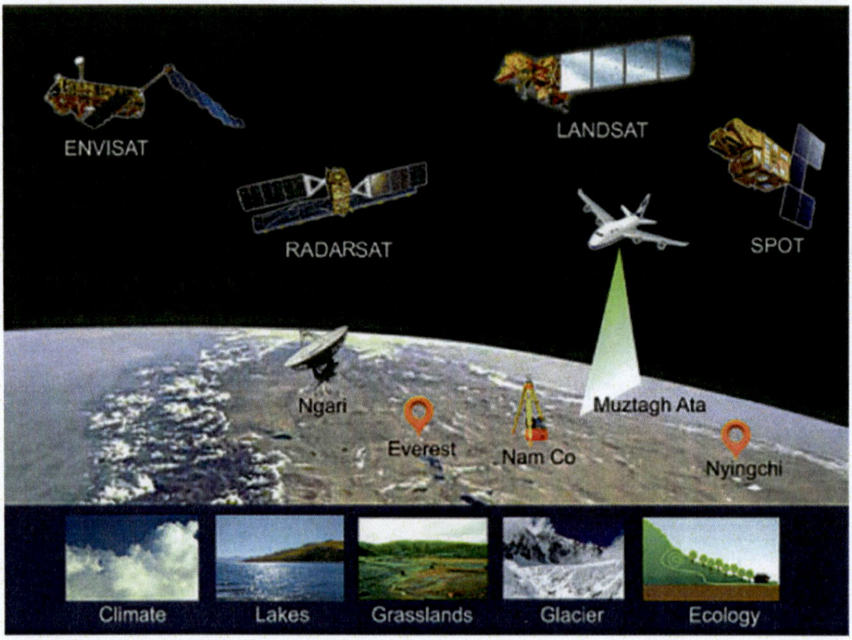

Fig. 4.17 Remote sensing techniques, including Earth observation satellites, aerial and ground-based observation stations. *Source* Guo et al. (2015)

can be estimated by integrating satellite measurements of the radiation budget and the AOD. Other processes which include water vapor, precipitation, and clouds are well-known products that contribute to climate change and fluctuations due to their varying rates, intensity, and concentrations. However, satellite remote sensing methods have revolutionized our capability in estimating and tracking these components, and are more accurate than the model estimates that were in place during a larger part of the twentieth century (Yang et al. 2013). Several state-of-the-art instrumentation techniques for ground-based measurements and satellite remote sensing technologies have been designed in the past three decades to assess optical and physiochemical properties of aerosols for a better understanding of radiative balance and feedback methods. The multi-angle imaging spectroradiometer (MISR), ozone monitoring instruments (OMI), and satellite retrievals of Moderate Resolution Imaging Spectroradiometer (MODIS) are utilized for this purpose. Ground-based observations of aerosol quantities also provide a basis for validation as well as atmospheric correction, and can be applied toward the validation of aerosol models used in atmospheric correction algorithms.

4.4.2 *Hydrological Processes*

Estimating hydrological variables using point-based observations has been the norm prior to the introduction of satellite remote sensing systems. However, with their introduction, unprecedented approaches and techniques in earth science observations emerged, providing new capabilities in several areas. Satellite remote sensing systems provide spatially continuous measurements of hydrological state variables, parameters, and processes, which include soil moisture content and snow cover characteristics. With the multiple stream of big satellite data, hydrological processes can now be better understood.

4.4.2.1 Hydrological State Variables

- Surface soil moisture plays a critical role in the partitioning of precipitation at the ground surface into latent heat exchanges with the atmosphere, as well as the partitioning between surface runoff and infiltration. Despite the importance of in-situ observations of soil moisture content, in calibrating and validating outputs from hydrological models and satellite-based soil moisture retrievals (Dorigo et al. 2011a, b), soil moisture monitoring stations appear to be limited and are only available at a few locations globally (Fig. 4.18). Even though the International Soil Moisture Network (ISMN, Dorigo et al. 2021, 2011b) was created to help organize and host in-situ soil moisture measurements from several operational networks across the globe, the locations of ISMN networks and sites across the world show that Asia, Africa, South America, and much of the Middle East lack in-situ soil moisture data (Fig. 4.18). The ISMN is a global network containing 71 networks that provides access to approximately 2842 stations, much of which are located in North America (USA), Europe, and Australia and some parts of Asia (Dorigo et al. 2021). The publicly and freely available ISMN soil moisture networks and data are still growing, and its database is regularly updated as evidence regular or irregular (Dorigo et al. 2021, 2011a, b). However, the point-based observations from these in-situ soil moisture networks are limited in providing reliable estimates of soil moisture content in time and space (Fig. 4.18). However, satellite hydrology systems, which provide observations at microwave wavelengths are susceptible to dielectric soil properties and moisture content, thereby providing details of only the top few centimeters of the soil column based on the sensor wavelength. The capability of these satellite systems (includes soil moisture missions, e.g., SMOS, ASCAT, and SMAP) in providing soil moisture estimates with varying wavelengths makes them more efficient, even in areas of low to moderate vegetation coverage (Margulis et al. 2002). This makes the evaluation of soil moisture retrieval algorithms an interesting research direction in this day and age. For instance, using in-situ soil moisture observation for 17 sites across Europe, some pioneering studies (e.g., Brocca et al. 2011) have validated the satellite-derived soil moisture from the Advanced Microwave

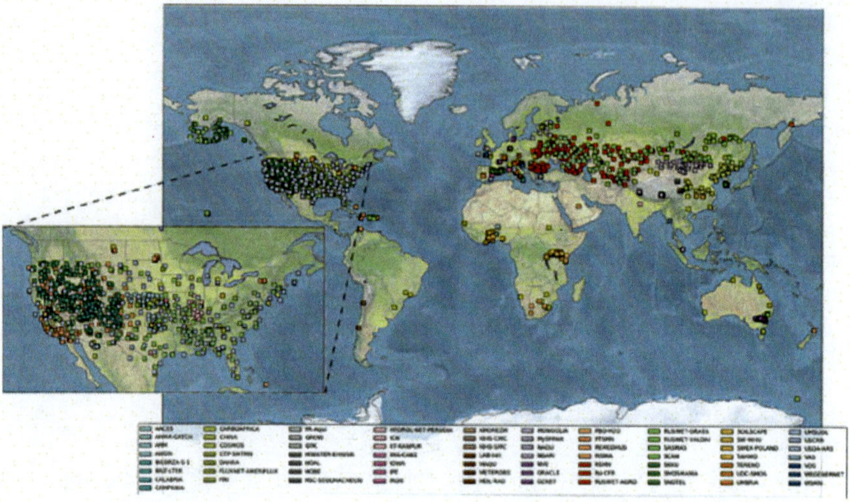

Fig. 4.18 Spatial distribution of globally available in-situ soil moisture stations based on the International Soil Moisture Network database. *Source* Dorigo et al. (2021)

Scanning Radiometer for the Earth observing system and the Advanced SCATterometer missions. Satellite evaluation of soil moisture is still ongoing for other regions of the world.

- Snow cover on the other hand can be characterized by the visible (VIS) and infrared (IR) sensors. However, this characterization is complicated by the existing cloud cover readings as a result of the proximity between snow and cloud VIS/IR spectral signatures. Snow water equivalent can be observed using passive microwave sensors, which are susceptible to the water equivalent of dry snowpacks and its microphysical properties under a wide range of weather conditions. In the review undertaken by Clifford (2010), they detailed the history, and identified challenges and future developments of snow water equivalent estimates from passive microwave sensors. They also highlighted that in comparison to independent model-based reanalysis datasets, snow water estimates derived from passive microwave sensors bear lower confidence levels compared to snow cover estimates derived from the VIS/IR measurements.

4.4.2.2 Moisture Fluxes

Generally, satellite remote sensors are not able to directly estimate water fluxes. However, these sensors are capable of measuring state variables including land surface reflectance, land cover, land surface temperature, cloud characteristics, and even vegetation properties. Moisture fluxes can then be deduced based on relationships between the state variables and the fluxes. The key and most dominant moisture fluxes at the land-atmosphere surface are precipitation and evapotranspiration. It is possible to infer precipitation rates from cloud top conditions with the use of VIS/IR satellite imageries, from microwave signals due to the strong interaction between microwave radiations and precipitation particles, or a combination of both (Tang et al. 2010). Precipitation extracts based on remote sensing platforms have continued to show promise as key input parameters to hydrological models which are the basis of estimating other moisture fluxes including evapotranspiration and stream flow estimates. The study undertaken by Yilmaz et al. (2010) reported a prior evaluation of a global flood monitoring system that leverages precipitation estimates derived from the Tropical Rainfall Measuring Mission (TRMM) multi-satellite precipitation analysis. The global flood monitoring framework was capable of simulating the onset of flood episodes generated by heavy precipitation and detected approximately 40% of global floods, though it suffered from region-dependent bias.

Evapotranspiration, which is equivalent to latent heat transfer, is basically inferred as a residual in the surface energy balance. Some of the factors influencing the surface energy balance include climatic parameters (such as wind speed, humidity gradient away from the leaf, and energy availability), environmental conditions (such as soil characteristics and water availability), and vegetation characteristics (such as vegetation height, stomatal resistance, and leaf area index). Land surface state parameters such as emissivity, reflectance, land cover, surface albedo, surface temperature, and vegetation properties are measurable by satellite remote sensing methods and can be related directly to the factors that influence surface energy balance in a given region.

4.4.2.3 Water Storage and Surface Water Budget

Terrestrial water storage (TWS) is the sum of changes in freshwater stored on and below the land surface. It is a critical component of the water cycle, yet was poorly observed globally up until 2002 (Tang et al. 2010). After the launch of the GRACE satellite (Tapley et al. 2004), it became possible to monitor the changes in these terrestrial water storage cycles at larger scales. Krogh et al. (2010) evaluated and compared TWS variations from global isotropic and regionally constrained GRACE mass concentrations and hydrological models over the Southern African region. Results from their experiments showed that the application of regional constraints generates more details than using the isotropic constraints. The GRACE mission possesses the capabilities of monitoring changes in terrestrial stored water, but empirical and statistical techniques will be required to partition the observed TWS variations into water storage components such as those in reservoirs, wetlands and lakes, groundwater, soil moisture, and snow packs (e.g., Ndehedehe 2022a). Radar altimetry methods have been employed in this regard to determine the dynamics of freshwater levels and to deduce the variations in water storage components such as surface water variability in lakes (e.g., Ndehedehe 2022d; Ndehedehe et al. 2017; Moore and Williams 2014). Given the coarse resolution of altimetric footprints (usually in the range of 1–10 km), most altimetric studies have emphasized surface water targets with surface areas larger than 300 km^2 (Tang et al. 2010). With the use of a water-detection algorithm, Zhang et al. (2010) explored the applications of satellite radar altimetry for the examination of relatively smaller inland water bodies. This experiment was hinged on the fact that the radar waveform returned from water is different from that returned from land. Outcome from their study showed that water elevations from their algorithms depict improvements over water elevations processed using standard radar altimetry techniques when compared with in-situ observations.

One of the limitations of current altimetry missions for inland water applications is that the instruments are nadir looking with 200–300 km spacing between orbital tracks. Therefore, the current platforms are only able to measure a limited fraction of the earth's largest lakes, wetlands reservoirs, and rivers, and do not possess the potential of simultaneously observing water elevations and inundated areas (Fig. 4.17). The SWOT (https://swot.jpl.nasa.gov/) mission is a swath mapping radar interferometer, which possesses the capabilities of providing simultaneous observations of water elevations and inundated areas for inland water bodies. Past literature that focused on hydrological models, processes, and applications are largely based on different constituent variables in the land surface water balance.

The terrestrial water balance is given as $\Delta\text{TWS} = P - E - R$, where ΔTWS represents changes in terrestrial water storage, P represents precipitation, E is evapotranspiration, and R is runoff. These components, which make up most of the hydrological fluxes, are now observable at varying spatio-temporal resolutions and accuracies due to the advent of satellite remote sensing techniques. For instance, the work undertaken by Gao et al. (2010) attempted to close the surface water budget using satellite-based ΔTWS, P and E products, and observed R estimates over major hydrological river basins in the United States. They assessed nine satellite-based estimates among which precipitation rates had the highest bias or error term. The deduced runoff, which was a residual of satellite-based hydrological fluxes, was generally overestimated because of excessive precipitation rates and the underestimation of evapotranspiration and TWS estimates from remote sensing platforms. Despite this outcome, the use of satellite remote sensing approach in the estimation of key hydrological units is a significant progress. However, there is still opportunity for fine-tuning and improving the estimates from these remote sensing platforms through their continued evaluation (Ndehedehe 2019). This is because retrievals from remote sensing are yet to provide consistent estimates of the land surface water budget (Gao et al. 2010).

4.4.3 Understanding Anthropogenic Influence

Anthropogenic influences, processes, materials, and objects are those from human activities, as opposed to those occurring naturally with human intervention. This intervention can be in the form of various human actions including wastes from biological or chemical substances and/or by-products (Fig. 4.19). For example, it is generally accepted that carbon dioxide is the key factor driving anthropogenic climate change and significant fluctuations in the global hydrologic systems. Anthropogenic

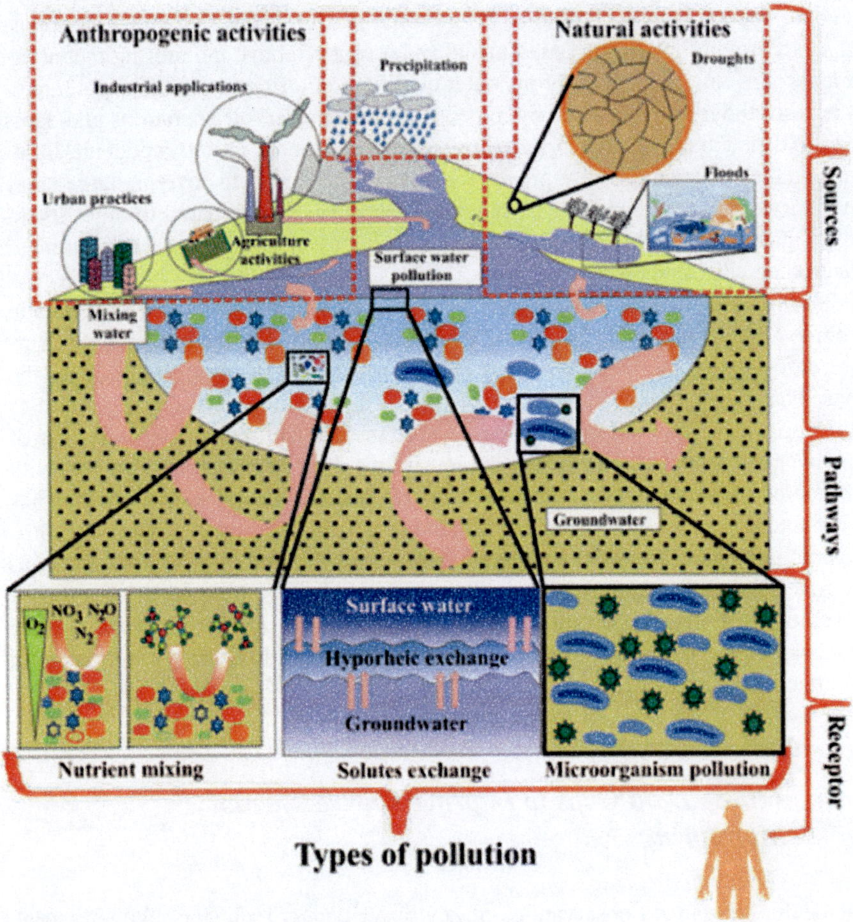

Fig. 4.19 The role of human impacts on water resources emphasizes the need for satellite-based techniques in monitoring. *Source* Akhtar et al. (2021)

impacts are well-known direct or indirect contributors to a large proportion of the tropospheric aerosol, even in relatively remote regions of the globe. As a result of the roles aerosols play in hydrology and cloud physics, there have been established links between anthropogenic activities and cloud/climate modification. It has been suggested that increased pollution from large industrialized and heavily urbanized sub-regions can increase the amounts of cloud condensation nuclei available for the formation of cloud downwind (Fig. 4.19). Overall, the higher the amount of cloud condensation nuclei active at typical cloud super-saturations, the more concentration of precipitation droplets are expected.

Prior to the advent of satellite hydrology systems, it was difficult to monitor the effects and impacts of anthropogenic influences on the climate system. However, in contemporary times, satellite-based geodetic platforms have provided an opportunity to provide long-term assessments and a cost-effective monitoring technique of anthropogenic influences. From water quality monitoring to quantifying changes in terrestrial hydrology caused by impacts of human reservoir operations and dams (e.g., Ndehedehe et al. 2017), satellite-based assessments of anthropogenic influence on the environment have proved to be less expensive, and covering large areas with varying extents. Additionally, they are less invasive compared to other methods of quantifying anthropogenic impacts, which may include field-based seismic or electromagnetic methods.

For instance, the InSAR technique (see Chap. 6) uses time series of SAR data to provide high spatial resolution (in a few meters) of monthly measurements of surface deformation and human impacts on the terrestrial ecosystem. InSAR uses the phase change of radar reflections of the Earth's surface to observe small variations in the position of the reflection points. Due to the availability of radar reflectors of suitable quality, InSAR can be applied toward the detection of surface motion with an accuracy of a few millimeters. Currently, the InSAR (see Chap. 6) technique has contributed immensely to understanding the short- and long-term effects of anthropogenic influences around the globe on public infrastructure like roads. It has also improved the understanding of how these anthropogenic influences affect climate and other hydro-geophysical quantities such as geothermal productions, groundwater changes, earthquake-related strain, tunneling and mining, and oil and gas productions (Bhattacharyya et al. 2019).

4.4.4 Filling Data Gaps to Improve Environmental Monitoring

The decline in gauged observations of key water budget indicators like terrestrial stored water, rainfall, river discharge, and evapotranspiration has led to increased reliance of satellite-based products. While other developed regions have sufficient gauged networks, emerging economies have extremely sparse distributions of such networks and in some areas they are not only discontinuous but also inadequate with their density falling far below the World Meteorological Organization guidelines (Ndehedehe 2019; Alsdorf et al. 2007). Earlier on, the routine reporting of these indicators, especially river discharge to the World Meteorological Organization, declined by 90% since 1990 mostly attributed to the costs and logistics required for the installation of gauged network (Ndehedehe 2019; Alsdorf et al. 2003). The impact of paucity of high-quality baseline data is that it creates knowledge gaps in water budget variations and the impacts of climate on terrestrial hydrology. Even though some regions have a highly densed network of gauged stations, the other non-trivial issue is the fact that gauge observations such as those from river stage are limited in large-scale monitoring of hydrological conditions (Alsdorf et al. 2007).

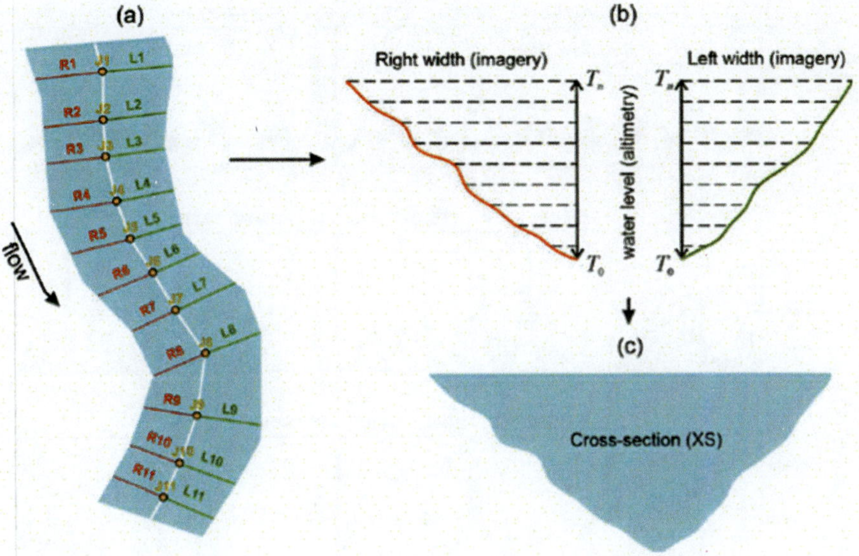

Fig. 4.20 Illustration of the lateral method in estimating river discharge. *Source* Lamine et al. (2021)

For instance, the physics of water flow across floodplains and wetlands provides one spatial dimension information about changes in surface water extent.

On the one hand, GRACE data has bridged the gap, providing new capability in large-scale assessment of terrestrial hydrology (e.g., groundwater) and water budget underpinned by the availability of satellite-based precipitation and soil moisture (Ndehedehe 2022a, c; Tapley et al. 2019). On the other hand, the emergence of novel techniques and algorithms that optimized the modeling of river discharge measurements have contributed to improving knowledge on the nature of climate change impacts on water resources (Ndehedehe 2022d). For instance, Fig. 4.20 illustrates the lateral method used by Lamine et al. (2021) to obtain river section and river discharge based on satellite imagery and altimetry data. By relying on the Alos PalsarRTC-DEM, they estimated the cross-sectional area and the average slope. Recently, the feasibility of estimating daily river discharge and runoff from the combination of GRACE data with satellite observations of soil moisture and precipitation (Camici et al. 2022) underscores benefits of satellites in hydrology-based monitoring to improve knowledge of global water cycle and managing water use. This capability was tested and calibrated in the Mississippi River basin (Fig. 4.21). Several performance metrics (e.g., Pearson correlation coefficient, root mean square error relative to the mean, and Kling–Gupta efficiency) employed to measure the goodness-of-fit between observed and predicted river discharge time series indicate strong agreement in the five calibrated sections of the basin between 2003 and 2016 (Fig. 4.21). This apparently attests to the importance of satellite hydrology missions in filling critical hydrological data gaps.

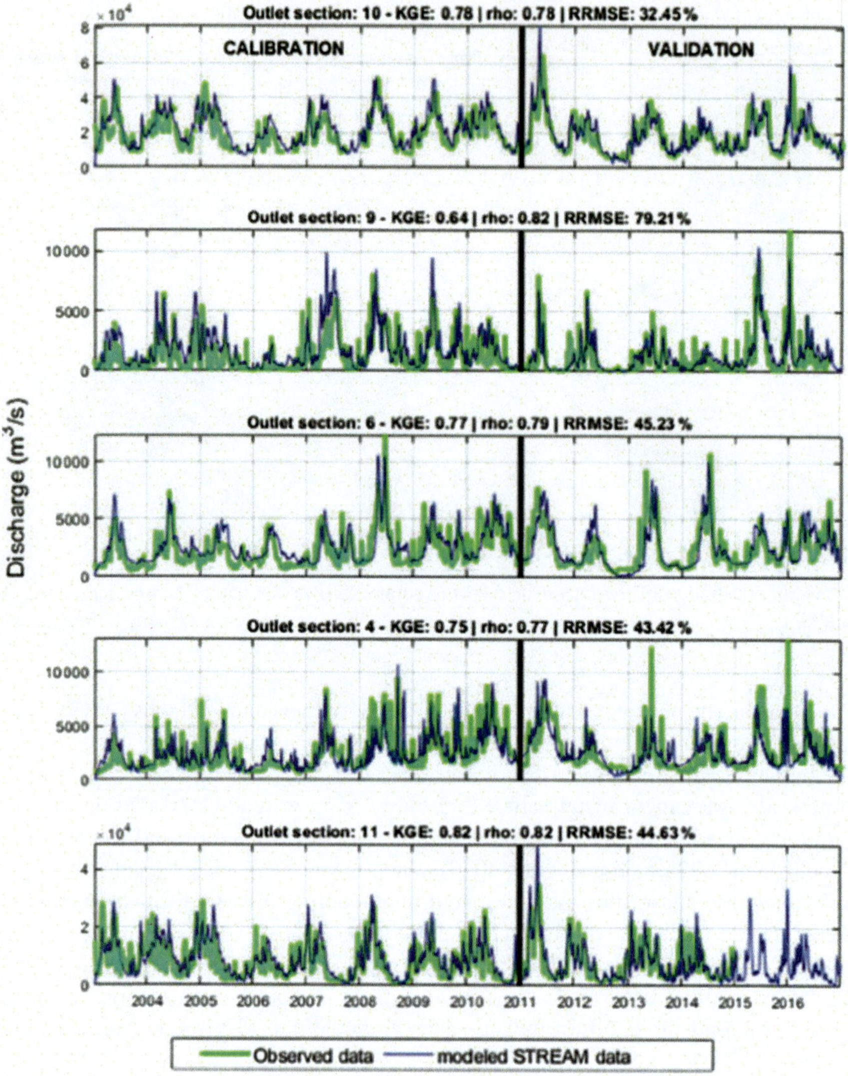

Fig. 4.21 Estimation of river discharge by combining GRACE data with satellite observations of soil moisture and precipitation. *Source* Camici et al. (2022)

The other benefit taken from this new river discharge monitoring is that it is scalable and can be implemented on a global scale to aid global runoff monitoring that underpins improved understanding of natural systems and the roles humans play to interrupt or disturb it (Camici et al. 2022). HydroSat, a repository for several hydrology data, is helping with providing information on surface water hydrology, including surface water extent, water level fluctuations, and river discharge, among

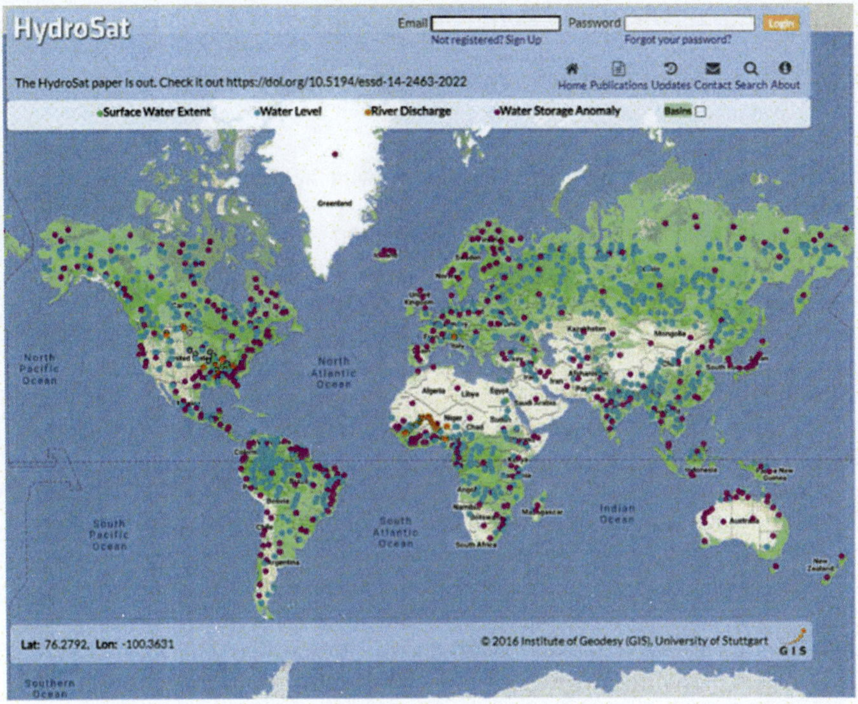

Fig. 4.22 Interface for the HydroSat database where several satellite hydrology products (surface water extent, water level, river discharge, and water storage anomaly) are hosted. *Source* Tourian et al. (2022)

others. As described by Tourian et al. (2022), it is a *database that contains geometric quantities of the global water cycle from geodetic satellites*. In the HydroSat package (Fig. 4.22), water volume anomaly for lakes and reservoirs was estimated by combining water level and surface water area variation from satellite altimetry observations. HydroSat is publicly available and can be accessed from this HydroSat database portal (http://hydrosat.gis.uni-stuttgart.de). The homepage of the HydroSat database is illustrated in Fig. 4.22. From the menu or buttons on the homepage (Fig. 4.22), time series data are available for surface water extent, water level, river discharge, and water storage anomaly across different parts of the world. Potential users will have to sign up through registration and as shown in the top right-hand corner of Fig. 4.22. Moreover, data can also be retrieved on basin scale. With limitations in publicly available global in situ database in terms of global coverage—spatial and temporal, HydroSat fills existing data gap, providing the opportunity to improve the monitoring of surface water hydrology to improve the understanding of changes in global water cycle (Tourian et al. 2022). However, the density of some data (e.g., river discharge) in the HydroSat database is limited in some regions (Fig. 4.22).

4.4.5 *Management and Policy Frameworks*

4.4.5.1 Understanding Impacts on Water Resources and Need for Management

Societies around the globe have long struggled with the management, planning, and distribution of freshwater resources amidst growing populations and increased used of water resources under a changing climate. Some of these impacts are evidenced in the drying of Lake Chad, the Aral Sea, as well as the formation of the Salton Sea in California. The impacts of extreme droughts on hydrological systems (e.g., loss of surface and groundwater) are well established in the literature (see, e.g., Scanlon et al. 2021; Ndehedehe et al. 2021; Ndehedehe 2022a; Pekel et al. 2016; Rodell et al. 2009; Tiwari et al. 2009). However, the continued depletion of groundwater in the Central Valley suggests such impacts can be further accelerated when these drought conditions persevere or their magnitude increases (Liu et al. 2022). Other consequences including the degradation of major river wetlands and deltas (e.g., Tigirs and Euphrates, Yellow, and Colorado), and the changes of stream flow through the damming rivers can be attributed to poor management and policy frameworks in these regions, among other factors (e.g., Petheram et al. 2018; Nilsson et al. 2005). Understanding the complex and socio-physical nature of increasing water challenges has resulted in the need for a more comprehensive management and policy framework. This is critical to help avoid poor decisions and for key stakeholders and actors to seek better options for the management of freshwater resources.

4.4.5.2 Water Resources System Analysis-WRSA

The implementation of water management policies are usually capital-intensive, expensive, and sometimes could be challenging due to several users having varied interests. It also requires approval from many political agencies and leaders. Therefore, policy and planning decisions are subject to discussions and processes, and are oftentimes subject to critical analysis to develop and relate alternative approaches to inform such discussions. The application of water resources system analysis (WRSA) for the prediction of the hydrologic, socio-economic, and environmental consequences of water management has long sought to provide better methods and techniques to replace the traditional water management systems (Brown et al. 2015). Recently, the WRSA has developed gradually to focus on the assessment of alternative frameworks in collaboration with policy makers and stake holders in a systematic approach, with the use of simulation and optimization techniques. Policy analysis within the WRSA basically consists of quantitative evaluation of policy impacts and alternatives, trade-off analysis, as well as development of policy frameworks. Policy frameworks are fundamentally deemed as the standards that prescribe, constrain, or affect water management decisions. Even though WRSA does not focus on how these policies came to be, the collaborations among professionals including, lawyers, economists, political scientists, and system analysts, however, tend to contribute to the emergence of new knowledge toward these policy development strategies (Brown et al. 2015).

The advent of simulation modeling techniques has created a new routine to aid the analysis of water policies. This approach has a universal application in the development of water resources plans. Simulation is applied toward the evaluation of plans or policies that are generated outside the model, typically referred to as "what if …?" analysis. Simulation applications vary between major water resources system models. This may entail environmental and economic performance indices to the proximal–universal use of hydraulic simulation models for the purpose of evaluating how water and flood distribution system options perform. These models include hydrological processes by accepting inflows as inputs, or groundwater dynamics to the analysis (Yates et al. 2005). Anthropogenic influence, which includes, water use and reservoir operations are usually stipulated in alternative or target objectives.

4.4.5.3 Collaborative Modeling Methods and Other Applications of the WRSA

Within a simulation-based planning protocol, some models are known to use optimization to simulate the decisions related to human actions rather than as an optimized basis for design decisions and developing plans. Several models rely on optimization processes to simulate water use and reservoir release decisions based on prescribed priorities (e.g., Riverware, WEAP, and CalSim). Sometimes, hydro-economic modeling utilizes optimization in a similar role, where anthropogenic influences are estimated as decision variables that maximize economic results (Rosenberg et al. 2007). The application of hydro-economic optimization equally extends to modeling water markets, market-like mechanisms, as well as creative applications of financial mechanisms with water supply frameworks (Harou et al. 2009). Identifying stakeholders' penchant to have control is vital. However, in cases where stakeholders are unlikely to agree to objectives, collaborative modeling has garnered significant attention (Langsdale et al. 2013). Collaborative modeling procedures entail the co-design of a water resources system model with various stakeholders having a focus on decision support for contributing to a more comprehensive understanding of the problem being addressed. They are known for the application of optimization or simulation techniques. Collaborative modeling methods function based on the hypothesis that better understanding of water problems obtained through an open process for model development can improve the chances that solutions will be achieved and accepted by stakeholders. Palmer et al. (2013) highlighted that the shared vision planning is a key example and is being applied in major river basins across the globe. The application of the WRSA in assessing proposed new infrastructural designs in river basins across the world has shown serious potentials especially in transboundary scenarios (Yang et al. 2014b; Ghile 2014). Other applications of the WRSA includes assessing the risks of climate change and planning for acclimatization to its projected effects since the 1990s, resourceful responses to uncertain probabilities, water demand modeling, and management, among others (Brown et al. 2015). Its relevance in water management and maintaining suitable policy frameworks makes it an important system in freshwater management across major sub-regions and river basins in the world (Brown et al. 2015).

4.4.5.4 Some Water Management Examples

Given the acceptability of simulations as an analysis approach to water management, water policy and planning applications entailing system optimization processes remain less popular outside academia. Generally, optimization models are prescriptive models that are utilized in the creation of plans or policies for the purpose of achieving key objectives. In flood management, for instance, risk-based optimization techniques are widely applied over the Netherlands and the UK in the establishment of flood safety levels for major level systems (Westerhof et al. 2022). One of the highest applications of optimization for policy and planning problems lies on the location where there is substantial political consensus on the performance objectives, especially one objective, which is the case with water distributions, levee systems, and hydro-power projects. Apart from the routine hydro-power applications, hydro-economic optimization techniques are popularly applied in academic research and policy studies. Some of the projects where they were applied include the reestablishment of Hetch Hetchy valley in Yosemite National Park (Brown et al. 2015). In the context of water resources planning, system optimization is increasingly being underpinned by information obtained from earth resources satellite and satellite hydrology systems like GRACE.

References

Abdelmohsen K, Sultan M, Ahmed M et al (2019) Response of deep aquifers to climate variability. Sci Total Environ 677:530–544. https://doi.org/10.1016/j.scitotenv.2019.04.316

Adamowski J, Fung Chan H, Prasher SO, Ozga-Zielinski B, Sliusarieva A (2012) Comparison of multiple linear and nonlinear regression, autoregressive integrated moving average, artificial neural network, and wavelet artificial neural network methods for urban water demand forecasting in Montreal, Canada. Water Resour Res 48(1). https://doi.org/10.1029/2010WR009945

Agutu N, Awange J, Ndehedehe C, Kirimi F, Kuhn M (2019) GRACE-derived groundwater changes over greater horn of Africa: temporal variability and the potential for irrigated agriculture. Sci Total Environ 693:133467. https://doi.org/10.1016/j.scitotenv.2019.07.273

Ahmed M, Sultan M, Yan E, Wahr J (2016) Assessing and improving land surface model outputs over Africa using GRACE, field, and remote sensing data. Surveys Geophys 37:529–556. https://doi.org/10.1007/s10712-016-9360-8

Ahmed M, Wiese DN (2019) Short-term trends in Africa's freshwater resources: rates and drivers. Sci Total Environ 695:133843. https://doi.org/10.1016/j.scitotenv.2019.133843

Akhtar N, Syakir Ishak MI, Bhawani SA, Umar K (2021) Various natural and anthropogenic factors responsible for water quality degradation: a review. Water 13(19). https://doi.org/10.3390/w13192660

Ali S, Liu D, Fu Q, Cheema MJM, Pal SC, Arshad A, Pham QB, Zhang L (2022) Constructing high-resolution groundwater drought at spatio-temporal scale using GRACE satellite data based

on machine learning in the Indus Basin. J Hydrol 612:128295. https://doi.org/10.1016/j.jhydrol.2022.128295
Aljazeera.com (2022) Europe's heatwaves, droughts put focus on climate change risks. Online. Retrieved from https://www.aljazeera.com/news/2022/8/13/europes-heatwaves-droughts-put-focus-on-climate-change-risks on 13th Aug 2022
Alsdorf D, Lettenmaier D, Vörösmarty C (2003) The need for global, satellite-based observations of terrestrial surface waters. Eos, Trans Am Geophys Union 84(29):269–276. https://doi.org/10.1029/2003EO290001
Alsdorf DE, Rodríguez E, Lettenmaier DP (2007) Measuring surface water from space. Rev Geophys 45(2):RG2002. https://doi.org/10.1029/2006RG000197
Beck HE, van Dijk AIJM, Levizzani V et al (2017) Mswep: 3-hourly 0.25deg global gridded precipitation (1979–2015) by merging gauge, satellite, and reanalysis data. Hydrol Earth Syst Sci 21(1):589–615. https://doi.org/10.5194/hess-21-589-2017
Becker M, Meyssignac B, Xavier L, Cazenave A, Alkama R, Decharme B (2011) Past terrestrial water storage (1980–2008) in the amazon basin reconstructed from GRACE and in situ river gauging data. Hydrol Earth Syst Sci 15(2):533–546. https://doi.org/10.5194/hess-15-533-2011
Becker M, Papa F, Frappart F, Alsdorf D et al (2018) Satellite-based estimates of surface water dynamics in the Congo River Basin. Int J Appl Earth Obs Geoinf 66:196–209. https://doi.org/10.1016/j.jag.2017.11.015
Bell JW, Amelung F, Ferretti A, Bianchi M, Novali F (2008) Permanent scatterer insar reveals seasonal and long-term aquifer-system response to groundwater pumping and artificial recharge. Water Resour Res 44(2). https://doi.org/10.1029/2007WR006152
Bhattacharyya A, Haldar A, Bhattacharyya M, Ghosh A (2019) Anthropogenic influence shapes the distribution of antibiotic resistant bacteria (ARB) in the sediment of Sundarban estuary in India. Sci Total Environ 647:1626–1639. https://doi.org/10.1016/j.scitotenv.2018.08.038
Biancamaria S, Durand M et al (2011) Assimilation of virtual wide swath altimetry to improve arctic river modeling. Remote Sens Environ 115(2):373–381
Brocca L, Hasenauer S, Lacava T, Melone F, Moramarco T, Wagner W, Dorigo W, Matgen P, Martínez-Fernández J, Llorens P, Latron J, Martin C, Bittelli M (2011) Soil moisture estimation through ascat and amsr-e sensors: an intercomparison and validation study across Europe. Remote Sens Environ 115(12):3390–3408. https://doi.org/10.1016/j.rse.2011.08.003
Brown CM, Lund JR et al (2015) The future of water resources systems analysis: toward a scientific framework for sustainable water management. Water Resour Res 51(8):6110–6124. https://doi.org/10.1002/2015WR017114
Broxton PD, van Leeuwen WJD, Biederman JA (2019) Improving snow water equivalent maps with machine learning of snow survey and lidar measurements. Water Resour Res 55(5):3739–3757. https://doi.org/10.1029/2018WR024146
Bürgmann R, Rosen PA, Fielding EJ (2000) Synthetic aperture radar interferometry to measure earth's surface topography and its deformation. Annu Rev Earth Planet Sci 28(1):169–209. https://doi.org/10.1146/annurev.earth.28.1.169
Burnett MW, Quetin GR, Konings AG (2020) Data-driven estimates of evapotranspiration and its controls in the Congo Basin. Hydrol Earth Syst Sci 24(8):4189–4211. https://doi.org/10.5194/hess-24-4189-2020
Camici S, Giuliani G, Brocca L, Massari C, Tarpanelli A, Farahani HH, Sneeuw N, Restano M, Benveniste J (2022) Synergy between satellite observations of soil moisture and water storage anomalies for runoff estimation. Geoscientific Model Dev 15(18):6935–6956
Castellazzi P, Martel R, Galloway DL, Longuevergne L, Rivera A (2016) Assessing groundwater depletion and dynamics using GRACE and InSAR: potential and limitations. Groundwater 54(6):768–780. https://doi.org/10.1111/gwat.12453
Chaussard E, Bürgmann R, Shirzaei M, Fielding EJ, Baker B (2014) Predictability of hydraulic head changes and characterization of aquifer-system and fault properties from InSAR-derived ground deformation. J Geophys Res Solid Earth

Chen J, Famigliett JS, Scanlon BR, Rodell M (2016) Groundwater storage changes: present status from GRACE observations. Surv Geophys 37(2):397–417. https://doi.org/10.1007/s10712-015-9332-4

Chen J, Wilson C, Seo K (2009) S2 tide aliasing in grace time-variable gravity solutions. J Geodesy 83:679–687. https://doi.org/10.1007/s00190-008-0282-1

Church J, White N (2011) Sea-level rise from the late 19th to the early 21st century. Surv Geophys 35:585–602. https://doi.org/10.1007/s10712-011-9119-1

Clifford D (2010) Global estimates of snow water equivalent from passive microwave instruments: history, challenges and future developments. Int J Remote Sens 31(14):3707–3726. https://doi.org/10.1080/01431161.2010.483482

Currell MJ, Ndehedehe CE (2022) The cambrian limestone aquifer, northern territory: Review of the hydrogeology and management rules to ensure protection of groundwater dependent values. Environment centre northern territory. Retrieved from http://www.researchrepository.griffith.edu.au/handle/10072/420013 on 13th December, 2022

Dannouf R, Yong B, Ndehedehe CE, Correa FM, Ferreira V (2022) Boosted regression tree algorithm for the reconstruction of grace-based terrestrial water storage anomalies in the yangtze river basin. Front Environ Sci 10. https://doi.org/10.3389/fenvs.2022.917545

Dorigo W, Himmelbauer I, Aberer D, Schremmer L, Petrakovic I, Zappa L, Preimesberger W, Xaver A, Annor F et al (2021) The International Soil Moisture Network: serving Earth system science for over a decade. Hydrol Earth Syst Sci 25(11):5749–5804. https://doi.org/10.5194/hess-25-5749-2021

Dorigo W, van Oevelen P, Wagner W, Drusch M, Mecklenburg S, Robock A, Jackson T (2011) A new International Network for in situ soil moisture data. Eos, Trans Am Geophys Union 92(17):141–142. https://doi.org/10.1029/2011EO170001

Dorigo WA, Wagner W, Hohensinn R, Hahn S, Paulik C, Xaver A, Gruber A, Drusch M, Mecklenburg S, van Oevelen P, Robock A, Jackson T (2011) The International Soil Moisture Network: a data hosting facility for global in situ soil moisture measurements. Hydrol Earth Syst Sci 15(5):1675–1698. https://doi.org/10.5194/hess-15-1675-2011

Durand M, Gleason CJ, Pavelsky TM et al (2023). A framework for estimating global river discharge from the surface water and ocean topography satellite mission. Water Resour Res 59(4):e2021WR031614. https://doi.org/10.1029/2021WR031614

Díaz-Alcaide S, Martínez-Santos P (2019) Advances in groundwater potential mapping. Hydrogeol J 27:2307–2324

Emery CM, Biancamaria S, Boone A, Ricci S, Rochoux MC, Pedinotti V, David CH (2020) Assimilation of wide-swath altimetry water elevation anomalies to correct large-scale river routing model parameters. Hydrol Earth Syst Sci 24(5):2207–2233. https://doi.org/10.5194/hess-24-2207-2020

Fasbender D, Peeters L, Bogaert P, Dassargues A (2008) Bayesian data fusion applied to water table spatial mapping. Water Resour Res 44(12). https://doi.org/10.1029/2008WR006921

Foroumandi E, Nourani V, Jeanne Huang J, Moradkhani H (2023) Drought monitoring by downscaling GRACE-derived terrestrial water storage anomalies: a deep learning approach. J Hydrol 616:128838. https://doi.org/10.1016/j.jhydrol.2022.128838

Fu L-L, Chelton DB, Le Traon P-Y et al (2010). Eddy dynamics from satellite altimetry. Oceanography 23(4). https://doi.org/10.5670/oceanog.2010.02

Galloway D, Burbey T (2011) Review: regional land subsidence accompanying groundwater extraction. Hydrogeol J 19:1459–1486. https://doi.org/10.1007/s10040-011-0775-5

Gao H, Tang Q, Ferguson CR et al (2010) Estimating the water budget of major us river basins via remote sensing. Int J Remote Sens 31(14):3955–3978. https://doi.org/10.1080/01431161.2010.483488

Getirana AC (2010) Integrating spatial altimetry data into the automatic calibration of hydrological models. J Hydrol 387(3):244–255. https://doi.org/10.1016/j.jhydrol.2010.04.013

Ghent D, Kaduk J, Remedios J, Balzter H (2011) Data assimilation into land surface models: the implications for climate feedbacks. Int J Remote Sens 32(3):617–632. https://doi.org/10.1080/01431161.2010.517794

Ghile YB, Taner MU, Brown C et al (2014) Bottom-up climate risk assessment of infrastructure investment in the niger river basin. Climatic Change 122:97–110. https://doi.org/10.1007/s10584-013-1008-9

Gong P, Wang J, Yu L, Zhao Y et al (2013) Finer resolution observation and monitoring of global land cover: first mapping results with Landsat TM and ETM+ data. Int J Remote Sens 34(7):2607–2654. https://doi.org/10.1080/01431161.2012.748992

Grace K, Davenport F (2021) Climate variability and health in extremely vulnerable communities: investigating variations in surface water conditions and food security in the West African Sahel. Popul Environ 42:553–577. https://doi.org/10.1007/s11111-021-00375-9

Guo H-D, Zhang L, Zhu L-W (2015) Earth observation big data for climate change research. Adv Climate Change Res 6(2):108–117. https://doi.org/10.1016/j.accre.2015.09.007

Gyawali B, Murgulet D, Ahmed M (2022) Quantifying changes in groundwater storage and response to hydroclimatic extremes in a coastal aquifer using remote sensing and ground-based measurements: The Texas Gulf Coast Aquifer. Remote Sens 14(3). https://doi.org/10.3390/rs14030612

Göttl F, Schmidt M, Seitz F (2018) Mass-related excitation of polar motion: an assessment of the new RL06 GRACE gravity field models. Earth Planets Space 70(195). https://doi.org/10.1186/s40623-018-0968-4

Hansen J, Sato M, Ruedy R et al (2007) Climate simulations for1880-2003 with GISS modelE. Climate Dyn 29:661–696. https://doi.org/10.1007/s00382-007-0255-8

Harou JJ, Pulido-Velazquez M, Rosenberg DE, Medellín-Azuara J, Lund JR, Howitt RE (2009) Hydro-economic models: concepts, design, applications, and future prospects. J Hydrol 375(3):627–643. https://doi.org/10.1016/j.jhydrol.2009.06.037

Hemati M, Hasanlou M, Mahdianpari M, Mohammadimanesh F (2021) A systematic review of landsat data for change detection applications: 50 years of monitoring the earth. Remote Sens 13(15). https://doi.org/10.3390/rs13152869

Henry C, Allen DM, Huang J (2011) Groundwater storage variability and annual recharge using well-hydrograph and GRACE satellite data. Hydrogeol J 19:741–755. https://doi.org/10.1007/s10040-011-0724-3

Hoffmann J, Leake SA, Galloway DL, Wilson AM (2003) Modflow-2000 ground-water model–user guide to the subsidence and aquifer-system compaction (sub) package. Geological Survey Washington DC. Retrieved from https://pubs.usgs.gov/of/2003/ofr03-233/ on 23rd Aug 2022

Huang C, Chen Y, Zhang S, Wu J (2018) Detecting, extracting, and monitoring surface water from space using optical sensors: a review. Rev Geophys 56(2):333–360. https://doi.org/10.1029/2018RG000598

Humphrey V, Gudmundsson L, Seneviratne SI (2017) A global reconstruction of climate-driven subdecadal water storage variability. Geophys Res Lett 44(5):2300–2309. https://doi.org/10.1002/2017GL072564

IMBIE (2018) Mass balance of the Antarctic Ice Sheet from 1992 to 2017. Nature 558:219–222. https://doi.org/10.1038/s41586-018-0179-y

Jing W, Zhang P, Zhao X, Yang Y, Jiang H, Xu J, Yang J, Li Y (2020) Extending GRACE terrestrial water storage anomalies by combining the random forest regression and a spatially moving window structure. J Hydrol 590:125239. https://doi.org/10.1016/j.jhydrol.2020.125239

Kalu I, Ndehedehe CE et al (2022) A new modelling framework to assess changes in groundwater level. J Hydrol Reg Stud 43:101185. https://doi.org/10.1016/j.ejrh.2022.101185

Kalu I, Ndehedehe CE, Okwuashi O, Eyoh AE (2021) Assessing freshwater changes over Southern and Central Africa (2002–2017). Remote Sens 13(13). https://doi.org/10.3390/rs13132543

Kalu I, Ndehedehe CE, Okwuashi O, Eyoh AE (2022) Integration of satellite geodetic observations for regional geoid modeling using remove-compute-restore technique. Earth Sci Inform 15:233–251

Khan SA, Wahr J, Bevis M, Velicogna I, Kendrick E (2010) Spread of ice mass loss into northwest greenland observed by GRACE and GPS. Geophys Res Lett

Kim B-H, Seo K-W, Lee C-K, Kim J-S, Chen J, Wilson CR (2022) Antarctic ice mass change (2003–2016) jointly estimated by satellite gravimetry and altimetry. J Geophys Res Solid Earth 127(9):e2021JB023297. https://doi.org/10.1029/2021JB023297

Krogh PE, Andersen OB et al (2010) Evaluating terrestrial water storage variations from regionally constrained grace mascon data and hydrological models over southern africa - preliminary results. Int J Remote Sens 31(14):3899–3912. https://doi.org/10.1080/01431161.2010.483483

Kumar SV, Wang S, Mocko DM, Peters-Lidard CD, Xia Y (2017) Similarity assessment of land surface model outputs in the North American Land Data Assimilation System. Water Resour Res 53(11):8941–8965. https://doi.org/10.1002/2017WR020635

Lamine B, Ferreira V, Yang Y, Ndehedehe C, He X (2021) Estimation of the Niger River cross-section and discharge from remotely-sensed products. J Hydrol Reg Stud 36:100862. https://doi.org/10.1016/j.ejrh.2021.100862

Langsdale S, Beall A, Bourget E, Hagen E, Kudlas S, Palmer R, Tate D, Werick W (2013) Collaborative modeling for decision support in water resources: principles and best practices. JAWRA J Am Water Resour Assoc 49(3):629–638. https://doi.org/10.1111/jawr.12065

Leuliette EW, Willis JK (2011) Balancing the sea level budget. Oceanography 24(2):122–129. https://doi.org/10.5670/oceanog.2011.32

Liu P, Famiglietti J, Purdy A et al (2022) Groundwater depletion in california's central valley accelerates during megadrought. Nature Commun 13:7825. https://doi.org/10.1038/s41467-022-35582-x

Margulis SA, McLaughlin D, Entekhabi D, Dunne S (2002) Land data assimilation and estimation of soil moisture using measurements from the Southern Great Plains 1997 field experiment. Water Resour Res 38(12):35–1–35–18. https://doi.org/10.1029/2001WR001114

Masood A, Tariq MAUR, Hashmi MZUR, Waseem M et al (2022) An overview of groundwater monitoring through point-to satellite-based techniques. Water 14(4). https://doi.org/10.3390/w14040565

McCabe MF, Rodell M, Alsdorf DE, Miralles DG, Uijlenhoet R et al (2017) The future of Earth observation in hydrology. Hydrol Earth Syst Sci 21(7):3879–3914. https://doi.org/10.5194/hess-21-3879-2017

Michailovsky CI, Milzow C, Bauer-Gottwein P (2013) Assimilation of radar altimetry to a routing model of the brahmaputra river. Water Resour Res 49(8):4807–4816. https://doi.org/10.1002/wrcr.20345

Moore P, Williams SDP (2014) Integration of altimetry lake lavels and GRACE gravimetry over Africa: inferences for terrestrial water storage change 2003–2011. Water Resour Res 50:9696–9720. https://doi.org/10.1002/2014WR015506

NASA (2002) Gravity recovery and climate experiment launch. National Aeronautics and Space Administration. Retrieved from http://www.jpl.nasa.gov/news/press-kits/gracelaunch.pdf Accessed 9th Feb 2017

Ndehedehe C (2022a) Groundwater from space, pp 211–230. Springer, Cham. https://doi.org/10.1007/978-3-030-99577-5_9

Ndehedehe C (2022b) Remote sensing hydrology, pp 3–17. Springer, Cham. https://doi.org/10.1007/978-3-030-99577-5_1

Ndehedehe C (2022c) Satellite geodetic missions, pp 53–70. Springer, Cham. https://doi.org/10.1007/978-3-030-99577-5_4

Ndehedehe C (2022d) Satellite observations of terrestrial water storage, pp 331–386. Springer, Cham. https://doi.org/10.1007/978-3-030-99577-5_14

Ndehedehe C, Awange J, Agutu N, Kuhn M, Heck B (2016) Understanding changes in terrestrial water storage over West Africa between 2002 and 2014. Adv Water Resour 88:211–230. https://doi.org/10.1016/j.advwatres.2015.12.009

Ndehedehe CE (2019) The water resources of tropical West Africa: problems, progress and prospect. Acta Geophysica 67(2):621–649. https://doi.org/10.1007/s11600-019-00260-y

Ndehedehe CE, Awange J, Kuhn M, Agutu N, Fukuda Y (2017) Analysis of hydrological variability over the Volta river basin using in-situ data and satellite observations. J Hydrol Reg Stud 12:88–110. https://doi.org/10.1016/j.ejrh.2017.04.005

Ndehedehe CE, Awange JL, Agutu NO, Okwuashi O (2018) Changes in hydro-meteorological conditions over tropical West Africa (1980–2015) and links to global climate. Global Planet Change 162:321–341. https://doi.org/10.1016/j.gloplacha.2018.01.020

Ndehedehe CE, Ferreira VG (2020) Assessing land water storage dynamics over Southern America. J Hydrol 580:124339. https://doi.org/10.1016/j.jhydrol.2019.124339

Ndehedehe CE, Ferreira VG, Agutu NO (2019) Hydrological controls on surface vegetation dynamics over West and Central Africa. Ecol Ind 103:494–508. https://doi.org/10.1016/j.ecolind.2019.04.032

Ndehedehe CE, Ferreira VG, Agutu NO, Onojeghuo AO, Okwuashi O, Kassahun HT, Dewan A (2021) What if the rains do not come? J Hydrol 595:126040. https://doi.org/10.1016/j.jhydrol.2021.126040

Neal J, Schumann G, Bates P, Buytaert W, Matgen P, Pappenberger F (2009) A data assimilation approach to discharge estimation from space. Hydrol Process 23(25):3641–3649. https://doi.org/10.1002/hyp.7518

Neely WR, Borsa AA, Burney JA, Levy MC, Silverii F, Sneed M (2021) Characterization of groundwater recharge and flow in California's San Joaquin Valley From InSAR-Observed surface deformation. Water Resour Res 57(4):e2020WR028451. https://doi.org/10.1029/2020WR028451

Nilsson C, Reidy CA, Dynesius M, Revenga C (2005) Fragmentation and flow regulation of the world's large river systems. Science 308(5720):405–408. https://doi.org/10.1126/science.1107887

Nordberg W (1965) Geophysical observation from nimbus i. Science 150(3696):559–572. https://doi.org/10.1126/science.150.3696.559

Palmer RN, Cardwell HE, Lorie MA, Werick W (2013) Disciplined planning, structured participation, and collaborative modeling - applying shared vision planning to water resources. JAWRA J Am Water Resour Assoc 49(3):614–628. https://doi.org/10.1111/jawr.12067

Pan B, Hsu K, AghaKouchak A, Sorooshian S (2019) Improving precipitation estimation using convolutional neural network. Water Resour Res 55(3):2301–2321. https://doi.org/10.1029/2018WR024090

Pekel JF, Cottam A, Gorelick N, Belward AS (2016) High-resolution mapping of global surface water and its long-term changes. Nature 540(7633):418–422. https://doi.org/10.1038/nature20584

Petheram C, Gallant J, Stone P, Wilson P, Read A (2018) Rapid assessment of potential for development of large dams and irrigation across continental areas: application to northern australia. Rangeland J 40(4):431–449. https://doi.org/10.1071/RJ18012

Rateb A, Scanlon BR, Pool DR, Sun A et al (2020) Comparison of groundwater storage changes from GRACE satellites with monitoring and modeling of major U.S. aquifers. Water Resour Res 56(12):e2020WR027556. https://doi.org/10.1029/2020WR027556

Rodell M, Velicogna I, Famiglietti JS (2009) Satellite-based estimates of groundwater depletion in India. Nature 460(7258):999–1002. https://doi.org/10.1038/nature08238

Rosenberg DE, Tarawneh T, Abdel-Khaleq R, Lund JR (2007) Modeling integrated water user decisions in intermittent supply systems. Water Resour Res 43(7). https://doi.org/10.1029/2006WR005340

Scafetta N, West BJ (2006) Phenomenological solar contribution to the 1900–2000 global surface warming. Geophys Res Lett 33(5). https://doi.org/10.1029/2005GL025539

Scanlon BR, Longuevergne L, Long D (2012) Ground referencing grace satellite estimates of groundwater storage changes in the California Central Valley, USA. Water Resour Res 48(4). https://doi.org/10.1029/2011WR011312

Scanlon BR, Rateb A, Pool DR et al (2021) Effects of climate and irrigation on GRACE-based estimates of water storage changes in major US aquifers. Environ Res Lett 16(9):094009. https://doi.org/10.1088/1748-9326/ac16ff

Scanlon BR, Zhang Z, Rateb A, Sun A, Wiese D, Save H et al (2019) Tracking seasonal fluctuations in land water storage using global models and GRACE satellites. Geophys Res Lett 46(10):5254–5264

Schmugge TJ, Kustas WP, Ritchie JC, Jackson TJ, Rango A (2002) Remote sensing in hydrology. Adv Water Resour 25(8):1367–1385. https://doi.org/10.1016/S0309-1708(02)00065-9

Seyoum WM, Kwon D, Milewski AM (2019) Downscaling GRACE twsa data into high-resolution groundwater level anomaly using machine learning-based models in a glacial aquifer system. Remote Sens 11(7). https://doi.org/10.3390/rs11070824

Smith RG, Knight R, Chen J, et al (2017) Estimating the permanent loss of groundwater storage in the southern san joaquin valley, california. Water Resour Res 53(3):2133–2148. https://doi.org/10.1002/2016WR019861

Sneed M, Brandt JT (2020) Mitigating land subsidence in the coachella valley, california, usa: an emerging success story. Proc Int Assoc Hydrol Sci 382:809–813. https://doi.org/10.5194/piahs-382-809-2020

Sośnica K, Jäggi A, Meyer U et al (2015) Time variable Earth's gravity field from SLR satellites. J Geodesy 89:945–960. https://doi.org/10.1007/s00190-015-0825-1

Sun AY, Scanlon BR, Zhang Z, Walling D, Bhanja SN, Mukherjee A, Zhong Z (2019) Combining physically based modeling and deep learning for fusing grace satellite data: can we learn from mismatch? Water Resour Res 55(2):1179–1195. https://doi.org/10.1029/2018WR023333

Tang Q, Durand M, Lettenmaier DP, Hong Y (2010) Satellite-based observations of hydrological processes. Int J Remote Sens 31(14):3661–3667. https://doi.org/10.1080/01431161.2010.483496

Tapley B, Bettadpur S, Watkins M, Reigber C (2004) The Gravity Recovery and Climate Experiment: Mission overview and early results. Geophys Res Lett 31:1–4. https://doi.org/10.1029/2004GL019920

Tapley B, Watkins M, Flechtner F, Reigber C, Bettadpur S et al (2019) Contributions of GRACE to understanding climate change. Nature Climate Change 9(5):358–369. https://doi.org/10.1038/s41558-019-0456-2

Thomas AC, Reager JT, Famiglietti JS, Rodell M (2014) A GRACE-based water storage deficit approach for hydrological drought characterization. Geophys Res Lett 41(5):1537–1545. https://doi.org/10.1002/2014GL059323

Tiwari VM, Wahr J, Swenson S (2009) Dwindling groundwater resources in northern India, from satellite gravity observations. Geophys Res Lett 36(18):L18401. https://doi.org/10.1029/2009GL039401

Tourian MJ, Elmi O, Shafaghi Y, Behnia S, Saemian P, Schlesinger R, Sneeuw N (2022) Hydrosat: geometric quantities of the global water cycle from geodetic satellites. Earth Syst Sci Data 14(5):2463–2486. https://doi.org/10.5194/essd-14-2463-2022

Vasco D, Harness P, Pride S, Hoversten M (2016) Estimating fluid-induced stress change from observed deformation. Geophys J Int 208(3):1623–1642. https://doi.org/10.1093/gji/ggw472

Vishwakarma B, Zhang J, Sneeuw N (2021) Downscaling GRACE total water storage change using partial least squares regression. Sci Data 8(95). https://doi.org/10.1038/s41597-021-00862-6

Wagner W, Naeimi V, Scipal K et al (2007) Soil moisture from operational meteorological satellites. Hydrogeol J 15:121–131. https://doi.org/10.1007/s10040-006-0104-6

Westerhof SG, Booij MJ, den Berg MCJV, Huting RJM, Warmink JJ (2022) Uncertainty analysis of risk-based flood safety standards in the Netherlands through a scenario-based approach. Int J River Basin Manage 1–16. https://doi.org/10.1080/15715124.2022.2060243

Wulder MA, White JC et al (2008) Landsat continuity: issues and opportunities for land cover monitoring. Remote Sens Environ 112(3):955–969. https://doi.org/10.1016/j.rse.2007.07.004

Yang J, Gong P, Fu R et al (2013) The role of satellite remote sensing in climate change studies. Nature Climate Change 3:875–883. https://doi.org/10.1038/nclimate1908

Yang Y, Long D, Guan H, Scanlon BR, Simmons CT, Jiang L, Xu X (2014) GRACE satellite observed hydrological controls on interannual and seasonal variability in surface greenness over mainland Australia. J Geophys Res Biogeosciences 119(12):2245–2260. https://doi.org/10.1002/2014JG002670

Yang Y-CE, Brown C, Yu W, Wescoat J, Ringler C (2014) Water governance and adaptation to climate change in the indus river basin. J Hydrol 519:2527–2537. https://doi.org/10.1016/j.jhydrol.2014.08.055

Yates D, Sieber J, Purkey D, Huber-Lee A (2005) WEAP21-A demand-, priority-, and preference-driven water planning model. Water Int 30(4):487–500. https://doi.org/10.1080/02508060508691893

Yilmaz KK, Adler RF, Tian Y, Hong Y, Pierce HF (2010) Evaluation of a satellite-based global flood monitoring system. Int J Remote Sens 31(14):3763–3782. https://doi.org/10.1080/01431161.2010.483489

Yin W, Hu L, Zhang M, Wang J, Han S-C (2018) Statistical downscaling of GRACE-Derived Groundwater Storage Using ET Data in the North China Plain. J Geophys Res Atmos 123(11):5973–5987. https://doi.org/10.1029/2017JD027468

Yin W, Zhang G, Han S-C, Yeo I-Y, Zhang M (2022) Improving the resolution of GRACE-based water storage estimates based on machine learning downscaling schemes. J Hydrol 613:128447. https://doi.org/10.1016/j.jhydrol.2022.128447

Yueh SH, Chaubell J (2012) Sea surface salinity and wind retrieval using combined passive and active l-band microwave observations. IEEE Trans Geosci Remote Sens 50(4):1022–1032. https://doi.org/10.1109/TGRS.2011.2165075

Zhang D, Zhang Q, Werner AD, Liu X (2016) GRACE-based hydrological drought evaluation of the Yangtze River Basin, China. J Hydrometeorology 17(3):811–828. https://doi.org/10.1175/JHM-D-15-0084.1

Zhang M, Lee H, Shum CK, Alsdorf D et al (2010) Application of retracked satellite altimetry for inland hydrologic studies. Int J Remote Sens 31(14):3913–3929. https://doi.org/10.1080/01431161.2010.483495

Chapter 5
Can the Wells Run Dry?

5.1 Introductory Remarks

Drought is a well known natural hazard that affects socio-economic systems and public health, as well as threaten the availability of freshwater. Drought can be slow, but when it intensifies and becomes severe over a substantially long period, its impacts on hydrological systems can be deleterious, leading to groundwater drought (Bierkens and Wada 2019; Thomas and Famiglietti 2019). Groundwater drought is a long-term decline in groundwater levels caused by reduced rainfall and recharge, low well yields, and long-term extraction (Van Lanen and Peters 2000). Groundwater drought (in other words, when groundwater wells run dry) can also occur when the water tables drop below their average levels, thus making it a class of drought different from agricultural, meteorological, and hydrological drought. The uniqueness of groundwater drought is that unlike other forms of drought (e.g., meteorological, hydrological, and agricultural), which are predominantly climate driven (Ndehedehe et al. 2021; Agutu et al. 2017; Sheffield and Wood 2007), it can be caused by either prolonged rainfall deficit or intensive human water extraction or both, making it complex and difficult to measure (Bloomfield and Marchant 2013; Mishra and Singh 2010). Additionally, groundwater drought impact negatively on water resources, including public water supply, water for industrial activities, and for agricultural irrigation. Because of the decline in water tables, groundwater drought is also responsible for constraining groundwater discharge to groundwater-dependent surface waters (e.g., springs, rivers, and streams) and ecosystems (e.g., vegetation and wetlands).

In the early stages of drought, water deficits are first noticed in surface waters and subsequently in other hydrological stores. At this stage, groundwater resources are relied upon for the provision of water supplies because during the early onset of drought, groundwater has the capability of sustaining surface flows through groundwater base flow (Hughes et al. 2012). However, groundwater appears to be susceptible to relatively prolonged or persistent drought events. This is because, when compared to surface water resources (e.g., rivers, lakes), a significantly longer time is required

C. Ndehedehe, *Hydro-Climatic Extremes in the Anthropocene*, Springer Climate,
https://doi.org/10.1007/978-3-031-37727-3_5

for groundwater storage to recover or be replenished as drought begins to terminate. Frequent monitoring of groundwater drought is thus encouraged and important, especially in semi-arid and arid regions where surface water resources appear to be apparently limited. This is due to the over-reliance of humans and aquatic ecosystem on groundwater resources. Given that they are usually driven by climatic factors (e.g., Kalu et al. 2022; Agutu et al. 2019) and regional hydro geologic conditions (Bloomfield et al. 2015), the identification and characterization of drought impacts on groundwater reservoirs are increasingly becoming complex.

Additionally, other region-specific factors, including changes in land-use land cover and human interventions (e.g., water diversion projects) contribute to the complexities that limits our understanding of groundwater processes (e.g., Ndehedehe 2022b). Water diversion projects, for instance, have slowed depletion rates in groundwater in China. Long et al. (2020) who reported this among other studies, argued that even though increased rainfall accounted for about 30% of groundwater recovery in Beijing (in north China Plain), water diversion to the region was a major driver (40%) of the rising water table or groundwater increase. This complexity in hydrological processes makes groundwater drought assessment crucial and past studies have attempted to design suitable monitoring indices for groundwater drought. For instance, Bloomfield et al. (2003) assessed groundwater drought based on minimum groundwater level and return period. They applied multi-linear regression with monthly rainfall observations to ascertain the minimum groundwater level. The proposed model was analyzed in the Jurassic limestone aquifer, Permo-Triassic sandstone, and Chalk aquifers, and yielded acceptable results. Another related study (Peters et al. 2005) attempted to monitor groundwater drought, leveraging a deviation from a specific level as an indicator. Ten 1000-years long periods of groundwater recharge and discharge via the nearest-neighborhood resampling were generated to facilitate their investigation of groundwater drought. This enabled the determination of groundwater drought episodes with various return periods, severity levels, durations, and extents. To efficiently manage groundwater resources and prevent groundwater drought, groundwater monitoring must be prioritized in large aquifers around the globe. In the sections that follow, global groundwater status as well as key challenges, constraints, and groundwater processes are detailed. This chapter also discusses the concept of groundwater drought as a new type of drought and how it can be measured. Sustainable limits and the importance of adaptation and resilience in the context of groundwater hydrology and climate change are also highlighted.

5.2 Groundwater Resources

Groundwater is a natural resource found in underground geologic units (Ndehedehe 2022b), better known as aquifers. The top 2 km of the Earth's crust comprises an estimated 22.6 million km^3 of groundwater, and out of this amount, about 0.1–5.0 × million km^3 were deposited within the last five decades (Gleeson et al. 2016). Several factors contribute to the highly heterogeneous nature of groundwater depth in many aquifers. Some of these include, hydrological conditions (e.g., rainfall

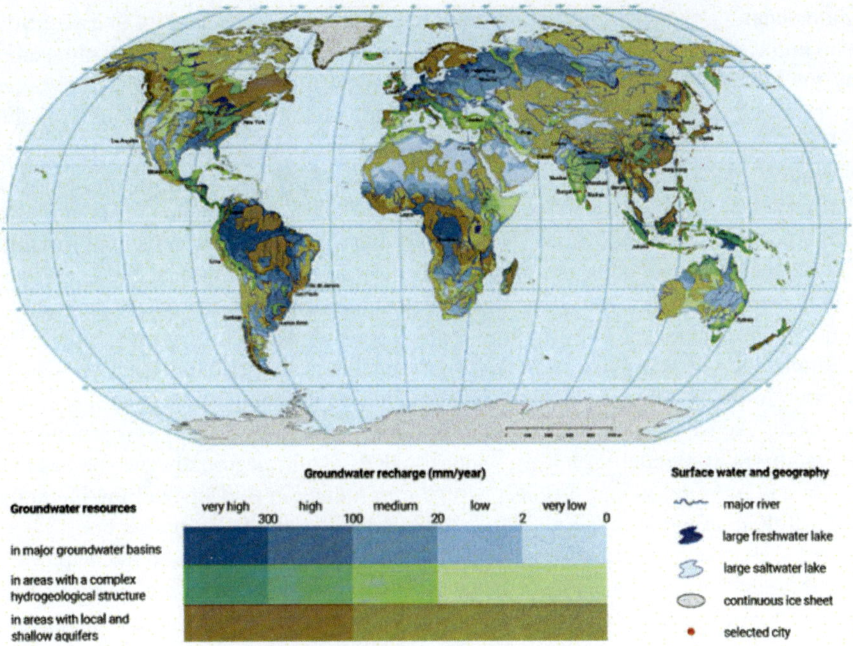

Fig. 5.1 Global groundwater recharge. *Source* https://www.unesco.org/reports/wwdr/2022/en

amount) leading to recharge, dynamic equilibrium between geology/geomorphology of the aquifers, regional to local-scale groundwater flow frameworks, and anthropogenic/human influence in terms of groundwater withdrawal all contribute to the varied global-scale distribution of groundwater depth. As captured in the global groundwater recharge map (Fig. 5.1), water availability depends largely on climatic factors, suggesting areas (e.g., Amazon and Congo basins) with large surface water and rainfall amount are critical hot-spots with the most freshwater. Normally, groundwater levels possess the capability of driving wetland extents, and patterns of vegetation distribution. It is believed that shallow groundwater resources influence an estimated 30% of global continental areas, which includes a contribution to an estimated 15% of surface water bodies. Even though precipitation replenishes groundwater through recharge, Fan et al. (2013) highlighted the contribution of shallow groundwater resources to seasonal or perennial rivers and much of the remainder impacting on plant root zone, through the capillary fringe.

Studies focusing on understanding groundwater resources from local to global scales, along with recharge and discharge volumes, and aquifer parameters, among several others are still emerging (e.g., Ndehedehe et al. 2023; Ndehedehe 2022b; Kalu et al. 2022; Ahmed et al. 2021; Long et al. 2020; Ferreira et al. 2020; Valencia et al. 2020). While developed regions, such as the United States and Europe, have comprehensive knowledge of groundwater abstraction rates and extent from the mid-last century, many other emerging economies have little or no data related to groundwater resources. This has limited the development of a global-scale picture and understanding of the resources and their regional-scale sustainability and security. The initial effort to develop a global-scale assessment of this resource was pioneered by the United Nations (UN) in the 1960s. However, it was not until recently that more robust and realistic estimates became available through interventions from the UN and Agricultural Organization's database (Fig. 5.1), United Nations Educational Scientific and Cultural Organization (UNESCO), which launched the World-wide Hydrological Mapping and Assessment Program's data, and detailed maps of major recharge areas around the globe (Döll and Fiedler 2008). With the advent of satellite remote sensing techniques, the application of advanced observational and numerical techniques for earth observations has gained significant attention due to their capabilities and accuracies in estimating hydrological units such as groundwater. For example, the GRACE (Gravity Recovery And Climate Experiment) mission (Chap. 6) and global modeling techniques have facilitated the monitoring of groundwater resources even at water management scales. The provision of continuous groundwater estimates from reliable platforms such as GRACE makes groundwater forecasting easier and more feasible, especially with the introduction of deep learning methods in hydrological modeling (Kalu et al. 2022; Rodell et al. 2018). In this book, deep learning approach to groundwater forecast and assessment is detailed in Chap. 8.

Groundwater recharge (Fig. 5.1) occurs when water infiltrates into the ground to replenish groundwater aquifers. The process, however, is heavily reliant on spatio-temporal variability in rainfall patterns as well as other hydrologic and land-based factors. Renewable groundwater resources are naturally recharged over time by rainfall (Bierkens and Wada 2019) and discharges occur by drainage to streams, springs, and by human extraction (Wada et al. 2010). When discharge rate exceeds recharge rate, groundwater storage decreases, and will be reflected as declining groundwater table. Due to the geological evolution of continental landmass, there are two major types of aquifer that characterize key groundwater systems that exist globally: (i) fractured, crystalline/ hard rock aquifers and (ii) sedimentary rock-sourced, porous fluvial aquifers (Fig. 5.2). However, the work undertaken by Fryar (2021) identified a third type, referred to as carbonate aquifers, which can either be porous or fractured aquifer systems. The latter play important roles in major parts of the world. Porous media aquifers, mostly dominated by alluvial aquifers, are typified by hundreds of meters of unconsolidated sediments that are weathered and accumulated in large sedimentary basins, thereby developing some of the most prolific groundwater aquifers in the world. These thick and unconsolidated alluvial aquifers (Fig. 5.2) appear conducive to recharge (Shamsudduha 2021). However, over extraction of groundwater in these aquifers result in a relatively slower decline of groundwater levels, this will lead to increased extraction as it warrants an unconsolidated extraction from greater depths. Notwithstanding, due to the large groundwater storage in many of these aquifers, they remain the most heavily depleted in several regions of the world (Ndehedehe 2022b; Mukherjee and et al. 2021).

In cratonic parts of the globe, hard rock aquifers are mostly prevalent. These types of aquifers are known to have limited groundwater storage in weathered soil and subsoil horizons of several meter thickness and are basal to unweathered and less-porous bedrock aquifers (Fig. 5.1). These capabilities make it possible for them to provide

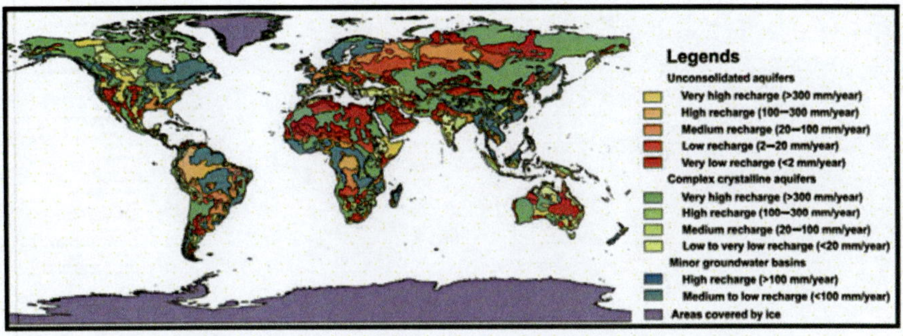

Fig. 5.2 Global aquifer map highlighting the characteristics and types of aquifer and their corresponding average annual recharge rates. *Source* Mukherjee and et al. (2021)

a complex hydrogeological formation. The bedrock is distinctively lacking of basic porosity but is, however, composed of variable quantities of secondary porosity fashioned by cracks, fissures, and fractures, where groundwater flows through desired routes. These hard rock aquifers produce groundwater characterized by lower productivity of extracting wells, poor quality, and less predictable spatiotemporal productivity and storage. This makes the availability of groundwater resource in hard rock regions discontinuous and limited, thus leading to unreliable estimations and predictions of groundwater availability for human use. Unsustainable groundwater extractions from these hard rock aquifers result in precipitous declining groundwater levels, thereby inflicting harsh limitations on groundwater extraction rates. Consequently, rapidly increasing groundwater needs and demands in many parts of these cratonic regions lead to significant groundwater storage decline, with extraction rates significantly surpassing recharge rates. Over time, this action causes a substantial increase in dry wells in the region.

It is true that efficient groundwater management should involve groundwater users, policy makers, water scientists, and stakeholders at different levels of society. But the management of groundwater could be difficult when large aquifers span through political boundaries and are shared across international borders. Sometimes, they are shared across state, national or even international borders, and when this is the case, such aquifers are termed transboundary aquifers (TBA) (Fig. 5.3). Presently, the International Shared Aquifer Resource Management initiative of the UNESCO-IHP has identified 592 TBAs. They include 318 in Europe, 129 in Asia and Oceania, 73 in America, and 72 in Africa. Despite the existence of several treaties to guide water sharing and management among member states, there is still a potential for international conflict in cases of exploitation and management of these groundwater resources by the riparian countries (Puri 2021). Some of these conflicts have been highlighted and key examples, include the Ganges–Brahmaputra–Meghna basins in South Asia, Guarani in South America, and the Nubian Sandstone in North Africa (Chakraborty et al. 2021). These treaties and guidelines also ensure that the world's largest aquifers in these TBA zones (Fig. 5.3), which may be potentially stressed

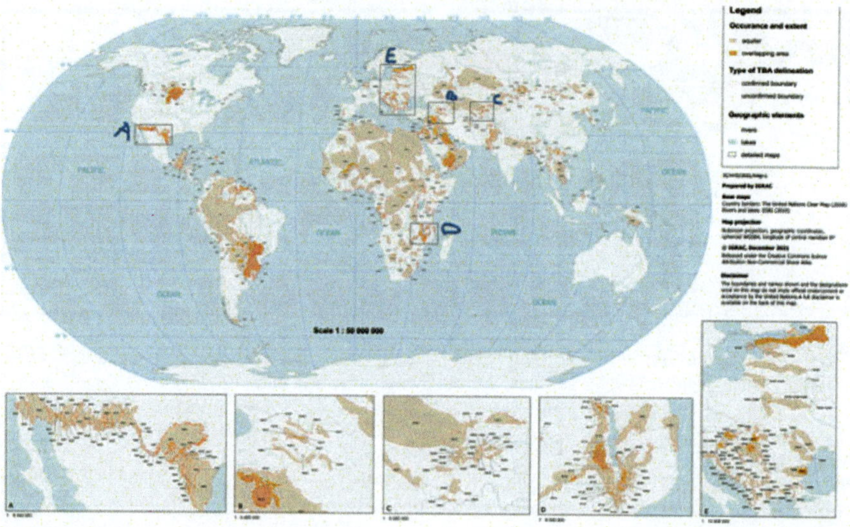

Fig. 5.3 Trans-boundary aquifer (TBA) map of the world (updated 2021) showing occurrence and extent as well as type of TBA delineation. The insert (polygons on map) labeled A to E are zoomed areas shown in the sub-panels below. *Source* https://unesdoc.unesco.org/ark:/48223/pf0000380193

are monitored and managed. Given the reliance on groundwater, large aquifers are experiencing stress due to water extraction and climate change, the TBAs need more attention to ensure sustainability and global water security.

5.2.1 *Global Groundwater Status*

The critical role of groundwater in sustaining freshwater-dependent ecosystems including human well-being is widely recognized. However, many of the world's largest aquifers (Fig. 5.2) are experiencing excessive groundwater abstractions for irrigation purposes. This has resulted in declines of groundwater tables and a reduction of groundwater outflow to surface water bodies (Gleeson and Richter 2018). With the increasing rate of climate change impacts, global groundwater demand is expected to rise further resulting from the increasing temporal variability of surface water flows (Taylor et al. 2013). Regions of the world where groundwater use is sustainable and critical thresholds of groundwater risk have not been transgressed appear to reduce by the day. Three key problems associated with global groundwater use in the Anthropocene include

- depletion due to overdraft,
- salinization and waterlogging due to inadequate drainage operations, and
- pollution due to industrial, agricultural, and other human activities.

In regions facing high population density and insufficient surface water, as well as dynamic tube-well-irrigated agriculture, several impacts of groundwater overdevelopment are becoming more evident. One of the most common symptoms of these consequences is the increasing decline in groundwater tables. For instance, in North China's Henan province, over 2 million hectares comprising of 52% of land are serviced by tube wells. As reported in Rajan et al. (2020), water table observations from 1975 to 1987 showed that 358 observation wells spanning 75,000 km^2 experienced declines of 0.75 to 3.68 meters during this period. The study undertaken by the International Water Management Institute reported that the Fuyang river basin of North China witnessed a reduction in surface water supplies to agriculture over a 20 year period for meeting industrial needs. This led farmers to resort to groundwater irrigation, causing an increase in the number of tube wells in the basin to about 91,000. The report showed a decline in the water table from 8 to 50 meters from 1967 to 2000 (Shah et al. 2000). Therefore, the existing aquifers over the Fuyang basin are arguably under severe stress as farmers are depleting the lower aquifers while industries are polluting the upper aquifers.

Since the 1980s, groundwater over-exploitation in North China has increased from 56 to 164 regions, resulting in a total area of 180,000 km^2 where this over-exploitation has been found. The review undertaken by Jiang (2009) on China's water scarcity articulated this widespread over-exploitation of groundwater and also reported km^2 that 70% of the region has experienced extensive groundwater use or abstraction. Long-term simulated groundwater depletion between 1960 and 2008 in the North China Plain was averaged at approximately 4 km^3/yr and Cumulative groundwater depletions in the central plain, piedmont district, and coastal plain, were 103 km^3, 50 km^3, and 5 km^3 (Cao et al. 2013). With the nature of accelerated use in the region's groundwater, the water table continues to fall and with about 30 m drop in the 1990s as opposed to about 4 m fall in the 1950s (Jiang 2009). Other regions in Asia nonetheless, appear to be facing similar problems or perhaps much greater groundwater problems (over-exploitation) compared to China.

For example, the arid regions of West and South Asia have reported extreme cases and a good example is Yemen. A World Bank memorandum on water sustainability and management in Yemen reported that groundwater extraction is a critical driver in groundwater change, contributing to threat in water security and wellbeing of the Yemeni people. Ward et al. (2007) reported a groundwater extraction rate of over 400% in the highland plains of Yemen. This aggressive extraction and water use makes Yemen the only country in the region where groundwater extraction exceeds recharge throughout the entire country.

Moreover, in North America, Mexico's aquifer's accounts for one of the most overdeveloped aquifers. In a study undertaken by Wester (1999), the water tables in ten understudied aquifers in the Mexican region are declining at a mean annual rate of 1.79 to 3.3 m/yr in recent times. Generally, satellite observations of groundwater have identified that North America (e.g., California in the US and Mexico cities) has critical hotspots of extensive groundwater depletion that have caused land subsidence rates (e.g., Castellazzi et al. 2021; Ojha et al. 2018, 2020). Despite the implementation of the groundwater management plan, California's Central Valley, continues to face excessive groundwater extraction, resulting in falling water tables, drying wells, and land subsidence, and recently (2019–2021), the losses in groundwater seem to have accelerated (8.57 km^3/yr) in the region due to mega-droughts (see, Liu et al. 2022).

In South Asia, the story is the same. In the western, northwestern, and peninsular Pakistan and India, over a million irrigation wells and counting have been identified in the region. Since the 1960s, the number of mechanized and tube wells in India has reached 19 million; there is also a substantial rise in irrigation wells in Pakistan Punjab and Bangladesh with an increase from 4% in 1972 to 70% in 1999 in the extent of irrigated areas by groundwater wells (Shah 2007). This expansion in groundwater wells and irrigated landscapes resulted in a significant increase in groundwater abstraction, causing annual groundwater withdrawals to exceed annual groundwater recharge in most parts of the region (Rajan et al. 2020). Groundwater withdrawal in India is one of the highest in the world, and the consequences are obvious, evidenced by groundwater drought and threat to food security for hundreds of millions of people.[1] In the Western Rajasthan and Haryana (Punjabs areas of India), the main groundwater problem is salinity.

In North Gujarat and Southern Rajasthan, the issue is fluoride pollution of groundwater, while in southern India, the problem is the diminishing well yields and intensifying groundwater pumping costs, resulting from competitive well deepening exercises. In parts of Bengal and western Bangladesh, arsenic contamination has been reported as the cause of groundwater stress (Khan 1994). In many coastal regions, one of the most significant effects of excessive groundwater pumping for irrigation is the saline ingress (salt water intrusion) into the coastal aquifers. This could create a significant challenge and affect the region's capability to provide for its growing population. Overall, groundwater decline in South Asia is a regional problem and in eastern Pakistan, northern India, and Bangladesh, about 54 cubic kilometers of groundwater loss every year was reported in intensively irrigated areas where about 600 million people are living (https://www.science.org/content/article/indias-groundwater-disappearing-alarming-rate).

One of the major challenges facing global groundwater in the world is sustainable management. As the problems and consequences of excessive groundwater depletion have been reported in different parts of the world (e.g., Ndehedehe 2022a; Taylor et al. 2013), it is thus important to design mitigation measures that can slow down or even reverse these actions to avoid further problems in future. For example, water harvesting is one of such measures. To this end, over 800,000 big and small dams have been constructed worldwide aimed at capturing and storing rain water. However, reservoirs meant for this rain water harvesting are unable to capture one-fifth of total precipitation at any given time because most of the water are absorbed into the sea due to runoff in the form of "rejected recharge" (Shah 2007). If a small percentage of this total runoff can be stored underground by reducing the speed of the runoff and creating time for groundwater recharge, this can result in a significant enhancement and replenishment of groundwater supplies. This action, nonetheless, needs proper aquifer management techniques and strategies where planned drawdown of the water

[1] (https://edition.cnn.com/2021/02/24/asia/india-groundwater-study-intl-hnk-scn/index.html).

table in dry months is a key factor in the strategy aimed at enhancing recharge from wet months and irrigation flows. This aquifer management technique may require time to be adopted in several emerging economies. However, some of the following methods have been tested over time to help with groundwater sustainability, and they include the following:

- recharge with imported surface water (Neely et al. 2021; Shah et al. 2000),
- recharge with rainwater (Rajasekhar 2020; Shah et al. 2000)
- vegetative treatment of the catchment (e.g., Zhao et al. 2012; Shah et al. 2000),
- domestic rainwater harvesting (marc Mwenge Kahinda et al. 2007; Shah et al. 2000).

5.2.2 Challenges and Constraints

Groundwater extraction is similar to mineral resource mining. With increased extraction rate, the relative quality and richness of the resource tends to diminish, especially in ares with fossil or non-renewable groundwater resources. For groundwater resources, the drop of the hydraulic head results in an increase in well depths and pumping lifts. This often leads to an increase in production costs, and in some cases, the deterioration of water quality as less desirable deep groundwater is produced. These among other issues contribute to challenges of groundwater over-exploitation and mismanagement. Groundwater over exploitation and contamination have become global problems (Gregory et al. 2013). Their impact on terrestrial ecosystem have increased due to climate change, which is an established critical driver of changes in groundwater resources (Ndehedehe 2022a; Bierkens and Wada 2019; Ojha et al. 2018; Kløve et al. 2014). But since we live in the Anthropocene, the composite effect of growing human population, urbanization, increased water use, land use change, unsophisticated pumping and drilling technology, industrialization, institutional changes, expansion of irrigated agriculture, more rigorous water quality standards, and the effect of climate change now contribute to the widespread, and often unregulated use of groundwater throughout the world.

Aquifers are regarded as the fundamental long-term reservoirs that possess the capacity to store freshwater resources. They are free from evaporation, regionally ubiquitous and are usually free from pathogens under normal conditions. For these and other reasons, aquifers are classified as very vital human and environmental resource. Unfortunately, groundwater over exploitation and contamination challenges are becoming so popular. This Chapter identifies six important challenges and constraints spanning this shared human-environmental resource based on the work of Gorelick and Zheng (2015). The six major dimensions highlight the fundamental interactions of groundwater resources with human beings and the natural environment. These six factors contributing to the challenge of over exploitation and contamination of groundwater resources are discussed in relation to historic drivers and key applications, while making suggestions for improved aquifer needs in each context.

- **Ecosystems:**
  The most significant contributor to the disturbance in the world's largest river systems have been traced to humans (Gorelick and Zheng 2015). As of 2005, over fifty percent of the world's largest river systems were impacted by dams, regardless of the presence of surface reservoirs and their benefits to the functioning of terrestrial and aquatic ecosystem (Nilsson et al. 2005). Due to the extent of managed surface waters around the world, the exploitation of groundwater resources is on a steady increase and the influence of groundwater use on rivers have intensified. Groundwater extraction from aquifers affects groundwater-dependent ecosystems (Fig. 5.5), thereby interrupting the balance between human and environmental needs. One major "environment-based" misconception in groundwater management is that the pre-development magnitude of aquifer recharge is presumed to be equivalent to the sustainable rate at which the aquifer can be explored. Popularly referred to as the "water budget myth" (Louwyck et al. 2023), this misconception results in a miscalculation that assumes sustainability equals pre-development of natural recharge, thus ignoring the fact that before pumping, this virgin recharge was already discharging to rivers as base flows, to estuaries, lakes, and the ocean as bottom discharge, and to naturally occurring springs (Louwyck et al. 2023). Presumably, this amount of groundwater recharge for human use via pumping can result in double counting the available water that can be safely extracted for use on in a long-term basis.
  Pumping can affect surface waters by minimizing groundwater discharge to rivers as base flows, thus constraining surface waters. When pumping begins water is basically taken from aquifer storage and as the pumping progresses, more amounts of water can come from the boundaries. The existing lag between the start of the pumping and the minimization in base flow from surface water can take several decades, while the achievement of a new equilibrium condition may take centuries (Gorelick and Zheng 2015). Factors affecting this lag time include magnitude of pumping, aquifer geometry, hydraulic properties, and other hydrologic quantities. Disregarding the ultimate source of pumped water can be harmful to groundwater dependent ecosystem and to downstream water users.
- **Hazards:**
  Engineering contributions to the quantitative assessment of hydrogeology, especially in response to the recognition of environmental hazards has grown immensely in recent times. A number of hazards associated with overexploitation of aquifers include land subsidence, groundwater mining, seawater intrusion, triggered earthquakes, etc. Groundwater mining has led to significant aquifer depletions on large scales such that its regional effects can be measured from space with the use of gravitational anomalies and radar techniques (e.g., GRACE, InSAR—Inteferometric Synthetic Aperture Radar discussed in Chap. 6). Major aquifers around the world are experiencing unprecedented exploitations with the net global groundwater depletion rate doubling since 1990 (Konikow 2011). This has also led to the

decline of regional hydraulic heads over wells with varying depths (Cao et al. 2013), increased cost of pumping and entrainment of deep saline water.

- **Human health**:
The threat of groundwater contamination became more conspicuous during the past five decades. With nuclear energy generated since the 1950s in the U.S., accumulating spend fuel led to an apprehension, culminating in a study by the National Academy of Sciences in the 1970's, recommending a more permanent storage for high-level nuclear waste (https://nap.nationalacademies.org/read/20484/chapter/1). Simultaneously, groundwater contamination at the Love canal in New York in the 1970's signified the inception of an era that recognized the human health consequences of groundwater contamination. The superfund program led to a study focused on the management and prediction model aimed at designing a field of contaminant hydrogeology with a significant emphasis on aquifer remediation.
- **Food security:**
Globally, over two-third of all groundwater extracted from aquifers are channeled into irrigation (Gorelick and Zheng 2015), and about half of the world's irrigated crops are heavily reliant on groundwater. In terms of land mass, out of the 301 million hectares set out for irrigation globally, over 35% rely on groundwater (Siebert et al. 2010). Basically, farmers have been efficient in their utilization of groundwater resources to an extent, especially in terms of supply. However, this is not the case when it comes to the contamination of groundwater resource and how it is managed. This has resulted in the development of simulation-optimization modeling procedure, which is aimed at curtailing this.
- **Groundwater for energy and resources:**
In the United States, groundwater abstraction accounted for less than 1% of total withdrawals for thermoelectric power in 2010 (Maupin et al. 2014). However, on a more localized scale, groundwater and energy have significant inter-connection as the exploitation, delivery, and treatment of water resources require energy, whereas almost all sources of energy need water for key aspects relating to production, such as cooling, extraction or even transportation. This strong link, which is basically referred to as the "water-energy nexus" has become a primary focus in the hydrological sciences and has been highlighted in some literatures (e.g., Bartos and Chester 2018).
- **Conflict:**
Several types of conflicts can arise from groundwater use from multiple regions, especially in trans-boundary aquifers (Fig. 5.3). Therefore, one of the biggest challenges for regional groundwater management is in resolving transboundary groundwater conflicts. The highlight of transboundary problems spans through intraregional, interstate, or international scales. The underling physical challenge, however, is the same in each case. That is, over exploitation in the part of an aquifer on one side of a political boundary tends to affect hydraulic heads and consequently pumping costs throughout the boundary regions. Since these groundwater systems in transboundary aquifers may be interconnected, another key issue is that the pumping on one end tends to draw clean or contaminated water

across such boundary, thus affecting water quality for the other region. Some past literatures have discussed at large the issues surrounding international transboundary groundwater aquifers and this has been receiving increasing attention over the years (Jarvis et al. 2005). Figure 5.3 depicts a comprehensive atlas showing transboundary aquifers which are known to form the basis for economic, hydrogeologic, legal, institutional, and environmental conflicts between several boundaries (Puri 2021).

5.3 Groundwater Drought: A New Type of Drought?

Droughts influence all the components of global water cycle, including groundwater. It is primarily caused by lack or long-term decline in precipitation. The nature of its propagation through the hydrologic cycle and other human factors could exacerbate impacts on hydrological systems (Ndehedehe et al. 2021; Ndehedehe 2022c). Even though drought phenomenon, concept and impact were discussed in the pioneering work of Wilhite and Glantz (1985) based on three classes of droughts identified, there are about six classification of droughts, which have been detailed and extensively discussed further by Ndehedehe (2022a). This include, meteorological, agricultural, hydrological, socio-economic, anthropogenic, and groundwater drought. Of these classes, groundwater drought is a unique type of drought and may not be well known. Groundwater drought is generally perceived as an extension of hydrological drought when below-normal groundwater levels are experienced, or type of hydrological drought characterized by prolonged low groundwater levels, minimized baseflow, and decreased flows to wetlands, springs, and groundwater-fed rivers (see, e.g., Uddameri et al. 2019; Sanginabadi et al. 2019; Liu et al. 2016; Bloomfield et al. 2015; Van Loon 2015). However, the pioneering work of Van Lanen and Peters (2000) on the subject provided a definition that established the distinctiveness of groundwater drought as a unique type of drought. They defined it as "*a lack of groundwater recharge, or a lack of groundwater expressed in terms of storages or groundwater heads in a certain area and over a particular period of time*". Ultimately, the groundwater drought concept acknowledges the role of climate (drought) and human action (extraction, abstraction or withdrawal) as critical drivers, and this type of drought is evidenced in declining groundwater level as well as reduced groundwater discharge and recharge over a significantly long period of time (Uddameri et al. 2019; Sanginabadi et al. 2019; Van Lanen and Peters 2000). When drought is prolonged, surface water is lost and groundwater recharge plummets and under such conditions human groundwater abstractions increase, resulting in reduced groundwater storage and discharges to streams and springs (Ndehedehe 2022b; Ndehedehe et al. 2021; Uddameri et al. 2019; Van Lanen and Peters 2000). Moreover, falling groundwater levels decrease transmissivity in unconfined aquifers as well as well yields (Uddameri et al. 2019; Van Lanen and Peters 2000). This is groundwater drought!

As a relatively new concept that is still not widely studied, the definition of groundwater drought could be varied but the idea remains the same—groundwater resources

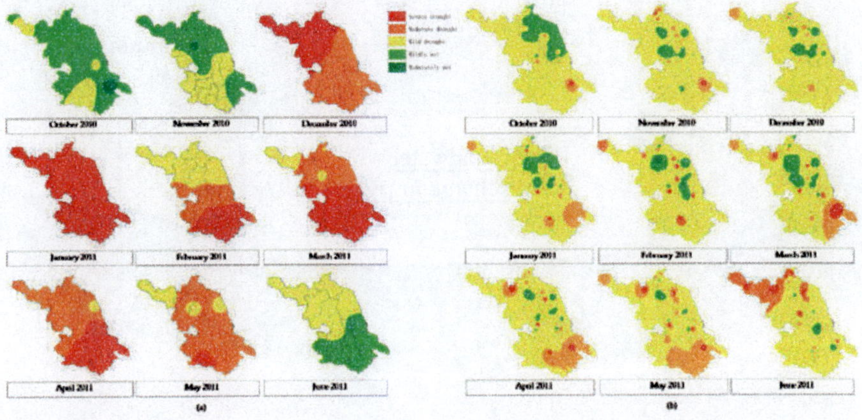

Fig. 5.4 Meteorological drought versus groundwater drought. **a** Spatial distribution of **a** Meteorological drought and **b** groundwater drought in 2010 and 2011 period in the Jiangsu Province of China. Both droughts were calculated from in situ or gauged data. *Source* Liu et al. (2016)

experience drought or can be restricted due to drought. However, groundwater is not only driven by droughts, as prolonged water withdrawals can trigger low well yields and considerable decline. For this reason, groundwater drought can also be simply defined as when the wells run dry. In other words, it is the result of decrease in groundwater recharge or when groundwater wells run dry (water table reduces) due to prolonged drought and human water abstraction (Ndehedehe 2022a). This understanding captures prolonged falling groundwater levels as the aftermath of limited rainfall and recharge, low well yields or long-term human water abstraction. As with other forms of drought (meteorological, agricultural, hydrological), groundwater drought also vary spatially. However, as illustrated in Fig. 5.4a, b, such variability are in sharp contrast to that of meteorological drought. For instance, in January 2011, when the entire Jiangsu Province was under severe meteorological drought (Fig. 5.4a), much of the Province experienced mild groundwater drought (Fig. 5.4b). However, when the severe meteorological drought persevered in March of the same year in the southern region of the Province, it escalated into a notable severe groundwater drought in the south east location of the Province (Fig. 5.4a, b). This further highlights the intrinsic response of groundwater to climatic conditions (Fig. 5.4b).

Groundwater drought also affects discharge to terrestrial, aquatic, and groundwater dependent ecosystems, including springs, streams, and vegetation (Fig. 5.5). When groundwater wells run dry or are substantially reduced due to over excessive pumping, it lowers the water table, thus reducing discharge to streams to support aquatic habitats (Fig. 5.5). While this leads to loss of aquatic species, the reduced water table can result in the demise of groundwater-dependent vegetation ecosystems (Fig. 5.5). Climate change and intensive human water extraction can lead to reduced water table and lower groundwater discharge, resulting in a drying well. The

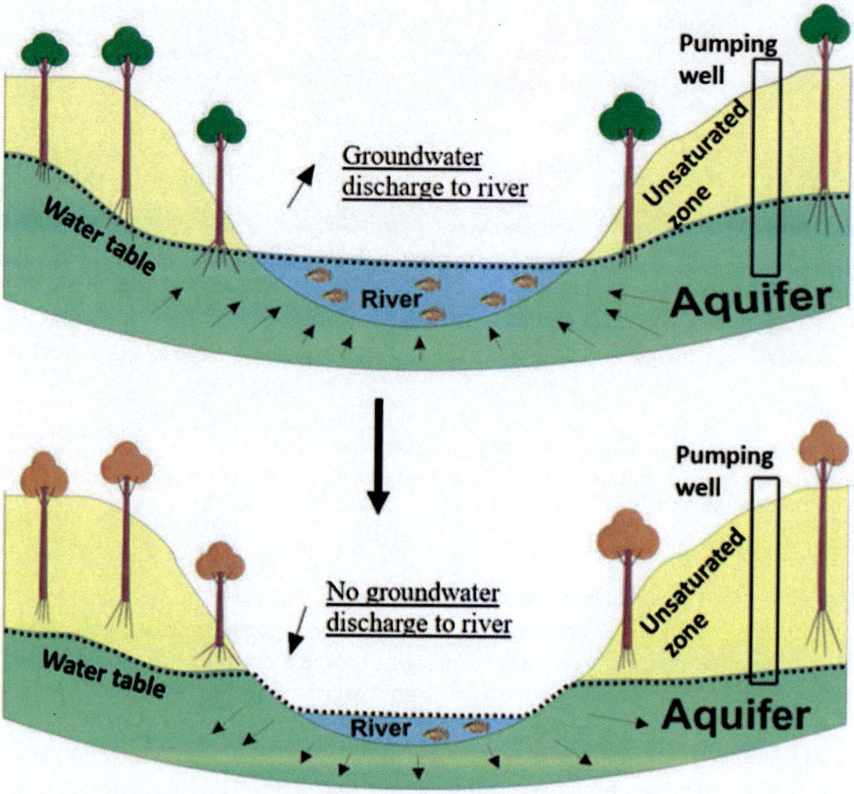

Fig. 5.5 Effects of groundwater pumping on groundwater-dependent ecosystems. Climate change and intensive human water extraction can lead to reduced water table, lower groundwater discharge and ultimately, groundwater drought or drying well

difficulty in monitoring groundwater due to lack of data in areas with the most need makes it challenging to predict. This is further complicated by the multiple dimensions of drivers of groundwater drought. For instance, as with soil moisture and hydrological drought, groundwater drought stems from meteorological drought, which gradually metamorphose to hydrological and subsequently groundwater drought when the former becomes prolonged (e.g., Ndehedehe 2022a). When integration of precipitation deficits and high latent heat transfer and evapotranspiration rates propagates through the hydrological cycle, the rapid loss of soil moisture and surface water will negatively impact groundwater recharge, which in turn may result in groundwater drought.

Another factor that could cause groundwater drought is extensive human groundwater abstraction to support agriculture through irrigation. Both factors can escalate the impacts of groundwater drought as is now the case in India. The attempt to analyze the propagation of meteorological droughts through the hydrologic systems to achieve a better understanding of the progression of groundwater drought is available in past literatures. For example, an earlier study undertaken by Peters et al. (2003) tracked drought in recharge through a linear reservoir by examining the propagation of a drought from groundwater recharge to discharge and the effect of aquifer characteristics on the propagation. Results from their study showed the progression of drought on many and short drought episodes in the recharge to fewer and longer drought episodes in the groundwater discharge.

Moreover, Peters et al. (2006) also investigated the spatial distribution and propagation of groundwater drought using hydraulic heads, simulated recharge, and groundwater discharge for a catchment in England. Results from their study showed the existence of several short term droughts in recharge and discharge patterns, with hydraulic head experiencing fewer and more severe droughts. It is important to note that as earlier mentioned, groundwater drought is different from hydrological and agricultural drought and research in this area is gaining significant attention (see, e.g., Huang et al. 2021; Pathak and Dodamani 2021; Seo and Lee 2019; Thomas et al. 2017; Liu et al. 2016; Bloomfield et al. 2015). Apart from natural factors like strong changes in climatic conditions, anthropogenic influence like pumping (Fig. 5.5) also impact groundwater levels, and where significant over exploitation of this freshwater resource is found, it may be difficult for groundwater to return to normal levels, thus potentially culminating in groundwater drought. In the context of drought, the synergy between climate change and human actions that contribute to water-deficit conditions is usually described as anthropogenic drought (Ndehedehe 2022a; AghaKouchak et al. 2021). This is a new type of drought, which describes water bankruptcy, attributing the latter to extensive human water abstraction from aquifers and climate change. With this in mind, groundwater drought is actually the forerunner of anthropogenic drought.

5.3.1 *Measuring Groundwater Drought*

There is a high tendency of groundwater drought exhibiting complex characteristics under the combined influence of natural and anthropogenic factors. Also, the lack of actively direct observational data on groundwater resources makes it difficult to quantitatively assess and measure groundwater drought. Regardless, since the issue of groundwater drought has received considerable attention from hydrologists, meteorologists, and geologists, it has become an emerging area of hydrological research, having witnessed a number of studies in the past decade. One of the most reliable tools for monitoring and forecasting drought conditions using a multi-analysis approach is the groundwater resource index developed by Mendicino et al. (2008). The use of groundwater drought index based on groundwater level or climate data (rainfall and evapotranspiration) and satellite data is emerging. While groundwater drought index is based on monitoring bores, such index is usually developed after the similitude of standardized precipitation index where long term accumulation scales of 12-month are presumed to capture drought impacts on groundwater variations (e.g., Ndehedehe et al. 2016; Bloomfield and Marchant 2013; McKee et al. 1995).

5.3.1.1 In situ and Data Assimilation

The illustration in Fig. 5.6 is the averaged time series of standardized groundwater index for each hydrological zones of the Lincolnshire, UK. Bloomfield et al. (2015), who evaluated this index (called it standardized groundwater level index) with the standardized precipitation index found a significantly good and strong (0.84) correlation between both indicators. Similarly, Li and Rodell (2015) utilized the groundwater drought index (GDI) in the assessment of groundwater drought over the central and northeastern United States. However, unlike Fig. 5.6, the groundwater drought indices in Li and Rodell (2015) were based on model-derived groundwater underpinned by GRACE data assimilation. The evaluation of model based groundwater drought indices with standardized precipitation index on a 12-month cumulation scale showed both indicators were strongly correlated in the US. The capability to monitor groundwater drought based on satellite assimilation of GRACE with outputs from land surface model provides a better opportunity for applications in water management, especially in monitoring water supplies and allocation. However, these are case specific examples and the performance of these groundwater indicators could be different in other locations (Fig. 5.6).

5.3.1.2 Satellite Remote Sensing Methods

The use of remote sensing methods like GRACE to assess groundwater status to support drought prediction is well established (e.g., Kalu et al. 2022; Ndehedehe 2022b; Agutu et al. 2019). The integration of GRACE with hydrological modeling to improve the assessment of groundwater drought is now increasing. For example, Thomas et al. (2017) pioneered the direct application of GRACE satellite by developing a groundwater drought indicator to measure groundwater drought over the central valley of California. They introduced a groundwater drought index based on GRACE observations. This index was based on normalized deviations of GRACE-derived groundwater storage and was called GRACE Groundwater Drought Index (GGDI). Even though nomenclatures might be different for these groundwater drought indices, the GGDI captured the properties of groundwater drought caused by the composite influence of climate change and human activities. As further reported for the California Central Valley, the considerable association of GGDI with in situ groundwater indices makes it an important tool in groundwater drought monitoring, especially to assess its characteristics and drivers (Thomas et al. 2017).

Additionally, the integration of GRACE with other remote sensing datasets to develop a machine learning framework (Fig. 5.7) to measure groundwater drought provided new opportunities for the use of satellite methods in groundwater drought assessment and monitoring. A case study by Seo and Lee (2019) over South Korea

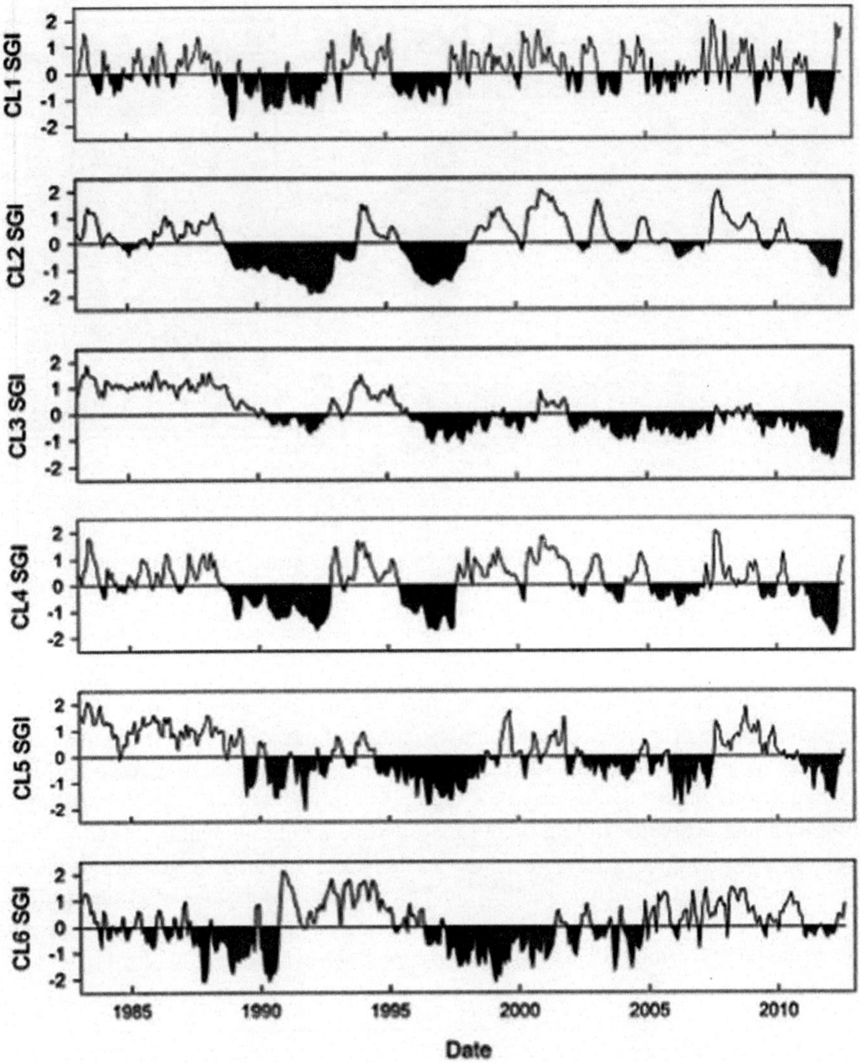

Fig. 5.6 Standardized groundwater index (SGI) based on monitoring bores in different areas or clusters, represented as CL on the y-axis of each plot. *Source* Bloomfield et al. (2015)

implemented an artificial neural network framework to quantify groundwater drought using GRACE, satellite rainfall, and output from models. Again, their approach demonstrated the feasibility of monitoring characteristics of groundwater drought. Machine learning framework that combines big satellite data with in situ observations (e.g., rainfall, temperature, etc.) to identify areas where groundwater resources are affected by human water extraction or drought (Fig. 5.7) can facilitate operational decisions to aid prioritization of management policies that limits threats to water

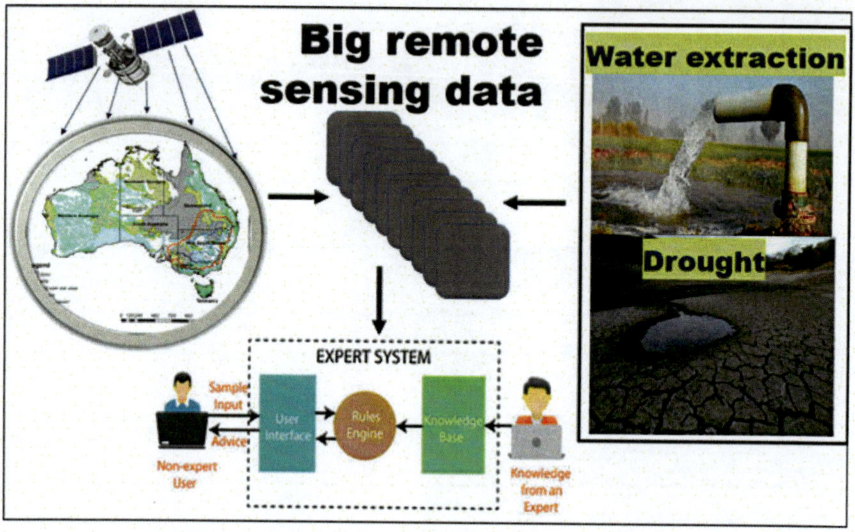

Fig. 5.7 Framework for groundwater drought assessment. The capability to quantify water use from space will benefit stakeholders in their operational decisions to aid prioritization of management policies that limits threats to water security and climate-induced environmental risks

security. This framework can provide the flexibility to conceptualize different representations of the problem and better understanding of the complexity of groundwater systems. As groundwater response to anthropogenic footprints and impacts of climate change could sometimes be complex, the ability to identify such impacts requires a multivariate problem-solving model that considers environmental, climatic, land surface, and geological conditions, among others. For instance, irrigation water use, aquifer dewatering, climate variability, and the geologic properties of some aquifers could create complexities in understanding groundwater systems such that sparsely distributed monitoring bores are unable to capture (Ndehedehe et al. 2021).

5.3.1.3 Groundwater Drought Metric

Drought indicators for monitoring groundwater drought are still developing and also an interesting area of research. As mentioned in the preceding section, these indicators can either be based on in situ groundwater data (Groundwater Drought Index) or the use of satellite methods like GRACE (GRACE Groundwater Drought Index). GRACE-derived groundwater drought indices are emerging as important indicators to monitor groundwater drought. Apart from identifying the trends and spatio-temporal distributions of drought, GRACE-based groundwater drought index is another reliable index to help detect the footprints of climate teleconnection patterns like the El Niño-Southern Oscillations demonstrated in the North China Plain North China plain (Wang et al. 2020). Moreover, the standardized water-level index

(SWI) was originally developed for hydrological drought assessment with the aid of groundwater level recharge deficit. However, the model appeared to be efficient in groundwater drought assessment. An example is the use of SWI as a groundwater drought index underpinned by satellite data to assess groundwater drought in Peddavagu watershed in the sub-basin of the Krishna river system over the Indian peninsula (Nagarajan and Sreedhar 2015).

Furthermore, Bloomfield et al. (2015) utilized monthly groundwater estimates to design the standardized groundwater level index (SGI), which is specifically applied in the assessment of groundwater drought. Results from their correlation analysis between SGI and SPI on multiple time scales formed the basis for adopting SGI in groundwater drought assessment. The comparison between SGI and SPI is still ongoing because of the varied response of groundwater drought to meteorological drought in some aquifers and the considerably non-linearity in groundwater systems. In this sense, a bivariate frequency analysis-based groundwater drought was compared with SPI in a tropical river basin in India by Pathak and Dodamani (2021). They implemented a cluster analysis of the long term monthly groundwater over the Gataprabha river basin in India. Results from their analysis showed that it was possible to classify observation wells so as to perform the Mann-Kendall test to analyze seasonal and annual groundwater level trends in the evaluation of the areal groundwater drought.

Furthermore, Liu et al. (2016) used monthly groundwater level estimates from 40 active wells over the Jiangus province in China (1989 – 2012), and applied the SGI to carry out a spatio-temporal groundwater drought analysis in the province. Their approach differed from that of Bloomfield et al. (2015) because the former considered both parametric and non-parametric cases. The monthly time series of groundwater level estimates were fitted with the gamma function; with little or no thought as to if the gamma function could fit with the monthly time series data of groundwater level estimates in other regions, which could affect the final assessment of groundwater level drought. This became the basis of the study undertaken by Lorenzo-Lacruz et al. (2017) who transformed the SGI based on the standardized runoff index and proposed an index according to various probability distributions of monthly groundwater level estimates. This approach ensures is adaptability of the estimated SGI to various water and climate-related conditions and has the potential of reflecting a more accurate depiction of groundwater level drought. Several other drought monitoring indices have been developed and implemented over the past decades. See Sect. 5 of Mishra and Singh (2010) for more details.

5.3.1.4 Multivariate Techniques in Groundwater Drought Characterization

Monitoring groundwater drought and identifying its characteristics is an important aspect of drought risk assessment and preparation for future droughts. The quantitative assessments of groundwater drought events, is thus critical to help characterize groundwater drought, similar to other forms of droughts (e.g., meteorological, soil moisture, and hydrological). Given the correlation and dependence between some characteristic variables, such as drought intensity, drought duration, and drought extent, traditional analysis using single variable (e.g., rainfall) may not be comprehensive enough to undertake a detailed assessment of groundwater drought episodes, as it can result in providing inaccurate and insufficient drought risk assessments. However, the advent of copula function has shown great promise in the area of drought characterization (Wu et al. 2018). The copula function possesses the capabilities to fit into the distribution functions of numerous drought-characteristic variables and has the potential of efficiently delineating existing correlation between the variables. Studies have shown that the copula function is a robust tool in the measurement and analysis of groundwater drought (e.g., Pathak and Dodamani 2021).

5.3.2 Impacts of Extraction and Drought

The impacts of excessive groundwater extraction and drought are all-encompassing and have critical consequences for society and ecosystems. For instance, industry and energy production, public water supply, terrestrial and aquatic ecosystems, agriculture and livestock farming, and recreation and tourism could be at risk of drought impacts (Stahl et al. 2016). Specifically, a reduction in groundwater levels may result in policies that limits groundwater abstractions for agricultural, industry as well as costly restrictions to public water supply. This can contribute to a huge economic stress on human population. An instance of this is the last major drought in the UK, which occurred between 2010 and 2012. This drought led to seven water companies in the east and south England imposing temporary bans on an estimated 20 million people. Another example was the drought in the farming sector in the spring of 2012 and the 2015 drought in Europe, resulting in the drying up of wells and boreholes (groundwater drought), which made the provision of drinking water for cattle extremely difficult (Van Lanen et al. 2016).

Droughts generate a convoluted web of impacts that extends to several sectors, including the economy. The negative impact of drought can go beyond the area experiencing it. Since half of the earth's terrestrial surfaces are susceptible to them, they can be classified as a widespread phenomenon (Mishra and Singh 2010). Of all the natural hazards that occurred in the 20th century, droughts were reported to have the most significant detrimental impacts on ecosystem (Obasi 1994). In contemporary times, a number of large scale intensive drought episodes have been observed in several regions of the world including Australia, Europe, Asia, Africa, etc. These drought events affected substantially large areas, and led to increased social and economic costs in most of these regions. Drought impact on these continents is discussed further in what follows.

- **Asia:**
  Due to increasing water stress in Asia, declines in the production of maize, rice, and wheat has been reported in the past few decades based on an Intergovernmental Panel on Climate Change (IPCC) study. Events such as increasing latent heat transfer, evapotranspiration rates and frequency of El-Nino events coupled with reduction in the number of rainy days are factors catalyzing the effect of drought on the region. During the drought of 1999 to 2000, over 60 million people living in central and southwest Asia were affected, and this was regarded as one of the largest drought impact on human population from a global perspective (IRI 2001). Countries that were most affected by this drought include Turkmenistan, Uzbekistan, Tajikistan, Western Pakistan, and Afghanistan and large-scale climate teleconnection patterns like the ENSO (La Niña) and variations from the Indian and Pacific Oceans contributed to the severity of these drought events. In another case, Xue et al. (2012) reported a severe economic and societal loss to northern China following the frequent severe drought episodes recorded in 1997, 1999 to 2002 in the region. In 2000, an estimation of over 40 million hectares of agricultural land areas were affected by drought in the region. Due to water scarcity and less precipitation in the region, some rivers began drying up. For example, the Yellow river experienced zero streamflow episodes from 1972 to 1997. This has become a more frequent and longer occurrence since the early 1990s. In 1997, this water scarcity in northern China led to a period of 226 days without streamflow in the Yellow river, which is recorded to be the longest drying up duration even in conventional times (Jiang 2009; Gou et al. 2007). With the progression of global warming and increase in atmospheric temperature, it has been observed that the risk of drought since the late 1970s has increased drastically (Mishra and Singh 2010).
  In 2022, China issued its first national drought alert of the year ("yellow alert") around August, as several cases of forest fires and high latent heat transfers were reported by the media. Apart from the sporadic wild fires in the region, the "yellow alert" was necessitated due to shrinkage of the Poyang lake, which is regarded as one of the most important lakes in Yangtze flood basin, central China's Jiangxi province, to a quarter of its normal size. Reports also showed that 66 rivers across 34 counties in the southwestern region of Chongquing have dried up with a 60% reduction in rainfall over the region thereby causing food and freshwater shortages in the Jiangxi province of China (Al-Jazeeral 2022). India is another region, which is a popular vulnerable drought-prone regions in Asia. In every three year period, at least one drought episode has been reported in the region for the past five decades. The increasing magnitude and frequency of these drought events in contemporary times are causes of concern as prolonged and wide spread drought continue to negatively affect the region (Bank 2003).
- **Australia:**
  Several cases of recurrent drought episodes have been reported in Australia, with the popularly known "Millennium Drought" lasting for almost ten years being the most notorious (e.g., Ndehedehe et al. 2021; van Dijk et al. 2013). It was characterized as one of the worst drought in the region since early European settlement

as it affected most of the Southern and Eastern Australia states, with many rivers experiencing record low flows during this period. The Australian Bureau of Agriculture and Resource Economics estimated that the 2006 drought resulted in a significant reduction in the national winter cereal crop by at least 36% and cost the rural Australian region an estimated AUD 3.5 billion, leaving many farmers in serious financial crisis (Wong et al. 2010).

- **Europe:**
  The severity of drought in many European countries is increasing significantly. While the northern to north-eastern European regions are more prone to a rise in flood frequencies, the southern and south-eastern European regions exhibit significant increases in drought frequencies. This means that the northern regions are more susceptible to flooding due to increased precipitation rates, while the southern parts are susceptible to drought due to prolonged dry spells and increased evaporation (Voss et al. 2002). Some studies suggest that the analysis for drought should be localized within a regional context due to its large scale characteristics so as to improve the accuracy of the results (e.g., Mishra and Singh 2010). During the past 30 years, Europe has been affected by some major drought episodes, with the most notable ones occurring in 1976 over Northern and Western Europe, 1989 over most of Europe, 1991 over most of Europe, and 2003 over large parts of Europe and was associated with the summer heat wave of 2003 (Feyen and Dankers 2009). The United Nations Environment Programme in 2005 reported one of the most serious droughts in 60 years over the Iberian Peninsula which led to a total reduction of EU cereal yields by an estimated 10%.
- **North America:**
  Over the past two decades, the United States has experienced an unprecedented increase drought impacts with more recorded droughts and heightened severity. For instance, the impact of the 1988 large area drought on the economy of the US was estimated to be USD 40 billion, an amount that is approximately three times the cost of losses associated with the 1989 San Francisco earthquake (e.g., Mishra and Singh 2010). According to the National Climatic Data Center, USA, an approximate 10% of the total land mass of the entire country experienced extreme drought at any given time in the last century (e.g., Mishra and Singh 2010). In the United States, drought episodes from 1980 to 2003 have been reported to account for ten out of the fifty eight weather related disasters (Ross and Lott 2003). Of the total USD 349 billion total cost of all weather-related disasters, drought alone accounted for USD 144 billion which was about 41.2% of the total budget (Ross and Lott 2003). This, therefore, makes drought the costliest natural disaster to have affected the United States. Most regions in Canada have also experienced drought episodes. However, the Canadian Prairies and the interior British Columbian region appear to be more susceptible to this disaster because of the high variability in precipitation recorded across the region (Canada 2004). Over the past two centuries, a minimum of 40 long-term droughts have occurred in Western Canada. In the southern region of Alberta, Manitoba, and Saskatchewan, multi-year droughts were recorded in the 1890s, 1930s, and 1980s (Wheaton 2000). The eastern part of Canada experiences the mildest drought events, as the episodes

are known to affect smaller areas, but less intense and less frequent. However, periodic spikes in drought intensity is noticed occasionally, such as the drought which occurred in the 20th century. Over a large portion of the Canadian prairies, various successive seasons of below average rainfall rates have resulted in one of the most severe prairie droughts ever recorded, severely damaging many water-dependent and water-related activities between 2001 and 2002 (Canada 2004). The average level of the Great Lakes in 2001 plummeted to their lowest points in over 30 years with Lake Huron and Lake Superior displaying proximate record lows (Mishra and Singh 2010).

- **Africa:**
  Since the late 1960s, series of reports and literatures have dwelled on the severe and unprecedented drought episodes over the Sahel (Ndehedehe et al. 2020). The drought has had a severely damaging impact on this ecologically vulnerable sub-region and was a major motivation toward the inauguration of the UN convention on combating desertification and drought (Zeng 2003). It is believed that drought frequency in this region has increased since the end of the nineteenth century. The droughts experienced in the 1910s, 1940s, 1960s, 1970s, and 1980s all resulted in serious famine in the region, though a partial recovery from the drought was reported (Lovejoy and Baier 1975). While a minimum of one severe drought has been confirmed in each century starting from the 1600's, the severity and frequency of the recent Sahelian drought attracted increasing attention and stood out. In the drought severity that lasted between 1960 and 1980, Batterbury and Warren (2001) reported the two major spikes between this period. This drought caused the famine that occurred from 1968 to 1974 and again in the early and mid 1980s.

5.4 Groundwater Processes

The direct and indirect effect of human water extraction and other factors such as climate, geology, and surface water on groundwater processes (Fig. 5.8) is an interesting subject in groundwater science. Given that groundwater has served as a historical buffer against climate variability over the past decades, our dependence on groundwater resources is expected to increase as water supplies around the world undergo stress based on population growth, urbanization, and industrialization effects. Groundwater systems around the world are also experiencing unprecedented climate change effects as a result of increase in temperature and climate variability. The composite influence of climate and human abstraction on groundwater makes it a threatened resource that needs to be managed effectively. Changes in global climate through rainfall deficits and higher evapotranspiration rates coalesce with water abstraction to influence the rates of groundwater recharge and discharge (Fig. 5.8), though such impacts could be different across various parts of the world. The rates, intensity and frequency of hydrological units such as precipitation, evapotranspiration, and runoff are significant contributors to groundwater variability. Several regions have

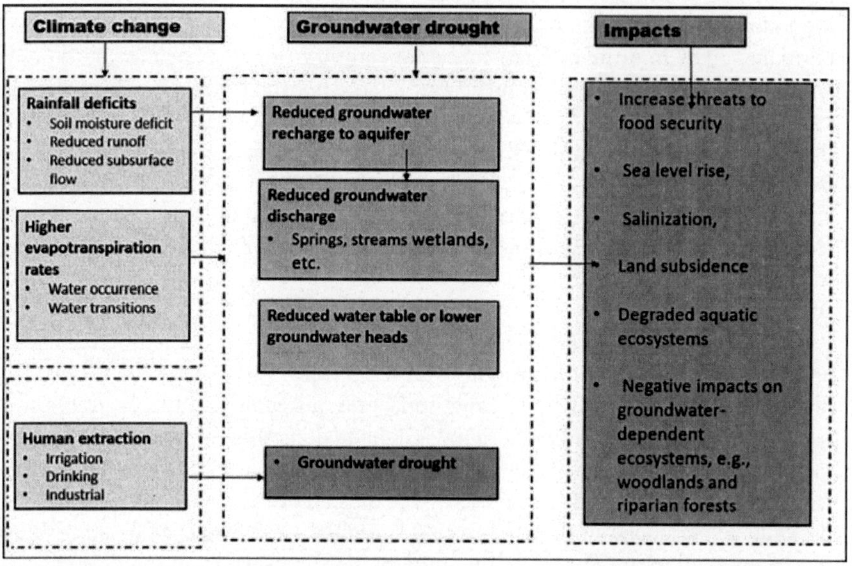

Fig. 5.8 Impacts of climate change and human activities on groundwater drought. Excessive groundwater pumping draws down the water table. When extreme drought is prolonged, groundwater depletion, higher evaporation, and low well yields will trigger groundwater drought. Groundwater drought reduces recharge and will eventually lead to several impacts, including deformation and loss of groundwater-dependent ecosystems (e.g., springs and streams) and collapse of forest ecosystems

experienced an increase in groundwater recharge due to impacts from increased precipitation patterns, more frequent large floods, or due to melting of permafrost.

In other regions, however, reduced precipitation and higher evapotranspiration rates are most likely to contribute to a decrease in aquifer recharge and when this prolongs, the water table is reduced, truncating discharge to springs and streams (Fig. 5.8). Changes in vegetation land cover affecting recharge rates and runoff will occur as a result of climate change and tends to worsen human and anthropogenic impacts such as deforestation. This impact on groundwater recharge will in turn influence the rates and quantities of groundwater discharge to stream base flows, springs, and the availability of groundwater for pumping (Fig. 5.8). A key challenge facing groundwater managers is the design of adaptation strategies that could account for uncertainties in such a way that can result in promise satisfactory outcomes, especially for water use under a specific range of climatic conditions. Some of the examples of the needed strategies include, the conservative allocation limits based on contingency policies that trigger other possible management arrangements predicated on laid out water availability conditions, and the intensification of storage through monitored aquifer recharge.

5.4.1 Groundwater Recharge

The water that crosses the water table and contributes to water stored in groundwater aquifers is called groundwater recharge (Currell and Ndehedehe 2022). Groundwater recharge is one of the key factors influencing hydrological balance and sustainability of subsurface regimes. Recharge can take various forms, including infiltration through diffuse rainfall over a wide area, and focused recharge occurring below specific landscape features (Currell and Ndehedehe 2022; Barron et al. 2012). In a general sense, recharge is related to the downward flow of water reaching the water table, and adding to the volume of already existing water in the groundwater reservoir. One mechanism of this action involves the downward percolation of soil water in excess of soil moisture shortage and infiltration excess. Despite the straightforwardness or simplicity of this definition, conceptual problems still exists. For example, the amount of groundwater recharge over any specific region is usually assumed to be the same as the amount of excess infiltration over the same region. In reality, however, not all water necessarily gets to the water table. Important controls on recharge processes include climate, hydrology, soils, topography, geology, land use, and vegetation and essential factors to be accounted for in the development of conceptual model at all spatial scales (Guillaumot et al. 2022; Currell and Ndehedehe 2022; Moeck et al. 2020; Barron et al. 2012). The relative importance of these predictors or controls can differ across regions. For example, in typically arid and temperate regions, climate parameters are less important in recharge estimation (Barron et al. 2012). However, in several other regions, recharge rates are considerably influenced by climatic parameters, i.e., seasonal changes and rates in precipitation and temperature (Guillaumot et al. 2022; Moeck et al. 2020).

The movement of water or infiltration to groundwater can be hampered by low-conductivity horizons and the water can vanish as interflow to nearby local depressions, where it evaporates or runs off instead of merging with the regional groundwater system. Also, another problem can emanate from the rise in water table of shallow aquifers. This can introduce a local groundwater system with associated local seepage discharge in the region of interest. The total downward flow to the water table is accordingly higher than the net recharge in the region. A comparable problem in regions with high water tables is related with time scales; water may enter the groundwater reservoir but through evapotranspiration and anthropogenic activities, this may likely be lost rather quickly. It is important to determine for a local groundwater system if the portion of the recharge flux is temporary or part of the total long term replacement or contribution to aquifers. A distinct discrepancy should therefore be made, both for modeling and conceptual purposes, between the potential volume of water available for recharge from the soil zone and the actual recharge volume getting to the water table.

5.4.1.1 Recharge as Evidence of Water Availability

Groundwater recharge is critical because the water flows to and fro aquifer sustains the water cycle and groundwater dependent ecosystems like springs, wetlands, rivers, as well as vegetation. Therefore, the task of quantifying the current rate of groundwater recharge is a prerequisite for proficient and sustainable groundwater resource management, especially in arid regions, where such resources are crucial to economic development. Primarily, recharge estimates are targeted at the aquifers required for development and are usually not from the uppermost phreatic source. In part, this recharge is controlled by atmospheric processes including precipitation, evapotranspiration and runoff, thus making any changes in the global hydrological and atmospheric process have a significant influence on groundwater in terms of quantity and quality. Therefore, understanding of the spatio-temporal processes and impacts on recharge rates in relation to groundwater sustainability and management is paramount. Since the mid-1980's, a notable upsurge in recharge studies has been reported in science journals, yet sources of uncertainties in the modeling of groundwater recharge have largely been ignored (Atawneh et al. 2021).

The emerging need for the adequate and reliable assessment of recharge estimates is equally supported by international and non-government agencies. Despite the increased publications on the subject in the past decades, there are still research gaps in the accurate assessment of recharge to improve understanding of groundwater systems. This is because recharge processes could be complex, especially in fractured aquifers. Quantifying recharge can thus be difficult due to differences in soil properties and the lack of data on relatively smaller scale. This was noted in the work of Guillaumot et al. (2022) who also identified weakness in the water table fluctuation (WTF) method, arguing that this approach

> *"neglects the impact of lateral groundwater redistribution in the aquifer; i.e., it assumes that all the WTFs are attributable to recharge."*

The freely available recharge data from more than 5000 locations around the world (Fig. 5.9) offer opportunities to assess recharge processes to gain understanding of critical drivers (climatic controls, soil-related parameters) that underpins recharge predictions (Moeck et al. 2020). However, given the spatial distribution of recharge data on a global scale (Fig. 5.9), this re-emphasize the need for more observations to improve understanding of recharge on a large scale, especially in critical groundwater regions where withdrawal is considerably high.

Even though recharge data and its accuracy, be it from numerical modeling or chloride mass balance, is still a challenge on global, regional and local scale (Guillaumot et al. 2022; Crosbie 2021; Moeck et al. 2020; Barron et al. 2012), the amount of recharge in a given region is an important metric of water availability (Fig. 5.10). This is because the amount of recharge will be evident in the rates of evapotranspiration and streamflow, which unfortunately are not well captured in hydrological models (Berghuijs et al. 2022). Areas with considerable large amount of recharge as indicated in Fig. 5.10 are notable and also show where most of the world's surface flow and large rates in evapotranspiration have been observed. Available water in

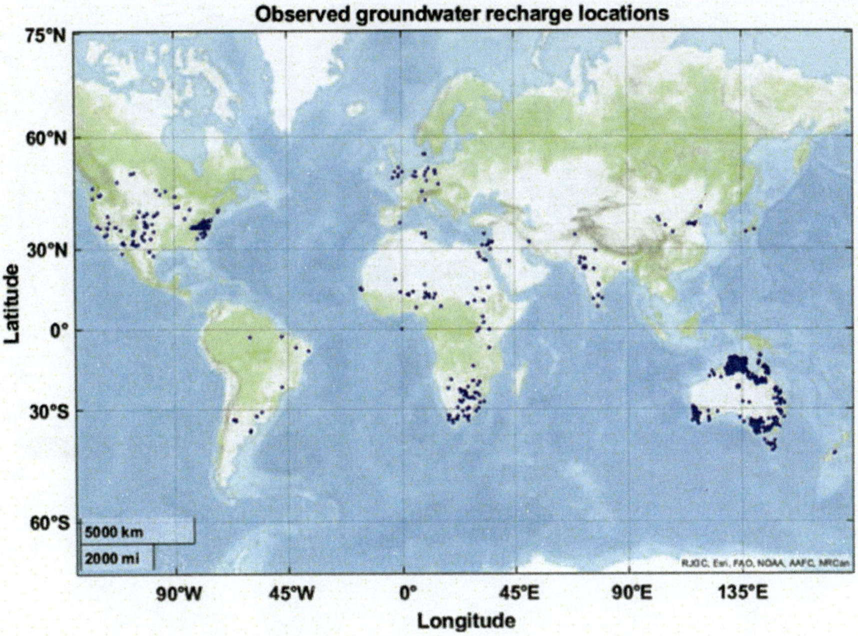

Fig. 5.9 Spatial distribution of direct natural groundwater recharge rates across the world. This map was prepared using data from https://opendata.eawag.ch/dataset/globalscale_groundwater_moeck based on the work of Moeck et al. (2020)

aquifer are determined by the balance between what is recharged (incoming flux) and the outgoing flux (discharge) to groundwater dependent ecosystems such as surface waters (i.e., springs and streams) and vegetation, as well as losses to human abstractions (Alley et al. 2002). For this reason, the plenitude of recharge can re-emerge as evapotranspiration and substantial streamflows (e.g., Berghuijs et al. 2022; Alley et al. 2002).

5.4.1.2 Types and Sources of Recharge

The different sources of recharge to groundwater systems are well known. According to de Vries and Simmers (2002), key recharge mechanisms from these sources are conceptually defined as (e.g., Currell and Ndehedehe 2022);

(i) Direct recharge: water supplemented to the sub surface water reservoirs in excess of soil moisture deficits and evapotranspiration by direct vertical percolation through the vadose zone.
(ii) Indirect recharge: percolation to the water table via the vadose zone.

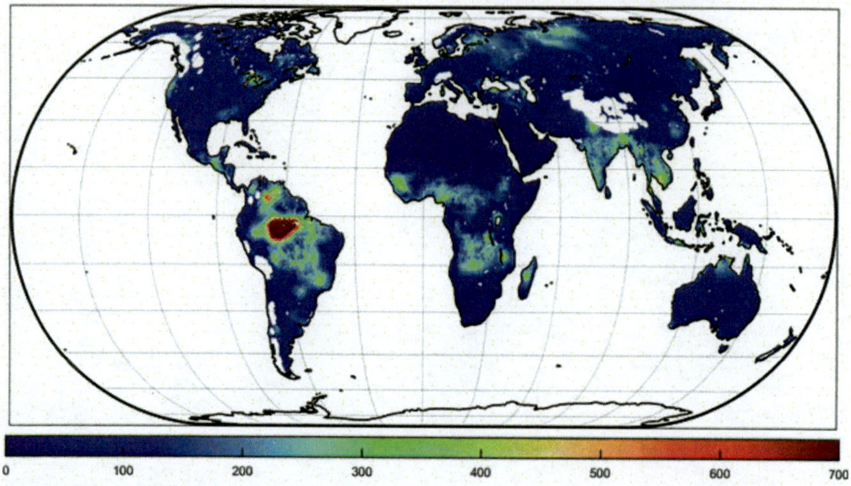

Fig. 5.10 Mean annual distribution of groundwater recharge depth (in mm/yr) or peak between 2003 and 2016. This recharge was estimated using data from the Gravity Recovery And Climate Experiment satellite

(iii) Localized recharge: an intermediate style of sub surface water recharge resulting from the horizontal surface concentrations of water in the absence of comprehensive channels.

In some locations, combinations of these various types of recharge are bound to occur. This can take place through the following processes:

(i) diffuse percolation,
(ii) macro-pore flow through root channels, fissures and desiccation cracks, and
(iii) preferential flow resulting from unstable wetting fronts and distinguished soil physical characteristics within the soil, notably between clayey and sandy sediments.

5.4.1.3 Techniques in Recharge Estimation

Recharge occurs to a reasonable extent even in the most arid regions and as it progresses, indirect and localized recharge tends to become more important than direct recharge, especially in terms of total aquifer replenishment. There are different methods to estimating recharge. The report by Currell and Ndehedehe (2022) on the hydrogeology of the Cambrian Limestone Aquifer in Australia's Northern Territory to support water management and important values sustained by its groundwater detailed several approaches to recharge estimation. Some of these include, the use of tritium data , chloride mass balance, chloride data from precipitation and groundwater, and numerical modeling, among several other methods (see, e.g., Guillaumot et al. 2022; Crosbie 2021; Moeck et al. 2020; Barron et al. 2012).

An example of numerical modeling is the work undertaken by Doble et al. (2014) to quantify overbank flood recharge using surface–subsurface flow model to evaluate the amount of infiltration that occurs through a floodplain. In each of the model scenarios, they found several limitations and uncertainties associated with the recharge process, e.g., differences in floodplain elevation played a key role in infiltration volume. Additionally, factors including, aquifer transmissivity and the clogging layer conductance restricted flood infiltration process (Doble et al. 2014). The water table fluctuation method to recharge estimation from the work of Guillaumot et al. (2022) is another modeling approach to quantifying recharge. As illustrated in Fig. 5.11, the technique requires several input including, precipitation, net precipitation (difference between rainfall and evaporation) data from boreholes. Furthermore, the process involved two steps-the calibration of a 1-dimensional analytical groundwater model to estimate hydrodynamic parameters from borehole data and the analytical inversion process (Fig. 5.11). The heterogeneous nature of some aquifers could be a challenge in this hybrid-modeling approach to groundwater recharge estimation (Fig. 5.11), but Guillaumot et al. (2022) argued that the level of confidence it offers makes it scalable and applicable in other hydrological domains, especially when long-term data (water table records) are available.

As these methods have uncertainties and limitations, groundwater recharge from satellite remote sensing using the Gravity Recovery and Climate Experiment (GRACE) has recently emerged (Fig. 5.10), thus serving as an independent line of evidence to complement other existing methods. This first estimates of recharge was based on changes in groundwater derived from GRACE and despite uncertainties associated with the use of soil moisture as an auxiliary data, these estimates align with those derived from other approaches in the literature. Apparently, water yield and recharge within the Amazon floodplain corridor is the highest and is followed by tropical regions in Africa (West and central) and south Asia, which receive substantially large amount of rainfall every year. (Fig. 5.10).

In contrast, semi-arid and typically dry regions have low recharge (Fig. 5.10) and it is possible that catchments located in these areas which show relatively high recharge could be benefiting from through-flows because of their interconnected systems. The study undertaken by Barron et al. (2012) emphasized the importance of rainfall as a key predictor of recharge in most tropical climates consistent with the recent work of Guillaumot et al. (2022). But Barron et al. (2012) also argued that even though changes in annual recharge depend on annual changes in rainfall, the proportion of rainfall that translates to recharge are unequal, that is, about 2 to 4 fold increase in recharge variations can be caused by changes in annual rainfall.

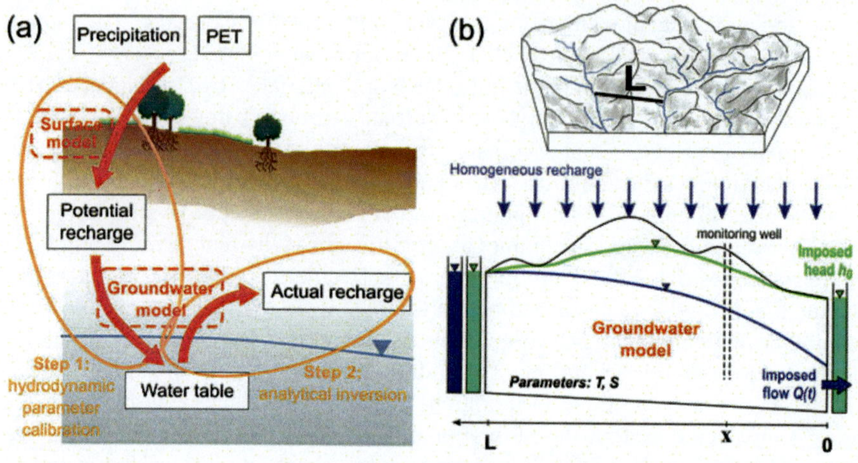

Fig. 5.11 Groundwater recharge using the water table fluctuation technique. *Source* Guillaumot et al. (2022)

5.4.2 *Discharge*

The loss of water from an aquifer to a surface-water body, the atmosphere, or extraction for human use is referred to as groundwater discharge. This action progresses to groundwater depletion when the rates of groundwater discharge are more than the rates of recharge. There have been significant groundwater depletion in the past five decades due to both direct and indirect impacts of climate change. Other factors including human activities, urbanization, and anthropogenic influences are also contributors to increased groundwater depletion rates. Over the past century, studies have attributed sea-level rise crises to changes in global groundwater discharge (Bierkens and Wada 2019; Taylor et al. 2013). Others projected that the rise in sea level would have been even greater if significant amounts of water had not been stored in hydrological systems (e.g., surface water and soil moisture) or channeled into aquifers by irrigation return-flow (Jakeman et al. 2016). Even if the present groundwater abstraction rates are not increased, some groundwater resources are still bound to be severely affected by climate change. Some of them include, the Chalk aquifer in eastern England (Yusoff et al. 2002), and the Edward aquifer in Texas, USA (Loáiciga et al. 2000).

5.4.2.1 Climate Change Impacts on Groundwater Discharge

Direct or indirect climate change effects on groundwater discharge include, changes in land-use practices or irrigation, changes in water demand, and soil degradation. The substantial rise in groundwater discharge observed from the mid-1900s has continued unabated. This increase is directly proportional to the growing population in many regions as well as the development of advanced well pumps that are used to support agricultural industries and public drinking water supplies. For instance, parts of the Ogallala aquifer in the US have recorded significant water-level declines since the 1950s which range from 5 to over 50 m depending on the relative scale of discharge and aquifer recharge (McMahon et al. 2007). Wang et al. (2009) also pointed out the strong correlation between the declining baseflow in the sand hills of Nebraska, USA and its soil texture, and projected that this could also be the case in other parts of the world. Given the varied amounts of precipitation rates influenced by some climate scenarios over the globe, many regions receive more precipitation than others. The study undertaken by Woldeamlak et al. (2007) pointed out that under severe wet-climate scenarios; runoff was the most vulnerable product and when integrated with the predicted surge in groundwater discharge, may lead to rising groundwater levels and winter precipitations, thus increasing flooding risks in the study region. Under dry-climate scenarios, recharge was the most vulnerable product and showed how it decreased over time in all seasons. The decrease which was projected to be within the range of 3 to 5m resulted in annual groundwater level declines. This can contribute adversely to local aquatic life in riverine ecosystems and local wetlands that rely on groundwater discharge to support base flow.

Groundwater is a key component of the global hydrological cycle and several other water resource projects. Therefore, the potential effects of climate change on both recharge and discharge activities deserve more attention by the scientific community. Currently, scientists lack the needed tools and datasets (such as long-term continuous monitoring of recharge and discharge estimates) to accurately forecasts recharge and discharge responses to future climate change in most environments, such as Africa and Asia. Green et al. (2007) pointed out the uncertainty of most regions of the world in projecting either the increase or decrease of recharge/discharge estimates especially under climate change. Under future climate change the location and timing of recharge and related impacts on groundwater supplies are poorly understood as pointed out by Gurdak et al. (2007). Yet, water resources, especially in several semi-arid and arid regions, are particularly prone to climate change effects and consequences and require more attention to enhance their chances of sustainability and usability in the near future.

5.4.2.2 Net Groundwater Discharge

Submarine groundwater discharge or the net groundwater discharge existing beneath the ocean is a large and integral part of the global hydrologic cycle, accounting for as much as 12,000 km^3 per year (Jakeman et al. 2016) and may otherwise supply freshwater for human consumption, agricultural irrigation, and industrial benefits. Determining the extent of groundwater discharge and its biogeochemical effects on the ocean has critical implications toward understanding climate change impacts on oceanic properties and processes (Windom et al. 2006). For instance, highly dissolved nitrogen phosphorus ratios in submarine groundwater discharge in relation to surface water tends to control the coastal oceans toward phosphorus limitations within the coming decades, thereby changing the current nitrogen-limited coastal primary production that has been the standard pattern for some decades now (Taniguchi et al. 2009).

5.5 Sustainable Limits

Groundwater depletion is regarded as one of the greatest sustainability challenges facing humanity in the twenty-first century. With sustainable development goals a decade away, the government, policy makers, and water agencies around the world are in pressing need for comprehensive and targeted measures to ensure that local communities adhere to groundwater management policies in an efficient and effective manner. This is a cause for concern as recent hydro-economic analysis show that unsustainable groundwater abstraction rates pose significant threat not just to local water and food security (Wada et al. 2012), but also international trade links globally (Dalin et al. 2017).

5.5.1 Addressing Scarcity Through Policies

In many stressed aquifer systems around the world, one of the most adopted options in mitigating groundwater stress has been to regulate groundwater abstraction by establishing and enforcing sustainable allocation limits. For these policies to be effective, it is required and expected that the "rules of the game" are socially accepted and sustainable, firmly supported and thoroughly enforced. Given that water users may decide to resist, ignore or violate the restrictions imposed on them, adherence to the rule cannot be totally assured by water managers or even the government. Although the introduction of strict monitoring and penalties may persuade people to play by the roles. However, such show of force could possibly erode trust between water users and their respective agencies. In emerging economies where most of the groundwater depletion is being experienced, the mechanism to monitor millions of water users could be an expensive endeavor and challenging. In addressing water scarcity problems in China, regulating water withdrawal requires well define legally enforceable water rights through the establishment of relevant institutions (Jiang 2009). Emerging economies can benefit from similar approach, these institutions and systems will coordinate the administration of such rights across different government agencies and organizations. Additionally, capacity building through training and data-driven decision support systems underpinned by research in water resource management can inform good policy framework in these regions (e.g., Jiang 2009).

Understanding the motivations for compliance and susceptibility to cultural factors and policy instruments remain a key challenge for even the most proficient water agencies in developed regions. Social research and empirical surveys usually take years to plan, design and develop while taking up a significant portion of the entire time, resources and effort in the process. Majority of the social research undertaken on the issues of compliance are site-specific and the results generated from them are usually difficult to generalize, extrapolate, and apply to other regions, given that every region has its own unique characteristic features. Other methods of enquiry into the views of water users toward groundwater policies and regulations are required to gain a prior understanding of how hydro-social systems could be directed toward local, national, and international socio-economic goals and objectives.

Reports from the United Nations Environment Programme (United Nations Environment Programme 2016) reveal that an estimated 30 to 50% of the global water supply is illegally gained, with water theft projected to increase as a result of drought and climate change impacts. If this trend continues, it is expected that a 40% shortfall in water availability will happen around 2030 (Castilla-Rho et al. 2019). However, to reverse this trend, it is expected that significant improvements be made on sustainable limits through compliance and enforcement. This is because regardless of how many new governance tools are generated, it will be inadequate, especially if compliance and enforcement are absent or insufficient. The role that adherence to groundwater management policies will play in the attempt of water agencies to attain the needed groundwater outcomes can be intentionally explored through computational social science models, which have served as valuable tools for synthesizing empirical and theoretical information on cultural values, human cooperation, and social norms (Castilla-Rho et al. 2019).

5.5.2 Human Population and Risk

In recent times, the increasing global population and urbanization in most groundwater aquifer locations are major drivers of global groundwater stress. This means that in many regions around the world, the sustainability of groundwater services is threatened due to population growth which is typical of the Anthropocene. Some of these threats are outlined below.

- **Artificial drainage:**
  Drainage activities aimed at evacuating sub surface water, either for shallow water-table control or for facilitating resource extraction, such as mining, affect groundwater services similar to groundwater abstractions for any intended use of water. The impact of artificial drainage and the displacement of sub and surface water, either for formal technical drainage provisions or by enlarging the regions of impermeable land surface could potentially minimize groundwater recharge rates.
- **Intensive groundwater abstraction:**
  As a result of the increasing human population, there is a significant increase in groundwater abstraction rates for irrigation, agriculture and human use. This has a direct effect on the hydrological regime of the aquifer in question. Initially, the abstracted flow is balanced by depletion of stored groundwater, but on a medium term, capture (i.e., the sum of intercepted natural groundwater discharge)

replaces this storage depletion and thereby induces increases in the recharge rates of groundwater. Rigorous groundwater abstraction has the potential of introducing significant negative side effects which includes increasing cost of groundwater abstraction, declining groundwater levels, degradation of wetlands, diminishing spring discharges and base flows, land subsidence, and the intrusion of brackish or other low quality water. It could also have positive side effects such as enhancing groundwater recharge, contributing to the drainage of water logged regions, and developing more storage capacity for buffering between dry and wet periods.

- **Salinization and Pollution:**
  The salinization or pollution of groundwater renders it unsuitable for certain groundwater services. This is a major concern for domestic water supply and water service delivery. Several processes can cause freshwater in aquifers to become saline or brackish. Some of them are due to natural causes (for e.g., flooding by sea water), while others are caused by human beings (for e.g., downward seepage of mineralized irrigation return flows). Some of the contributors and mechanisms to this process may include households, mining, industries, agriculture produce, and inadequate expulsion of waste water.

5.5.3 *Adaptation and Resilience to Climate Change*

Increased freshwater demand caused by population growth, economic prosperity, and change in dietary requirements has been further exacerbated by climate change (Bierkens and Wada 2019). While it remains difficult to estimate the direct impact of climate change impacts on groundwater resources, it is certain that proper climate change adaptation and mitigation policies will significantly contribute to cushioning such impacts on groundwater resources. In some cases, however, shift in policy solutions and water planning guidelines can contribute to worsening the challenges associated with groundwater use and management. For instance, in the Cambrian Limestone Aquifer of Australia's Northern Territory (NT), there are two rules for water allocation that appears to be contradictory and not supportive of groundwater management in the region.

An excerpt from the report of Currell and Ndehedehe (2022) to illustrate this in the three major sedimentary basins (Daly, Georgina, and Wiso river basins) of the Cambrian Limestone Aquifer is indicated below:

> *"...The Top End rules allow for allocation of groundwater licenses up to a fraction (20%) of estimated recharge. This approach is broadly in line with other groundwater management jurisdictions in Australia and worldwide, which calculate a Sustainable Yield as a fraction of recharge and/or discharge with the intention of balancing extraction with a need to maintain long-term groundwater access and limit impacts on GDEs.*
>
> *...The Arid Zone rules, which apply to the south of a line delineating northward and southward flowing surface water catchments of the NT, contain two clauses relevant to groundwater.*

The first states that licensed extractions should have no detrimental impacts on GDEs; the second allows for depletion of 80% of total pre-development groundwater storage, over a 100-year time frame..."

The Top End rules apply to the relatively wetter and hydrologically more productive northern section of the Cambrian Limestone Aquifer unlike the Arid Zone rules designed for the dryer and semi-arid portion of this hydrogeological groundwater aquifer. As further argued by Currell and Ndehedehe (2022), the two aspects of the rules in the Arid Zone case are contradictory and that

"...It is not possible for the first aim to be achieved if the second part is permitted...".

This is a good example of how policies on sustainable management of groundwater can be self-defeating, thus the need to adopt management approaches frameworks that aligns well with contemporary best practice (Currell and Ndehedehe 2022) and underpins effort to increase resilience to the impacts of climate change on groundwater resources.

Climate adaptation and resilience measures are designed to manage the consequence of a dynamic climate as well as minimize future potential risks. Adaptation and resilience strategies encompass national, regional, and practical measures taken at all political levels and by individuals. In highly populated cities and regions of the world, groundwater is regarded as the most critical resource in terms of sustainable development through provision of low cost, high quality and reliable water supplies. An estimated 70% of the drinking water in the European Union, 60% of agricultural irrigation and 80% of rural water supply in sub-Saharan Africa is largely dependent on groundwater services (Hiscock 2011). Groundwater resources also sustain ecosystem and landscapes in humid regions by replenishing riparian areas, wetlands, and unique aquatic ecosystems in coastal environments. Due to the hidden nature of groundwater resources, its development is usually unaccounted for and not monitored, thus making it difficult to incorporate such into overall water resource management. This may contribute to contamination and over-exploitation. Therefore, even without the adverse effects of climate change, sustainable management of groundwater resources is a major challenge. Groundwater is a widely distributed resource at regional, local, and basin scales. Stakeholders that benefit from this freshwater reservoir including large industries, municipalities, and farmers are responsible for influencing its policies by government and water managers. Generally, governance systems, incentives, resource policies, data collection and available information need to relate to a wider scale range with various adaptive management strategies in both urban and rural areas to ensure best practice in groundwater management.

Intensifying climate change impacts strong variability and trends in historic hydrological data. During periods of surface water shortages and droughts, one of the surest responses is the abstraction of groundwater resources as an emergency supply. However, it is possible for this abstraction and groundwater use under conditions of climate change to become unsustainable, especially in regions expected to experience an increase in drought duration and frequency. Also, the perceived rise in sea levels due to climate change impacts will further threaten coastal freshwater aquifers and reservoirs, especially those reservoirs already experiencing salinisation as a result of over-exploitation.

The effects of discharge and groundwater development take several years to become evident (Alley 2006). Thus, responsible water managers and even the government tend to neglect the data collection and analysis needed to sustain informed groundwater management until problems ensue. This type of reactionary stance to groundwater management is a significant problem that needs to be addressed. Even though some groundwater systems are renewable, effective monitoring of groundwater resources is critical because they can also exist as fossil groundwater thus making them a non-renewable natural resource on human time scale. For instance, the groundwater that is abstracted from the aquifers in many arid and semiarid regions was recharged during wetter periods under paleoclimate conditions (Green 2016). Adaptation and resilience measures can be reactive or preventive and this applies to both natural and social systems. Confirming the sustainability of investments in groundwater resource planning and development over considerable periods and lifetime of schemes while taking explicit account of the dynamics of climate is referred to as climate proffing. In the absence of dependable projections of future changes in hydrological and climate models, adaptation measures, processes and methods can be employed. This includes, enhanced water use efficiency, management of water demand and offering no-regret choices to cope with climate change.

Having outlined the implication of climate change mitigation on groundwater resources, it is pertinent to discuss how groundwater could be applied and sustained through climate change adaptation standards. It was established earlier in this Chapter that climate change plays a key role in affecting surface water supplies in several ways. Some of this include, changing seasonality of river flows and snow melt, strong variations in precipitation patterns, increasing evapotranspiration rates, intensification of storm episodes, and the increased frequency of drought and flood episodes. Groundwater resources have the potential to complement surface water shortages to deliver key services. Some examples are given below.

(i) Urban water supply: Sustaining a reliable supply of fresh drinking water to urban areas is vital for health and total well-being of the inhabitants. Apart from good health being dependent on fresh drinking water, economic, and community health are also dependent on this resource. The impacts of climate change tend to increasingly endanger cities that depend on surface water catchments. For instance, in the mid-1970s, inflows into Perth's water storage initiated a sequence of "step dynamics" such that a decline in the order of 70% of the prior long-term-mean was experienced (Petrone et al. 2010). During the Millennium Drought of 2002 to 2010, five other cities in southern Australia also experienced water shortages reduced to critical low levels. The impacted states adapted to this anomaly by diversifying the water supplies for these cities by introducing reuse, groundwater and desalination sources. The cities of Perth, Adelaide, and Sydney designed new groundwater resources, implemented monitored aquifer recharge routines. Additionally, some aquifers were set aside and marked as reserves in case of future occurrences.

(ii) Irrigated food production: Over fifty percent of the globally accessible freshwater resources are already redirected for human use, while seventy percent of the world's water consumption is in agricultural production (Jakeman et al. 2016). It is being estimated that food demand will double over the next five to eight decades, and if adequate measures are not taken toward increased productivity, water use for food production will increase by 70-90% under a changing climate.

(iii) Freshwater biodiversity conservation: Freshwater biodiversity has undergone significant alterations as a result of the over-exploitation of surface and groundwater resources. Current and emerging techniques aimed at conserving freshwater biodiversity, especially for climate change adaptation have focused on making available surface environmental flows, otherwise referred to as environmental water demand management in other regions (Richter 2010). In Australia, the main emphasis of environmental flow programs (Kennard et al. 2010) is on the conservation of large wetland systems as well as restore and conserve riverine areas, especially over the lower reaches of river systems. It is presumed that surface water environment which flows under the conditions of short-term changes and long-term climate impacts will be enough to sustain the ecological characteristics of these wetlands. However, studies have shown that desiccation and water quality impacts caused by drought episodes and worsened by climate change are not suitably improved by environmental water policies and frameworks (Pittock 2013). These strategies, however, presume that large wetlands in downstream reaches of ecosystems and river basins can be conserved in a similar state to the present.

References

AghaKouchak A, Mirchi A, Madani K, Baldassarre GD (2021) Anthropogenic drought: Definition, challenges and opportunities. Rev Geophys e2019RG000683. https://doi.org/10.1029/2019RG000683

Agutu N, Awange J, Zerihun A, Ndehedehe C, Kuhn M, Fukuda Y (2017) Assessing multi-satellite remote sensing, reanalysis, and land surface models' products in characterizing agricultural drought in East Africa. Remote Sens Environ 194:287–302. https://doi.org/10.1016/j.rse.2017.03.041

Agutu N, Awange J, Ndehedehe C, Kirimi F, Kuhn M (2019) GRACE-derived groundwater changes over Greater Horn of Africa: temporal variability and the potential for irrigated agriculture. Sci Total Environ 693:133467. https://doi.org/10.1016/j.scitotenv.2019.07.273

Ahmed A, Alrajhi A, Alquwaizany AS (2021) Identification of groundwater potential recharge zones in Flinders Ranges, South Australia using remote sensing, gis, and mif techniques. Water 13(18). https://doi.org/10.3390/w13182571

Al-Jazeeral (2022) China issues first national drought alert, battles to save crops. *Al Jazeera*. Retrieved from https://www.aljazeera.com/news/2022/8/19/china-issues-first-national-drought-alert-battles-to-save-crops on Ist September 2022

Alley WM (2006) Tracking u.s. groundwater: Reserves for the future? Environ: Sci Policy Sustain Develop 48(3):10–25. https://doi.org/10.3200/ENVT.48.3.10-25

Alley WM, Healy RW, LaBaugh JW, Reilly TE (2002) Flow and storage in groundwater systems. Science 296(5575):1985–1990. https://doi.org/10.1126/science.1067123

Atawneh DA, Cartwright N, Bertone E (2021) Climate change and its impact on the projected values of groundwater recharge: A review. J Hydrol 601:126602. https://doi.org/10.1016/j.jhydrol.2021.126602

Bank W (2003) Report on financing rapid onset natural disaster losses in India: a risk management approach. World Bank. Report No. 26844-IN, Washington, DC

Barron OV, Crosbie RS, Dawes WR, Charles SP, Pickett T, Donn MJ (2012) Climatic controls on diffuse groundwater recharge across Australia. Hydrol Earth Syst Sci 16(12):4557–4570. https://doi.org/10.5194/hess-16-4557-2012

Bartos M, Chester M (2018) The conservation nexus: valuing interdependent water and energy savings in arizona. Environ Sci Technol 48:2139–49. https://doi.org/10.1021/es4033343

Batterbury S, Warren A (2001) The African Sahel 25 years after the great drought: assessing progress and moving towards new agendas and approaches. Glob Environ Change 11(1):1–8. https://doi.org/10.1016/S0959-3780(00)00040-6

Berghuijs WR, Luijendijk E, Moeck C, van der Velde Y, Allen ST (2022) Global recharge data set indicates strengthened groundwater connection to surface fluxes. Geophys Res Lett 49(23):e2022GL099010. https://doi.org/10.1029/2022GL099010

Bierkens MFP, Wada Y (2019) Non-renewable groundwater use and groundwater depletion: a review. Environ Res Lett 14(6):063002. https://doi.org/10.1088/1748-9326/ab1a5f

Bloomfield JP, Gaus I, Wade SD (2003) A method for investigating the potential impacts of climate-change scenarios on annual minimum groundwater levels. Water Environ J

Bloomfield JP, Marchant BP (2013) Analysis of groundwater drought building on the standardised precipitation index approach. Hydrol Earth Syst Sci 17(12):4769–4787. https://doi.org/10.5194/hess-17-4769-2013

Bloomfield JP, Marchant BP, Bricker SH, Morgan RB (2015) Regional analysis of groundwater droughts using hydrograph classification. Hydrol Earth Syst Sci 19(10):4327–4344. https://doi.org/10.5194/hess-19-4327-2015

Canada E (2004) Threats to water availability in canada. National Water Research Institute, Burlington, Ontario, 1:128. NWRI Scientific Assessment Report Series No. 3 and ACSD Science Assessment Series

Cao G, Zheng C, Scanlon BR, Liu J, Li W (2013) Use of flow modeling to assess sustainability of groundwater resources in the North China Plain. Water Resour Res 49(1):159–175. https://doi.org/10.1029/2012WR011899

Castellazzi P, Garfias J, Martel R (2021) Assessing the efficiency of mitigation measures to reduce groundwater depletion and related land subsidence in Querétaro (Central Mexico) from decadal InSAR observations. Int J Appl Earth Obs Geoinf 105:102632. https://doi.org/10.1016/j.jag.2021.102632

Castilla-Rho J, Rojas R, Andersen M, Holley C, Mariethoz G (2019) Sustainable groundwater management: How long and what will it take? Glob Environ Change 58:101972. https://doi.org/10.1016/j.gloenvcha.2019.101972

Chakraborty M, Mukherjee A, Ahmed KM (2021) Transboundary groundwater of the ganges–brahmaputra–meghna river delta system. In: Mukherjee A, Scanlon BR, Aureli A, Langan S, Guo H, McKenzie AA (eds) Global groundwater. Elsevier, pp 129–141. https://doi.org/10.1016/B978-0-12-818172-0.00010-4

Crosbie RS, R. P. (2021) Constraining probabilistic chloride mass-balance recharge estimates using baseflow and remotely sensed evapotranspiration: the Cambrian Limestone Aquifer in northern Australia. Hydrogeol J 29:1399–1419. https://doi.org/10.1007/s10040-021-02323-1

Currell MJ, Ndehedehe CE (2022) The cambrian limestone aquifer, northern territory: Review of the hydrogeology and management rules to ensure protection of groundwater dependent values. Environment centre northern territory. Retrieved from https://research-repository.griffith.edu.au/handle/10072/420013 on 13th December, 2022

Dalin C, Wada Y, Kastner T et al (2017) Groundwater depletion embedded in international food trade. Nature 543:700–704. https://doi.org/10.1038/nature21403

de Vries J, Simmers I (2002) Groundwater recharge: an overview of processes and challenges. Hydrogeol J 10:5–17. https://doi.org/10.1007/s10040-001-0171-7

Doble R, Crosbie R, Peeters L, Joehnk K, Ticehurst C (2014) Modelling overbank flood recharge at a continental scale. Hydrol Earth Syst Sci 18(4):1273–1288. https://doi.org/10.5194/hess-18-1273-2014

Döll P, Fiedler K (2008) Global-scale modeling of groundwater recharge. Hydrol Earth Syst Sci 12(3):863–885. https://doi.org/10.5194/hess-12-863-2008

Fan Y, Li H, Miguez-Macho G (2013) Global patterns of groundwater table depth. Science 339(6122):940–943. https://doi.org/10.1126/science.1229881

Ferreira V, Yong B, Tourian M, Ndehedehe C, Shen Z, Seitz K, Dannouf R (2020) Characterization of the hydro-geological regime of Yangtze River basin using remotely-sensed and modeled products. Sci Total Environ 718:137354. https://doi.org/10.1016/j.scitotenv.2020.137354

Feyen L, Dankers R (2009) Impact of global warming on streamflow drought in europe. J Geophys Res: Atmos 114(D17). https://doi.org/10.1029/2008JD011438

Fryar AE (2021) Groundwater of carbonate aquifers. In: Mukherjee A, Scanlon BR, Aureli A, Langan S, Guo H, McKenzie AA (eds) Global groundwater. Elsevier, pp 23–34. https://doi.org/10.1016/B978-0-12-818172-0.00002-5

Gleeson T, Richter B (2018) How much groundwater can we pump and protect environmental flows through time? presumptive standards for conjunctive management of aquifers and rivers. River Res Appl 34(1):83–92. https://doi.org/10.1002/rra.3185

Gleeson T, Befus K, Jasechko S et al (2016) The global volume and distribution of modern groundwater. Nature Geosci 9:161–167. https://doi.org/10.1038/ngeo2590

Gorelick SM, Zheng C (2015) Global change and the groundwater management challenge. Water Resour Res 51(5):3031–3051. https://doi.org/10.1002/2014WR016825

Gou X, Chen F, Cook E, Jacoby G, Yang M, Li J (2007) Streamflow variations of the Yellow River over the past 593 years in western China reconstructed from tree rings. Water Resour Res 43(6). https://doi.org/10.1029/2006WR005705

Green TR (2016) Linking climate change and groundwater. Springer International Publishing, Cham, pp 97–141. https://doi.org/10.1007/978-3-319-23576-9_5

Green TR, Bates BC, Charles SP, Fleming PM (2007) Physically based simulation of potential effects of carbon dioxide-altered climates on groundwater recharge. Vadose Zone J 6(3):597–609. https://doi.org/10.2136/vzj2006.0099

Gregory JM, White NJ, Church JA, M. F. P. B et al (2013) Twentieth-century global-mean sea level rise: Is the whole greater than the sum of the parts? J Clim 26(13):4476–4499. https://doi.org/10.1175/JCLI-D-12-00319.1

Guillaumot L, Longuevergne L, Marçais J, Lavenant N, Bour O (2022) Frequency domain water table fluctuations reveal impacts of intense rainfall and vadose zone thickness on groundwater recharge. Hydrol Earth Syst Sci 26(22):5697–5720. https://doi.org/10.5194/hess-26-5697-2022

Gurdak JJ, Hanson RT, McMahon PB, Bruce BW, McCray JE, Thyne GD, Reedy RC (2007) Climate variability controls on unsaturated water and chemical movement, High Plains Aquifer, USA. Vadose Zone J 6(3):533–547. https://doi.org/10.2136/vzj2006.0087

Hiscock KM (2011) Groundwater in the 21st Century–Meeting the challenges. Springer Netherlands, Dordrecht, pp 207–225. https://doi.org/10.1007/978-90-481-3426-7_13

Huang J, Cao L, Yu F, Liu X, Wang L (2021) Groundwater drought and cycles in Xuchang City. China. Front Earth Sci 9. https://doi.org/10.3389/feart.2021.736305

Hughes J, Petrone K, Silberstein R (2012) Drought, groundwater storage and stream flow decline in southwestern Australia. Geophys Res Lett 39(3). https://doi.org/10.1029/2011GL050797

IRI (2001) The drought and humanitarian crisis in central and southwest Asia: A climate perspective. International research institute (IRI) for climate and society. Retrieved from https://www.preventionweb.net/publication/drought-and-humanitarian-crisis-central-and-southwest-asia-climate-perspective on 22nd September, 2022

Jakeman AJ, Barreteau O, Hunt RJ, Rinaudo J.-D, Ross A, Arshad M, Hamilton S (2016) Integrated groundwater management: An overview of concepts and challenges. Springer International Publishing, pp 3–20. https://doi.org/10.1007/978-3-319-23576-9_1

Jarvis T, Giordano M, Puri S, Matsumoto K, Wolf A (2005) International borders, ground water flow, and hydroschizophrenia. Groundwater

Jiang Y (2009) China's water scarcity. J Environ Manag 90(11):3185–3196. https://doi.org/10.1016/j.jenvman.2009.04.016

Kalu I, Ndehedehe CE, Okwuashi O, Eyoh AE, Ferreira VG (2022) A new modelling framework to assess changes in groundwater level. J Hydrol: Reg Stud 43:101185. https://doi.org/10.1016/j.ejrh.2022.101185

Kennard MJ, Pusey BJ, Olden JD, Mackay SJ, Stein JL, Marsh N (2010) Classification of natural flow regimes in Australia to support environmental flow management. Freshwater Biol 55(1):171–193. https://doi.org/10.1111/j.1365-2427.2009.02307.x

Khan H (1994) Management of groundwater resources for irrigation in Bangladesh. Food and agricultural organization. Retrieved from https://agris.fao.org/agris-search/search.do?recordID=XF9548947 on on 18th September 2022

Kløve B, Ala-Aho P, G. B et al (2014) Climate change impacts on groundwater and dependent ecosystems. J Hydrol 518:250–266. https://doi.org/10.1016/j.jhydrol.2013.06.037

Konikow LF (2011) Contribution of global groundwater depletion since 1900 to sea-level rise. Geophys Res Lett 38(17). https://doi.org/10.1029/2011GL048604

Li B, Rodell M (2015) Evaluation of a model-based groundwater drought indicator in the conterminous U.S. J Hydrol 526:78–88. https://doi.org/10.1016/j.jhydrol.2014.09.027

Liu P, Famiglietti J, Purdy A et al (2022) Groundwater depletion in California's Central Valley accelerates during megadrought. Nat Commun 13:7825. https://doi.org/10.1038/s41467-022-35582-x

Liu P, Famiglietti J, Purdy A et al (2022) Groundwater depletion in California's Central Valley accelerates during megadrought. Nat Commun 13:7825. https://doi.org/10.1038/s41467-022-35582-x

Loáiciga H, Maidment D, Valdes J (2000) Climate-change impacts in a regional karst aquifer, Texas, USA. J Hydrol 227(1):173–194. https://doi.org/10.1016/S0022-1694(99)00179-1

Lorenzo-Lacruz J, Garcia C, Morán-Tejeda E (2017) Groundwater level responses to precipitation variability in Mediterranean insular aquifers. J Hydrol 552:516–531. https://doi.org/10.1016/j.jhydrol.2017.07.011

Lorenzo-Lacruz J, Garcia C, Morán-Tejeda E (2017) Groundwater level responses to precipitation variability in Mediterranean insular aquifers. J Hydrol 552:516–531. https://doi.org/10.1016/j.jhydrol.2017.07.011

Louwyck A, Vandenbohede A, Heuvelmans G, Van Camp M, Walraevens K (2023) The water budget myth and its recharge controversy: linear vs nonlinear models. Groundwater 61:100–110. https://doi.org/10.1111/gwat.13245

Lovejoy PE, Baier S (1975) The desert-side economy of the central sudan. Int J Afr Hist Stud 8(4):551–581. https://doi.org/10.2307/216696

Marc MKJ, Taigbenu AE, Boroto JR (2007) Domestic rainwater harvesting to improve water supply in rural South Africa. Phys Chem Earth, Parts A/B/C 32(15):1050–1057. https://doi.org/10.1016/j.pce.2007.07.007

Maupin M, Kenny J, Hutson S, Lovelace J, Barber N, Linsey K (2014) Estimated use of water in the United States in 2010. U.S. Geol Surv Circular 1405:55. https://dx.doi.org/10.3133/cir1405

McKee TB, Doeskin NJ, Kieist J (1995) Drought monitoring with multiple time scales. In: Conference on applied climatology, American meteorological society. Boston, Massachusetts, pp 233–236. Retrieved from: www.southwestclimatechange.org/node/911. Accessed 13 July, 2014

McMahon P, Dennehy K, Bruce B et al (2007) Water-quality assessment of the High Plains Aquifer, 1999–2004. US geological survey, US Geological Survey Professional Paper, p 1749

Mendicino G, Senatore A, Versace P (2008) A groundwater resource index (gri) for drought monitoring and forecasting in a mediterranean climate. J Hydrol 357(3):282–302. https://doi.org/10.1016/j.jhydrol.2008.05.005

Mishra AK, Singh VP (2010) A review of drought concepts. J Hydrol 391:202–216. https://doi.org/10.1016/j.jhydrol.2010.07.012

Moeck C, Grech-Cumbo N, Podgorski J, Bretzler A, Gurdak JJ, Berg M, Schirmer M (2020) A global-scale dataset of direct natural groundwater recharge rates: A review of variables, processes and relationships. Sci Total Environ 717:137042. https://doi.org/10.1016/j.scitotenv.2020.137042

Mukherjee A, Scanlon BR et al (2021) Global groundwater: from scarcity to security through sustainability and solutions. In: Mukherjee A, Scanlon BR, Aureli A, Langan S, Guo H, McKenzie AA (eds) Global groundwater. Elsevier, pp 3–20 https://doi.org/10.1016/B978-0-12-818172-0.00001-3

Nagarajan R, Sreedhar MG (2015) Micro-level drought vulnerability assessment using standardised precipitation index, standardised water-level index, Remote Sensing and GIS. ACRS 2015: The 36Th Asian Conference On Remote Sensing. Retrieved from https://www.researchgate.net/publication/283121546_Micro-Level_Drought_Vulnerability_Assessment_Using_Standardised_Precipitation_Index_Standardised_Water-Level_Index_Remote_Sensing_and_GIS on 20th September, 2022

Ndehedehe C (2022a) Drought events. Springer International Publishing, Cham, pp 249–280 https://doi.org/10.1007/978-3-030-99577-5_11

Ndehedehe C (2022b) Groundwater from space. Springer International Publishing, Cham, pp 211–230 https://doi.org/10.1007/978-3-030-99577-5_9

Ndehedehe C (2022c) Hotspots of climatic influence. Springer International Publishing, Cham, pp 629–688 https://doi.org/10.1007/978-3-030-99577-5_20

Ndehedehe CE, Agutu NO, Okwuashi OH, Ferreira VG (2016) Spatio-temporal variability of droughts and terrestrial water storage over Lake Chad Basin using independent component analysis. J Hydrol 540:106–128. https://doi.org/10.1016/j.jhydrol.2016.05.068

Ndehedehe CE, Agutu N, Ferreira VG, Getirana A (2020) Evolutionary drought patterns over the Sahel and their teleconnections with low frequency climate oscillations. Atmos Res 233:104700. https://doi.org/10.1016/j.atmosres.2019.104700

Ndehedehe CE, Ferreira VG, Agutu NO, Onojeghuo AO, Okwuashi O, Kassahun HT, Dewan A (2021) What if the rains do not come? J Hydrol 595:126040. https://doi.org/10.1016/j.jhydrol.2021.126040

Ndehedehe CE, Ferreira VG, Adeyeri OE, Correa FM, Usman M, Oussou FE, Kalu I, Okwuashi O, Onojeghuo AO, Getirana A, Dewan A (2023) Global assessment of drought characteristics in the anthropocene. Resour Environ Sustain 12:100105. https://doi.org/10.1016/j.resenv.2022.100105

Neely WR, Borsa AA, Burney JA, Levy MC, Silverii F, Sneed M (2021) Characterization of groundwater recharge and flow in California's San Joaquin Valley From InSAR-Observed surface deformation. Water Resour Res 57(4):e2020WR028451. https://doi.org/10.1029/2020WR028451

Nilsson C, Reidy CA, Dynesius M, Revenga C (2005) Fragmentation and flow regulation of the world's large river systems. Science 308(5720):405–408. https://doi.org/10.1126/science.1107887

Obasi GOP (1994) WMO's role in the international decade for natural disaster reduction. Bull Am Meteorol Soc 75(9):1655–1662. https://doi.org/10.1175/1520-0477(1994)075<1655:WRITID>2.0.CO;2

Ojha C, Shirzaei M, Werth S, Argus DF, Farr TG (2018) Sustained groundwater loss in california's central valley exacerbated by intense drought periods. Water Resour Res 54(7):4449–4460. https://doi.org/10.1029/2017WR022250

Ojha C, Werth S, Shirzaei M (2020) Recovery of aquifer-systems in Southwest US following 2012–2015 drought: Evidence from InSAR, GRACE and groundwater level data. J Hydrol 587. https://doi.org/10.1016/j.jhydrol.2020.124943

Pathak AA, Dodamani BM (2021) Connection between meteorological and groundwater drought with copula-based bivariate frequency analysis. J Hydrol Eng 26(7):05021015. https://doi.org/10.1061/(ASCE)HE.1943-5584.0002089

Peters E, van Lanen H, Torfs P, Bier G (2005) Drought in groundwater-drought distribution and performance indicators. J Hydrol 306(1):302–317. https://doi.org/10.1016/j.jhydrol.2004.09.014

Peters E, Bier G, van Lanen H, Torfs P (2006) Propagation and spatial distribution of drought in a groundwater catchment. J Hydrol 321(1):257–275. https://doi.org/10.1016/j.jhydrol.2005.08.004

Peters E, van Lanen H, Torfs P, Bier G (2005) Drought in groundwater-drought distribution and performance indicators. J Hydrol 306(1):302–317. https://doi.org/10.1016/j.jhydrol.2004.09.014

Petrone KC, Hughes JD, Van Niel TG, Silberstein RP (2010) Streamflow decline in southwestern Australia, 1950–2008. Geophys Res Lett 37(11). https://doi.org/10.1029/2010GL043102

Pittock J (2013) Lessons from adaptation to sustain freshwater environments in the Murray-Darling Basin. Australia. WIREs Clim Change 4(5):429–438. https://doi.org/10.1002/wcc.230

Puri S (2021) Transboundary aquifers: a shared subsurface asset, in urgent need of sound governance. In: Mukherjee A, Scanlon BR, Aureli A, Langan S, Guo H, McKenzie AA (eds) Global groundwater. Elsevier, pp 113–128 https://doi.org/10.1016/B978-0-12-818172-0.00009-8

Rajan A, Ghosh K, Shah A (2020) Carbon footprint of India's groundwater irrigation. Carbon Manag 11(3):265–280. https://doi.org/10.1080/17583004.2020.1750265

Rajasekhar M, Gadhiraju SR, Kadam A et al (2020) Identification of groundwater recharge-based potential rainwater harvesting sites for sustainable development of a semiarid region of southern India using geospatial, AHP, and SCS-CN approach. Arab J Geosci 13(24). https://doi.org/10.1007/s12517-019-4996-6

Richter BD (2010) Re-thinking environmental flows: from allocations and reserves to sustainability boundaries. River Res Appl 26(8):1052–1063. https://doi.org/10.1002/rra.1320

Rodell M, Famiglietti JS, Wiese DN, Reager JT, Beaudoing HK, Landerer FW, Lo MH (2018) Emerging trends in global freshwater availability. Nature 557(7707):651–659. https://doi.org/10.1038/s41586-018-0123-1

Ross T, Lott N (2003) A climatology of 1980-2003 extreme weather and climate events. National climatic data center. http://www.ncdc.noaa.gov/ol/reports/billionz.html

Sanginabadi H, Saghafian B, Delavar M (2019) Coupled groundwater drought and Water Scarcity Index for intensively overdrafted aquifers. J Hydrol Eng 24(4):04019003. https://doi.org/10.1061/(ASCE)HE.1943-5584.0001764

Seo JY, Lee S.-I (2019) Spatio-temporal groundwater drought monitoring using multi-satellite data based on an artificial neural network. Water 11(9). https://doi.org/10.3390/w11091953

Shah T (2007) The groundwater economy of south asia: an assessment of size, significance and socio-ecological impacts. In: Opportunities and threats to development, The agricultural groundwater revolution, pp 7–36

Shah T, Molden D, Sakthivadivel R, Seckler D (2000) The global groundwater situation: Overview of opportunities and challenges. In: International water management institute, Colombo, Sri Lanka. Retrieved from https://publications.iwmi.org/pdf/H025885.pdf on 18th September 2022

Shamsudduha M (2021) Groundwater storage dynamics in the himalayan river basins and impacts of global change in the anthropocene. In: Mukherjee A, Scanlon BR, Aureli A, Langan S, Guo H, McKenzie AA (eds) Global groundwater. Elsevier, pp 47–63 https://doi.org/10.1016/B978-0-12-818172-0.00004-9

Sheffield J, Wood EF (2007) Characteristics of global and regional drought, 1950–2000: Analysis of soil moisture data from off-line simulation of the terrestrial hydrologic cycle. J Geophys Res: Atmos 112(D17). https://doi.org/10.1029/2006JD008288

Siebert S, Burke J, Faures JM, Frenken K, Hoogeveen J, Döll P, Portmann FT (2010) Groundwater use for irrigation-a global inventory. Hydrol Earth Syst Sci 14(10):1863–1880

Stahl K, Kohn I, Blauhut V, Urquijo J et al (2016) Impacts of european drought events: insights from an international database of text-based reports. Nat Hazards Earth Syst Sci 16(3):801–819. https://doi.org/10.5194/nhess-16-801-2016

Taniguchi M, Burnett WC, Ness GD (2009) Integrated research on subsurface environments in asian urban areas. Sci Total Environ 407(9):3076–3088. https://doi.org/10.1016/j.scitotenv.2009.02.002

Taylor R, Scanlon B, Döll P et al (2013) Ground water and climate change. Nat Clim Change 3:322–329. https://doi.org/10.1038/nclimate1744

Thomas B, Famiglietti J (2019) Identifying climate-induced groundwater depletion in GRACE. Sci Rep 9:4124. https://doi.org/10.1038/s41598-019-40155-y

Thomas B, Famiglietti J (2019) Identifying climate-induced groundwater depletion in GRACE. Sci Rep 9:4124. https://doi.org/10.1038/s41598-019-40155-y

Uddameri V, Singaraju S, Hernandez E (2019) Is Standardized Precipitation Index (SPI) a useful indicator to forecast groundwater droughts?-insights from a Karst aquifer. JAWRA J Am Water Resour Assoc 55(1):70–88. https://doi.org/10.1111/1752-1688.12698

United Nations Environment Programme ICPO (2016) Strategic report: Environment, peace and security: A convergence of threats. https://wedocs.unep.org/20.500.11822/17008

Valencia LO, Johansen K, Solorio B et al (2020) Mapping groundwater abstractions from irrigated agriculture: big data, inverse modeling, and a satellite-model fusion approach. Hydrol Earth Syst Sci 24(11):5251–5277. https://doi.org/10.5194/hess-24-5251-2020

van Dijk AIJM, Beck HE, Crosbie RS, de Jeu RAM, Liu YY, Podger GM, Timbal B, Viney NR (2013) The millennium drought in southeast Australia (2001–2009): Natural and human causes and implications for water resources, ecosystems, economy, and society. Water Resour Res 49(2):1040–1057. https://doi.org/10.1002/wrcr.20123

Van Lanen HA, Laaha G, Kingston Daniel G et al (2016) Hydrology needed to manage droughts: the 2015 european case. Hydrol Process 30(17):3097–3104. https://doi.org/10.1002/hyp.10838

Van Loon AF (2015) Hydrological drought explained. WIREs. Water 2:359–392. https://doi.org/10.1002/wat2.1085

Van Loon AF (2015) Hydrological drought explained. WIREs. Water 2:359–392. https://doi.org/10.1002/wat2.1085

Voss R, May W, Roeckner E (2002) Enhanced resolution modelling study on anthropogenic climate change: changes in extremes of the hydrological cycle. Int J Clim 22(7):755–777. https://doi.org/10.1002/joc.757

Wada Y, van Beek LPH, Bierkens MFP (2012) Nonsustainable groundwater sustaining irrigation: A global assessment. Water Resour Res 48(6). https://doi.org/10.1029/2011WR010562

Wada Y, van Beek LPH, van Kempen CM et al (2010) Global depletion of groundwater resources. Geophysical Res Lett 37(20). https://doi.org/10.1029/2010GL044571

Wang F, Wang Z, Yang H, Di D, Zhao Y, Liang Q (2020) Utilizing GRACE-based groundwater drought index for drought characterization and teleconnection factors analysis in the North China Plain. J Hydrol 585:124849. https://doi.org/10.1016/j.jhydrol.2020.124849

Wang T, Istanbulluoglu E, Lenters J, Scott D (2009) On the role of groundwater and soil texture in the regional water balance: An investigation of the Nebraska Sand Hills, USA. Water Resour Res 45(10). https://doi.org/10.1029/2009WR007733

Ward C, Beddies S, Hariri K, Yaffiei SO, Sahooly A, Gerhager B (2007) Yemen's water sector reform program–a poverty and social impact analysis (PSIA). World Bank Group, Washington, D.C. Retrieved from http://documents.worldbank.org/curated/en/503681468335429999/Main-report on 17th September 2022

Wester P (1999) Institutional responses to groundwater depletion : the aquifer management councils in the State of Guanajuato, Mexico. In: The international symposium on integrated water management in agriculture, 16–18 June, Gomez Palacio, Mexico. https://cir.nii.ac.jp/crid/1573668924793687808

Wheaton E (2000) Canadian prairie drought impacts and experiences. In: Wilhite D (ed) Drought: A global assessment, vol 1. Routledge Press. London, UK, pp 312–330

Wilhite DA, Glantz MH (1985) Understanding: the drought phenomenon: The role of definitions. Water Int 10(3):111–120. https://doi.org/10.1080/02508068508686328

Windom HL, Moore WS, Niencheski LFH, Jahnke RA (2006) Submarine groundwater discharge: A large, previously unrecognized source of dissolved iron to the South Atlantic Ocean. Marine Chem 102(3):252–266. https://doi.org/10.1016/j.marchem.2006.06.016

Woldeamlak S, Batelaan O, De Smedt F (2007) Effects of climate change on the groundwater system in the grote-nete catchment, belgium. Hydrogeol J 15:891–901. https://doi.org/10.1007/s10040-006-0145-x

Wong G, Lambert MF, Leonard M, Metcalfe AV (2010) Drought analysis using trivariate copulas conditional on climatic states. J Hydrol Eng 15(2):129–141. https://doi.org/10.1061/(ASCE)HE.1943-5584.0000169

Wu J, Liu Z, Yao H, Chen X, Chen X, Zheng Y, He Y (2018) Impacts of reservoir operations on multi-scale correlations between hydrological drought and meteorological drought. J Hydrol 563:726–736. https://doi.org/10.1016/j.jhydrol.2018.06.053

Xue X, Wang T, Guo J, Yang Y, Peng F, Liu L (2012) Impact of global warming on drought in china. Sci Cold Arid Reg 4(3):201–210. https://doi.org/10.3724/SP.J.1226.2012.00201

Yusoff I, Hiscock KM, Conway D (2002) Simulation of the impacts of climate change on groundwater resources in eastern England. Geol So London Special Publ 193(1):325–344. https://doi.org/10.1144/GSL.SP.2002.193.01.24

Zeng N (2003) Drought in the sahel. Science 302(5647):999–1000. https://doi.org/10.1126/science.1090849

Zhao F, Xu Z, Zhang L (2012) Changes in streamflow regime following vegetation changes from paired catchments. Hydrol Process 26(10):1561–1573. https://doi.org/10.1002/hyp.8266

Part IV
Environmental Sensing

Chapter 6
Remote Sensing of the Environment

6.1 Introductory Remarks

The application of Earth resources satellites in environmental monitoring provides the opportunity to make informed decisions at different spatial scales that lead to a more resilient and sustainable world. The use of these satellites to support environmental assessment and decision-making in various capacities is experiencing an astronomical increase due to sophistication in new sensors and imaging capabilities (Ndehedehe 2022c). As part of the key strategic initiative to help foster the 2030 Agenda for Sustainable Development, experts across multidisciplinary fields of remote sensing are now working with end users in the application of remote sensing products to provide insights that underpin policies across several areas including, food security, urban resilience, preservation of wetlands, and conservation of biodiversity, among others (Kavvada et al. 2022). Optical missions like the Landsat programs and suite of Sentinel systems and those of hyperspectral platforms have been widely used in different disciplines within environmental science. Remote sensing data from these satellite missions have provided strong capability in global-regional analysis and local-scale analysis of the environment, thus facilitating our understanding of global environmental change through enhanced environmental monitoring, change detection, and understanding key drivers and processes surrounding human civilization. This monitoring leads to several benefits in several areas of society, including risk and resource management, urban development, as well as assess options that underpin planning, policy decisions and targeted government interventions. For instance, public policy framework on drought response has benefited from Earth observation data to aid the allocation of funds and water distribution to small scale farmers affected by extreme droughts (e.g., Ndehedehe et al. 2021; Brito et al. 2018). Additionally, consistent and repetitive monitoring of the global environment using remote sensing data are important for developing a robust approach to reducing risks of impacts from extreme events and identifying new priorities to drive socio-economic progress and promote sustainable development (Kavvada et al. 2022).

C. Ndehedehe, *Hydro-Climatic Extremes in the Anthropocene*, Springer Climate,
https://doi.org/10.1007/978-3-031-37727-3_6

6.2 New Capabilities in Satellite Remote Sensing Systems

Satellite Earth observation data has provided complementary source of information to environmental sensing and monitoring of the land surface, sometimes supporting in situ monitoring networks and in other cases, the only available source of data for large-scale assessment of the environment. New satellite sensors have emerged and they can now measure either directly or indirectly several components of the hydrological cycle, e.g., evapotranspiration, rainfall, and terrestrial water storage (e.g., Ndehedehe 2022d, f; Sheffield et al. 2018). Even though satellite data provide capabilities to measure a broad range of features, phenomena, and objects (e.g., water quality, floods, vegetation, biomass, among others), there are multitude of challenges to be considered when using these data for water resources management. These challenges have been summarized and categorized by Sheffield et al. (2018) as

- quality: how well the satellite measures object of interest or target and compares to observed data.
- resolution: this is the temporal (frequency of observation) and spatial (smallest object that can be measured) scales at which the data is collected.
- sampling: how often and where the satellite collects the data on land surface. Some satellites have global coverage but still limited in extent compared to others.
- legacy: this is the length and continuity of the mission and the availability of long-term records. Some satellite missions have ended and others are still ongoing.
- latency: this is how readily the data are transmitted and made available to the public. Some data are available in almost near real time while others take a while.

The sensor design and their orbital characteristics are responsible for some of these challenges as well as play important role in the applications of these satellite systems in various fields including their use in hydrology and water resources (Ndehedehe 2022c, f; Sheffield et al. 2018). Based on the uniqueness, the caveats identified above, and capability of these sensors, the work undertaken by Sheffield et al. (2018) identified the relevant remote sensing data for several environmental applications, as well as detailed suitable satellite missions that can provide data for water resources management and monitoring of hydrological hazards (e.g., floods, drought, etc.). Some of these data and missions include, precipitation (e.g., TMPA and GPM/IMERG), soil moisture (e.g., AMSR-E, SMOS, Sentinel-1, etc.), surface water height and extent (e.g., Jason-2/3, Sentinel-3, Landsat, Sentinel-2, etc.), land surface temperature (e.g., Landsat, Sentinel-3, MODIS, ASTER, etc.), snow (e.g., MODIS SCE, VIIRS SCE, etc.), evapotranspiration (e.g., Global ESI, MODIS ET, RS-PM, etc), groundwater (GRACE and GRACE-FO), and vegetation (e.g., Landsat, Sentinel-2, MODIS, VIIS, SPOT, etc.) (Sheffield et al. 2018). However, some of the inherent challenges limiting the use of some of these satellite missions have been addressed in recently launched satellite missions (e.g., SWOT) and emerging future missions (NISAR), thus boosting the capabilities of these sensors in remote sensing of the environment, including applications in water resources management. Overall, this chapter discusses the potentials of these new and emerging systems (optical, radar, and geodetic satellite missions) in environmental sensing.

6.2.1 Optical Satellite Systems

Well known traditional remote sensing systems like the multispectral Landsat series have been in used for several decades now, supporting a broad range of environmental applications. But new Landsat sensors like the recently launched Landsat 9 (https://landsat.gsfc.nasa.gov/satellites/landsat-9/) comes with innovations that can improve these capabilities further, including applications in water quality, crop mapping, water resources management, disaster management, land cover land use, climate change (e.g., droughts, wildfire, and floods), urban studies, among several others. These applications are apparently underpinned by the open data policy of the relevant agencies (e.g., US Geological Survey/ National Aeronautics and Space Administration-NASA and European Space Agency-ESA) warranting easy access to these data. Notably, the Landsat mission has been in operation for about 50 years now and its legacy is evident in the increasing use of data from this program at multiple temporal and spatial scales relevant to management decisions (Ndehedehe 2022c; Hemati et al. 2021). Similarly, the litany of ESA Sentinel missions, which commenced in 2016 ensure the continuity of global scale environmental monitoring of the Earth's atmosphere, oceans, and land on different spatial scales, thus complementing the Landsat program and other existing multi-satellite missions. However, with only few months after the launch of Landsat 9 in September 2021, a new Landsat mission, called *Landsat Next* is planned for launch later in 2030. Landsat Next is more than a follow-up mission to Landsats 8 and 9 sensors 9 (Operational Land Imager 1/2) as its sensor design and technology are expected to provide new capabilities for the next generation of Landsat users (https://landsat.gsfc.nasa.gov/satellites/landsat-next/).

The summary of its spectral characteristics in Fig. 6.1 shows Landsat Next is not just a follow-up mission but a strategic investment to improve capabilities of satellite observation in Earth resources monitoring. Landsat Next is a complete deviation from the suite of existing Landsat missions as it is expected to have 26 bands. Five of these bands are expected to have similar characteristics (spatial and spectral) to the European Space Agency's multispectral Sentinel-2 spectral channels. Notably, the similarity in these bands is critical to facilitate data merging and synchronization with other missions like Sentinel-2 while the additional spectral channels are expected to broaden Landsat applications underpinned by its capability to collect more data than its predecessor due to improved temporal resolution[1]. The data merging between multi-spectral Sentinel-2 mission and the planned Landsat Next is possible in some bands because of the similarity in ground sampling distance and spectral channels,

---

[1] (https://landsat.gsfc.nasa.gov/satellites/landsat-next/).

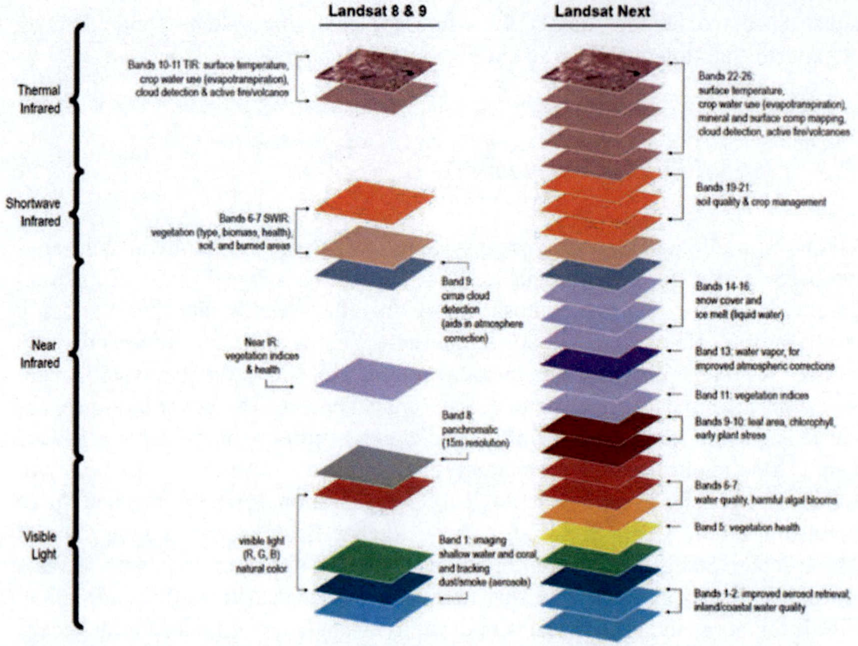

Fig. 6.1 Spectral characteristics of the proposed Landsat Next in 2030 in relation to ongoing Landsat 8 and 9 missions. *Source* NASA (https://landsat.gsfc.nasa.gov/satellites/landsat-next/)

e.g., Red Edge 1 and 2, NIR Broad, (https://landsat.gsfc.nasa.gov/satellites/landsat-next/). Based on the new spectral channels in Landsat Next, some anticipated emerging applications of this new satellite mission will include, soil quality and crop management (bands 19–21); snow cover and ice melt mapping (bands 14–16); early plant stress, leaf area, and chlorophyll (bands 9–10); and water quality and harmful algal blooms (bands 6–7), among others (Fig. 6.1). Satellite programs like the Landsat and Sentinel missions remain very important tools in advancing understanding of environmental changes in the Anthropocene. As opposed to the moderate spatial (30 m for all Landsat series and 10 and 20 m for Sentinel-2) and temporal (16 days for Landsat) resolutions of these past missions, Landsat Next is a new generation of satellite missions that will enhance multi-disciplinary science applications in monitoring environmental changes. Unlike the relatively new Landsat 9 OLI, this capability of Landsat Next will be underpinned by improved temporal, spectral, and spatial resolutions (10 and 20 m). Improvements in spatial resolution apparently will lead to extracting features and mapping of land surface conditions more accurately due to finer details.

6.2.2 Radar Technology

Radar imaging systems are emerging as excellent substitutes and in other cases complementary to passive remote sensing systems. This is because radar systems operate independently of lighting conditions and largely independently of weather (e.g., Dong and Chen 2018). This makes them very useful in several fields, including oceanography, land surface temperature mapping, crop health and wetland monitoring, disaster monitoring (e.g., floods, droughts, landslides, etc.), water resources, meteorology, climatology, desertification, surface albedo, geology, forestry, and mapping oil spill, among others (see, e.g., Ndehedehe 2022c; Okwuashi et al. 2021; Hemati et al. 2021; Oussou et al. 2020; Lavender and Andrew 2016; Beck et al. 2006; Rango 1994). Importantly, because radar observations are not limited by cloud cover and can collect data at any time of day, they are considered as more sophisticated and their applications in environmental monitoring are much broader compared to multispectral data like the Landsat and hyperspectral systems. Additionally, new satellite radar technology like the NASA-ISRO SAR Mission (NISAR) has emerged as a sophisticated SAR mission with broadened applications compared to other well known radar missions (e.g., Sentinel-1, Radarsat-2). In this chapter, the focus is on radar missions because of their sophistication, all-weather, day and night imaging capabilities unlike the passive systems. This uniqueness of the radar mission is important to help consolidate on the gains of satellite remote sensing in tracking environmental changes in the Anthropocene. With this in mind, the novel NISAR mission is discussed in detail being the latest of radar missions. Its mission objectives, scope, and broad multidisciplinary capabilities in environmental assessment are highlighted. Among the capabilities discussed, the potential of its combination with satellite gravity missions like GRACE to improve the assessment of groundwater status and impacts associated with over extraction is stressed.

6.3 The NISAR Mission: A Blessing to the Anthropocene?

One of the key motivation for investments in radar program is because of the limitations of passive remote sensing systems. They are limited by cloud cover and sun illumination conditions because of their dependence on external source of energy (i.e., the Sun). Based on imaging capabilities and sensor design characteristics, the data collected by remote sensing systems are different. Remote sensing platforms, be it satellite, aerial photography or close range like remotely piloted aircraft can be passive or active systems. The former include optical sensors or optical cameras and scanners on-board satellite (or space-based) and airbone platforms. Passive remote sensing methods can be classified as multispectral (e.g., Landsat, Sentinel, SPOT, MODIS, Quick Bird, etc.) and hyperspectral remote sensing, which uses imaging spectrometers that measure near-laboratory-quality spectra in narrow spectral bands (Ndehedehe 2022c). Unlike the former in which images are collected by recording the reflection of solar radiation from the earth using airborne and spaceborne sensors (e.g., Lavender and Andrew 2016; Camps-Valls et al. 2012), hyperspectral images are highly innovative remote sensing data that enable a better distinct discrimination of earth resources or targets (e.g., Okwuashi and Ndehedehe 2020; Lein 2012). However, as opposed to the passive systems, the active systems or radar remote imaging systems generate their own electromagnetic energy in the microwave region to "illuminate" the terrain and detects the reflected energy or signal from the Earth surface (Ndehedehe 2022c).

Synthetic Aperture Radar (SAR) data is one example of radar remote sensing. Through transmitting and receiving in multiple polarizations, the radar system can detect unique and distinct features of both natural (e.g., water bodies) and man-made targets (e.g., buildings). Another example of radar remote sensing is the Polarimetric Synthetic Aperture Radar (PolSAR). This is an optimized version of the well known SAR system. PolSAR can collect information regardless of weather conditions by transmitting and receiving electromagnetic waves in diverse polarimetric states (e.g., Okwuashi et al. 2021). However, the NASA-ISRO SAR Mission (NISAR, Fig. 6.2) is a new satellite radar technology involving partnership between US and India that will provide more robust sophisticated capability in environmental sensing. The new mission is designed to measure changes in the earth system including ice masses, biomass, groundwater, natural hazards, sea level rise, among other important applications. As will be discussed in the next section, all of the important science applications of NISAR are in key environmental areas where major changes have been observed in the Anthropocene. Some of these include, groundwater, coastal wetlands, extreme events and hazards, snow and ice sheets, glaciers, forest ecosystems, soil moisture, permafrost, pollution, etc. NISAR will provide information on these critical environmental and hydro-climatic variables, including monitoring changes in land surface motions, thus making it the first multidisciplinary SAR mission. Given these capabilities, the NISAR mission is indeed a blessing to the Anthropocene. One would consider it a blessing because, on the one hand, it fills the existing data gaps, especially areas where gauged observations are lacking. On other hand, information from NISAR is potentially critical to help advance human societies in terms of helping to facilitate dialog among knowledge providers and stakeholders. This information can also feed directly into action plans and policy that seek mitigation from climate change and human influence as well as protection of valuable natural and ecosystem resources.

6.3.1 NISAR Mission Science Overview and Scope

NISAR is an acronym for **N**ASA-**ISRO** **S**ynthetic **A**perture **R**adar (SAR). Unlike past SAR missions like those of Sentinel-1, NISAR is a new radar mission with multidisciplinary capabilities and applications that will collect data across the globe on land surface changes to help understand drivers and consequences of such changes (NISAR 2018). This mission is the result of collaboration and partnership between

Fig. 6.2 Artistic expression of the **N**ASA-**IS**RO **S**ynthetic **A**perture **R**adar (NISAR) concept. NISAR is a joint Earth-observing mission between NASA and the Indian Space Research Organization (ISRO). The multidisciplinary mission will measure the causes and consequences of land surface changes using advanced radar imaging technique. *Source* NASA-ISRO SAR Mission (https://nisar.jpl.nasa.gov/mission/quick-facts/)

the National Aeronautics and Space Administration (NASA) and the Indian Space Research Organisation (ISRO). This is why the mission is coined NISAR is a highly innovative and sophisticated SAR mission consisting of a dual frequency with fully polarimetric radar system that has an imaging swath greater than 240 km (NISAR 2018). Because of this design (Fig. 6.2), NISAR will have a temporal frequency of 12-days (Table 6.1), making it possible to generate interferometric time series and perform systematic global mapping of changes across different regions of the Earth. The NISAR platform consists of two fully capable synthetic aperture radar instruments. These are the 24 cm wavelength L-SAR and 10 cm wavelength S-SAR with each of these instruments designed as an array-fed reflectors to work as SweepSAR (a method that achieves wide swath at full resolution) scan-on-receive wide swath mapping systems (NISAR 2018). NISAR is designed to measure ice mass and changes in land surface motions, ecosystem disturbances, and biomass to improve understanding of earth system processes that underpin our capability to improve forecasts and assessment of landscape changes and among others. Table 6.1 summarizes the features that will make this possible, including the availability of time

Table 6.1 Overview of NISAR capabilities and characteristics as detailed in the NISAR Mission science handbook. *Source* (NISAR 2018)

Item	NISAR	Capabilities
1	L-band (24 cm wavelength)	Foliage penetration and interferometric persistence
2	S-band (12 cm wavelength)	Sensitivity to light vegetation
3	SweepSAR technique with imaging swath greater than 240 km	Global data collection
4	Polarimetry (single/dual/quad)	Surface characterization and biomass estimation
5	12-day exact repeat	Rapid sampling
6	3–10 m mode-dependent SAR resolution.	Small-scale observations
7	3 years science operations (5 years consumables)	Time series analysis
8	Pointing control less than 273 arcseconds	Deformation interferometry
9	Orbit control less than 350 m	Deformation interferometry
10	Greater than 30% observation duty cycle	Complete land/ice coverage
11	Left/right pointing capability	Polar coverage, north and south
12	Nominal mission pointing: left only	Time series continuity
13	SAR resolution- 3–10 m	

series of radar observations that will reliably aid the assessment of temporal changes of the earth's landscape every 12 days (Table 6.1). As further detailed in NISAR (2018), the science focused areas include Solid Earth, Ecosystems, and Cryospheric science disciplines. Specific science applications of NISAR and how its potential to meet the needs of a broad end-user cross-disciplinary community are detailed in the next section.

6.3.2 *Environmental Applications of NISAR*

As a multidisciplinary geodetic and polarimetric radar-imaging mission, NISAR will serve several scientific objectives including the following:

- determining the probability of geodetic pertubations and earth movements like earthquakes, volcanic eruptions, landslides, and land subsidence, which may also be triggered by groundwater depletion,
- understanding changes in carbon storage and uptake in, e.g., agricultural, permafrost systems, wetland, etc., and
- understanding the impacts of climate change on ice sheets and sea level rise.

Some specific science focused areas and multi-disciplinary environmental applications of NISAR are summarized in what follows.

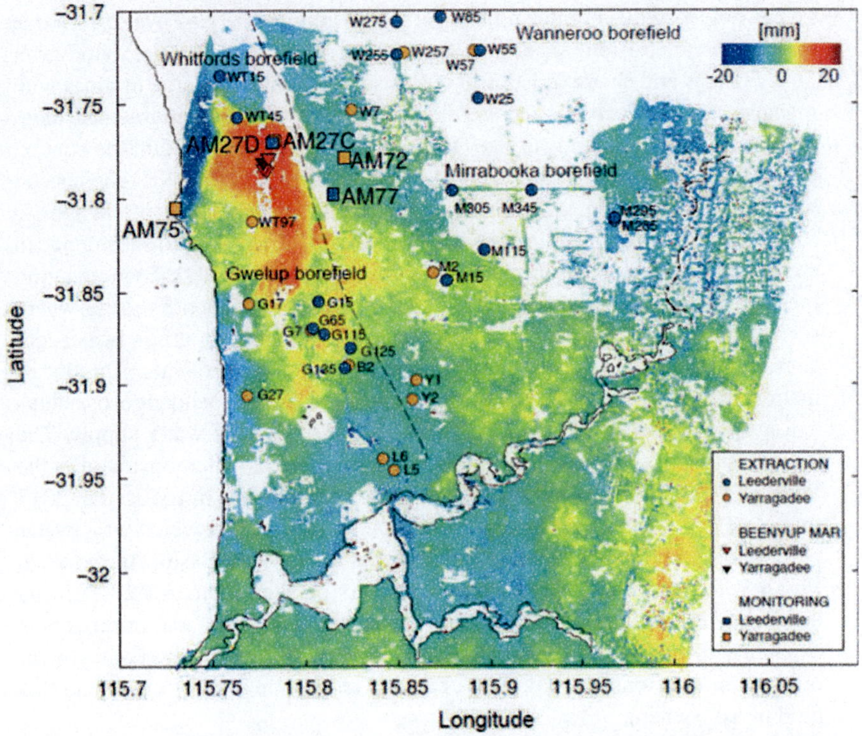

Fig. 6.3 Cumulative groundwater-related ground displacements in Perth basin since 2016. The points (circles and squares) indicate locations of extraction, injection, and monitoring bores while the acronym MAR in the legend is Managed Aquifer Recharge—a water management program targeted to reduce extraction and sustain levels of hydraulic head in relevant aquifers *Source* Parker et al. (2021)

(i) **Groundwater**: Groundwater remains an important source of water for agriculture and other uses. During prolonged droughts, groundwater extraction increases and with expansion of agriculture into semi-arid regions, substantial groundwater depletion is inevitable (Ndehedehe et al. 2021). Several regions including in India, northern Africa, California Central Valley, Middle East, North China Plain, China, and Canning Basin, Australia, are experiencing groundwater loss due to these reasons (Ndehedehe 2022a; Rateb and Kuo 2019; Ojha et al. 2018). The combined impacts of groundwater droughts (substantial and prolonged decline in groundwater aquifer) and water extraction can cause a gradual shift of the land surface (Fig. 6.3), or land subsidence (e.g., Parker et al. 2021; Castellazzi et al. 2021; Bierkens and Wada 2019). When this happens, it leads to the collapse of public infrastructure like causing damages to buildings, bridges, cracks in roads and can alter runoff trajectory, resulting in potentially higher risk of flooding (Ojha et al. 2018). In the recent past, the use of SAR data from previous missions (e.g., Radarsat-2, ALOS-PALSAR, and Sentinel-

1) has helped with monitoring such impacts and quantifying water loss due to withdrawal (Castellazzi et al. 2016, 2018). Using SAR data, groundwater flow changes and characteristics in response to artificial recharge or areas with managed aquifer recharge can be monitored by observing patterns and magnitudes of uplift and land surface displacements (Fig. 6.3). Within the context of water resources and assessment of groundwater status, the NISAR mission will provide a radar imaging capability to aid the measurement of changes in surface elevation to generate time series of the surface uplift and subsidence in areas affected (https://nisar.jpl.nasa.gov/mission/quick-facts/). By generating these through InSAR techniques, NISAR will enable us identify areas where surface uplift and subsidence are associated with changes in groundwater loss caused by drought or slow recharge, and extensive withdrawals. This information will enhance our ability to protect groundwater dependent ecosystems and avoid consequences resulting from permanent loss of water supply. The unique capability of SAR data in this respect was already demonstrated in the Perth basin where managed aquifer recharge program was introduced in 2017 to replenish groundwater through the injection of treated recycled wastewater two important confined groundwater aquifers in the Perth Basin. As shown in Fig. 6.3, these confined groundwater aquifers or sites (bore AM27C on the map) showed substantial uplift due to the injection. NISAR will improve this capability through improved spatial and temporal resolution and complement other radar missions by providing extended radar time series to facilitate this kind of assessment.

(ii) **Extreme events and hazards**: As was discussed in Chap. 1 of this book, extreme events like floods, fire, and drought are very typical of the Anthropocene. The impacts resulting from these events have damaging consequences including loss of lives and properties, diminished water supply, low crop yield, poor water quality, loss of hydro-power potential, etc. These can lead to social and economic problems on various levels (e.g., household, local regional, and national). With a varying spatial resolution of 5–10 m and relatively reasonable temporal frequency, NISAR will provide the capability to map extreme events (e.g., floods) on a relatively fine scale that will underpin important management decisions and operations. For instance, it will be possible to map floods in open and forested landscapes around the globe twice every twelve days to provide timely information for accurate flood assessment. NISAR data is expected to be publicly available after two days of observation and within hours in response to disasters (http://nisar.jpl.nasa.gov/applications). This open data policy and speedy delivery of products will ensure actionable, timely data for several applications, including rapid emergency response to extreme floods and other forms of environmental disasters. Moreover, with the all-weather, day/night imaging capability of NISAR, areas affected by hurricanes, tornadoes, and volcanic hazards will be mapped to facilitate the assessment of damages caused by the magnitude and severity of these disasters. There is also potential in flood forecasting. Maps of changes in water levels will be derived from NISAR to predict potential of downstream flooding and enhance the mapping of major tropical river systems. In this context, NISAR will fill a major data gap of poorly distributed stream gauges by making it possible to generate water levels in flooded riverine areas from radar images. NISAR is also set to help with quickly generating hazard maps of natural disasters including major hurricanes, earthquakes, landslides, and tsunamis.

(iii) **Coastal resources, forest, and wetlands**: Much of the world's human population are clustered around the coast. This has resulted in the emergence of big cities and highly populated urban centers like New York, Mumbai, Shangai, Los Angeles, Lagos, and Sáo Paulo, among others. These regions are at the interface of land and water, thus usually at risk of exacerbating land loss caused by changing environmental conditions. For example, estimates of substantial land loss in the Mississippi Delta revealed that about 25% of the delta's land area has been lost since 1932 (http://nisar.jpl.nasa.gov/applications). Satellite imaging of the Earth is helping with the efficient monitoring of this kind of changes. In particular, radar observations of forest and vegetation changes, including quantifying distribution and amount of above-ground woody biomass can now be improved by relying on L-band PALSAR and LiDAR data (Ndehedehe 2022e). This is largely underpinned by the sensitivity of radar to forest disturbance, also making it an important source of independent information for below-canopy changes. The NISAR microwave radar will strengthened this capability and beyond by enabling accurate measurements of coastal and inland surface water changes to inform on potential areas that should benefit from coastal restoration projects. As highlighted in the NISAR science application documents (http://nisar.jpl.nasa.gov/applications), monitoring changes in land loss and gain is important to understanding variations in deltas to help with protection through reversing land loss in the shrinking deltas. The NISAR mission will do this efficiently by monitoring the status of low elevation coastal areas.

(iv) **Land Subsidence**: When groundwater withdrawal is extensive due to prolonged drought or resource extraction, this can lead to surface deformation or landslides (Ndehedehe 2022a). This is because SAR-derived vertical ground displacements are assumed to be associated with aquifer compaction in areas with extensive groundwater extraction (e.g., Smith and Majumdar 2020; Castellazzi et al. 2018). Land subsidence can occur naturally and caused by human activities like groundwater extraction (http://nisar.jpl.nasa.gov/applications). Case-specific examples in the US and Mexico show how excessive groundwater extraction from compressible aquifers can cause land subsidence (Ojha et al. 2018; Castellazzi et al. 2018). As was mentioned previously, this causes damage to buildings and public infrastructures like roads, etc. The use of InSAR techniques from radar observations have helped with detailed measurements of large-scale land subsidence associated with groundwater over extraction or prolonged drought, thus complementing the lack of observational data in this area (e.g., Smith and Majumdar 2020; Ojha et al. 2020). NISAR data will measure changes in land elevation and collect SAR images that will improve our capability to detect and quantify localized land subsidence.

(v) **Snow and ice sheets, glaciers, and terrestrial permafrost**: Sea level rise is another typical problem in the Anthropocene caused by ocean thermal expansion and melting of ice and glaciers, among other factors (See Chap. 1). Satellite remote sensing technique, either independent or in combination with other methods is helping with tracking ice sheets and glacier ablations (Ndehedehe 2022g). Satellite missions like the ICESat measure ice sheet mass balance, cloud, and aerosol heights (https://icesat.gsfc.nasa.gov/). They collect multi-year data on the elevation of the Earth's ice sheets, clouds, vegetation, and the thickness of sea ice that underpin our capability to track climate change influence on hydrology. Even the European Space Agency Sentinel-3 mission is also equipped with the capability to monitor sea and land ice topography, generally, ongoing satellite missions are limited in monitoring the rapid depletion of some glaciers. Satellite radar data from the NISAR mission will fill this gap by providing global maps of ice flow that underpins impact assessments of sea level rise (http://nisar.jpl.nasa.gov/applications). Measurements from NISAR will be systematic, revealing both short-term variations and long-term trends as well as marked changes in the behavior of glaciered regions (http://nisar.jpl.nasa.gov/applications). This will provide an opportunity to optimize predictive models in assessing impacts of sea level rise. Additionally, the NISAR mission will observe global changes in permafrost landscapes on a more frequent basis (bi-weekly) to help with assessing major hazard to infrastructure.

(vi) **Soil moisture**: Radar observations have made it possible to retrieve a 3-km grid global surface soil moisture every 8-day period using the L-band dual-copolarized Soil Moisture Active Passive SAR data (Kim and van Zyl 2017). Even though NISAR mission will provide similar capability, radar imaging of soil moisture from NISAR will be optimized to generate field scale or high-resolution product. Additionally, the frequency at which global soil moisture will be mapped is set to improve, occurring every 6 to 12 days at the field or water management scale. Given the importance of soil moisture in regulating interactions between land surface and the atmosphere and serving as a direct indicator of water availability for crops and agricultural drought (Agutu et al. 2021), the level of unprecedented spatial detail offered by a NISAR-derived soil moisture will enable an accurate assessment of agricultural forest water resources. Satellite soil moisture products retrievals suggest that improvements are required due to the influence of vegetation in some areas and measurement errors (Agutu et al. 2021; Kumar et al. 2018). The further assessment of these uncertainties and measurement errors will be underpinned by NISAR data and can potentially help improve future model simulations and forecasts of soil moisture through data assimilation.

(vii) **Environmental pollution**: Advancements in satellite technology now makes the monitoring of flaring stations possible (Faruolo et al. 2021; Anejionu 2019). The emission of gas flares occurs in both visible and infrared radiation at different wavelengths. Therefore, one can observe the visible radiation easily from satellite imagery with very high spatial resolution (for e.g., 30cm) of the visible spectrum. Diverse nighttime lights include fires, gas flares, human settlements, and fishing boats based on the position, frequency, and emergence of the lights. With high-resolution satellite imagery and with the capability to identify different light sources at nighttime using satellite data, it is now possible to distinguish between similar signals from gas flares and human settlements amidst other factors ranging from parallax visions, omission/commission, OLS pixel size, surface effects, as well as geo-location when viewed from space (Faruolo et al. 2021). However, apart from other environmental hazards like landslides, wildfire, and floods, the NISAR data will provide a better capability to track changes caused by oil spill. To cushion the impacts of damage, rapid detection of areas in inland open waters affected by oil spill is critical and NISAR data can provide this information. NISAR-radar imaging of oil spill or slicks on surface water will be easy to detect because of their uniqueness as they usually appear as dark features. The capability to quickly detect slicks or oil-related effects on surface water is because amount of backscattered signals into radar antenna is attenuated due to oil damps (http://nisar.jpl.nasa.gov/applications).

6.4 Multi-satellite Systems to Improve Understanding of Water Storage Components

6.4.1 Remote Sensing and Geospatial Methods in Groundwater Assessment

The capability of optical remote sensing to assess groundwater potential is emerging, making it possible to improve understanding of recharge and discharge mechanisms in areas where monitoring bores are lacking or limited. Several new techniques using satellite remote sensing to assess the status of groundwater aquifers that underpins reasonable management of groundwater resources are emerging. Landsat 7 ETM, for example, has been applied to isolate hydrothermally altered rocks in eastern Tauride belt (SE Turkey) using the Crosta techniques (Aydal et al. 2007). Even though this technique is not detailed further here, the combination of Landsat 8 with digital elevation models supports the identification of fractured aquifers (Fig. 6.4). This was demonstrated by Oussou et al. (2020) who employed digital image processing techniques (e.g., image enhancement and directional Sobel filters) to analyze Landsat 8 and PALSAR data in the Zou basin of Western Africa. Following their identification of main orientations of fractures, the productivity of some aquifers' based on productivity of some bores (e.g., boreholes located near the main orientations) were detected. This was possible by generating accurate fracture maps and derived products (fracture density, coincidence map and cross-points) from the use of Landsat 8 (visible and infrared bands) and PALSAR-derived digital elevation models as well as in situ bore well data. As an illustration, the coincidence (left) and fracture density (right) maps over the Zou basin in Benin republic, West Africa is

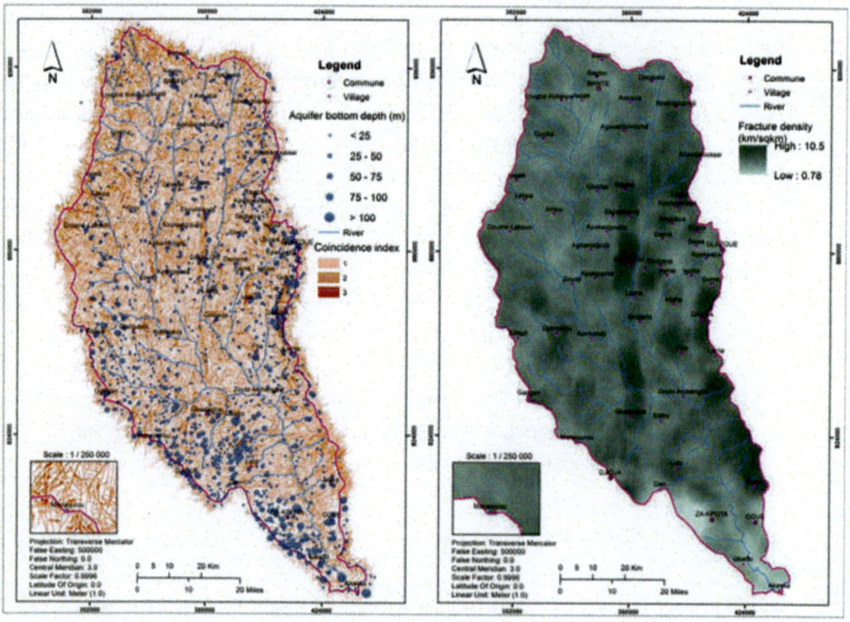

Fig. 6.4 Remote sensing and geospatial technique to identify fractured aquifers. On the left and right panels are the Coincidence and fracture density maps, respectively, over the Zou basin in Benin republic, West Africa. *Source* Oussou et al. (2020)

indicated in Fig. 6.4. The coincidence map is an important tool that supports better and more precise fracture identification Oussou et al. (2020). Fractures beneath the weathered layer may be difficult to detect. However, optical remote sensing methods (Fig. 6.4) provides opportunities to extract important information regarding their hidden geometries (Oussou et al. 2020). In the Hawker Area located in South Australia, remote sensing and geospatial methods were combined with the multi-influencing factors to identify groundwater recharge potential areas (Ahmed et al. 2021). The processing of remote sensing data like Landsat using band ratio, enhancement methods and multivariate methods will lead to lineament density and land cover maps while the processing of digital elevation data will result in drainage density and slope features. These geospatial layers can then be integrated using weighted approach (e.g., multi-influencing factors, analytical hierarchy method, etc.) and ranked to delineate groundwater recharge potential zones (e.g., (Ahmed et al. 2021)). The example in Fig. 6.5a shows groundwater recharge potential for the Hawker Area in South Australia. Based on the ranking system adopted, three classes of recharge potential were identified: good, moderate, and low Fig. 6.5b. The different groundwater potential zones show different proportion of areas ranging from greater than 25% (good) to above 50% (moderate).

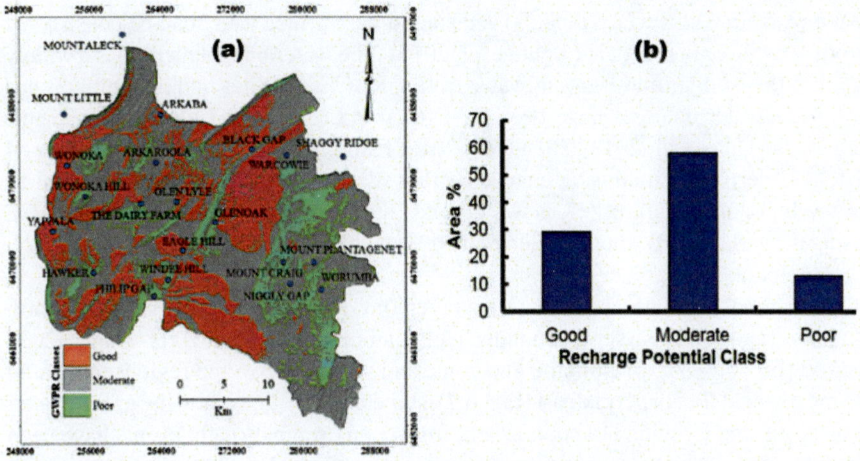

Fig. 6.5 Delineating recharge potential areas in South Australia by combining remote sensing and geospatial methods. *Source* Ahmed et al. (2021)

6.4.2 Groundwater from GRACE Satellite

As was mentioned earlier in Chap. 4, satellite observations of changes in terrestrial water storage from the Gravity Recovery and Climate Experiment (GRACE) has enabled the assessment of groundwater status and is further detailed in Ndehedehe (2022a). The GRACE mission has helped with assessing and quantifying groundwater variations across several regions of the world, thus contributing to our understanding of climate change impacts on the water cycle (Tapley et al. 2019). Large aquifer systems and how they are affected by climate change and extensive water extraction can now be tracked using GRACE satellites in conjunction with auxiliary data from models (Ferreira et al. 2020; Ojha et al. 2018; Chen et al. 2016). To achieve this, soil moisture and surface water information are usually taken from models and sometimes in situ. They are then subtracted from the terrestrial water storage observed by GRACE using the water budget method (e.g., Ferreira et al. 2020). This approach is based on the assumption that GRACE observed variations in terrestrial water storage (**T**) comprise changes in groundwater (**G**), soil moisture (**S**), ice and snow water as: (**I**), inland surface water/reservoir storage (**R**), groundwater (**G**), and canopy water storage (**C**) as (Ndehedehe 2022a; Ferreira et al. 2020),

$$\mathbf{T}_{i,j} = \mathbf{G}_{i,j} + \mathbf{S}_{i,j} + \mathbf{I}_{i,j} + \mathbf{R}_{i,j} + \mathbf{C}_{i,j} \tag{6.1}$$

were the sub-scripts i and j are the number of rows (time), and columns (spatial locations), respectively. In practice, groundwater storage anomalies can then be estimated by rearranging the water budget equation (Eq. 6.1). Soil moisture, surface water, canopy water storage, and snow are usually sourced from models (e.g., out puts from land surface models) and sometimes surface water can be constructed from satellite or in situ data (Agutu et al. 2019). The water budget method is a widely used in the estimation of groundwater from GRACE satellites and declining trends in groundwater changes have been reported in past literatures that employed this approach (Ojha et al. 2018; Chen et al. 2016; Famiglietti 2014). The consistency of GRACE-derived groundwater changes with well observations have been established in the literatures. Figure. 6.7 shows groundwater changes estimated from GRACE using the widely used water budget method Water budget method (Ndehedehe 2022a; Ferreira et al. 2020).

As with aforementioned studies, some regions, including north-east Brazil, north America, and Asia also show groundwater depletion for the April 2002–August 2021 period (Fig. 6.7). Apart from the Patagonia and ice/snow covered regions that could be the result of lost in permafrost (Fig. 6.7), some of these areas showing groundwater loss are strong climatic hot spots where frequent extreme drought events have been reported or well known. For instance, frequent extreme drought in California (e.g., 2006–2010 and 2011–2015), Eastern Europe, and southeastern Brazil (e.g., 2014–2017) led to critical water crises in these regions. Assessing groundwater status from GRACE satellites offers new capability in freshwater reporting. By integrating this satellite-based monitoring framework with national groundwater database architecture, this can lead to a better understanding of groundwater hydrology and freshwater accounting. GRACE-derived terrestrial water storage has enabled the monitoring of freshwater changes globally (Rodell et al. 2018). By subtracting other relevant water storage components (e.g., soil moisture and surface water) from terrestrial water storage, changes in groundwater can be estimated (Fig. 6.7). With this capability, important contributions to our understanding of global groundwater and its responses to changes in climate are now feasible. The change in water availability in time and space and groundwater loss in some regions can potentially limit access to drinking water as well as reduce the availability of freshwater for irrigation. As these water issues can lead to socio-economic problems and crises, satellite measurements of freshwater can serve as a tool to assess water security on a regional and local scale.

6.4.3 Satellite Data Fusion for Groundwater Monitoring

While this is an important tool in the Anthropocene to aid large-scale assessment of groundwater status, the new framework for freshwater water accounting introduced by Valencia et al. (2020) combined information from optical satellite systems with ancillary and model data to quantify changes in irrigated crop water use and groundwater abstraction rates. This approach was based on a data fusion and inversion approach that combined multi-satellite information with model outputs. The framework (Fig. 6.6) that was developed from this study showed satellite products or its derivatives (Landsat, NDVI Leaf Area Index, Albedo, etc.) as essential inputs to quantifying groundwater abstraction rates. Using this framework, Valencia et al. (2020) estimated approximately 5.5 billion cubic meters for the year 2015. Annual

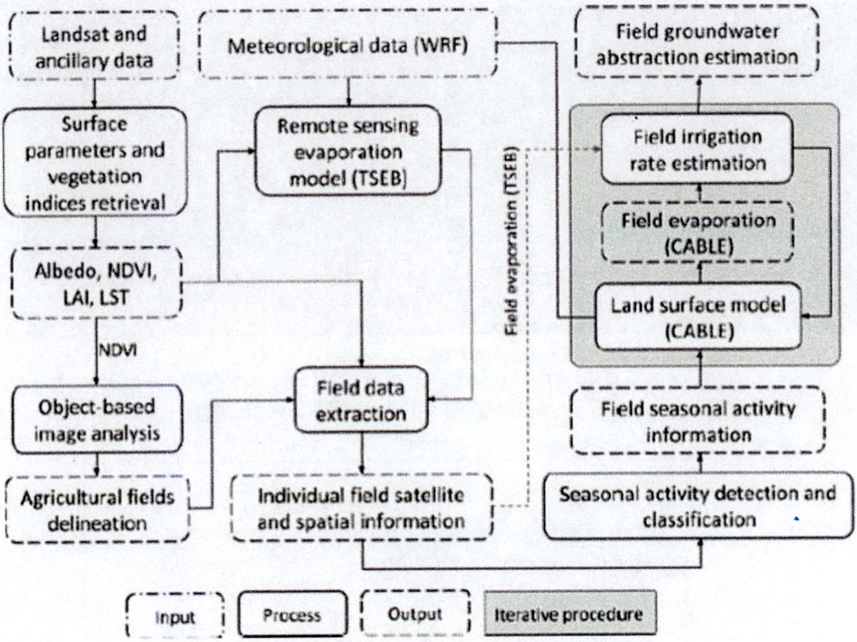

Fig. 6.6 Satellite data fusion techniques to quantify groundwater abstraction using. *Source* Valencia et al. (2020)

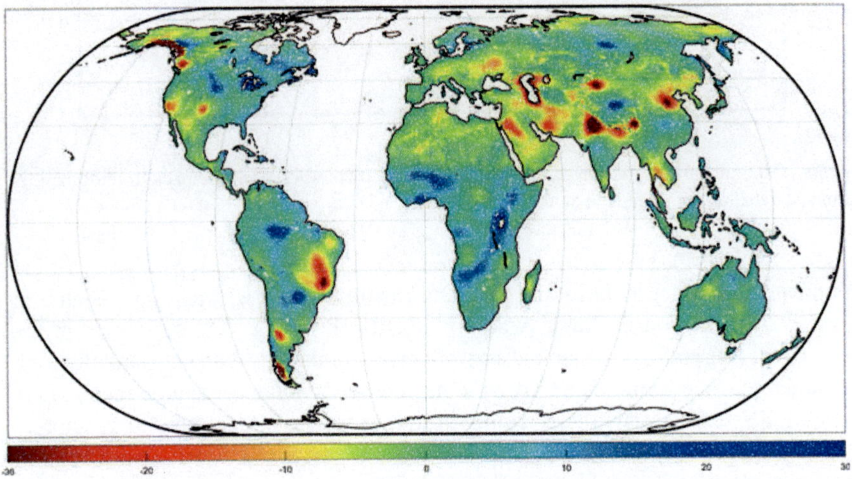

Fig. 6.7 Global groundwater (mm/yr) changes estimated from the Gravity Recovery and Climate Experiment satellite (April 2002–August 2021). *Source* Ndehedehe (2022a)

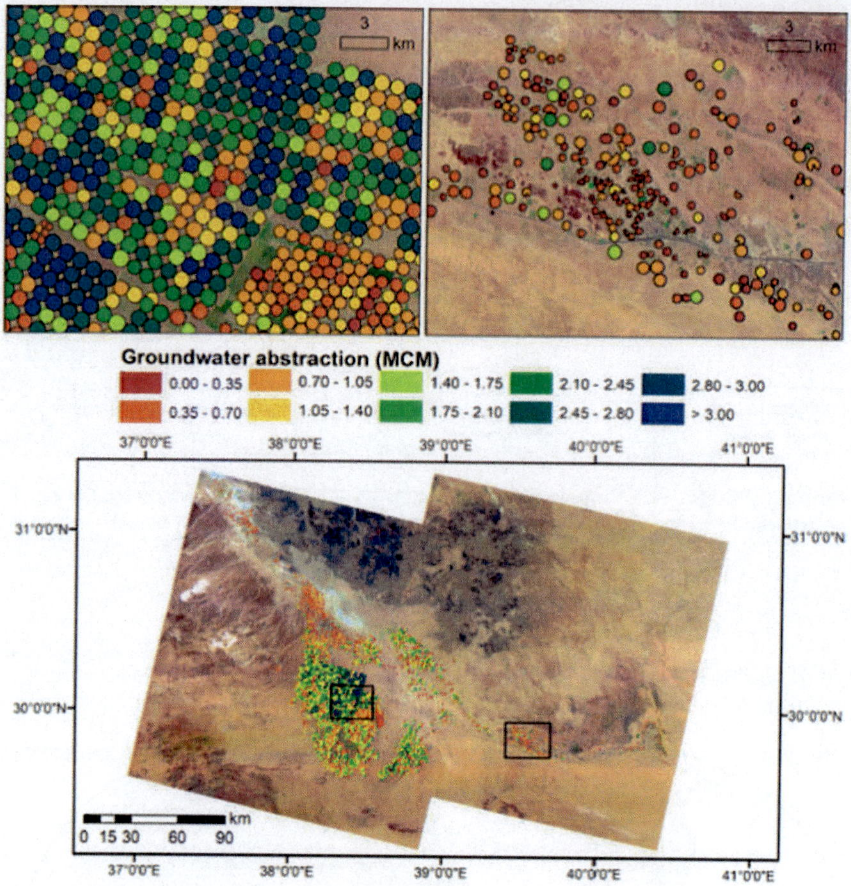

Fig. 6.8 Groundwater extraction rates (in million cubic meters-MCM). The background image is a Landsat data. *Source* Valencia et al. (2020)

groundwater abstraction based on this data fusion method as was implemented in an agricultural region in Saudi Arabia if illustrated in Fig. 6.8. As shown in the upper panel (Fig. 6.8), there are clusters of agricultural areas with substantially high abstraction rates of more than 3 MCM/yr as opposed to other areas with much lower abstraction rates. Clearly, there are opportunities to use optical data to accurately define and quantify irrigation estimates and groundwater extraction to aid agricultural water management on a local scale. These opportunities are particularly underpinned by the continuous availability of satellite information though the presence of cloud cover can limit the availability of satellite data in some locations due to environmental conditions. The use of radar data in this sense circumvents such problems. As discussed in the next section, satellite radar data like, Radarsat-2, ALOS-PALSAR, and Sentinel-1 have been widely used in groundwater assessment. However, there is even greater and improved potential with NISAR in groundwater assessment on a much higher resolution, be it spatially or temporally. This is discussed further in the next section.

6.4.4 Integrating SAR with GRACE Follow-on to Assess Groundwater Status: The NISAR Prospect

In Chap. 1, groundwater decline was identified as a major characteristics of the Anthropocene relative to the Holocene conditions. The depletion of global groundwater resources is well established in the literature with regions like Northwest India, U.S California's Central Valley, Middle East, Australia, Mexico, and the Northern China Plain, among others being some of the major hydrological hotspots where groundwater losses have been reported (see, e.g., Parker et al. 2021; Castellazzi et al. 2018; Ojha et al. 2018; Chen et al. 2016; Famiglietti 2014; Voss et al. 2013). Given that 70% of global groundwater withdrawals are used for irrigation and agricultural purposes, groundwater will continue to remain an important source of freshwater in the future, especially in arid and semi-arid domains where surface water is limited. As mentioned earlier, over pumping of groundwater that results in depletion causes land subsidence. Several urban centers around the world, for example, Mexico, Chennai, Perth, Sao Paulo, Jakarta, Melbourne, and Ontario, among others have experienced groundwater loss that resulted in substantial land subsidence and negative impact on public infrastructure. The use of SAR data independently or in combination with GRACE satellite data has helped with the assessment of subsidence rates and quantifying groundwater losses in these regions (e.g., Castellazzi et al. 2021; Ojha et al. 2018, 2020). On a water management scale, SAR data provides the capability to help track groundwater depletion and manage aquifer recharge volumes that helps avoid infrastructure damage (Parker et al. 2021). As illustrated in Fig. 6.3 for the Perth Basin, areas with maximum relative uplift (bore AM27C on the map) and regions with subsidence (bore AM75 on the map), which shows negative ground displacement where identified by generating interferograms from successive SAR satellite images. An interesting aspect of SAR data applications in groundwater management is its capability to track the response of artificial recharge of aquifers using treated wastewater has demonstrated in the Perth Basin (Parker et al. 2021). This is obvious in Fig. 6.3, which shows manage aquifer recharge (MAR on the map) locations (Leederville and Yarragadee) experiencing an uplift of up to 20 ± 3 mm due to injection.

The use of SAR data alone to map groundwater overestimation in Central Mexico is another case study demonstrating the utility or radar time series to identify aquifers under threat. The study undertaken by Castellazzi et al. (2021) used SAR to track how change in water management slowed down groundwater depletion in the Querétaro Valley of Mexico. Using interferometry analysis of radar time series (Radarsat-2 and Sentinel-1) covering the 2004–2020 period, they reported a five-times decrease in maximum ground deformation. Figure 6.9a illustrates InSAR-derived vertical

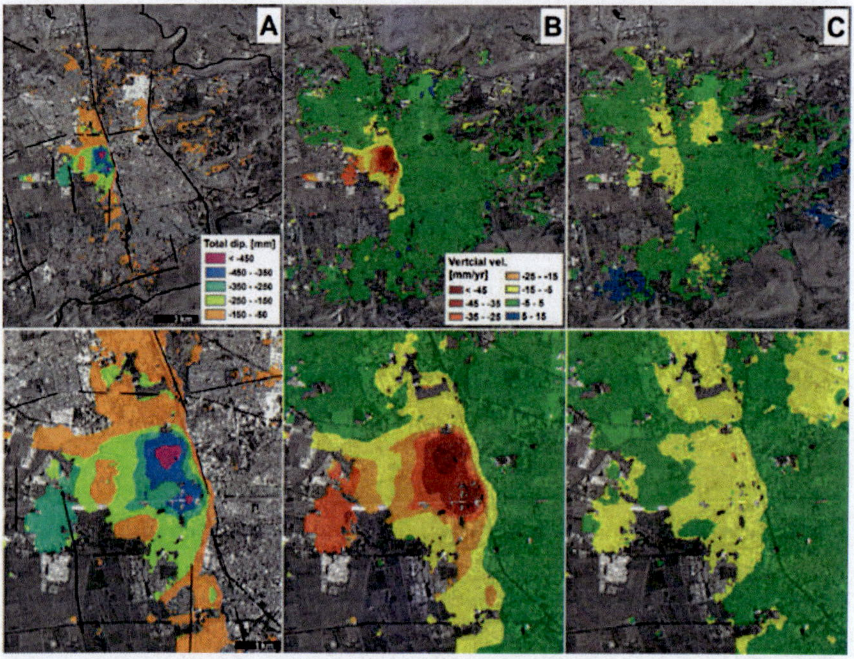

Fig. 6.9 InSAR-derived vertical ground displacement due to groundwater changes for **a** the entire period (2006–2020) **b** before February 2011 **c** after February 2011. The panels below **a–c** are zoomed locations over the "5 de Febrero" fracture. *Source* Castellazzi et al. (2021)

ground displacement capability for the entire period. Deformation rates associated with groundwater depletion prior to the water management change implemented in 2011 indicates vertical displacement velocity of up to -45 mm/yr (Fig. 6.9b). However, this was slowed down after the change in water management scheme (when water import scheme was implemented as discussed in Castellazzi et al. (2021)) came into effect (Fig. 6.9c) because of the decrease in extraction rates.

The integration of GRACE and SAR-based interferometric analysis of large sets of SAR images with in situ groundwater has strengthened the capability to characterize groundwater changes and their impacts on a much larger scale (e.g., Ojha et al. 2020; Castellazzi et al. 2018). GRACE-derived groundwater changes can either be assessed independently with those of InSAR (Interferometry of Synthetic Aperture Radar) to evaluate groundwater depletions or trends or integrated to improve water management plans on a broader scale. However, it can also aid the characterization of aquifers and improve understanding of how they respond to extreme events like droughts and flood as well as assess pressure equilibrium between fine-grain and coarse-grain aquifer aquitard (e.g., Ojha et al. 2020). By helping to constrain depletion rates and displacements in river basins with considerable groundwater use, the integration of GRACE and its Follow On (FO) mission with SAR data will provide a new capability that underpin water management in data deficient regions

(Rateb and Kuo 2019; Castellazzi et al. 2018). Arguably, similar capability is also expected when NISAR is integrated with the GRACE-FO mission. Existing multi-temporal InSAR data from Sentinel-1A/B satellites C-band and Radarsat-2 have been combined with GRACE to assess land surface deformation related to water storage changes in major aquifers around the world, including groundwater have been implemented (e.g., Li et al. 2020; Ojha et al. 2020; Rateb and Kuo 2019; Castellazzi et al. 2018). The NISAR data will also demonstrate this capability and the availability of data from the GRACE-FO will support its continuous assessment of groundwater changes in the Anthropocene.

6.4.5 Remote Sensing of Soil Moisture

6.4.5.1 Measuring Soil Moisture

Surface soil moisture is the total amount of water available in the unsaturated zone. It is also a key factor in water resource management and its assessment is critical to improve understanding of land surface interactions with atmosphere, carbon cycles and underpins capability to predict weather accurately as well as in agricultural water management for food security (Li et al. 2022; Seneviratne et al. 2010). Vegetation, for instance, depends on water available in the root zone layer and apart from influencing the life cycle of insects through moisture availability on the soil surface, soil moisture regulates runoff. The importance of understanding soil moisture effect on earth surface assessment is clear from its inclusion into the leading 50 essential climate variables by the Global Climate Observing Systems (Bojinski et al. 2014). Additionally, studies (e.g., Ndehedehe et al. 2019; Agutu et al. 2017) that have assessed the water driven changes in vegetation and agricultural drought (lack of soil moisture availability in the root zone or a period characterized by decline in soil moisture) have identified soil moisture as an ideal indicator for these purposes. Given these importance and numerous roles, including crop yield and productivity and understanding how prevailing climatic conditions affect plants biological properties and their development (Agutu et al. 2021), soil moisture retrievals from optical, radar, and microwave satellite sensor have been the focused of several hydrology research. Apparently, the numerous investments in these satellite technologies (Table 6.2) have contributed to the progress in this area and also underpinned by the role of soil moisture in partitioning rainfall into surface runoff and groundwater and regulating evapotranspiration (Gao et al. 2022; Agutu et al. 2021; Ndehedehe et al. 2018).

Soil moisture can be quantified in several ways. This include, ground-based measurements and model simulations, satellite remote sensing (Table 6.2), and close range remote sensing like the use of remotely piloted aircraft systems or unmanned aerial vehicles. However, ground-based sensor networks are considerably limited or completely lacking in most regions of the world (Fig. 6.10), making the accurate assessment of soil moisture content difficult. Obviously, the US has the highest gauge stations for soil moisture monitoring in the world and only sparsely distributed in

Table 6.2 A synopsis of some satellite remote sensing soil moisture sensors/products. *Source* Ndehedehe (2022b)

Sensor/missions	Spatial resolution (km)	Temporal resolution	Channel	Period
Soil Moisture Active Passive (SMAP)	3, 9, 36	2–3 days	1.4 GHz	2015–Present
Advanced Microwave Scanning Radiometer2 (AMSR2)	25	Daily	Multiple from 6.9 GHz	2012–Present
Soil Moisture Ocean Salinity Satellite (SMOS)	35	2–3 days	1.4 GHz	2009–Present
Advanced Scatterometer (ASCAT)	25	1–2 days	5.3 GHz	2006–Present
Advanced Microwave Scanning Radiometer (AMSR-E)	25	Daily	Multiple from 6.9 GHz	2002–2011
Scatterometer ERS	50	3–4 days	5.3 GHz	1992–Present
Microwave Imager TRMM	25	Daily	Multiple from 10.7 GHz	1997–2015
Special Sensor Microwave Imager (SSM/I)	25	Daily	Multiple from 19.4 GHz	1987–Present
Scanning Multichannel Microwave Radiometer (SMMR)	140	Daily	Multiple from 6.6 GHz	1978–1987

western Europe and lacking across most continents (Fig. 6.10). Despite the overwhelming importance of soil moisture in agriculture, interestingly, soil moisture observational networks are lacking in much of the world's croplands and agricultural landscapes where they are typically needed the most (Fig. 6.10). After more than a decade of the existence of new satellite-based soil moisture platforms, improving soil moisture estimates from satellite using novel retrieval algorithm, including machine learning is increasingly becoming an important aspect and focus in soil hydrology research (e.g., Li et al. 2022; Kerr et al. 2016).

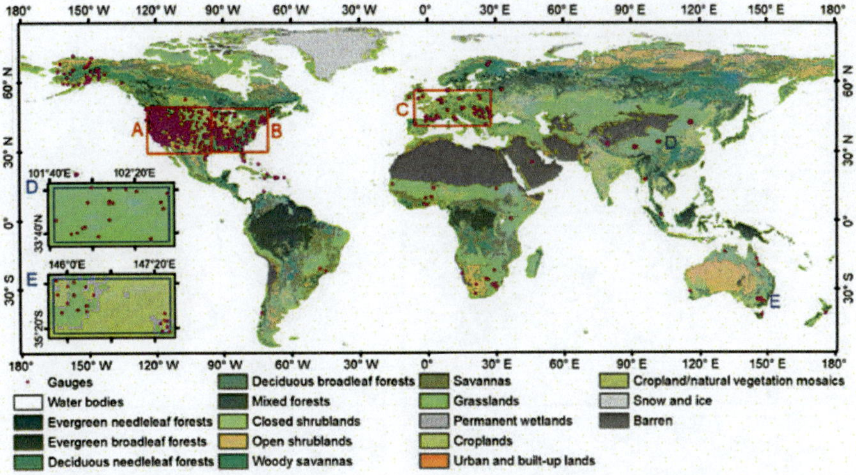

Fig. 6.10 Spatial distribution of global soil moisture gauged stations. Soil moisture observational networks are lacking in much of the world's croplands and agricultural landscapes where they typically needed the most *Source* Gao et al. (2022)

6.4.5.2 New Space-Based Sensors for Soil Moisture Observations

To improve the monitoring of soil moisture across the globe, remotely sensed surface soil moisture platforms are now available (Table 6.2). There are two spaceborne soil moisture L-band radiometer missions currently in operation. They are the European Space Agency-owned Soil Moisture and Ocean Salinity (SMOS) and Soil Moisture Active Passive (SMAP) developed by the National Aeronautics and Space Administration (NASA). The presence of an interferometric radiometer, which allows dual-polarized and multi-angular brightness temperature observations makes it possible for SMOS to concurrently retrieve soil moisture and vegetation optical depth (Li et al. 2022). Moreover, some independent (not synthesized from other products) soil moisture products like the SMOS-INRA-CESBIO were developed to optimize applications in soil hydrology (Fernandez-Moran et al. 2017). With passive microwave remote sensing providing an unprecedented opportunity to quantify surface soil moisture and vegetation water content using the vegetation optical depth (VOD), new retrieval algorithm of soil moisture and L-band vegetation optical depth from the dual-channel SMAP radiometric observations have been reported by (Li et al. 2022; Wang et al. 2021). The mean and seasonal spatial variation of different soil moisture products across the globe between April 2015 and March 2019 show the consistency of this new SMAP retrieval algorithm (referred to as SMAP-INRAE-BORDEAUX) (Fig. 6.11).

Microwave satellites, like the *C*, *X*, and *L* bands generate important retrievals of soil moisture content given their ability to easily penetrate cloud cover. With an adequate amount of incident rays and the corresponding back scatter, they can

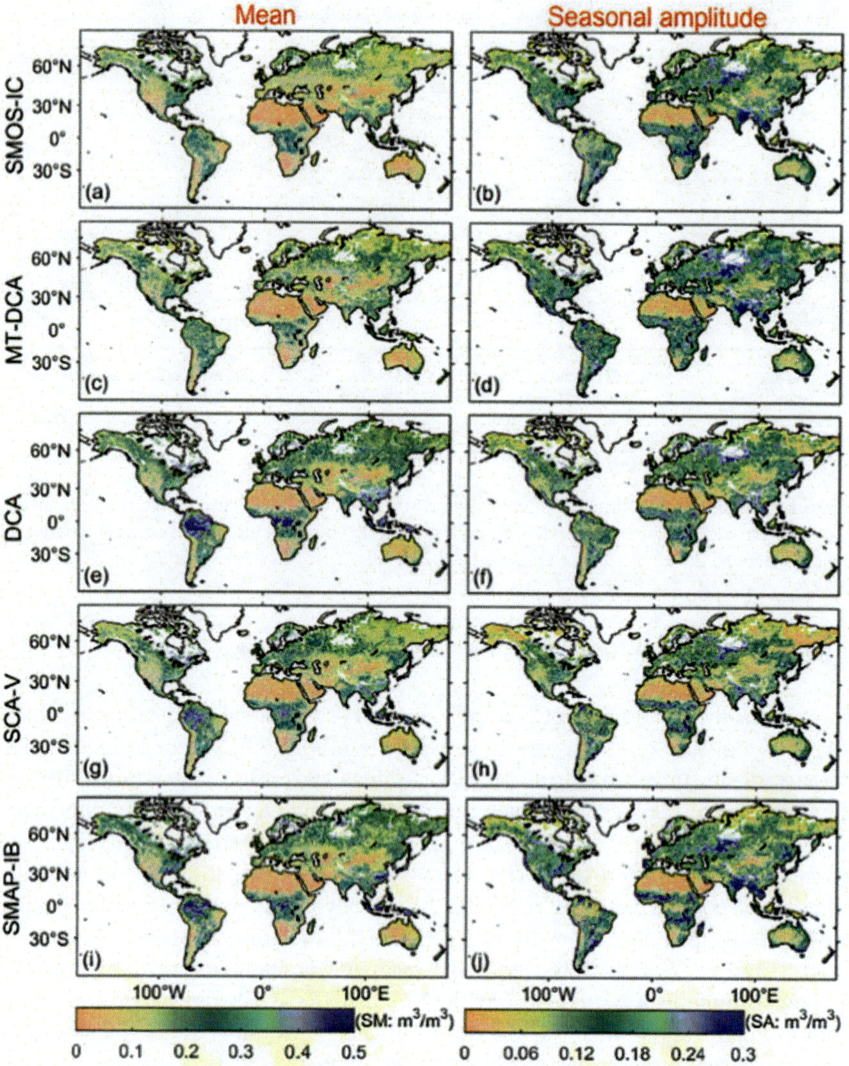

Fig. 6.11 The mean and seasonal spatial variation of different global soil moisture products including SMAP between April 2015 and March 2019. SMAP-IB retrieval algorithm is compared with others including, multi-temporal dual-channel algorithm (MT-DCA), dual-channel algorithm (DCA), and SMOS product (SMOS-IC). *Source* Li et al. (2022)

accurately assess the top soil (Wang et al. 2021; Njoku et al. 2003). The basic principle of microwave retrievals is that it relies on the sensitivity of the passive brightness temperature, as well as its active backscattering to the soil moisture content. Several sensors have been made available for adequate soil moisture detection for more than four decades. As detailed in Table 6.2, they include, the Soil Moisture

Active Passive (SMAP), Advanced Microwave Scanning Radiometer2 (AMSR2), Soil Moisture Ocean Salinity Satellite (SMOS), Advanced Scatterometer (ASCAT), the Advanced Microwave Scanning Radiometer, European Remote Sensing Scatterometer, Special Sensor Microwave Imager (SSM/I), and Scanning Multichannel Microwave Radiometer (SMMR) (Wagner et al. 2013; Njoku et al. 2003; Njoku and Entekhabi 1996). In contrast to the single sensor soil moisture products, Liu et al. (2011) improved on the original concept and developed a blended soil moisture product through the integration of both active and passive soil moisture from various satellite sensors. This product is useful for effective drought monitoring and trend analysis to assess impact of climate change. This is particularly so due to the availability of long-term blended soil moisture products. Several of these satellite soil moisture products described in Table 6.2 can improve knowledge on changes in actual soil water content and volumetric soil content. For instance, based on the SMOS satellite data, a Soil Water Deficit Index was constructed to measure agricultural drought by Martínez-Fernández et al. (2016). The SMOS soil moisture-based indicator in their study showed strong potential in agricultural drought monitoring. Even though the assimilation of satellite soil moisture like SMAP into model has been evaluated to assess potential for agricultural drought monitoring. The use of satellite remote sensing based soil moisture products like SMAP and SMOS provides an opportunity to directly measure agricultural drought (Ndehedehe 2022b). By merging AMSR-E and AMSR2 products (Table 6.2), consistent long-term vegetation optical depth product can now be monitored, thus enhancing the feasibility for monitoring changes in vegetation optical depth as an above-ground biomass surrogate (Wang et al. 2021).

6.4.6 Satellite-Derived Changes in Surface Water Storage

Rivers, water stored in lakes, reservoirs, wetlands provide essential services, and ecological functions such as regulating nutrient cycle, generation of hydropower, maintaining fishery production, water supply for agriculture, recreation and tourism, as well as support for a broad range of terrestrial and aquatic biodiversity (Ndehedehe et al. 2022). Process-based knowledge of the cascading impacts of droughts and human water extraction on surface water hydrology is thus crucial and can directly feed into management and policy frameworks. To improve the understanding of surface water changes, a plethora of satellite missions, including satellite radar-altimetry and the recently launched Surface Water Ocean Topography-SWOT (discussed in Sect. 4.3.3 of Chap. 4). Prior to the launch of SWOT, several remote sensing techniques and data assimilation methods that combine satellite with gauged observation have been used to quantify changes in surface water hydrology (e.g., surface water extents/volume and total drainable water storage) (e.g., Elmi and Tourian 2022; Kitambo et al. 2022; Tourian et al. 2018; Getirana et al. 2017; Pekel et al. 2016). These methods and approaches include, the use of Global Inundation Estimate from Multiple Satellites (GIEMS-2 product) and the combination of in situ with data from multi-mission satellite missions, e.g., GRACE, satellite altimetry, MERIT Hydro data, Sentinel-3A, and Sentinel-3B, among several other remote sensing data (Elmi and Tourian 2022; Kitambo et al. 2022; Prigent et al. 2020; Tourian et al. 2018). The GIEMS-2 product provides monthly estimates of surface water extent on a 25 km spatial resolution for the period between 1992 and 2015, and is suitable for climatological studies that seek to improve understanding of the impacts of climate variability on surface water changes (Prigent et al. 2020). The combination of in situ with satellite observations to quantify changes in surface hydrology has been demonstrated in large humid river basins like the Amazon and Congo basins. For example, using a multi-satellite approach, that included the use of GIEMS data with water level observations from the ENVISAT altimeter heights, Becker et al. (2018) estimated water level maps and surface water storage in the Congo river basin floodplains, lakes, rivers, and wetlands during the 2003–2007 period. The GIEMS data are available and can be requested using this link (https://www.estellus.fr/index.php?static13/giems-d15).

A related study in the Congo basin by Kitambo et al. (2022) also assessed surface water storage changes in the Congo basin, showcasing the importance of satellite data as an important source of data for hydrological modeling and improving understanding of hydrological processes underpinned by its capability to better characterize surface water hydrology. To achieve this, they combined ground-based observation with several remote sensing data, including GIEMS-2 product, satellite altimetry the European Remote Sensing-2 satellite the Environmental Satellite, Jason-2 and 3, the Satellite with ARgos and AL tiKa, and Sentinel-3A and Sentinel-3B missions (Fig. 6.13). River networks and their channel characteristics (e.g., width, flow, etc.) need to be properly captured to improve the accurate monitoring of changes in surface water extents or volumes (e.g., Elmi and Tourian 2022; Lamine et al. 2021; Dewan et al. 2017). The assessment and evaluation of channel changes of the Ganges–Padma River system in Bangladesh benefited from multitemporal Landsat data underpinned by a long-term river flow data. Dewan et al. (2017) who studied this river system, analyzed the channel width, sinuosity and braiding index, and used a soil–vegetation limit approach, which improves channel planform mapping to identify river boundary in Landsat data. They reported a significant variability in the width of the Ganges (between 5.36 and 3.23 km) and showed that, over a 38-year window, both the Ganges and the Padma experienced an expansion, contraction, and readjustment in configuration. The realistic representation of river networks and properties is lacking for several rivers and increasingly less available for many gauged rivers, monitoring freshwater changes using satellite methods that integrates other empirical analysis is thus a useful alternative (e.g., Elmi and Tourian 2022; Kitambo et al. 2022; Ndehedehe et al. 2022; Prigent et al. 2020).

Furthermore, using remote sensing techniques (e.g., GIEMS data combined with ENVISAT and GRACE observations), Becker et al. (2018) reported a mean annual variation of 81 ± 24 km^3 in surface water storage over the Congo basin that contributes to $19 \pm 5\%$ of observed annual changes in terrestrial water storage as observed by GRACE satellite. An example of the surface water storage estimated from satellite remote sensing methods is illustrated in Fig. 6.14. The spatial distribution of surface water storage in the Congo basin shows large storage variations of 1.3 km^3/year in some grid cells along the major river channels, while most areas show surface water changes ranging from 0.1–0.2 km^3/year per grid cell (Fig. 6.14). Satellite methods like GRACE have also been used in combination with altimetry based measurements to quantify the total amount of water that drains a river basin. Tourian et al. (2018) pioneered this over the Amazon basin, defining total drainable water storage (TDSW) as the *"total stored water that can exit or drain the river basin through natural hydrologic runoff generation as time approaches infinity, or as the basin would asymptotically approach zero storage given no additional inputs"*. Combining discharge, altimetric water level, and surface water extent from GIEMS repository with GRACE-based terrestrial water data, they estimated that TDWS for the Amazon basin as 1,766 km^3 and that TDWS predominantly (about 65%) exists as surface water in the basin. These capabilities of remote sensing provide a broad range of opportunities to keep track of the impacts of climate change and human water management on surface water hydrology in ways that can directly feed into management policies and support water planning and allocation.

6.5 Monitoring to Improve Decision and Policies

6.5.1 Freshwater Management

Improving the resilience of water supplies, especially in large urban communities where most of the world's human population reside is critical because of the growing concerns of extreme drought and climatic extremes. In some areas, human induced changes on groundwater are gradually surpassing the impacts of climate change. In the Tigris–Euphrates basin, falling groundwater storage at the rate of 7.56 km^3/year resulted in -10 mm/yr subsidence rate near Baghdad (Rateb and Kuo 2019). Multi-temporal SAR data showing varying degrees of subsidence rates in Southwest US (e.g., Las Vegas, Albuquerque, Colorado Springs, etc.) will benefit water managers and policymakers to help drive or draft sustainable and meaningful groundwater management plans (Ojha et al. 2020). The feasibility of satellite observations, including data from the Global Navigation Satellite System—GNSS (e.g., Global Positioning System, Galileo navigation system, GLObal NAvigation Satellite System, etc.) to assess the response of hydrological cycle and extreme events like drought, further helps water managers to predict freshwater availability. Following the work of Ferreira et al. (2021), this feasibility was demonstrated in Southeast Brazil, indicating that GNSS-predicted changes in terrestrial stored water considerably agreed with those of GRACE.

As was highlighted earlier, artificial recharge was implemented in response to rising water demand and reduced rainfall in the Perth basin to slow down water loss from groundwater aquifers. The study by Parker et al. (2021) demonstrated how SAR observations can be employed to measure displacement within the vicinity where this artificial recharge (injection of recycled and treated wastewater) occurred. Replenishing groundwater storage can happen in different ways, including regulating withdrawal rate, and improving recharge as mentioned earlier. To regulate withdrawal rates, groundwater pumping can be decreased and available terrestrial water recycled for other applications. Water transfer from considerably humid regions to the arid regions (e.g., Arizona and California) (Scanlon et al. 2012), and alternative water sources can also be developed (e.g., desalinating sea water) as part of mitigation strategies that underpin resilience to (Schiermeier 2008; Muller 2018). Climate change and over extraction will continue to impact on water supply systems, including reservoir storage and aquifers. The development of risk-based decision-making and modeling frameworks for water resource management will be underpinned by our capability to track and monitor changes of how these systems change in space and time.

6.5.2 Disaster and Risk Management

In some regions, surface deformations and other geophysical phenomena such as earthquakes, volcanic eruptions, earth fissuring and fracturing, and sinkholes are physical dynamic processes that pose a great deal of danger to public infrastructure and human lives. These hazards are usually associated with surface subsidence and their monitoring can assist with future preparation and increasing resilience of potential future hazards (e.g., Aldaajani et al. 2022). While these geophysical phenomena can be natural, vertical crustal displacements and subsidence have been linked to prolonged drought events and water extraction as was earlier mentioned (e.g., Ojha et al. 2020; Ferreira et al. 2018, 2019a). These water related disasters are increasing being monitored by integrating GRACE with vertical crustal displacements obtained from Global Positioning System (GPS) and model data (Fig. 6.12). The work undertaken by Ferreira et al. (2019a) demonstrated the feasibility of combining GPS data and vertical crustal displacements estimated from model outputs with GRACE observations. As illustrated in Fig. 6.12, there is a strong agreement between GRACE and GPS series and a considerable amount of observed amplitudes of vertical crustal displacements of the annual signals are concentrated mainly around the Amazon rainforest and Brazilian areas. The relationship between time series of GPS with GRACE revealed that approximately 45% of the GPS stations had correlation coefficients greater than 0.60 (Fig. 6.12a–c).

Given the changes in hydrological cycle caused by climate change impacts on water availability as well as the Earth's surface elastic response to the surface mass loading/unloading (Ferreira et al. 2018, 2019a), the combined use of GRACE and GPS data to assess crustal displacements and quantify land water storage is thus crucial to improve understanding on the impacts of extreme events like droughts on water availability (Ndehedehe 2022f; Ferreira et al. 2019b). Drought monitoring capability using GPS and GRACE as illustrated in Fig. 6.15 comes with benefits that complements other drought indicators derived from precipitation, soil moisture,

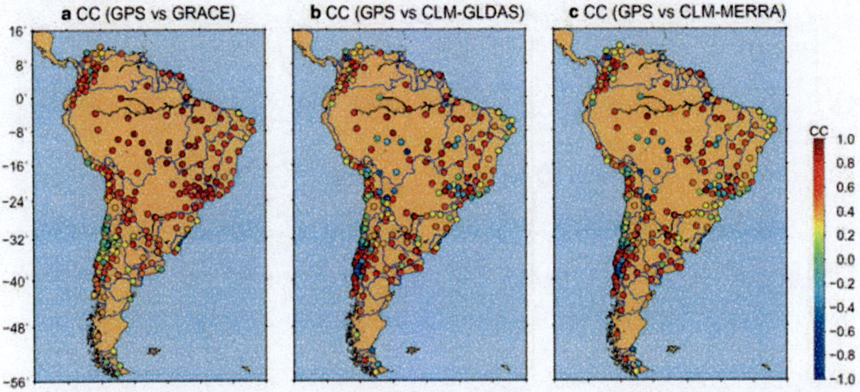

Fig. 6.12 Correlation coefficients (CCs) for **a** Global Positioning Systems-GPS vs GRACE-Gravity Recovery and Climate Experiment, **b** GPS versus land surface models (CLM-GLDAS), and **c** GPS versus reanalysis data (CLM-MERRA). CLM-GLDAS is Community Land Model-Global Land Data Assimilation System and MERRA is Modern-Era Retrospective Analysis for Research and Applications. *Source* Ferreira et al. (2019a)

and climate date (Ndehedehe 2022b). For instance, detected changes in the Earth's shape as a response to water mass loading and unloading were possible with geodetic observations like GRACE (Fig. 6.15). This capability offers alternative in monitoring changes in extreme events (droughts and floods) of humid regions like coastal West Africa and some parts of Asia and south America where observational data is lacking. And the frequency (daily and/or weekly basis) in which such monitoring occurs can help with timely assessment of impacts of climate variability on hydrology. Some case-specific examples of radar data supporting disaster and risk management are further detailed in the next subsections.

6.5.2.1 Groundwater-Induced Subsidence

Radar observations can deliver detailed measurements of large-scale ground deformations associated with prolonged groundwater withdrawal due to drought, aquifer systems, and resource extraction (Smith and Majumdar 2020). And InSAR-based assessments of regional surface displacements in western Arabia as a function time was recently reported by Aldaajani et al. (2022). They found that a proportion of the subsidence was associated with surface fractures in western Arabia that developed simultaneously with severe rainfall events in regions with alluvial sediments at the surface. However, there is some part of the observed subsidence that was linked to groundwater withdrawal (Aldaajani et al. 2022). Excessive groundwater extraction from compressible aquifers leads to land subsidence that will significantly cause damage to buildings and infrastructure and other environmental disasters and the use of GRACE with InSAR techniques to assess the sensitivities of aquifer systems

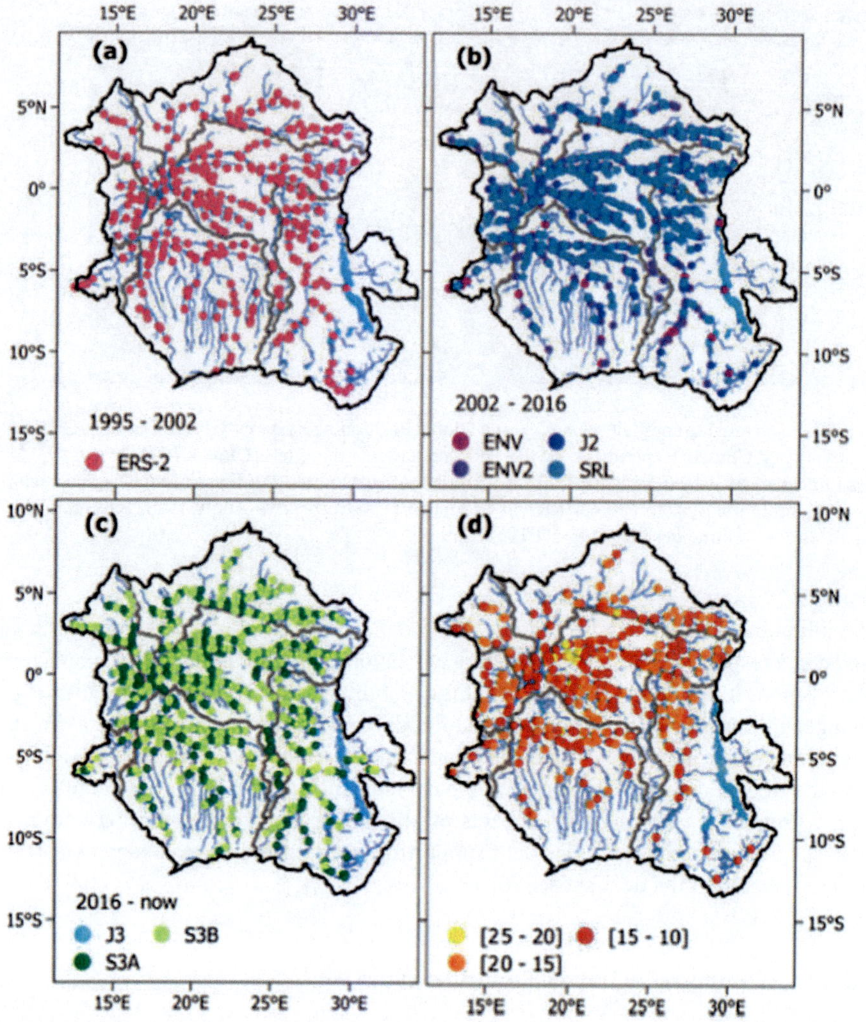

Fig. 6.13 Satellite altimetry virtual stations (VS) in the Congo basin. **a** ERS-2 (European Remote Sensing-2) VSs. **b** ENV (ENVISAT), ENV2 (ENVISAT-2), J2 (Jason-2), and SRL (Satellite with ARgos and ALtiKa) VSs **c** J3 (Jason-3), S3A (Sentinel-3A), and S3B (Sentinel-3B) VSs **d** VSs with an actual long time series ranging from 25 and 20 years (yellow), 20 and 15 years (orange), and 15 and 10 years (red). These VS are also available for other regions of the world. *Source* Kitambo et al. (2022)

to climate processes (e.g., inter-annual variations in rainfall) and water extraction provides an opportunity to track impacts from drought and water extraction. By generating maps and time series of surface displacements in areas where water planning is in place, including hotspots of groundwater extraction, tracking these impacts can inform on policy that underpins mitigation strategies and resilience.

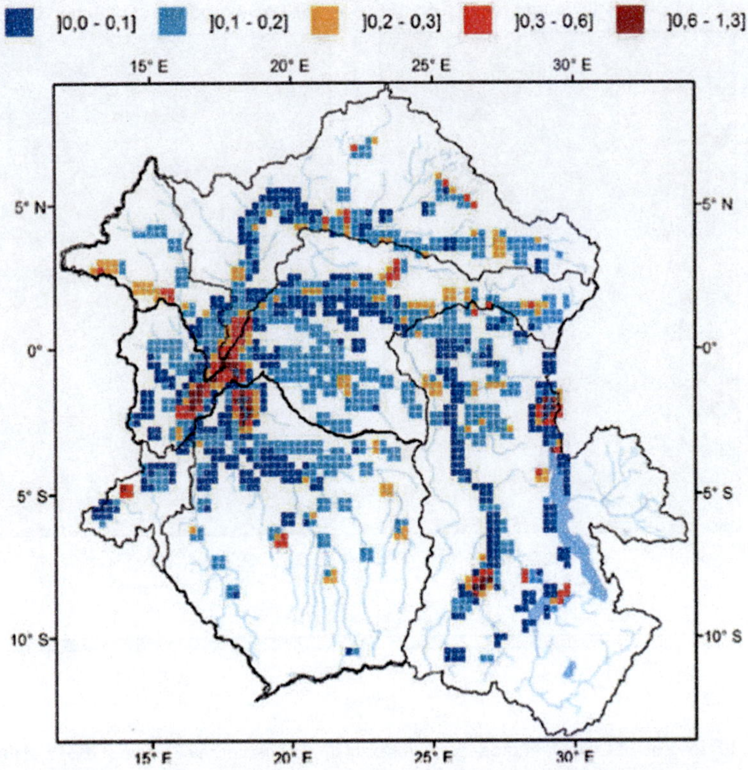

Fig. 6.14 Surface water storage (SWS) in the Congo basin estimated from remote sensing methods. The maximum annual amplitudes in surface water storage are indicated in km^3 during the 2003–2007. Deep blue shows the lowest SWS values (0–0.1 km^3), while deep red indicates the highest SWS amount ranging from 0.6–1.3 km^3. *Source* Becker et al. (2018)

6.5.2.2 Earthquake Response

On February 6, 2023, the southern region of Turkey and western Syria experienced a magnitude 7.8 and 7.5 earthquakes that led to widespread destruction in both countries (https://www.jpl.nasa.gov/images/pia25564-satellites-assess-earthquake-damage-in-turkey). Several online media channels confirmed this unprecedented earthquake is one of the worst in human history, killing more than 33,000 people in total (29,695 in Turkey and 3,574 in Syria).[2] As with quantifying impacts of floods and wildfire as well as assessing land surface displacements associated with groundwater depletion, satellite radar observations from Sentinel-1 and ALOS missions are helping with assessing the impact of earthquakes in large human settlements (Fig. 6.16). Facilitated by the The Advanced Rapid Imaging and Analysis

[2] https://www.reuters.com/world/middle-east/survivors-ever-fewer-earthquake-rubble-turkey-syria-2023-02-12/.

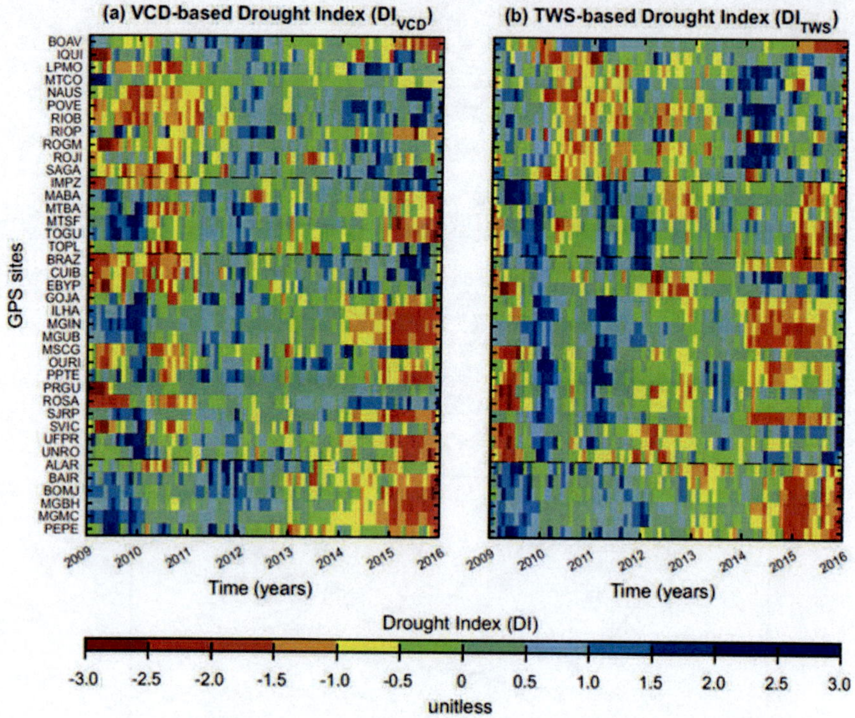

Fig. 6.15 Illustration of Drought indices derived from Global Positioning System (GPS) and Gravity Recovery and Climate Experiment data (GRACE) to assess impact of climate change on hydrology. The GPS sites are located in important river basins within Brazil and includes the São Francisco, Tocantins, La Plata, and Amazon river basins. Vertical crustal deformations were based on GPS data while GRACE was used for the terrestrial water storage (TWS)-based drought index. *Source* Ferreira et al. (2018)

team at NASA's Jet Propulsion Laboratory and California Institute of Technology, areas affected by the unprecedented large earthquakes in Turkey and Syria, including cities of Islahiye, Kahramanmaras, and Nurdagi were identified. This was achieved by comparing radar data (using the In-SAR technique) on February 8, 2023 to observations from the same satellite before the earthquake occurred (on April 7, 2021 and April 6, 2022). The illustration in Fig. 6.16 is a preliminary damage proxy map that shows key areas with increasing surface change where damage to infrastructure and buildings were rather severe. On the disaster map shown in Fig. 6.16, the color variation from yellow to red in the three locations (Iskenderum, Kahramanmaras, and Antakya), which have been zoomed out or elaborated further on the left indicates the rapid and considerable change in surface. That is, areas with deep red are locations with the most impact (regions of increasing change) where earth movements have damaged buildings and public infrastructures. There could be uncertainties in some areas due to higher elevation areas (lack of colored pixels) with snow vege-

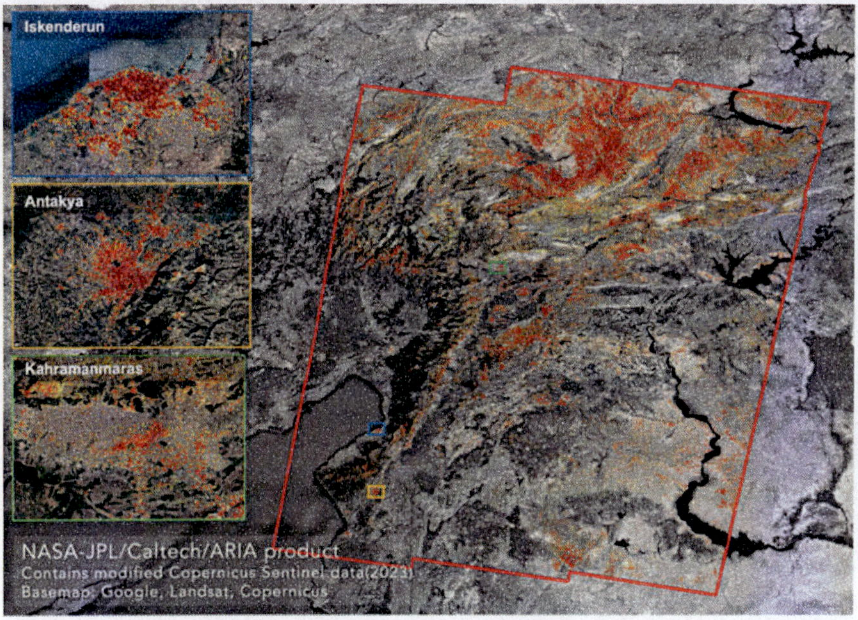

Fig. 6.16 Radar derived recent earthquake impact and spread over Turkey. The colors from yellow to deep red show areas of increasing surface change. *Source* NASA-JPL

tated landscapes (scattered colored pixels). Overall, this radar-derived disaster map highlights critical areas sensitive to building damage that does not only reveal the scale of devastation but can help disaster response.

6.5.3 Sustainable Development Goals

Satellite remote sensing is emerging as an important technology and innovation that will drive understanding of critical environmental science geared toward achieving the 17 Sustainable Development Goals of the United Nations (UN) (https://sdgs.un.org/goals). As part of a major step toward meeting critical geospatial data needs that underpins the UN 2030 Agenda for Sustainable Development, the Sustainable Development Goals Science Satellite 1 (SDGSAT-1) was launched on 5th November (Guo et al. 2022). Developed and managed by the Big Earth Data Science Engineering Project of the Chinese Academy of Sciences, the SDGSAT-1 mission is first of its kind comprising three sensors dedicated for the implementation of UN 2030 Agenda for Sustainable Development Goals (https://sdg.casearth.cn/en/mobile/datas/SDGSAT). The SDGSAT-1 satellite mission will accurately collect data on various human activities and measure changes on land surface conditions (http://en.people.cn/n3/2021/1112/c90000-9918780.html) that align with these Sustainable

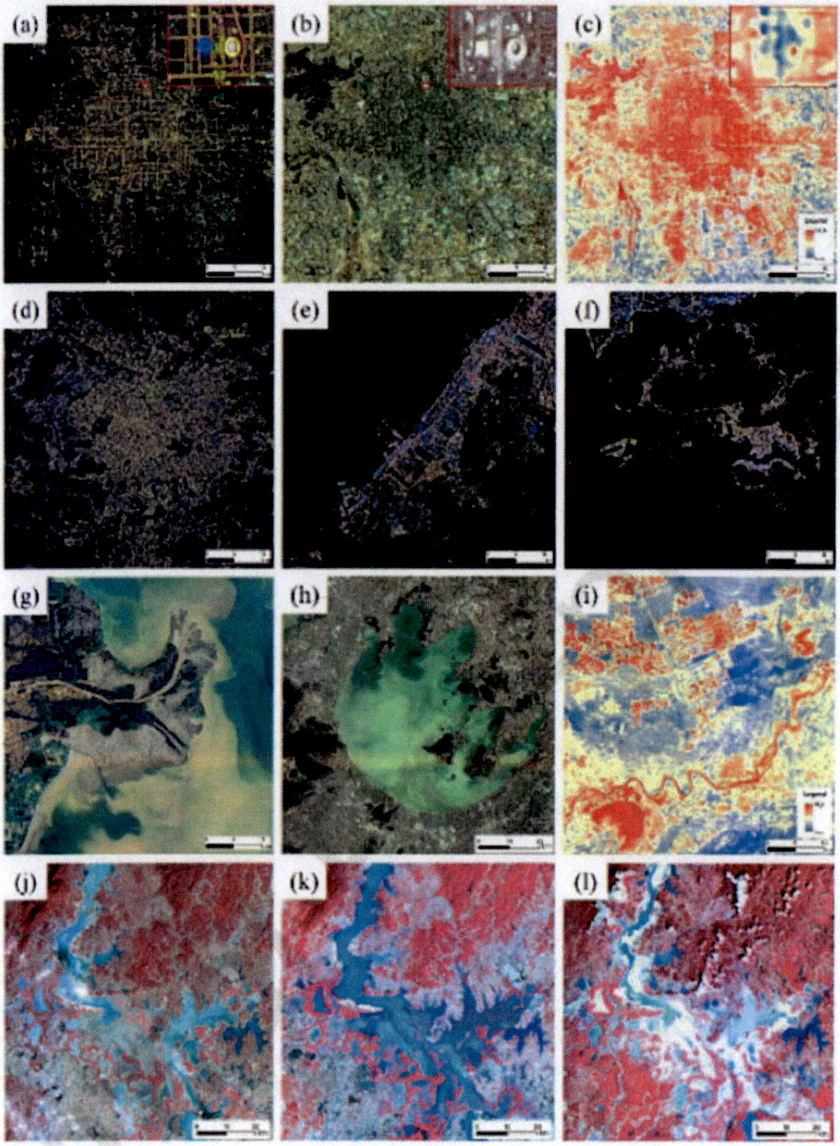

Fig. 6.17 First results from the Sustainable Development Goals Science Satellite 1 (SDGSAT-1) mission. *Source* Guo et al. (2022)

Development Goals. This capability will help fill major geospatial data gaps and drive engagement within SDG goals and sustainable development through creation of reliable stream of geoinformation data that underpins 17 Sustainable Development Goals of the UN 2030 Agenda (Guo et al. 2022). SDGSAT-1 is designed to orbit at an altitude of around 505 km and is a dedicated mission for the acquisition and evaluation of SDGs related data and relevant indicators such as climate and human impacts on urban, residential and coastal activities and nature.[3] The SDGSAT-1 sensors are highly innovative, providing different data for various SDG indicators at high spatial and temporal resolutions to support applications in several areas, including *"population, agriculture, well-being estimation economic livelihoods, water quality, informal settlements, climate related hazards and natural disasters (SDG 13), and coastal environments and terrestrial"* (Guo et al. 2022). The first results (multispectral images) from the SDGSAT-1 system have recently been reported and is illustrated in (Fig. 6.17).

References

Agutu N, Awange J, Ndehedehe C, Kirimi F, Kuhn M (2019) GRACE-derived groundwater changes over Greater Horn of Africa: temporal variability and the potential for irrigated agriculture. Sci Total Environ 693:133467. https://doi.org/10.1016/j.scitotenv.2019.07.273

Agutu N, Awange J, Zerihun A, Ndehedehe C, Kuhn M, Fukuda Y (2017) Assessing multi-satellite remote sensing, reanalysis, and land surface models' products in characterizing agricultural drought in East Africa. Remote Sens Environ 194:287–302. https://doi.org/10.1016/j.rse.2017.03.041

Agutu N, Ndehedehe C, Awange J, Kirimi F, Mwaniki M (2021) Understanding uncertainty of model-reanalysis soil moisture within Greater Horn of Africa (1982–2014). J Hydrol 603:127169. https://doi.org/10.1016/j.jhydrol.2021.127169

Ahmed A, Alrajhi A, Alquwaizany AS (2021) Identification of groundwater potential recharge zones in Flinders Ranges, South Australia using remote sensing, gis, and mif techniques. Water 13(18). https://doi.org/10.3390/w13182571

Aldaajani T, Simons M, Yunjun Z, Bekaert D, Almalki KA, Liu Y-K (2022) Using InSAR time series to monitor surface fractures and fissures in the Al-Yutamah Valley, Western Arabia. Remote Sens 14(8). https://doi.org/10.3390/rs14081769

Anejionu OCD (2019) Rationale, historical developments and advances in remote sensing of gas flares. Int J Remote Sens 40(17):6700–6719. https://doi.org/10.1080/01431161.2019.1590880

Aydal D, Arda1 E, Dumanlilar Ö (2007) Application of the Crosta technique for alteration mapping of granitoidic rocks using ETM+ data: case study from eastern Tauride belt (SE Turkey). Int J Remote Sens 28(17):3895–3913. https://doi.org/10.1080/01431160601105926

Beck PS, Atzberger C, Høgda KA, Johansen B, Skidmore AK (2006) Improved monitoring of vegetation dynamics at very high latitudes: a new method using MODIS NDVI. Remote Sens Environ 100(3):321–334. https://doi.org/10.1016/j.rse.2005.10.021

Becker M, Papa F, Frappart F, Alsdorf D, Calmant S, da Silva JS, Prigent C, Seyler F (2018) Satellite-based estimates of surface water dynamics in the Congo River Basin. Int J Appl Earth Obs Geoinf 66:196–209. https://doi.org/10.1016/j.jag.2017.11.015

Bierkens MFP, Wada Y (2019) Non-renewable groundwater use and groundwater depletion: a review. Environ Res Lett 14(6):063002. https://doi.org/10.1088/1748-9326/ab1a5f

[3] https://www.space.com/china-long-march-6-rocket-launches-sdgsat-1.

Bojinski S, Verstraete M, Peterson TC, Richter C, Simmons A, Zemp M (2014) The concept of essential climate variables in support of climate research, applications, and policy. Bull Am Meteorol Soc 95(9):1431–1443. https://doi.org/10.1175/BAMS-D-13-00047.1

Brito SSB, Cunha APMA, Cunningham CC, Alvalá RC, Marengo JA, Carvalho MA (2018) Frequency, duration and severity of drought in the semiarid Northeast Brazil region. Int J Climatol 38(2):517–529. https://doi.org/10.1002/joc.5225

Camps-Valls G, Tuia D, Gómez-Chova L, Jiménez S, Malo J (2012) Remote sensing image processing. Morgan and Claypool Publishers

Castellazzi P, Garfias J, Martel R (2021) Assessing the efficiency of mitigation measures to reduce groundwater depletion and related land subsidence in Querétaro (Central Mexico) from decadal InSAR observations. Int J Appl Earth Obs Geoinf 105:102632. https://doi.org/10.1016/j.jag.2021.102632

Castellazzi P, Longuevergne L, Martel R, Rivera A, Brouard C, Chaussard E (2018) Quantitative mapping of groundwater depletion at the water management scale using a combined GRACE/InSAR approach. Remote Sens Environ 205:408–418. https://doi.org/10.1016/j.rse.2017.11.025

Castellazzi P, Martel R, Rivera A et al (2016) Groundwater depletion in Central Mexico: use of GRACE and InSAR to support water resources management. Water Resour Res 52(8):5985–6003. https://doi.org/10.1002/2015WR018211

Chen J, Wilson C, Tapley B, Scanlon B, Güntner A (2016) Long-term groundwater storage change in Victoria, Australia from satellite gravity and in situ observations. Global Planet Change 139:56–65. https://doi.org/10.1016/j.gloplacha.2016.01.002

Dewan A, Corner R, Saleem A, Rahman MM, Haider MR, Rahman MM, Sarker MH (2017) Assessing channel changes of the Ganges-Padma River system in Bangladesh using Landsat and hydrological data. Geomorphology 276:257–279. https://doi.org/10.1016/j.geomorph.2016.10.017

Dong P, Chen Q (2018) Lidar remote sensing and applications. Taylor & Francis, CRC Press, US

Elmi O, Tourian MJ (2022) Retrieving time series of river water extent from global inland water data sets. J Hydrol 128880. https://doi.org/10.1016/j.jhydrol.2022.128880

Famiglietti JS (2014) The global groundwater crisis. Nature 4:945–948. https://doi.org/10.1038/nclimate2425

Faruolo M, Caseiro A, Lacava T, Kaiser JW (2021) Gas flaring: a review focused on its analysis from space. IEEE Geosci Remote Sens Mag 9(1):258–281. https://doi.org/10.1109/MGRS.2020.3007232

Fernandez-Moran R, Al-Yaari A, Mialon A, Mahmoodi A et al (2017) SMOS-IC: an alternative SMOS soil moisture and vegetation optical depth product. Remote Sens 9(5). https://doi.org/10.3390/rs9050457

Ferreira V, Montecino H, Ndehedehe C, Heck B, Gong Z, Westerhaus M, de Freitas S (2018) Space-based observations of crustal deflections for drought characterization in brazil. Sci Total Environ 644:256–273. https://doi.org/10.1016/j.scitotenv.2018.06.277

Ferreira V, Montecino H, Ndehedehe C, Yuan P, Xu T (2021) Chapter 14 - the versatility of gnss observations in hydrological studies. In: Petropoulos GP, Srivastava PK (eds) GPS and GNSS technology in geosciences, pp 281–298. Elsevier. https://doi.org/10.1016/B978-0-12-818617-6.00004-4

Ferreira V, Yong B, Tourian M, Ndehedehe C, Shen Z, Seitz K, Dannouf R (2020) Characterization of the hydro-geological regime of Yangtze River basin using remotely-sensed and modeled products. Sci Total Environ 718:137354. https://doi.org/10.1016/j.scitotenv.2020.137354

Ferreira VG, Montecino HD, Ndehedehe CE, del Rio RA, Cuevas A, de Freitas SRC (2019a) Determining seasonal displacements of Earth's crust in South America using observations from space-borne geodetic sensors and surface-loading models. Earth Planets Space 71(1):84. https://doi.org/10.1186/s40623-019-1062-2

Ferreira VG, Ndehedehe CE, Montecino HC, Yong B, Yuan P, Abdalla A, Mohammed AS (2019b) Prospects for imaging terrestrial water storage in South America using daily GPS observations. Remote Sens 11(6). https://doi.org/10.3390/rs11060679

Gao L, Gao Q, Zhang H, Li X, Chaubell MJ, Ebtehaj A, Shen L, Wigneron J-P (2022) A deep neural network based SMAP soil moisture product. Remote Sens Environ 277:113059. https://doi.org/10.1016/j.rse.2022.113059

Getirana A, Kumar S, Girotto M, Rodell M (2017) Rivers and floodplains as key components of global terrestrial water storage variability. Geophys Res Lett 44(20):10,359–10,368. https://doi.org/10.1002/2017GL074684

Guo H, Dou C, Chen H, Liu J, Fu B, Li X, Zou Z, Liang D (2022) SDGSAT-1: the world's first scientific satellite for Sustainable Development Goals. Sci Bull. https://doi.org/10.1016/j.scib.2022.12.014

Hemati M, Hasanlou M, Mahdianpari M, Mohammadimanesh F (2021) A systematic review of landsat data for change detection applications: 50 years of monitoring the earth. Remote Sens 13(15). https://doi.org/10.3390/rs13152869

Kavvada A, Cripe D, Friedl L (2022) Introduction to global sustainability frameworks and the role of earth observations, chapter 1, pp 1–11. American Geophysical Union (AGU). https://doi.org/10.1002/9781119536789.ch1

Kerr Y, Al-Yaari A, Rodriguez-Fernandez N, Parrens M, Molero B et al (2016) Overview of SMOS performance in terms of global soil moisture monitoring after six years in operation. Remote Sens Environ 180:40–63. https://doi.org/10.1016/j.rse.2016.02.042

Kim S-B, van Zyl JJ et al (2017) Surface soil moisture retrieval using the L-Band synthetic aperture radar onboard the soil moisture active-passive satellite and evaluation at core validation sites. IEEE Trans Geosci Remote Sens 55(4):1897–1914

Kitambo B, Papa F, Paris A, Tshimanga RM, Calmant S, Fleischmann AS, Frappart F, Becker M, Tourian MJ, Prigent C, Andriambeloson J (2022) A combined use of in situ and satellite-derived observations to characterize surface hydrology and its variability in the Congo River basin. Hydrol Earth Syst Sci 26(7):1857–1882. https://doi.org/10.5194/hess-26-1857-2022

Kumar SV, Dirmeyer PA, Peters-Lidard CD, Bindlish R, Bolten J (2018) Information theoretic evaluation of satellite soil moisture retrievals. Remote Sens Environ 204:392–400. https://doi.org/10.1016/j.rse.2017.10.016

Lamine B, Ferreira V, Yang Y, Ndehedehe C, He X (2021) Estimation of the Niger River cross-section and discharge from remotely-sensed products. J Hydrol Reg Stud 36:100862. https://doi.org/10.1016/j.ejrh.2021.100862

Lavender S, Andrew L (2016) Practical handbook of remote sensing. Taylor and Francis, CRC Press, New York

Lein JK (2012) Environmental sensing; analytical techniques for earth observation. Springer, New York

Li J, Wang S, Michel C, Russell HA (2020) Surface deformation observed by InSAR shows connections with water storage change in Southern Ontario. J Hydrol Reg Stud 27:100661. https://doi.org/10.1016/j.ejrh.2019.100661

Li X, Wigneron J-P, Fan L, Frappart F, Yueh SH, Colliander A et al (2022) A new smap soil moisture and vegetation optical depth product (smap-ib): Algorithm, assessment and inter-comparison. Remote Sens Environ 271:112921. https://doi.org/10.1016/j.rse.2022.112921

Liu YY, Parinussa RM, Dorigo WA et al (2011) Developing an improved soil moisture dataset by blending passive and active microwave satellite-based retrievals. Hydrol Earth Syst Sci 15(2):425–436. https://doi.org/10.5194/hess-15-425-2011

Martínez-Fernández J, González-Zamora A, Sánchez N, Gumuzzio A, Herrero-Jiménez C (2016) Satellite soil moisture for agricultural drought monitoring: assessment of the SMOS derived Soil Water Deficit Index. Remote Sens Environ 177:277–286. https://doi.org/10.1016/j.rse.2016.02.064

Muller M (2018) Lessons from cape town's drought. Nature 559:174–176

Ndehedehe C (2022a) Groundwater from space, pp 211–230. Springer, Cham. https://doi.org/10.1007/978-3-030-99577-5_9
Ndehedehe C, (2022b) Hotspots of climatic influence, pp 629–688. Springer, Cham. https://doi.org/10.1007/978-3-030-99577-5_20
Ndehedehe C (2022c) Optical remote sensing systems, pp 35–52. Springer, Cham. https://doi.org/10.1007/978-3-030-99577-5_3
Ndehedehe C (2022d) Remote sensing hydrology, pp 3–17. Springer, Cham. https://doi.org/10.1007/978-3-030-99577-5_1
Ndehedehe C (2022e) Remote sensing of surface vegetation, pp 131–176. Springer, Cham. https://doi.org/10.1007/978-3-030-99577-5_7
Ndehedehe C (2022f) Satellite geodetic missions, pp 53–70. Springer, Cham
Ndehedehe C (2022g) Satellite observations of terrestrial water storage, pp 331–386. Springer, Cham. https://doi.org/10.1007/978-3-030-99577-5_14
Ndehedehe CE, Ferreira VG, Agutu NO (2019) Hydrological controls on surface vegetation dynamics over West and Central Africa. Ecol Ind 103:494–508. https://doi.org/10.1016/j.ecolind.2019.04.032
Ndehedehe CE, Ferreira VG, Agutu NO, Onojeghuo AO, Okwuashi O, Kassahun HT, Dewan A (2021) What if the rains do not come? J Hydrol 595:126040. https://doi.org/10.1016/j.jhydrol.2021.126040
Ndehedehe CE, Ferreira VG, Getirana A, Agutu NO (2022) Understanding the influence of climate variability on surface water hydrology in the Congo Basin, chapter 5, pp 63–81. American Geophysical Union (AGU)
Ndehedehe CE, Okwuashi O, Ferreira VG, Agutu NO (2018) Exploring evapotranspiration dynamics over Sub-Sahara Africa (2000–2014). Environ Monit Assess 190(7):400. https://doi.org/10.1007/s10661-018-6780-6
NISAR (2018) NASA-ISRO SAR (NISAR) mission science users' handbook. NASA Jet Propulsion Laboratory, pp 261 pp
Njoku E, Jackson T, Lakshmi V, Chan T, Nghiem S (2003) Soil moisture retrieval from AMSR-E. IEEE Trans Geosci Remote Sens 41(2):215–229. https://doi.org/10.1109/TGRS.2002.808243
Njoku EG, Entekhabi D (1996) Passive microwave remote sensing of soil moisture. J Hydrol 184(1):101–129. https://doi.org/10.1016/0022-1694(95)02970-2
Ojha C, Shirzaei M, Werth S, Argus DF, Farr TG (2018) Sustained groundwater loss in California's central valley exacerbated by intense drought periods. Water Resour Res 54(7):4449–4460. https://doi.org/10.1029/2017WR022250
Ojha C, Werth S, Shirzaei M (2020) Recovery of aquifer-systems in Southwest US following 2012–2015 drought: Evidence from InSAR, GRACE and groundwater level data. J Hydrol 587. https://doi.org/10.1016/j.jhydrol.2020.124943
Okwuashi O, Ndehedehe CE (2020) Deep support vector machine for hyperspectral image classification. Pattern Recognit 103:107298. https://doi.org/10.1016/j.patcog.2020.107298
Okwuashi O, Ndehedehe CE, Olayinka DN, Eyoh A, Attai H (2021) Deep support vector machine for PolSAR image classification. Int J Remote Sens 42(17):6498–6536. https://doi.org/10.1080/01431161.2021.1939910
Oussou FE, Yalo N, Ndehedehe CE et al (2020) Fracture aquifers identification in the Zou basin (West Africa) using remote sensing and GIS. Geocarto Int 1–24. https://doi.org/10.1080/10106049.2020.1852613
Parker A, Pigois J-P, Filmer M, Featherstone W, Timms N, Penna N (2021) Land uplift linked to managed aquifer recharge in the Perth Basin, Australia. Int J Appl Earth Obs Geoinf 105:102637. https://doi.org/10.1016/j.jag.2021.102637
Pekel J-F, Cottam A, Gorelick N, Belward AS (2016) High-resolution mapping of global surface water and its long-term changes. Nature 540:418–422. https://doi.org/10.1038/nature20584
Prigent C, Jimenez C, Bousquet P (2020) Satellite-derived global surface water extent and dynamics over the last 25 years (giems-2). J Geophys Res Atmos 125(3):e2019JD030711. https://doi.org/10.1029/2019JD030711

Rango A (1994) Application of remote sensing methods to hydrology and water resources. Hydrol Sci J 39(4):309–320. https://doi.org/10.1080/02626669409492752

Rateb A, Kuo C-Y (2019) Quantifying vertical deformation in the Tigris–Euphrates Basin due to the groundwater abstraction: insights from GRACE and Sentinel-1 satellites. Water 11(8). https://doi.org/10.3390/w11081658

Rodell M, Famiglietti JS, Wiese DN, Reager JT, Beaudoing HK, Landerer FW, Lo MH (2018) Emerging trends in global freshwater availability. Nature 557(7707):651–659. https://doi.org/10.1038/s41586-018-0123-1

Scanlon BR, Faunt CC, Longuevergne L, Reedy RC et al (2012) Groundwater depletion and sustainability of irrigation in the US High Plains and Central Valley. Proc Natl Acad Sci 109(24):9320–9325. https://doi.org/10.1073/pnas.1200311109

Schiermeier Q (2008) Water: purification with a pinch of salt. Nature 452:260–261. https://doi.org/10.1038/452260a

Seneviratne SI, Corti T, Davin EL et al (2010) Investigating soil moisture-climate interactions in a changing climate: a review. Earth-Sci Rev 99(3):125–161. https://doi.org/10.1016/j.earscirev.2010.02.004

Sheffield J, Wood EF, Pan M, Beck H, Coccia G, Serrat-Capdevila A, Verbist K (2018) Satellite remote sensing for water resources management: potential for supporting sustainable development in data-poor regions. Water Resour Res 54(12):9724–9758. https://doi.org/10.1029/2017WR022437

Smith RG, Majumdar S (2020) Groundwater storage loss associated with land subsidence in western United States mapped using machine learning. Water Resour Res 56(7):e2019WR026621. https://doi.org/10.1029/2019WR026621

Tapley B, Watkins M, Flechtner F et al (2019) Contributions of GRACE to understanding climate change. Nature Climate Change 9:358–369. https://doi.org/10.1038/s41558-019-0456-2

Tourian MJ, Reager JT, Sneeuw N (2018) The total drainable water storage of the amazon river basin: a first estimate using GRACE. Water Resour Res 54(5):3290–3312. https://doi.org/10.1029/2017WR021674

Valencia LO, Johansen K, Solorio B et al (2020) Mapping groundwater abstractions from irrigated agriculture: big data, inverse modeling, and a satellite-model fusion approach. Hydrol Earth Syst Sci 24(11):5251–5277. https://doi.org/10.5194/hess-24-5251-2020

Voss KA, Famiglietti JS, Lo M, de Linage C, Rodell M, Swenson SC (2013) Groundwater depletion in the Middle East from GRACE with implications for transboundary water management in the Tigris-Euphrates-Western Iran region. Water Resour Res 49(2):904–914. https://doi.org/10.1002/wrcr.20078

Wagner W, Hahn S, Kidd R, Melzer T, Bartalis Z, Hasenauer S, Figa J, De Rosnay P, Jann A, Schneider S, Komma J (2013) The ASCAT soil moisture product: a review of its specifications, validation results, and emerging applications. Meteorologische Zeitschrift. https://doi.org/10.1127/0941-2948/2013/0399

Wang M, Wigneron J-P, Sun R et al (2021) A consistent record of vegetation optical depth retrieved from the AMSR-E and AMSR2 X-band observations. Int J Appl Earth Obs Geoinf 105:102609. https://doi.org/10.1016/j.jag.2021.102609

Part V
Machine Learning in Hydrology

Chapter 7
Assimilated Deep Learning to Assess Terrestrial Hydrology

7.1 Introductory Remarks

Hydrology deals with the quantitative assessments of the occurrence and dynamics (in space and time) of water on and beneath the earth's surface. It also encompasses the properties of water (e.g., changes in water quality), and its inter-relationship with the living and material component of the ecosystem. Freshwater sustains biodiversity, terrestrial, and aquatic life and plays critical roles in driving agricultural production and peaceful coexistence in regions where shared water resources could trigger rivalry and agitations (e.g., Ndehedehe 2022a; Odorico et al. 2020). Assessing changes in freshwater is therefore critical, particularly given its contributions to national income and gross domestic product (e.g., Ndehedehe 2019; OECD 2015). The combination of several geospatial methods with new satellite hydrology missions like the Gravity Recovery and Climate Experiment (GRACE, Tapley et al. 2004) to quantify changes in freshwater to assess impacts from climate and human activities are, therefore, targeted efforts to improve understanding of changes in terrestrial hydrology (e.g., Ndehedehe 2022c; Rodell et al. 2018a, 2019).

The movement of freshwater around our planet contributes to changes in the earth's gravity field and in the hydrological cycle. This idea is the foundation of modern-day remote sensing techniques like GRACE used in the global monitoring of freshwater resources (i.e., terrestrial water storage). In the global water cycle, water evaporates from rivers, lakes, and the ocean into the atmosphere, where it resides for a short time (an average of eleven days in tropical regions), before returning back to the earth's surface as precipitation. Some proportion of the water falling on land runs off in rivers, and sometimes can lead to a temporary formation of lakes, due to runoff intensity and landscape accumulation tendencies. Apart from surface runoff, there is a proportion of water that infiltrates through the soil to the aquifers below, and sometimes, emerges or discharges later to lakes, rivers or even the oceans. The remaining portion of water falling on land is returned back to the atmosphere through evaporation as a result of latent heat transfer. The rate of evaporation is higher in the oceans than on land. The difference in the amount of evaporation rates over land and

C. Ndehedehe, *Hydro-Climatic Extremes in the Anthropocene*, Springer Climate,
https://doi.org/10.1007/978-3-031-37727-3_7

ocean is caused by surface runoff, which leads to water on land flowing back into the ocean. This results in a significant lag between the evaporation rates over land and the oceans when induced by the same amount of precipitation.

In the conservation of total water, there must be a balance between the two most significant drivers of hydrological fluxes (i.e., precipitation and evapotranspiration). This is particularly so, given that the human planet does not gain or lose water. However, past studies have reported that evapotranspiration accounts for about 70% of total precipitation globally (Good et al. 2015). On the land, or on a continental scale, precipitation having a rate of 0.108×106 km^3/yr exceeds evapotranspiration having a rate of 0.071×106 km^3/yr, whereas, over the oceans, evapotranspiration having a rate of 0.435×106 km^3/yr dominates precipitation having a rate of 0.398×106 km^3/yr (Berner and Berner 2012). The difference in each scenario is 0.037×106 km^3/yr and comprises water transported from the oceans to the land at atmospheric water vapor (case 1), or that returned to the oceans as river runoff (case 2) (Berner and Berner 2012). This water cycle and dynamic process that take place both on, below, and above the earth's surface gives rise to a renewable and variable quantity referred to as "freshwater".

Freshwater resources are needed to meet the demands of humans, agriculture, livestock, industry, mining, hydroelectric power, and thermodynamics. Apart from livestock and human consumption of freshwater either for food production, or household, drinking, etc., the remaining users of freshwater hardly return the water to the atmosphere. For example, thermoelectric power plants need large amounts of water from a river for cooling, but then return a larger percentage of that water to the stream, albeit at a much higher temperature. Without a more efficient and applicable water resource management technique, the demand for freshwater at this current rate arguably, may not go beyond 2040 (Ahmadi et al. 2020). The work of de Marsily and Abarca-del Rio (2016) offers some insight into the nature of water scarcity and the expected severity of impacts on food security. They identified decreasing groundwater stocks and increasing global water consumption for food from 1300 m^3/year per capita in 2000, to about 1400 m^3/year and 1500 m^3/year in 2050 and 2100, respectively, as few indicators that may lead to challenges in food production and or deficit.

Water problems and the expected increase in water consumption of this magnitude underpinned by rising global human population can lead to unprecedented changes in the environment and atmosphere due to changes in drought characteristics and global climate (e.g., Ndehedehe et al. 2023). The aftermath of such problems can lead to an impact on food production and the generation of energy, thus resulting in global food shortage and poor economic growth. The need for enhanced management of water resources, based on empirical data and scientific knowledge will therefore be critical in the coming years. With agriculture accounting for almost 85% of global consumption of freshwater, a 50% rise in global food production is anticipated in the coming years (Jury and Vaux 2007). The use of machine learning techniques for the assessment and modeling of freshwater resources is thus warranted. With the complexities in hydrological processes that limit the capabilities of traditional machine learning techniques, deep learning methods are emerging not only in feature extraction from remotely sensed data (e.g., Okwuashi and Ndehedehe 2020), but also to track the impacts of global climate on freshwater changes (Kalu et al. 2022a). The novel capabilities of such machine learning and the implementation of assimilated or hybrid deep learning techniques to identify critical drivers of climate-induced changes on land water storage is articulated in this chapter. Specifically, an experiment and a case specific study of South America to showcase the potential and utility of this method is detailed in this chapter.

7.1.1 Climatic Impacts on Freshwater Availability

The spatio-temporal distribution of groundwater recharge and river flows are significantly influenced by changes in hydrological fluxes such as land surface temperature, evapotranspiration, and precipitation (Chiew 2006). The uneven or abnormal changes in these fluxes are due to climatic impacts, and in large sub-regions, they affect hydrological processes and systems, including snow, ice, permafrost (frozen groundwater), and rising glaciers in lakes located in mountainous regions (Rosenzweig et al. 2007). Some of these changes, which vary between regions, seasons, and in between years are expected to last for some years (Collins et al. 2012). For example, projections from the Coupled Model Intercomparison Project Phase 6 climate models suggest drought duration will increase and for regions with strong climate–groundwater interactions, such changes could directly impact on groundwater availability (Ndehedehe et al. 2023). The impacts of climate change on freshwater resources pose a significant threat to sustainable development. For example, the study undertaken by Kundzewicz et al. (2008) reported that runoff is projected to increase in some regions and decrease in other regions, thereby causing water problems in some catchments. By the mid-century, an unprecedented decrease of about 10–30% of total annual river runoff and water availability is expected, especially over dry regions at mid-latitudes and in the dry tropics (Milly et al. 2005). Conversely, this impact will equally result in a 10–40% increase at high latitudes and in some wet tropical regions, while more pronounced changes are expected by the end of this century (Milly et al. 2005). Most of the sub-regions that are currently going through water stress are likely to suffer from decreasing water resource availability as a result of climate change, as both groundwater recharge and river flows are on a continuous decline (e.g., Döll and Flörke 2005; Usman et al. 2021).

As global warming keeps intensifying, resulting in rising land surface temperature, a shift in winter precipitation from snow to rain is expected. This will equally result in a change in the timing of the peaks of stream flow in a wide range of mountainous and continental regions. The spring snowmelt peak is eliminated or diminished, while the winter flows increase. As the polar ice caps melt due to warming, river flows increase in the short term, but decline once the ice caps disappear. One sixth of the global population reside in river basins supplied by water from snowmelt or glacier, which emanates from major mountain ranges, including the Hindukush, Himalaya, and Andes (Vergara et al. 2007). Additionally, the dynamics in the timing of streamflow in these regions may contribute immensely to water resources' availability, especially in the reduction of low flows in summer and autumn seasons. Changes in the frequency and intensity of climatic extremes, such as flooding and drought have equally been projected. According to an intergovernmental expert consensus reported by De Pryck (2021), the Intergovernmental Panel on Climate Change projected increase in the proportion of total rainfall from heavy precipitation in tropical and high latitude areas by 2090. This rise in rainfall is expected to increase flood frequency and magnitude in these regions. Where flood frequency is expected to intensify in regions experiencing high rainfall rates, drought frequency is also anticipated to increase for regions where reduced precipitation is projected.

By 2090, it is expected that drought-affected areas will increase in extent, while the proportion of the land surface in extreme drought will increase ten-fold from the present. From the experiment undertaken by Burke et al. (2006), results from the Palmer Drought Severity Index reveals an overall drying trend of 0.3 and 0.56 per decade, respectively, for the first and second half of the century in a global analysis. Due to the potential increase in the frequency of both flood and drought in the coming years, some drainage basins are also expected to experience these extreme climatic conditions. The benefit of projected increase in annual runoff in some regions, such as the eastern and southeastern Asia will be reduced by hostile impacts of increased variability and seasonal runoff shifts on flood risk and water supply, especially in densely populated low-lying river deltas (Kundzewicz et al. 2008). In addition, increased precipitation during the wet season in those areas may not lessen dry season problems, especially if the additional water from rainfall cannot be conserved. It is also expected that areas where runoff is projected to decline will likely face a decrease in the value of the service provided by water resources, e.g., as energy sources, or habitat for freshwater flora and fauna.

The potential effect of climate change on freshwater quality is another issue. Much of the assessment of water quality issues has only been undertaken in developed countries. Similar assessments in emerging economies are largely unreported. The countries in sub-Saharan Africa are some regions where an increase in water temperature along major river basins has been documented (Muringai et al. 2021). The impact of increased temperature in this area has reportedly altered the rate of operation of some key chemical processes in water. But in terms of variability in freshwater, the projected decrease in groundwater recharge and streamflow is expected in the south and north American countries (Kundzewicz et al. 2008). Moreover, changes in the intensity of precipitation events impact the rate at which materials are transported to groundwater and surface waters, and alteration in flow volumes influences dilution of loads. Apart from climate change-induced impacts such as water stress and reduced freshwater availability in largely semiarid domains, water quality will also be affected, evidenced in increasing pathogen loads caused by large rainfall amounts in locations with poor sanitation infrastructure (Kundzewicz et al. 2008). Some of the major repercussions of declining water quality due to climate change, include immense pollution loads from diffuse sources as a result of heavy precipitation through higher runoff rates and infiltrations; increase in water withdrawals from low-quality sources; overloading the capacity of water and wastewater treatment plants noticed during extreme rainfall; and the malfunctioning of water infrastructure during flood events (Delpla et al. 2009).

Rapid population growth has been reported in coastal regions having less than 10% of the global renewable water supply (Gormsen 1997). These regions house one quarter of the global population and will be among the most vulnerable in the advent of extreme climatic conditions. Another challenge is the rate of saline intrusion caused by excessive water withdrawals from aquifers. This is expected to exacerbate due to rising sea level effects, and will equally have a significant effect on freshwater resources (e.g., groundwater) and availability. These potential changes in timing, volume, and quality of surface and groundwater could impact the reliability of safe water supplies. Their impact will be further underpinned by damaging flood events, affecting agricultural and industrial operations, and causing water-related ailments. Other impacts include transporting water-borne diseases, as well as limiting aquatic ecosystem, and the services they provide in biodiversity functioning.

Due to the expected impacts of changes in water availability in the coming years, global-scale studies are intensifying, and new algorithms are being applied to further assess the danger of water scarcity (Huang et al. 2021; Ndehedehe et al. 2023; Robinson et al. 2020). Food insecurity resulting from water deficit in time and space is expected to be driven by limited rainfall caused by changes in atmospheric patterns like strong El-Ninõ events, lasting several years (e.g., de Marsily and Abarca-del Rio 2016; Ndehedehe et al. 2021). For example, using the water stress indicator, it has recently been shown that about 61.1% of global population experienced higher water stress (during 1971–2010 period) due to increased water withdrawal for agriculture (Huang et al. 2021). While growing water withdrawal is the reason for rising water scarcity for regions with about 21% of global population, the plenitude of studies suggest that billions of people living in water-stressed areas are adversely affected as a result of climate change (e.g., de Marsily and Abarca-del Rio 2016; Delpla et al. 2009; Hayashi et al. 2013; Huang et al. 2021; Ndehedehe et al. 2021). However, the actual numerical estimation will depend on key factors such as,

- the interaction between population growth and climate change in the coming years
- how water managers and policy makers adapt to this circumstance,
- understanding the role and impact of non-linear climatic variables in driving climate change.

The modeling of the latter requires more advanced machine learning techniques because traditional methods appear obsolete in modeling non-linear climatic quantities of this degree. Thus, the motivation behind this chapter is to exemplify the capabilities of these advanced machine learning techniques in terrestrial hydrology. In particular, the duple-based deep learning method has been implemented here to demonstrate how it can provide further understanding of the role of global climate indices on continental freshwater availability.

7.1.2 Terrestrial Water Storage

Terrestrial water storage (TWS) can be defined as the sum of water stored on and beneath the surface of the earth. This includes five major freshwater components, namely, surface water, soil moisture, canopy water, groundwater, wetlands, and snow. TWS is a basic component of the terrestrial and global hydrological cycle (Fig. 7.1), exerting significant control over the energy, water, and biogeochemical fluxes, thereby playing a key role in global climatology (Kalu et al. 2021; Ndehedehe 2022a; Rodell et al. 2018a). For instance, soil water storage influences the partitioning of energy fluxes and water resources over the land surface. This suggests its role in contributing to precipitation recycling, and hydrological extremes, such as floods and drought. Surface water storage plays a key role in freshwater dynamics, especially over sub-tropical regions, as well as nutrient and sediment transports, and also contributes significantly to the balance of greenhouse gas emissions to the atmosphere. Groundwater serves as a repository for a large percentage of freshwater available for human use. They are stored in aquifers beneath the earth's surface, and play a key role in supplying human consumption, supporting aquatic and terrestrial ecosystem functioning, and agricultural water demands in different regions of the world (e.g., Ndehedehe 2022b). In the estimation of current rates of global mean sea level rise, TWS is a key unknown, but contributes significantly to earth rotation variations such as length of day (Church and Gregory 2001). As an integrated estimate of surface and groundwater resources, TWS has significant implications for freshwater availability and management. Despite its significance and importance in ecosystem functioning, changes in TWS was only recently, in the last decade being adequately measured at sub-regional and continental scales. This was due to the lack of comprehensive global network for routine TWS monitoring. While the existing ground and satellite-based techniques only provided estimates for soil moisture and surface water, there was no integrated measurement and estimate for TWS until after April 2002 when satellite gravity missions made this possible.

The limitation of global-scale assessment of changes in TWS through direct observation was resolved by the launch of the Gravity Recovery and Climate Experiment (GRACE) mission in March 2002 (Tapley et al. 2004). The GRACE mission was launched under the auspices of the National Aeronautics and Space Administration (NASA) in the United States in collaboration with their German counterparts, Deutsches Zentrum fur Luft, and Raumfahrt (DLR). The main aim of GRACE is to measure monthly aggregated estimates of changes in the earth's gravity field. This will enable the tracking of mass changes in the hydrosphere, oceans, and cryosphere. GRACE uses two satellites (popular referred to as *Tom and Jerry*), which move in tandem, in near-circular orbits with an altitude of about 500 km and separated by approximately 220 km (e.g., Ndehedehe 2022d). These twin satellites orbit the Earth one after the other, with an embedded measurement system (Global Positioning System receivers and sensors as well as high-precision accelerometers), as opposed to the usual single satellite approaches with one dedicated sensor (Tapley et al. 2004). The basic measurement is determined from micron level tracking of the satellite-to-

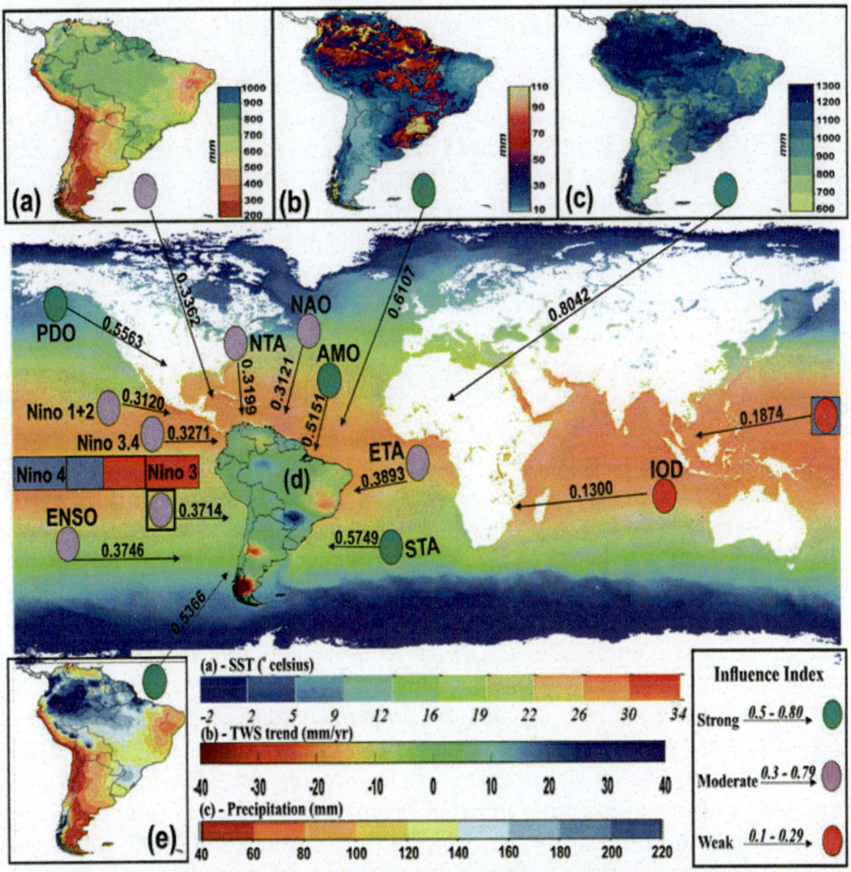

Fig. 7.1 Illustrating the collinearity between the predictors and predictand showing products with strong, moderate, and weak influence on the land water storage of the study region from April 2002 to June 2017. Predictors are grouped into (i) global teleconnection patterns, which include climate indices and averaged SST parameters, such as Nino 1+2, Nino 3.0, Nino 3.4, Nino 4, NTA, STA, ETA, ENSO, NAO, AMO, PDO, and IOD, and (ii) localized hydrological products, which include **a** evapotranspiration rates, **b** runoff estimates, **c** soil moisture estimates, and **e** precipitation patterns. The predictand is **d** GRACE-derived TWS estimates. SST units are in degree Celsius, while the rest are in mm

satellite space, whose variances is based on the individual gravitational attractions on the satellites as they pass over the Earth's surface (Tapley et al. 2004). After a month, the obtained measurement is used to compute an estimated global spherical harmonic model of the Earth's gravity field, which is subsequently used in the estimation of the mass changes of the Earth's surface. Accurate measurements of these mass changes are then related to equivalent water heights (land water storage) or global or continental changes in TWS fields (Fig. 7.2). As illustrated over South America, TWS or equivalent water heights varies across seasons in different regions or countries and for the Amazon basin area, the amplitudes of water storage changes are strongest between January and March compared to other periods (Fig. 7.2). It is possible to get minute differences between the computed mass series determined from the GRACE measurements due to the processing choices made by different vendors or the Science Data Systems (SDS) centers and their users. To increase the utility of GRACE data in diverse applications in earth science, SDS centers now provide quality-controlled, gridded, and basin-integrated mass change products.

Since its launch the GRACE data derived from gravity anomalies have been able to provide hydro-geodesists with new insights into TWS fluctuations that are difficult to measure from in situ data alone. Given the difficulty in the global assessment and monitoring of variations in hydrological patterns, this remote sensing technique has made it possible to track glacial ablations and ice sheets in the polar regions, monitor groundwater depletions, provide estimates of canopy water in dense rain-forested regions, and assess complex hydrological processes, by collecting and archiving time-variable monthly gravity observations (Cazenave and Chen 2010; Tapley et al. 2004). The GRACE satellite deciphers monthly gravity fields with an accuracy index of approximately 200,000 km^2 at low latitudes to 90,000 km^2 near the poles (Swenson et al. 2006). However, the GRACE mission facilitated a comprehensive evaluation of all available secondary data to create confidence in the use of GRACE to measure fluctuations in TWS on regional and water management scales (Rodell et al. 2018b). For the first time, GRACE provided the opportunity to quantify surface mass trends and changes in TWS, continental aquifers, glaciers, and ice sheets. To date, GRACE and its Follow-On mission continue to broaden our understanding of large-scale mass-redistribution associated with isostatic adjustments, earth movements, and earthquakes as well as changes in river runoff, TWS, Barystatic sea level, groundwater aquifers, and other forms of hydrological stores like Total Drainable Water Storage (see, e.g., Chen et al. 2020, 2022; Humphrey et al. 2023; Ndehedehe 2022b; Tourian et al. 2018).

Without GRACE, satellite observations are constrained to observing changes in the surface height of the ice sheets using radar and laser altimetry. These technologies are however influenced by variations in the properties of the surface layer and firn compaction, and only provide indirect measurements of the net mass change (e.g., runoff, ice discharge, evaporation, evapotranspiration, and precipitation). With the advent of GRACE, we are able to obtain the first direct measurement of ice mass change. Where altimetry-derived estimates are basically restricted to multi-annual trends of height-changes due to sampling and sensitivity issues at the ice sheet surface, the GRACE- derived ice sheet and glacier mass changes are obtained with an unprecedented temporal resolution of one month. The long-term mass trends determined using GRACE for the ice sheets are minutely manipulated by sampling methods, or unknown surface processes. However, the Arctic and Antarctica still have this correction problem due to observed poor mass-redistribution induced by the glacial-isostatic adjustment (GIA).

With the advent of satellite remote sensing and altimetry, extensive data collection and observation in the monitoring of water storage and fluxes in a dynamic world has been reported, thus enabling a global scale analysis of freshwater (Famiglietti et al.

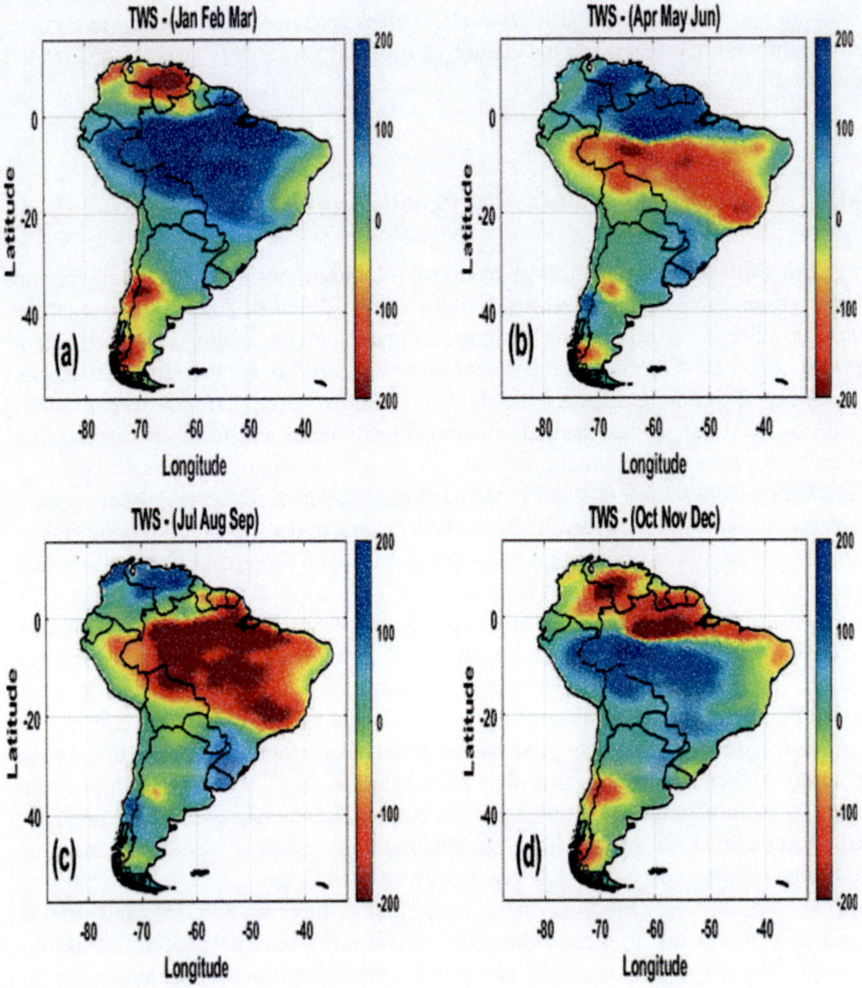

Fig. 7.2 Seasonal changes in terrestrial water storage over the study region from 2002 to 2017

2015). The GRACE-derived terrestrial water storage changes have been extensively used in various applications and assessments (e.g., Kalu et al. 2021; Ndehedehe and Ferreira 2020a; Rodell et al. 2018a) and were fully detailed in Chap. 4 of this book. In an attempt to better monitoring of our earth, the need to replace the model-based gravity parameters which have been in use (e.g., EGM2008) with continuous real-time gravity values (e.g., GRACE) is imminent. However, the direct estimates provided by the GRACE mission are inadequate when it comes to fully understanding the influence of other global climate modes on continental freshwater resources. This has warranted the increased use of artificial intelligence and machine learning algorithms, in a bid to provide sufficient answers as to the rate, intensity, and effects of non-linear atmospheric products, such as global climate modes, sea surface temperature components, and localized hydrological parameters on terrestrial water storage.

7.1.3 Deep Learning in Hydrological Modeling

In recent times, the application of machine learning theories to hydrological and climatic modeling has increased steadily (Kalu et al. 2022a, b, 2023). These machine learning models are algorithms designed to learn patterns and/or improve system performance based on the data they consume with little or no human intervention. The ability of these algorithms to think and act on their own as if they were humans (artificial intelligence) has resulted in increasing efficiency and accuracy in providing information about their interactions between non-linear climatic datasets, and the way they affect our environment, and by extension our daily lives. Other important aspects of machine learning operations lie in its ability to make decisions and predictions of numerical models and climatic quantities even without being explicitly programmed to do so (that's the unsupervised case).

Modern-day machine learning techniques have two major objectives. The first is to classify data based on models, which have been previously developed, and the last is to make notable predictions for future outcomes based on the previously learnt model. For instance, a hypothetical algorithm specific to classifying data may utilize computer vision of moles coupled with supervised learning to train it to classify cancerous moles. A machine learning algorithm for stock trading can inform the traders of future potential profit and loss outcomes depending on their previous trading experience. In hydrological studies, machine learning has the potential of providing notable outcomes of climate extremes which are dependent on various hydrological and meteorological processes, such as precipitation, ET, and runoff. This kind of machine learning uses the independent variable as the predictor, while the corresponding dependent variables serve as the predictand for the machine learning operations (Fig. 7.3). A simple architectural framework for back propagation neural network adopting this style of machine learning for hydrological applications is illustrated in Fig. 7.3. The concept of machine learning operations, however, limits system performance to exclusive machine operations, thus allowing them to still think and act like machines.

With the advent of deep learning operations, a more sophisticated approach to machine learning emerged. This innovation and sophistication enhances the feasibility to address the challenge of "machine-laden" operations because they have been specifically modeled after the human brain. Some deep structured architectures such as deep belief networks, deep neural networks, deep reinforcement learning, and convolutional and recurrent neural networks have been widely applied in multi-disciplinary fields. These include computer vision (Ponti et al. 2017), speech vision (Shahamiri 2021), machine translations (Singh et al. 2017), bioinformatics (Li et al. 2019), natural language processing (Young et al. 2018), medical image anal-

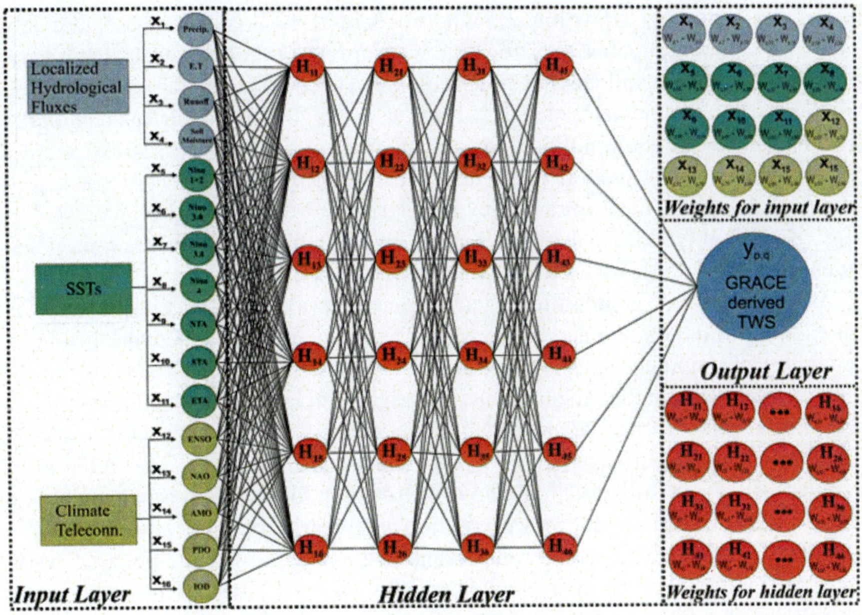

Fig. 7.3 Architectural framework of the adopted back propagation neural network used for the experiment depicting one input layer (consisting of the input parameters, i.e., hydrological fluxes (Precipitation, Evapotranspiration, Runoff, and Soil moisture), SSTs (Nino1+2, Nino 3.0, Nino 3.4, Nino 4, NTA, STA, and ETA), and climate teleconnections (ENSO, NAO, AMO, PDO, and IOD)); four hidden layers with six sets of neurons each (depicting a deep neural network training mode); the output layer as well as the varying weight parameters for all the nodes

ysis (Litjens et al. 2017), and even climate science (Mudigonda et al. 2021), have reported improved performances when compared to their machine learning counterparts. Based on these reports, this chapter attempts to refine predictions of deep learning systems by assimilating both machine and deep learning operations. This integration promises to show notable enhancement in the assessment of multi-scale climatic influences on continental hydrology, using coupled interactions between surrounding oceans and land water storage. In this chapter, the assimilated deep learning approach is employed to learn and understand land-ocean-atmosphere interaction, hydrological processes, and drivers of changes in TWS.

The modern-day application of deep learning algorithms has been compared to data-driven stochastic and physically-based techniques in hydrological assessments. Xu and Valocchi (2015) posited that data-driven techniques have enhanced accuracy over physically-based techniques in base flow predictions of regional groundwater models. This argument is easily corroborated by the application of data-driven techniques in calibration, parameter estimations, and its efficiency in handling a variety of uncertainties better than their physically-based counterparts. The robustness and simplicity of the data-driven models in hydrological studies have been widely reported in past literature (e.g., Yu et al. 2021). Physically-based groundwater flow models are characterized by important hydrogeological parameters that do not only vary spatio-temporally, but are difficult to measure. Data-driven models that rely on statistical learning techniques have been employed to correct uncertainties associated with calibrated groundwater model (Xu and Valocchi 2015). Data-driven techniques such as gene expression programming (Shah et al. 2021), boosted regression trees (Naghibi et al. 2016), support vector machines, random forest and M5 trees, as well as neural networks, have been widely applied in hydrological and climate studies (e.g., Kalu et al. 2023; Ndehedehe et al. 2020; Nourani et al. 2019; Okwuashi and Ndehedehe 2017; Sun 2013).

A neural network is an artificial intelligence algorithm that was inspired by the functioning of the human nervous system. The efficiency of the network lies in its capability to comprehensively operate disorderly but all-inclusive datasets even without needing strong model assumptions. The algorithm is well-known to have strong self-learning modes, self-adaptive abilities, and a good non-linear approximation quotient. The back propagation (BP) neural network algorithm is one of the most popular techniques for predictions and is well-known for its robust problem-solving ability. When the BP neural network is implemented in any predictive model, the connected factors, which are correlated and contain a significant amount of noise embedded in the systems are usually of concern, especially toward the compromise of data integrity (Okwuashi et al. 2021). Given the potential over-fitting and under-fitting problems prevalent in this algorithm, it is usually necessary to eliminate the noise and correlation among the input data as much as possible. This action of pre-processing model inputs can help to improve the performance and quality of the BP model, as well as its prediction accuracy.

Some studies have applied neural networks in GRACE and other water-related assessments. These include climate prediction analysis, groundwater prediction, water resource management research, and stream flow forecasting (House-Peters and Chang 2011; Ilin et al. 2005; Maier et al. 2010; Sun 2013). They have also been applied in precipitation and temperature analysis, population and irrigation studies, and simulation of groundwater pumping and weather conditions (Chu and Chang 2009). The study undertaken by Feng et al. (2008) simulated groundwater conditions to assess human influence using neural networks. They demonstrated the modeling capability of neural networks in capturing human footprints, including water extraction patterns on different spatial scales, thus improving understanding of critical drivers of regional groundwater conditions. Most of these studies (e.g., Ahmed et al. 2019) have employed machine/deep learning techniques to model TWS changes simply by using only hydrological products, including rainfall, evapotranspiration, runoff, etc. without accommodating the effects of global climate indices (sea surface temperature and teleconnection patterns) in their analysis. However, emerging studies (e.g., Kalu et al. 2022b) are beginning to value the role of the latter in improving groundwater modeling using deep belief network. The deep belief network groundwater modeling experiment is fully demonstrated in Chap. 8 of this book.

Climate teleconnection patterns influence regional meteorological processes. The knowledge of the influence of climate teleconnection on continental TWS is considerably important, especially given the fact that TWS is a hydrologic state and water cycle variable that includes groundwater storage. A thorough analysis of this concept provides insights on water resource managements, heat waves, wet conditions, and even potential flood and drought scenarios. Past attempts in modeling this process have dwelled on the influence of these global climate teleconnection indices on the dynamics of sea level rise, rainfall rates and intensity, and even changes in the distribution of TWS (DiNezio 2013; Merrifield et al. 2012; Ndehedehe and Ferreira 2020a). The need for an improved and notable prediction of TWS to adequately guide our understanding of future impacts of extreme climate conditions on water availability, mitigate harsh weather patterns, and guide hydrologists in decision-making over large sub-regions is thus critical.

In this chapter, a complex and multi-layered deep neural pattern that enable data to be moved between nodes in a highly connected manner is demonstrated (Fig. 7.3). This multi-layered deep neural network results in a non-linear transformation of the input hydrological and climatic datasets. The latter is becoming increasingly complex due to climate change and strong anthropogenic influence affecting natural processes (e.g., hydrology) as well as their respective cycles. Prior to the implementation of this multi-layered deep neural network, the original datasets, which consists of hydrological fluxes, sea surface temperature (SST), and global teleconnection patterns were pre-processed using unsupervised statistical decomposition techniques. The overarching goal of this pre-processing is to extract the dominant climatic features of the original predictors and use them for TWS predictions. The decomposition techniques to be used are the principal component analysis (PCA) and the independent component analysis (ICA). These techniques play a key role in highlighting the impacts of climate variability, land surface processes, hydro-climate drivers, and even atmospheric circulation patterns in the modeled TWS estimates. Therefore, in this chapter, the aim is to design an assimilated deep learning framework to predict the impact of SST-based indices on continental TWS using an improved BP-based model (Kalu et al. 2022a). The improved BP-based models, which are named, PCA-back propagation neural network (BPNN) and ICA-BPNN are significant in the assessment of climatic drivers, their impacts, and relative contributions on TWS dynamics.

The objectives of this study are (i) building a deep learning framework to reconstruct climate-driven TWS using the PCA-BPNN and ICA-BPNN models, with localized hydrological products (e.g., rainfall), and global SST and teleconnection datasets, (ii) examine the association of global climate teleconnection indices on continental TWS, and draw attention to the influence and impacts of localized hydrological fluxes and global climate indices in modeling TWS, and (iii) assess the influence of key climatic drivers on the spatio-temporal distribution of GRACE-TWS from 2002 to 2017. This method pioneers deep learning applications in diagnosing the effect of hydrological fluxes and climate teleconnection patterns on terrestrial hydrology on a continental scale by (a) making predictions of GRACE-TWS using an integrated statistical and BP neural network model, and (b) highlighting the impacts of global climate indices and hydrological fluxes (e.g., rainfall, evapotranspiration, etc.) as important drivers of terrestrial hydrology and continental TWS modeling.

7.2 South America: Hydrology and Climatology

South America (SA) is one of the most physiographically diverse wet-dry tropics composed of several aquifer types, highlands, lowlands, and highly productive watersheds. Its unique hydrological and hydrodynamic characteristic has attracted the attention of the global hydro-community, as it drains about 30% of the continental freshwater that reaches the oceans (Ferreira et al. 2019b). The continent houses the extremely wet Amazon River basin, which in turn hosts at least 10% of the world's known biodiversity, including endemic and endangered flora and fauna. It also plays home to considerable drier regions where geodetic perturbations and the impacts of other physical processes such as land subsidence and topographic deformation have been reported (e.g., Ferreira et al. 2019a, b). Some of the other major hydrogeological river basins and physiographic domains within the South America region include, the Orinoco, which is a tropical basin that occupies one-third of the Parana basin area but maintains a higher annual mean discharge. Also, the Sao Francisco and Parnaiba rivers, which cross the Brazilian semiarid are essential for regional subsidence (Brêda et al. 2020). Ndehedehe and Ferreira (2020b) reported that water availability is high in the region's tropical basins, including the Amazon, Orinoco, and Magdalena, and demonstrates significant distribution of annual precipitation as opposed to other hydrological hotspots in the world. They further stressed on the recurrence of water scarcity in the region experienced by most residents in northeast Brazil, despite being a major water tower, holding approximately 12% of global freshwater resources and accounting for an estimated 20% of global freshwater discharge. This dearth in precipitation rates are attributed to limited rainfall and frequent extreme hydrological drought conditions.

Precipitation patterns in SA are basically regulated by atmospheric circulation patterns and inter-tropical convergence zone, which defines the rates, intensity and seasonal characteristics of precipitation in most of the tropical wet-dry climatic belts. Using the MSWEP precipitation model over SA from 1990–2010, Brêda et al. (2020) reports a low annual precipitation rate observed in the Pantagonia, south Andes, and Brazilian semiarid regions, and a higher annual precipitation occurring in the Pampas and majorly across the sub-regions surrounding the equatorial line, especially in the west Amazon. This heterogeneous precipitation pattern over the continent reflects aridity and serves as a significant pointer to the role of global climate indices and teleconnection patterns on the climatology of the continent (Ndehedehe and Ferreira 2020b).

7.2.1 *Impact of Climate Modes on South American Weather Systems*

Global teleconnection patterns are important components of the climate system. They are large-scale atmospheric variability indices, occurring over months to years, as a result of the variations in circulation patterns, including, and large anomalies in SSTs. These patterns are closely linked to stationary planetary waves and zonal non-linear inter-relationships. The unprecedented oceanic warming in the Northeast pacific uncovered the linked teleconnections between the North Pacific and the weak 2014/2015 El-Nino to the atmospheric forcing patterns (Bond et al. 2015). The ocean warming resulted in major ecological consequences over the eastern pacific (Letícia M. Cavole 2016). An understanding of these teleconnection patterns and their relationship to climate variability at different time scales is important. This is because they provide the framework that supports planning schemes for water resources and sustainable strategies for freshwater conservations.

Over the years, the SA region has witnessed significant weather and climate fluctuations pattern as a result of teleconnection patterns, linking atmospheric and/or oceanic patterns from great distances to the continent. A notable example of the oceanic patterns disturbing the SA weather patterns is the El-Nino Southern Oscillation (ENSO). The ENSO is a natural oceanic-atmospheric coupled phenomenon highlighted by SST anomalies in the equatorial Pacific Ocean (ocean component) and the circulation pattern of its adjacent atmosphere (atmospheric component). They are known to impact climate globally, either directly or indirectly from the tropics to the extra-tropics. Over SA, reports have shown that these impacts are connected with the anomalous walker circulation descending a branch position (Cai et al. 2020), influencing the tropical regions, through Rossby wave trains via the Pacific-South America patterns (Grimm and Ambrizzi 2009). During El-Nino events, the walker circulation ascending branch is strengthened over the warmer waters of the eastern-central Pacific Ocean, and its eastern descending branch relocates to the tropical Atlantic ocean and northeastern SA, thereby incorporating the eastern Amazon region and northeastern Brazil in the process (Ambrizzi et al. 2004). This process has been linked to the reduction in precipitation over Northeast Brazil and east of the Amazon (Cai et al. 2020; Grimm and Ambrizzi 2009), with the opposite scenario occurring in La-Nina years (Grimm 2003). On the other hand, SA regions like Peru and Ecuador are getting increased amount of precipitation (Reboita and Ambrizzi 2021). This ENSO impact over the eastern portion of the Amazon basin still remains a topical issue, given the importance of the region to global climatology. The study undertaken by Lu and Dong (2005) also underlined the essential role of SSTs of the surrounding oceans on precipitation dynamics over the SA region.

7.2.2 *Evidence of Climatic Footprint on South American Terrestrial Hydrology*

Just like other highly productive freshwater ecosystems, precipitation drives TWS and changes in hydrological fluxes in the tropical regions of SA (Ndehedehe 2019). This means that regions that receive more annual rainfall are most likely to experience stronger amplitude and TWS dynamics. This claim was corroborated by Ndehedehe and Ferreira (2020b) who reported a 4500 mm mean annual precipitation in the Amazon river basin, which translated into a rise in TWS during the same period as a result of the precipitation-induced interchange of fluxes within the floodplain corridors (See Fig. 1a and b in Ndehedehe and Ferreira (2020a)). Generally, land water storage dynamics over the SA region from 2000 to 2017 is an outcome of several factors. These include, climate teleconnection patterns, recovery from dry periods, geodetic disturbances and deformations of the nearby oceans, global warming impacts, and anthropogenic influences, among other drivers (Rodell et al. 2018b). However, there is an overwhelming and a somewhat conclusive evidence of the impacts of climate teleconnections on changes in continental freshwater in SA and the tropical north of continent is a major hotspot of climate-induced (e.g., ENSO) changes in TWS (e.g., Ndehedehe and Ferreira 2020b; Ni et al. 2017; Phillips et al. 2012). This makes SA a unique case study for this experiment to demonstrate the potential of assimilated deep learning in assessing influence of climate variability on terrestrial hydrology to further improve understanding of global climate on TWS.

7.3 Data Types and Processing Methods

The original input parameters used for this experiment are climate drivers, which play the role of independent neurons when injected into the BPNN system model (Table 7.1). They are functions of four localized hydrological fluxes (i.e., precipitation, evapotranspiration, runoff, and soil moisture), seven prominent SST products of the nearby oceans (i.e., North Tropical Atlantic (NTA), South Tropical Atlantic (STA), East Tropical Atlantic (ETA), Nino 1+2, Nino 3.0, Nino 3.4 and Nino 4), and five global climate teleconnection patterns (e.g., ENSO, North Atlantic Oscilla-

Table 7.1 Fit based on Pearson's correlation between the original sixteen input variables used in developing the BPNN model. Predictor modes that exhibit high interplay between each other are represented in shaded cells

Input variables (i.j)	TRMM	ENSO	NAO	AMO	PDO	IOD
Evapotranspiration	0.7559	0.0340	0.2649	−0.2691	0.1547	0.0506
Soil moisture	0.6467	0.1657	0.1243	−0.3822	0.0930	−0.1205
Runoff	0.8063	0.1933	0.1883	−0.4205	0.1388	−0.0630
Nino 1+2	−0.0564	−0.3925	−0.0755	0.0246	0.3243	0.3136
Nino 3.0	−0.1417	−0.6451	−0.0285	0.0550	0.4754	0.3463
Nino 3.4	−0.1499	−0.7539	0.0032	0.0448	0.5904	0.2271
Nino 4	−0.1202	−0.7531	0.0156	0.0621	0.6286	0.1285
NTA	−0.0790	0.0201	−0.2738	0.6949	0.0288	0.0542
STA	0.1081	0.0718	−0.0998	0.0121	0.0778	0.0719
ETA	0.1228	0.1167	−0.1077	0.2314	0.0476	0.1198

tion (NAO), Atlantic Multi-decadal Oscillation (AMO), Pacific Decadal Oscillation (PDO), and the Indian Ocean dipole (IOD)). The GRACE-TWS on the other hand, served as a dependent parameter for our deep learning method.

7.3.1 Independent Data Types

7.3.1.1 Precipitation

Monthly TRMMv7 3B43 precipitation estimates (April 2002–June, 2017) obtained from the National Aerospace and Space Administration (NASA) Goddard Space Flight Center (GSFC) combined with the Integrated Multi-satellitE Retrievals for Global Precipitation Measurements (IMERG) were used for this experiment. This satellite-based precipitation was applied to assess the spatio-temporal inconsistencies in the SA region, particularly given the impacts of SST and climate teleconnection on dominant rainfall, which has been widely reported in the region. The TRMM 3B43/IMERG-based precipitation is used to assess monthly distribution of rainfall in time and spaces at 0.25 by 0.25 degrees (Huffman et al. 2007). The TRMM/IMERG rainfall dataset was rescaled to maintain and preserve a consistent spatial resolution and rainfall grids or cells consistent with the other datasets (e.g., GRACE-derived TWS) used in this study. The TRMM 3B43/IMERG-satellite precipitation is freely and publicly available for download (https://disc.gsfc.nasa.gov/datasets/TRMM_3B43_7/summary).

7.3.1.2 Evapotranspiration

The most practical and feasible means of obtaining spatially distributed evapotranspiration estimates over any terrestrial surface are through remote sensing techniques. The evapotranspiration (ET) dataset utilized in this study is (MODIS16A2) provided by the Numerical Terradynamic Simulation Group (NTSG), University of Montana. As part of the NASA/EOS project, this ET product can be used to calculate regional water and energy balance, soil water status to provide information that underpins water resource management on regional and global scales. The data set is available in NASA's repository and data portal (https://modis.gsfc.nasa.gov/data/dataprod/mod16.php). MODIS16A2 global ET monthly datasets from 2002 to 2017 were generated over a spatial resolution of 0.5 km pixel and a temporal period of 8 days. This ET estimates are pre-processed products based on the Penman-Monteith equation, and incorporates the effects of evaporation from the soil, wet canopy, and plant transpiration at dry canopy surfaces (Mu et al. 2011).

7.3.1.3 Runoff

Monthly runoff estimates plays a key role in the regulating TWS dynamics in tropical regions, as shown in the water budget equation. The global runoff data center which provided the monthly runoff dataset used in the experiment (www.bafg.de/GRDC) operates under the auspice of the world meteorological organization, and provides river discharge data of nearly 9000 gauging stations from all over the world. It is updated continuously and consists of daily and monthly discharge data which can even be used for the verification of atmospheric and hydrologic models. This product is generated with an underlying precipitation climatology, which is readjusted when the measured runoff exceeds the input precipitation and contradicts the terrestrial water budget equation. By including monthly runoff estimates as part of the predictors (Figs. 7.1 and 7.3), the aim is to strengthen the performance of the proposed deep learning framework in analyzing hydrological conditions over the tentative test bed identified in this chapter.

7.3.1.4 Soil Moisture

To improve understanding of the influence of soil moisture changes in the hydro-meteorological interactions of multi-scale climatic system, as well as the role they play in regulating TWS, we also included soil moisture datasets in the original predictors (Figs. 7.1 and 7.3). Soil moisture plays a key role in the studies of land surface hydrological processes and monitoring of climatic events in tropical regions (Ndehedehe et al. 2018). Their impact on TWS, however, varies based on soil type, low-level wind, temperature, and moisture. Apart from land water storage interactions; these products also play primary roles in the surface model behavior of soil moisture contents in tropical regions. Monthly terraclimate soil moisture data at 0.50 grid for the period between April, 2002 to June, 2017 was used in this experiment to enhance our deep learning estimation of hydrological impacts/contributions to continental TWS. The Terraclimate data is a monthly climatic water balance product that uses climatically aided interpolation, combining multi-resolution climate data from the WorldClim dataset (https://www.worldclim.org/) and the Climate Research Unit Ts4.0 data (https://data.ceda.ac.uk/badc/cru/data/cru_ts/) and other Reanalysis product (https://jra.kishou.go.jp/JRA-55/index_en.html).

7.3.1.5 Multivariate ENSO Index

ENSO is a periodic climatic pattern that encompasses the dynamics in the temperatures of waters in the central and eastern tropical Pacific Ocean (Trenberth 1997). It describes the presence of abnormally warm (El-Nino) and cold (La-Nina) SST anomalies in the eastern pacific. Given the interaction of the Multivariate ENSO Index (MEI) with inter annual changes in water availability and its assimilation of six other parameters and anomalies describing the Pacific, it was preferred in this experiment to other available indices. However, the comparison between MEI and other ENSO indices (such as the Nino 3.4 and 4.0) in this study found negligible difference based on their correlation coefficients.

7.3.1.6 North Atlantic Oscillation Index

The North Atlantic Oscillation (NAO) index is composed of a north-south dipole anomalies. The first one is located in Greenland, while the second, having an opposite sign spans through the central latitudes of the North Atlantic region (i.e., 350N and 400N). These opposing phases of the NAO are primarily responsible for basin-wide dynamics, especially in the location and intensity of the North Atlantic jet stream and storm track, as well as the in large-scale harmonization of the mean meridional and zonal heat forms and moisture transmissions. This results in the high-level dynamics of global precipitation and temperature patterns in regions of notable hydro-meteorological interactions, such as the SA region.

7.3.1.7 Atlantic Multi-decadal Oscillation

The Atlantic Multi-decadal Oscillation (AMO) is an aesthetically consistent variability pattern in the North Atlantic SSTs having duration of 720–960 months and comprehensively computed from the average SST anomalies of the Northern. They have been shown to be significantly correlated with multi-decadal variations of tropical summer precipitation rates, and contribute substantially to global warming (Chylek et al. 2016; Ndehedehe et al. 2017). It is still expected that the relative influence of the AMO activities to global warming postulated by Chylek et al. (2016) will increase during the second half of the 21st century. Therefore, the need to understand the potential contributions of this climatic index to dynamics in continental water storage is warranted (Ndehedehe et al. 2017).

7.3.1.8 Pacific Decadal Oscillation

The Pacific Decadal Oscillation (PDO) is a periodic pattern of ocean-atmosphere climate dynamics concentrated over the mid-latitude Pacific basin. It is perceived as either warm or cool surface waters in the Pacific ocean, north of 200N. The PDO can be likened to a long-lived El-Nino-like pattern of Pacific climate dynamics having extremes epitomized by rife oscillations in the North America climate and the Pacific basin. As a result of its heterogeneous oscillating patterns characterized by low frequency oscillations, it serves as a potential tool for future to explore future changes in climatic conditions. The influence of PDO on land water storage is gradually emerging and such evidenced suggests the footprint of multi-Decadal changes in hydrology (Ndehedehe and Ferreira 2020b).

7.3.1.9 Indian Ocean Dipole

The Indian Ocean Dipole (IOD) is largely portrayed by irregular oscillation of SSTs, which allows the interchanging patterns of warm and cold temperatures between the eastern and western part of the Indian Ocean. It is basically illustrated by the anomalous SST gradient between the southeastern equatorial Indian Ocean and the western equatorial Indian Ocean, which is commonly known as the Dipole Mode Index (DMI). This study explored the monthly correlation between this dataset and TWS for the period of 2002–2017, as well as their contribution to the fluctuations in hydro-meteorological patterns of the SA region. This data can be accessed from the European Climate Assessment and Data repository (http://climexp.knmi.nl).

7.3.1.10 Sea Surface Temperature Products

The SST (sea surface temperature) products which formed part of the predictors for the deep learning analysis play a key role in providing an all-encompassing effects of surrounding oceans on the hydrology of the SA region, and they include Ninos 1+2 (0-10S, 90W-80W), 3.0 (5N-5S, 150W-90W), 3.4 (5N-5S, 170W-120W), and 4.0 (5N-5S, 160E-150W). Other SST products such as the NTA, ETA, and STA from the Atlantic Ocean were also used in our experiment. The mean SST datasets from 2002-2017 used in this study was downloaded from https://www.esrl.noaa.gov/psd/data/gridded/data.noaa.oisst.v2.html.

7.3.2 Dependent Data Types

7.3.2.1 GRACE-Derived TWS

The application of GRACE-derived TWS in hydrological research and assessment has gained substantial attention over the years. Its role as a vital tool in sub-surface water storage monitoring and groundwater evaluations and processes has been overly explored in recent studies (Kalu et al. 2021; Ndehedehe and Ferreira 2020a, b; Ndehedehe et al. 2017). In this chapter, the monthly GRACE products from the University of Texas Center for Space Research (CSR-SH) was employed to serve as the dependent variable to be predicted.

The adequate prediction of this component not only provides a robust algorithm for the prediction and allocation of freshwater resources, but equally serves as a tool to account for data gaps in the GRACE monthly record. This missing observations have impeded accurate estimations and analysis of hydrological processes in several data-deficient regions, especially in their application of climate change studies. This study used GRACE mascons and the use of this product in global freshwater analysis is increasingly becoming popular in hydrological research, especially given its simplicity in the implementation of different geophysical constraints to help filter out noise from the level 2 GRACE dataset (Wiese et al. 2016). The mascon solutions, arguably provide results consistent with other global GRACE products even without implementing complex pre-processing processes like smoothing and de-stripping (Ndehedehe and Ferreira 2020b).

7.3.3 *Deep Learning Framework*

To ascertain the performance index of algorithm used for regression, it is necessary to measure the trade-off amount between calculation interval and performance. Over the years, the application of the BPNN has been well recognized for its popularity as an optimal training algorithm. This is because its mode of back propagation creates an avenue to fine-tune its predictions so as to record very minimal bias between the observed and predicted parameters (Maier and Dandy 2000). Its efficiency, robustness, and ease in implementation makes it often preferred to other neural network models for regression analysis. The BPNN appears to be a superior learning algorithm when a sufficient number of optimized and relatively noise-free parameters are used as original input predictor values (Fig. 7.4), regardless of the convolution of the specific domain problems (Walczak and Cerpa 1999). Even though Goethals et al. (2007) reported that the BPNN has a knack for withstanding noise in the training dataset during the training procedure, too many errors in the training dataset can impede its fine-tuning capability by demanding several iterations, and in some cases, prevent the neural network from learning the desired model. In case of few extremely noisy training datasets, some studies have recommended algorithms such as the radial basis function networks, and the counter propagation networks apart from the BPNN model, given that they have performed well in domains having insufficient and noisy input parameters as training datasets (Barnard and Wessels 1992; Fausett and Elwasif 1994).

It is worthy of note that the operation of input functions such as the momentum term, learning rate, error, and transfer function influence the performance of the optimized PCA- and ICA-BPNN if they are superlative. One of the most frequently used transfer functions is the sigmoid transfer function, which utilizes the logistic and hyperbolic tangent function in their operations. Other transfer function, including the linear and hard limit functions have equally been reported to have fast convergence rates in past studies (Maier and Dandy 2000). The mean square error function plays a key role in minimizing the training process, and is very common in error propagation of neural networks. In this experiment, however, an independent and normal error distribution function which has been recommended on the basis of obtaining optimal results was implemented.

In training datasets laced with significant number of outliers, Liano (1996) proposed the application of the least mean log squares error function in the attempt to countermand the impact of the outliers in the system model. The learning rate of this function is directly proportional to the size of the steps taken in the weight index. Essentially, learning rates are established during the training process, and by

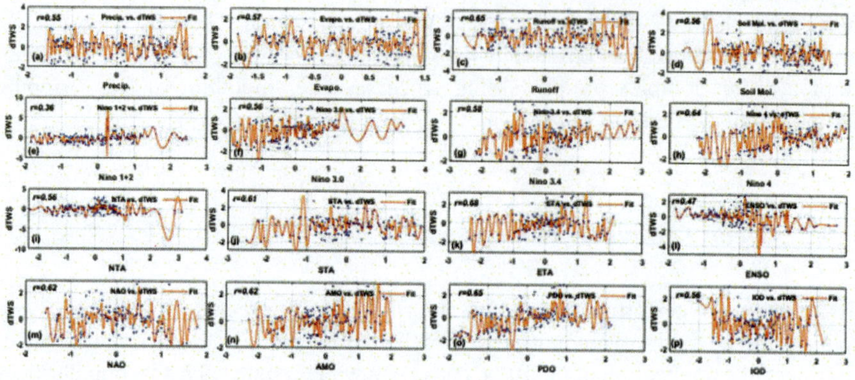

Fig. 7.4 Using the smoothing spine algorithm (with smoothing parameter, p = 0.8) to estimate the functional influence between the predictors and the response variable. The response variable (TWS) has been deaseasonalized (see Sect. 7.4.4) to eliminate harmonic constituents and trends; therefore, the residual (dTWS) was used as the dependent variable in this evaluation **a** dTWS versus precipitation. **b** dTWS versus Evapotranspiration **c** dTWS versus Runoff **d** dTWS versus Soil Moisture **e** dTWS versus Nino1+2 **f** dTWS versus Nino3.0 **g** dTWS versus Nino3.4 **h** dTWS versus Nino 4 **i** dTWS versus NTA **j** dTWS versus STA **k** dTWS versus ETA **l** dTWS versus ENSO **m** dTWS versus NAO **n** dTWS versus AMO **o** dTWS versus PDO **p** dTWS versus IOD

utilizing the trial-and-error method, the optimal learning rates are established. To improve quantities such as the convergence rate and the learning speed of the system mode, the momentum term is usually assimilated into the training algorithm. If the momentum term is greater than 1.0, then the training procedure will not converge. This brought about the suggestion of a learning rate of 0.7 with a momentum range of 0.8 and 0.9 or a learning rate of 0.6 with a momentum of 0.9 by Dai and MacBeth (1997). Other studies on the other hand, have suggested the determination of the bounds for the convergence on momentum parameters and learning rates, and have demonstrated that the momentum term can improve the learning rate indexes over which the model converges (Goethals et al. 2007).

7.3.4 Weight and Bias Initialization

The proposed BPNN used in our study consist of three fully interconnected layers, including the input, hidden, and output layers. The input layer houses the optimized input parameters, the hidden layer houses the six sets of neurons each depicting a deep neural pattern, and the output layer consists of the predictand (GRACE-TWS). The last neuron of the input and hidden layer is considered a bias node, given its resultant output equivalent of 1. Assuming the input layer is illustrated by available P patterns for the training procedure, the totality of the original input datasets can be represented as an $N + 1$ matrix with P rows and $N + 1$ column, given that all records of the last column equals 1. Similarly, the target dataset in the output layer can be characterized by a matrix W, having P rows and M columns. The weights between the neurons in the input and hidden layers generate a matrix w having entries, $w^1_{j,j}$ ($i = 1, N, j = 1, N + 1$), given that each discrete $w^1_{j,j}$ communicates with the ith neuron of the input layer with the jth neuron of the hidden layer.

As a result of this propagation, the hidden layer output can be represented as $A^{hidden,OUT} = f1(W^1 A^{input,IN})$ given that $A^{input,IN}$ illustrates the input matrix transpose. Also, the interface between the neurons weights in the hidden and output layer results in a matrix $w^2_{i,j}$ ($i = 1, M, j = 1, N + 1$), given that each weight $w^2_{i,j}$ associates the jth neuron of the hidden layer with the jth neuron of the output layer. This depicts that the interaction of the entire system model, ranging from the input layer to the output layer can be denoted by the matrix $A^{0,IN}$ with $N + 1$ rows and P columns. The optimal initial weights of the BPNN with four hidden layers basically result in a linear least squares problem, and can be further examined through the minimization procedure. The solution to this linear least square problem containing the optimal weights liking the four hidden layers to the output layer neuron is obtained by implementing the QR factorization technique based on the singular value decomposition algorithm (Qadir 2020).

7.3.5 Number of Hidden Layers

To improve the closeness of fit between a predicted neural network model and the corresponding predictand, an increase in the number of hidden layers is pertinent. However, a fewer number of hidden layer neurons improves the extrapolation efficiency or smoothness of the network, but may lack the desired accuracy needed (Walczak and Cerpa 1999). If a neural network is composed of one hidden layer, it can be likened to perform linearly with the capability of predicting any function as long as the hidden layer consist of sufficient as pointed out in the previous section above. However, for optimal results, some studies suggest that machine learning operations of non-linear constituents should be initiated with at least two hidden layers (Okwuashi and Ndehedehe 2020). For this particular experiment, however, the trial-and-error method to ascertain the optimal number of hidden layers to be used in our BPNN network was implemented, and this was based on the experience of the original input datasets. This strategy was adopted because of the highly problematic reliance of many optimal network geometries.

7.3.6 *Number of Hidden Neurons*

The choice of the number of neurons is largely a function of the total original input parameters used in our training, whereas the number of the neurons in the output layer is equivalent to the number of model outputs, which is 1 in our case. However, the choice of the number of neurons in the hidden layer is an important aspect of the deep learning framework, as it plays a key role in the accuracy index of the neural training. Bebis and Georgiopoulos (1994) suggested that a few hidden neurons provide faster training but results in scarce feature indicators, whereas more hidden neurons result in longer training periods, and may improve fitting between the predicted and the observed datasets. Some studies suggest that any continuous function having X_i inputs in the range [0 1] and X_o outputs can be designated as a feed forward procedure having $2X_i + 1$ hidden neurons. However, another popular procedure for optimal hidden layer neuron selection is the rule of thumb technique used in the determination of the number of neurons in the hidden layer for an optimal training experience. While some are based on the number of original predictors in the input layer, others are focused on the total number of input and output layer neurons in the network. We represent the pioneering rules that suggest the number of hidden neurons as a function of the number of input and/or output nodes in Table 7.2.

In this experiment, however, we will attempt to base the number of hidden layer neurons on the available connection weights from the input layer. Given that the training sample quantity has already been fixed to 16, which serves as an indication of the optimal parameters of connection weights, the rule depicted in Table 7.3 was applied to avoid model over-fitting, thus arriving at six neurons for each set. With the number of connection weights, and the number of input and output neurons made available by the original input predictors and the predictand, respectively, it was possible to estimate the number of hidden neurons. That is, if only one hidden layer is to be used, then the number of connection weights should not exceed $N_{TS} \times 1/10$, and the number of hidden neurons should not be less than $(X_i + X_o) \times 1/2$.

Table 7.2 Pioneering suggestions for number of hidden neurons; using the number of input (Xi) and output (Xo) nodes as a determining function

Suggestion	Source
$2 \times Xi$	Kanellopoulos and Wilkinson (1997)
$3 \times Xi$	Kanellopoulos and Wilkinson (1997)
$2 \times Xi + 1$	Fletcher and Goss (1993), Patuwo et al. (1993)
$\frac{1}{2} \times (Xi + Xo)$	Piramuthu et al. (1994)
$\frac{3}{4} \times Xi$	Lenard et al. (1995)
$2/3 \times Xi$	Wang (1994)

Table 7.3 Attempting to avoid over fitting by using the number of training samples (NTS) to estimate the maximum number of connection weights

Max. number of connection weights	Source
NTS	Rogers and Dowla (1994)
NTS × 1/2	Masters (1993)
NTS × 1/4	Walczak and Cerpa (1999)
NTS × 1/10	Weigend et al. (1990)
NTS × 3/100	Amari et al. (1997)

7.4 Methodology

7.4.1 Statistical Decomposition of Predictors

A justification for utilizing the statistical decomposition approach in the original input predictors is:

- attempting to explain the maximum amount of variance in our predictor datasets,
- reducing the impact of noise and outliers in our system model, and
- artificially generating new data samples for missing features.

Basically, the statistical decomposition technique allows our set of input predictors to be re-arranged into a new set of predictors that are orthogonal to each other and maximizes the amount of variance in the dependent sample that can be explained with the smallest number of Empirical Orthogonal Function (EOF) predictors, which is six in our case for both PCs and ICs. This advantage promises to optimize the predictability of the BPNN network, and reduce the lag between the observed and predicted neuron in the output layer. Additionally, it will also aid in understanding the influence of the predictor parameters on the predictand (Fig. 7.5). The PCA and ICA-decomposition technique was used in this experiment.

7.4.1.1 Principal Component Analysis

The principal component analysis is a well-known statistical decomposition tool which has been widely applied in feature extraction and dimensionality reduction (Ndehedehe et al. 2016a). It works by splitting a given time series into various components or signals through statistical rotation, while retaining interpretability and diminishing data loss. The relevance and prevalence of this statistical method in climate analysis can be largely attributed to its robustness, simplicity, and proficiency in the interpretation of optimized multivariate data, with the least amount of principal components. The technique provides the best illustration of a p-dimensional dataset in q-dimension by attempting to augment the variance in q dimensions. Advancement in technology has resulted in the acquisition of high-resolution datasets and the

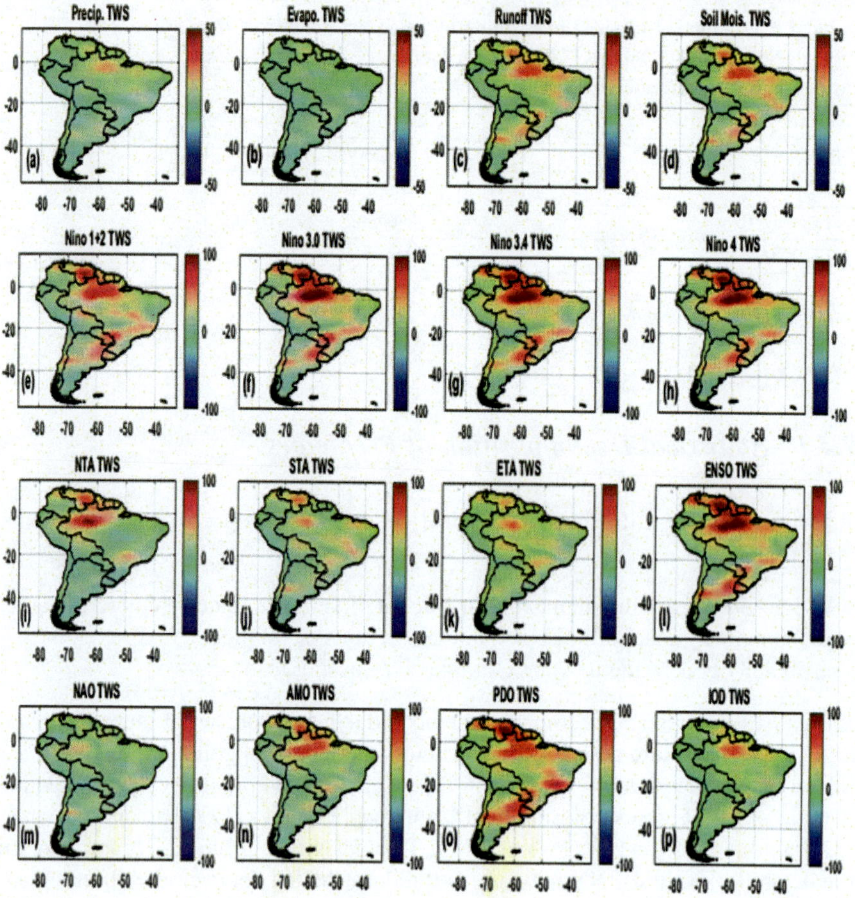

Fig. 7.5 Impact of the independent parameters illustrated in Fig. 7.3 on the deseasonalized TWS of the South America region highlighting the strength of each product to the region's water fluxes as shown in Fig. 7.1. **a** precipitation. **b** evapotranspiration **c** runoff **d** soil moisture **e** Nino1+2 **f** Nino3.0 **g** Nino3.4 **h** Nino 4 **i** NTA **j** STA **k** ETA **l** ENSO **m** NAO **n** AMO **o** PDO **p** IOD

application of big data in modern-day research. The need to integrate these datasets to provide accurate results through data mining and deep learning techniques has contributed to the increasing use of statistical decomposition techniques to assess signals in big-datasets. The well-known PCA technique has been regularly used for this. The relevance of its application in key research disciplines including bioinformatics, web data analysis, image processing, machine leaning, and climate analysis has been reported in past literatures (Ivits et al. 2014; Montazerolghaem et al. 2016). The motivation behind integrating the PCA constituents into the BPNN stems from the ability of the PCA model to linearly combined centered variables given the units of the matching parameters are homogenous. Also, its role in the rotation of space coordinates makes it have minimal impact on data structure. Therefore, the generated PCs which are the linear combination of parameters are known to reproduce the elemental datasets to the greatest degree and exhibit heterogeneity between each other. Assuming that our data matrix has m variables, (x_1, x_2, x_3, x_m) and

$$X_{m,n} = \begin{pmatrix} x_{1,1} & x_{1,2} & \cdots & x_{1,n} \\ x_{2,1} & x_{2,2} & \cdots & x_{2,n} \\ \vdots & \vdots & \vdots & \vdots \\ x_{m,1} & x_{m,2} & \cdots & x_{m,n} \end{pmatrix}$$

n times observations, we normalized the original input parameters by using the following method:

$$\frac{X_{ti} - \overline{X_t}}{S_t}. \tag{7.1}$$

Where $\overline{X_t} = (1/n \sum_{t=1}^{n} X_{ti})$ and $S_t = \sqrt{(1/n - 1) \sum_{t=1}^{n} (X_{ti} - \overline{X_t})^2}$. For expediency, let's continue to represent the normalized Y_{ti} as X_{ti}. Let the eigenvalues of covariance matrix of normalized data be given by $\lambda_1 \geq \lambda_2 \geq \ldots \lambda_m \geq 0$. Also, we assume $_{1},\ _{2},\ _{m}$ to be the corresponding eigenvector. The ith principal component is such that $F_i = {}_i^T X$, where $i = 1, 2, m$. Usually, $\lambda_k / \sum_{i=1}^{m} \lambda_k$ is known as the contribution rate of the kth principal component and $\sum_{i=1}^{k} \lambda_i / \sum_{i=1}^{m} \lambda_i$ is known as the cumulative contribution rate of the first k principal components. The first k principal component presumably, consist of the most information m original variables, if the cumulative contribution rate goes beyond 85%.

7.4.1.2 Independent Component Analysis

The independent component analysis (ICA) is a fourth order statistic, which applies statistical moments higher than the PCA. It builds on the capability of the PCA but seeks to determine or identify the directional vectors in the form of independent patterns in a data matrix. The algorithm evaluates the undetermined subtleties (or unknown dynamics or hidden patterns) embedded in multivariate system through the rotation of empirical orthogonal functions (Aires et al. 2002). Pioneering studies based on the work of Cardoso and Souloumiac (1993) described it as blind source separation in signal processing. Its application in several fields have been reported including Ndehedehe et al. (2017), who demonstrated the basic decomposition technique of ICA on the analyses using a hydrological time series. Using ICA, they decomposed the hydrological data matrix into statistically significant independent patterns. For instance, assuming the measured indicator $x(t)$ is a Gaussian data vector observed at time t, and the root indicator $s(t)$ is a Gaussian vector having independently conjoint parameters, such that $x(t) = Zs(t)$, where Z is a full rank linear mixing matrix. A linear transformation model is adopted to have

$$Y = W^T x(t) = W^T Zs(t) = A^T s(t). \tag{7.2}$$

The goal is to determine the solution of the ICA model. Where, $A = Z^T W$, W depicts an estimator of a row of the matrix Z^{-1}, and Y is a linear permutation of S_i with A_i showing the weight. Given that the two random independent parameters incline to the Gaussian distribution than the original parameters, $A^T s$ depicts a better fit to the Gaussian distribution than any S_i. Assuming that W is a maximizing quotient of the non-Gaussian distribution of $W^T x$; by implication, we have that $W^T x = A^T s$ is equal to one of the independent parameters. Negentropy is employed as a quantitative assessment of the degree of the non-Gaussian of a random parameter Y. This idea has an important rapport between the original entropy which has been used significantly in information theory. Liu and Wang (2011) defined the entropy of a random parameter as the information degrees of a given observe variable. The negentropy J is defined by

$$(Y) = H(Y_{gauss} - H(Y). \tag{7.3}$$

where $H(Y)$ is the entropy, given by $H(Y) = -\int p(Y) \log p(Y) dy$, $p(Y)$ represents the probability density, while Y_{gauss} represents a Gaussian random vector which has an equal covariance matrix with Y. The negentropy is always less than zero. However, the only condition for the negentropy to be zero is if and only if Y follows a Gaussian distribution. For the estimation of negentropy, the estimated computation technique is usually applied, which is given as

$$J(Y) \approx c_i[E\{G_i(Y)\} - E\{G_i(v)\}]^2 \tag{7.4}$$

Where c_i represents some positive constants, v represents a Gaussian parameter with unit variance and zero mean, while G_i represents some non-quadratic functions. However, this procedure does not warrant an accurate estimation index. Therefore, with the assimilation of a non-quadratic function G, we transform the approximation to

$$J(Y) \propto c_i[E\{G(Y)\} - E\{G(v)\}]^2 \tag{7.5}$$

Where $G(Y) = e(-Y^2/2)$. If the indicators are repeatedly observed, they are designated as original indicator matrix, X_{IC}. When the discrete matrix W represents the inverse of the mixing matrix Z, we now have the independent component matrix as

$$Y = WX_{IC} \tag{7.6}$$

having the likelihood of evaluating root indicators S, where one row of matrix Y depicts independent vectors. This chapter, however, utilizes the ICA technique based on a fixed-point model to decipher the independent matrix (Y^1). Extra details about the mathematics formulations, numerical steps, and computational routines of the ICA are available in the literature (Theis et al. 2005).

7.4.2 Back Propagation Neural Network

The back propagation neural network (BPNN) is a class of neural networks whose practice is based on fine-tuning the weights of a neural net based on the error rate (i.e., loss) obtained in the preceding epoch (i.e., iteration). The principle of the BPNN is based on the "supervised procedure" and can be significantly applied in the generation of predictive models, and how they influence the response variable (Fig. 7.6). Its utilization of non-linear transmission functions such as the logarithm S type (logsig) function and the hyperbolic tangent *S* type (tansig) function gives it credence in the prediction of the corresponding dependent datasets against other neural network models. The BPNN is popular for its front feedback neural pattern when dealing with neurons in training, as well as providing a learning framework characterized under supervision study. The learning process is a duple-phase training procedure, involving a forward and backward propagation.

The first phase is the forward propagation, which attempts to maintain a positive step from the input layer to the output layer. Using the iterative properties of the algorithm, the weighted values and threshold of each layer (Fig. 7.3) are calculated and disseminated into the BPNN network. Unlike the forward propagation phase, the second phase is the back propagation. The latter involves a revision of the originally achieved weighted and threshold parametric values (Fig. 7.7). Its operation in the system model is largely influenced by the total error generated during the estimation of each weight and threshold value from the output layer to the input layer. The operations of these two phases are interchangeable and iterative till convergence is achieved in the system model. When the output layer of the BPNN is simulated by the properties of the prior functions in the initial learning process, it generates an output layer promising a good fit between observed and predicted parameters. Okwuashi and Ndehedehe (2020) recounted the neural network as the only algorithm that exhibits the special technique of memory, learning, and discernment.

Apart from the predictive capabilities of the BPNN, another justification of its application in this experiment is its potential to be used as a technical framework to assess the inter-relationship between GRACE-TWS and hydrological variables, including fluxes, teleconnections, and SST products. This can be accomplished due to the BPNN's flexible topological framework, explainable learning process, and potential of minimizing under and over-fitting problems in similitude to the artificial neural network (ANN). Using a multi-layer perceptron procedure, the BPNN utilizes a set of successively arranged hidden layer frame composed of non-linear elements called neurons; this framework links the independent variables (predictors) to the dependent variable (predictand). These neurons existing in the hidden layer of the neural model instigates new variables based on the predictors from the previous layer, which is multiplied to a dynamic-specific weight index and the addition of a bias term for fine-tuning. The application of major hydrological fluxes that define the dynamic climatic response of the South America was intended to reflect the rapid response of surface (flood related) and groundwater (drought related) to hydrological extremes over the study region.

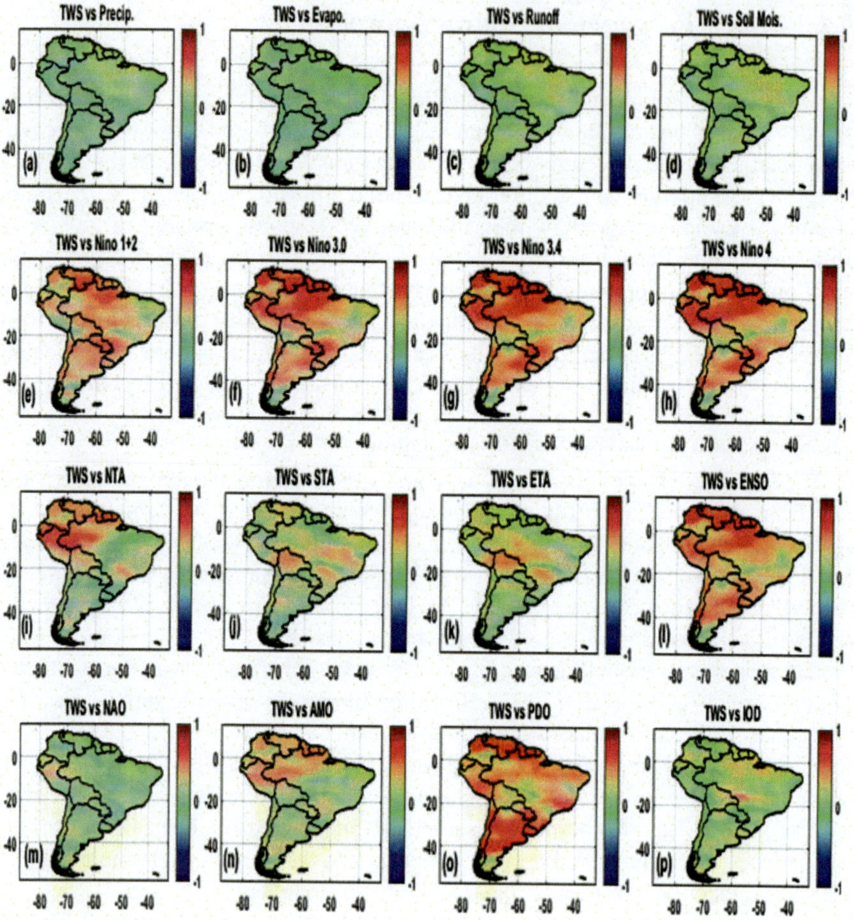

Fig. 7.6 Pearson's correlation coefficient between the deseasonalized terrestrial water storage and all independent parameters (**a**–**p**) used in this experiment

The hydrological fluxes inputs were integrated with teleconnection patterns and SST components as predictors, whose influence on continental TWS has been severely corroborated by past studies (e.g., Ndehedehe and Ferreira 2020a), thus their use as predictors in the neural network is justified. The three set of predictors input used in the BPNN of this experiment are:

- Hydrological fluxes (i.e., precipitation evapotranspiration, runoff and soil moisture products),
- Climate teleconnection parameters (i.e., ENSO, NAO, AMO, PDO, and IOD),
- Sea surface temperature products (i.e., Nino1+2, Nino 3.0, Nino3.4, Nino4, NTA, STA, and ETA).

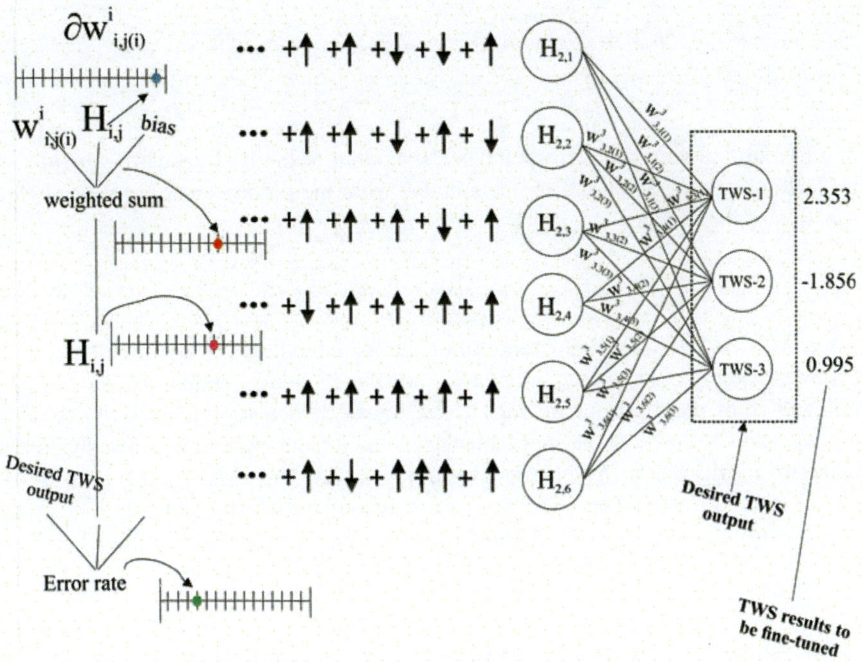

Fig. 7.7 Achieving the desired terrestrial water storage output by iterative fine-tuning procedure of the back propagation framework

The BPNN used in this experiment is a classic three-neuron layer framework, consisting of one input layer, four sets of hidden layers, and one output layer, each of these layers have 16 neurons, 6-set neurons, and 1 neuron, respectively (Fig. 7.3). Apart from the input neurons (i.e., $x_1 - x_{16}$), which only connect one input value with its parallel weights index ($W_{p,1} - W_{p,96}$), the total input for each discrete neuron represents the totality of all input values, each multiplied by its weight, and a bias term (zp) and can be described as a unit weight from a supplementary input.

$$H_{P1} = \sum_{i=1}^{n}(w_{p,1}.X_i + Z_p \rightarrow H_{q1} = f(H_{P1}) \tag{7.7}$$

$$H_{q1} = \sum_{i=1}^{n}(w_{q,1}.H_{P1} + Z_q \rightarrow H_{r1} = f(H_{q1}) \tag{7.8}$$

$$H_{r1} = \sum_{i=1}^{n}(w_{r,1}.H_{q1} + Z_r \rightarrow H_{s1} = f(H_{r1}) \tag{7.9}$$

$$H_{s1} = \sum_{i=1}^{n} (w_{s,1}.H_{r1} + Z_s \rightarrow y_{p,q,r,s} = f(H_4) \tag{7.10}$$

where $p, q, r, s = 1, 6;\ \ n = 16$ with Eqs. 7.7–7.10 representing the models of the first to fourth hidden layers, respectively; the output value $y_{(p,q,r,s)}$ can be estimated by feeding the net input and the parameters from the hidden layers to the transfer function of the neurons.

$$y_{p,q,r,s} = f(H_4), \tag{7.11}$$

where H_{P1}, H_{q1}, H_{r1}, and H_{s1} represent neurons in the first to fourth hidden layers, respectively, $w_{(p,1)}$, $w_{(q,1)}$, $w_{(r,1)}$, and $w_{(s,1)}$ are the dynamic weight index parameters initiated from the input parameters to the fourth hidden layers, respectively. Z_p, Z_q, Z_r, and Z_s represent the supplementary weight index parameters introduced for better fits for the first to last hidden layers, respectively, H_4 portrays a net total of the first to fourth hidden layers, while $y_{p,q,r,s}$ is the output, which is directly a function of the fourth hidden layer but indirectly depends on the entire BPNN model.

7.4.3 Implementation of Assimilated Deep Learning

7.4.3.1 Principal Component Analysis-BPNN Model

The PCA statistical procedure was applied to the original input predictors which have been grouped into hydrological fluxes, teleconnection patterns and SSTs to produce new parameters for the prediction model using the following steps:

1. **Step 1**: Normalize the original predictors
   Simply put, normalization process is a scaling technique in which data points are shifted and rescaled so that they end up in a range of 0 and 1. Therefore, the input layer which consists of the original predictors with dynamic magnitude index was normalized prior to the application of the PCA technique on them

$$x_i' = \frac{x_i}{\max(x_i)} \tag{7.12}$$

2. **Step 2**: The correlation coefficient matrix (R) is obtained using the normalized data from step one.

$$R = (r_{ij})_{16\times16} \tag{7.13}$$

$$r_{ij} = \frac{1}{n}\sum_{i=1}^{n} x_{ii}' x_{ij}'\ \ i, j = 1, 2, 16 \tag{7.14}$$

The function expressed in Eq. 7.13 seeks the generation of the R-values.

3. **Step 3**: Using the generated R-values from Eq. 7.13, the characteristic vector and characteristic value are estimated, which leads to the determination of the corresponding principal components. The input parameters yielded 183 characteristic values (April 2002–June 2017), given that they all maintained a temporal resolution of 1 month which means $\lambda_i (i = 1, 2, \ldots, 183)$ where λ_i represents the variance of the principal components. The λ_i values are arranged or ordered in a decreasing sequence ($\lambda_i \geq \lambda_2 \geq\geq \lambda_{183} \geq 0$) together with their corresponding characteristic vectors $(l_1, l_2, l_3, \ldots, l_{183})$, so that the ith principal component can be achieved by applying the function;

$$Y_{(PC,i)} = l_1' X = l_{i1} X_1 + l_{i2} X_2 \cdots + l_{i16} X_{16} \tag{7.15}$$

Where $i = 1, 2, 16$.

4. **Step 4**: Total the orderly sets of principal components to allow the accumulated input rate of the variance a = $\sum_{i=1}^{m} \lambda / \sum_{i=1}^{m} \lambda_i$ to go beyond a certain threshold (we choose 99% for this experiment). The principal components were computed as $Y_{(PC,1)}, Y_{(PC,2)}, Y_{(PC,3} \ldots, Y_{(PC,m)}, m < 184$ for every discrete value $\lambda_i (i = 1, 2, \ldots, 183)$, and $Y_{(PC,i)}$ $(i = 1, 2, m)$ as a linear integration of x_i. We now take the m principal components as the input parameters for our BPNN model (for this study, it was restricted to 6). This results in injecting the newly determined m principal components into the BP network and then with the aid of the PCA-BPNN model, we make our optimized predictions.

7.4.3.2 Independent Component Analysis-BPNN Model

In the proposed ICA-BPNN model, the independent components (IC) of the original input layer neurons using the ICA technique were first extracted. The discrete parameter IC's are depicted by the feature of the original predictors. The noisy ICs have the least effective information of original predictor dataset and were also identified in the statistically decomposition exercise. After the identification and removal of the noise IC, the remaining ICs are used as the input of the BPNN mode. Each of the input parameters is considered as a row; given the 183-monthly dataset of the original 16 input parameters, we derive a matrix X_{IC} of size 16×183. Using the ICA technique, the separate matrix W, and the independent component matrix Y can be developed. Each row of Y, the yi of size 1×400 represents an IC. Each IC represents a product, which represent different features of the original input layers from previous studies (e.g., Ndehedehe et al. 2016b). The noise-removal phase is very important in the ICA-BPNN modeling procedure after the latent IC's has been generated. We therefore, adopt the well-known testing and acceptance (TnA) procedure in solving this problem (Liu and Wang 2011). The obtained m ICs are taken into account prior to the primer of the TnA technique. We excluded one IC for each iteration step and utilized the remaining m-1 ICs for the reconstruction process of the original input matrix. Let us represent the reconstructed original input variables as X^R and the excluded ICs as y_k,

$$X^R = \sum_{i=1, i \neq k}^{m} a_i y_i, \; 1 \leq k \leq m \tag{7.16}$$

Where, $X^R = [x_1^R, x_2^R, x_m^R]^T$, x_i^R depicts the ith reconstructed parameter, a_i represents the ith column vector of mixing matrix A which is the inverse of the separate matrix W (Eq. 7.2) , while y_i is the ith IC. Considering the cases, $k = 1, 2, m$, which involves the reiteration of m ICs with the exclusion of one IC, we introduce the relative hamming distance (RHD) in quantifying the reconstruction error between the original) and reconstructed parameters which is given as

$$RHD = \frac{1}{n-1} \sum_{i=1}^{n-1} [R_i(T) - \hat{R}_i(T)]^2, \tag{7.17}$$

where $R_i(t) = sign[A_i(T+1) - A_i(T)]$; $\hat{R}_i(T) = sign[Pr_i(T+1) - Pr_i(T)]$. In this modeling, sign[param]= 1 if param > 0, sign[param]= 0 if param $= 0$, and sign[param]= -1 if param < 0. Also, A_i, Pr_i and n represents the actual value, the predicted value and the total number of data points, respectively.

With the RHD reconstruction error, it becomes possible to evaluate the similarity between the original input and reconstructed parameters. The RHD shows a higher similarity between the original and reconstructed parameters if the value tends to zero. However, when the RHD value is farther from zero, it means that the eliminated IC consist of more effective and vital information of the original dataset while depicting the poor collinearity between the original and reconstructed parameters. It is important to investigate the reconstruction which guarantees the RHD value closest to zero as it tends to make the best predictions with minimal bias. Table 7.4 represents the RHD reconstruction errors of each iterate.

Table 7.4 shows that the value of RHD, which is reconstructed by using IC1, IC2, IC3, IC4, IC5, and IC6 and eliminating IC7, is the smallest. It is deduced that the

Table 7.4 Relative Hamming Distance (RHD) reconstruction errors

Included ICs	Excluded ICs	RHD value
IC1 IC2 IC3 IC4 IC5 IC6	IC7	0.014
IC1 IC2 IC3 IC4 IC5 IC7	IC6	0.276
IC1 IC2 IC3 IC4 IC6 IC7	IC5	0.085
IC1 IC2 IC3 IC5 IC6 IC7	IC4	0.422
IC1 IC2 IC4 IC5 IC6 IC7	IC3	0.856
IC1 IC3 IC4 IC5 IC6 IC7	IC2	1.021
IC2 IC3 IC4 IC5 IC6 IC7	IC1	1.006

IC7 contains the least information and IC7 represents the noise IC. Therefore, IC1, IC2, IC3, IC4, IC5, and IC6 were conducted as the input of the proposed ICA-BP model, while IC7 was removed from our training.

7.4.4 Deseasonalization of the Dependent Data Type

For proper evaluation of the footprints of global climate modes on GRACE-TWS over South America, the trends and harmonic components (e.g., annual and semi-annual signals) of the GRACE hydrological signals, were removed through deseasonalization. The residual from the deseasonalization process, which we refer to as dTWS was then used in the individual assessment of each independent component and framework. The compartmentalization of the harmonic components and TWS trends from the GRACE time series was undertaken following procedures adopted from Ndehedehe et al. (2017). It was, however, implemented using the parameterization technique of the multi-linear regression algorithm. This approach is summarized as

$$\begin{aligned} Y(m,n,T) &= \rho_0 + \rho_1 T + \rho_2 \sin(2\pi T) + \rho_3 \cos(2\pi T) + \rho_4 \sin(4\pi T) \\ &\quad + \rho_5 \cos(4\pi T) + \rho_6 N(T + \varphi_N) + e(T) \end{aligned} \tag{7.18}$$

where the grid locations are depicted by m, n, the time in years is represented by T, the constant offset and linear trend is represented by ρ_0 and ρ_1, respectively, the annual signals are depicted by ρ_2 and ρ_3, and the semi-annual signals are represented by ρ_4 and ρ_5. The amplitude of TWS dynamics illustrating the impact of large-scale ocean-atmospheric oscillations is represented by ρ_6. The normalized time series developed after the elimination of the long- term mean of each climatic mode is depicted by N, the phase lag between the TWS time series and each climate index is shown as φ_N while the random error term with respect to time is represented as $e(T)$. The residual component generated after the elimination of the linear rates Eq. 7.19 and harmonic components (see Eq. 2 in Ndehedehe and Ferreira 2020b) is assumed to contain prevalent slow dynamic climatic signals and other sub-regional forcings. The application of the local least-squares polynomial approximation-based filter for Savitzky and Golay (1964) was employed in smoothing the effects of low frequency noise for both dependent (dTWS) and independent (precipitation and climate modes) parameters. The deseaonalized (X_{dTWS}) signal parameters is represented by

$$\begin{aligned} Y_{dTWS} &= Y(m,n,T) - [\rho_1 T + \rho_2 \sin(2\pi T) + \rho_3 \cos(2\pi T) + \rho_4 \sin(4\pi T) \\ &\quad + \rho_5 \cos(4\pi T)] \end{aligned} \tag{7.19}$$

7.5 Results and Discussion

As a result of the recursive optimization of weights and bias in the BPNN network, it is capable of learning with significantly higher learning rates and accuracies. The back propagation technique was particularly helpful in the fine-tuning of the predicted variable on our multi-layer feed forward neural network. Given that the back propagation neural pattern attempts to learn a set of weights to achieve an optimal weight and bias term so as to generate the best fit, the process was appropriately performed for the PCA (loss = 0.5%) and ICA (loss = 0.2%) models. An advantage of the good fit achieved from the PCA- and ICA-BPNN algorithms was the prior optimization through statistical decomposition of the original input parameters. For the PCA-BPNN mode, the original datasets were normalized and the correlation coefficient obtained from the normalized dataset. This was used to generate the characteristic vector values and subsequently, the corresponding PCs to be injected into the BPNN framework. The number of dominant PC patterns to serve as predictor input was restricted to six. Prior to their integration into the neural pattern, they were averaged to allow the accumulated input rate of the variance exceed the 99% set threshold value of the set model. This then led to the development of the PCA-BPNN model which we used to predict TWS.

For the ICA-BPNN mode, the individual ICs of the original input parameters was first extracted given that ICA uses a filtering system in its decomposition technique. Also, the identification and removal of noise ICs was vital to the efficiency and robustness of our proposed model. The TnA procedure was applied for this process. The RHD algorithm served as a tool to estimate the comparison between the original input parameters and the reconstructed parameters. This was important to ascertain the strength of the decomposed predictors, which will play a key role in the efficiency of the proposed ICA-BPNN model. The first 146 monthly dataset of the derived independent (PCA-BPNN, PC1-PC6 and ICA-BPNN, IC1-IC6) and dependent (PCA: TWS-PC1, TWS-PC2, TWS-PC3 and ICA: TWS-IC1, TWS-IC2, TWS-IC3) parameters served as the training samples, while the remaining 37 served as the testing dataset.

This 8–2 ratio of training and testing datasets has been reported to be extremely efficient by allowing the network enough dataset to train and provide it with capabilities to uncover hidden patterns in the neural pattern for adequate and more accurate predictive capabilities (Okwuashi and Ndehedehe 2017). Before the training procedure, it was necessary to authenticate the synchronization of the original input datasets with the CSR-TWS time series, especially in terms of spatial and temporal homogeneity. The developed model was useful in predicting the time series of three dominant independent modes of the GRACE-CSR-TWS. During the testing phase using the remaining 37 monthly dataset for both predictor and predictand, an average fit index of 86% was achieved. The details and results of the assimilated deep learning procedure is discussed further in subsequent sections.

7.5.1 Impacts of Global Climate Modes on GRACE Hydrological Signals

The delineation of the contributions and impacts of global climate indices and SST products on the land water storage dynamics of the South African region was a key objective in this study. This was achieved using an assimilated PCA- and ICA-BPNN model, and was useful in the identification of significant modes of deseasonalized TWS, which was correlated with the normalized climate teleconnection indices as shown in Fig. 7.6. The large spatial extent of the study region (South America) affected the fitting index between the SST and TWS, given that global indices perform best when tested over relatively smaller localized regions, such as river basins (Sajjad and Ghaffar 2019). The Atlantic indices showed the strongest averaged correlation with TWS with an r index of 0.642. This analysis also shows the strong collinearity between the Pacific indices and the SA's deaseasonlaized TWS, which suggests that proximity of the nearby oceans is a strong factor in the regulation of continental TWS cahnges (Figs. 7.5, 7.6 and 7.8). However, the strength of their influences is mostly felt in the northern region of the continent housing the Amazon river basin (Figs. 7.5, 7.6, and 7.8). Prior to deaseasonalization, the Pearson's correlation (r) was used to determine the extent of relationships between global climate indices and TWS (Fig. 7.1, Table 7.1), the Pacific and Atlantic indices as well as SSTs from these oceans, show the highest association with the spatial evolutions of TWS over the entire SA region ($r = 0.556, 0.515$, and 0.575 for PDO, AMO, and STA, respectively, Table 7.1, Figs. 7.8, 7.9). Generally, independent patterns of TWS dynamics over the study region are largely influenced with the PDO signals due to their proximity to the study region (Fig. 7.1). The IOD and the Nino 4.0 indices showed a weak relationship with the region's TWS ($r = 0.1300$ and 0.187 respectively), while ENSO ($r = 0.375$), Nino 1+2 (0.3120), Nino 3.0 ($r = 0.371$), Nino 3.4 ($r = 0.3271$), NTA($r = 0.03199$), NAO ($r = 0.3121$), and ETA ($r = 0.3893$) exhibited moderate relationships on the region's TWS (Fig. 7.1).

The relationships of these climate modes with the spatio-temporal patterns of the TWS over Africa (Fig. 7.1) highlights the footprint of global climate over the continent (Fig. 7.10). From the localized spatio-temporal patterns generated from cumulant decomposition (dTWS), we notice higher TWS amplitude toward Peru, Brazil, Venezuela, and Argentina, which are the regions with the most dynamic hydrological response in the continent (Boulanger et al. 2005). These countries are key hydrological hotspots, housing the river basins in the region, including the well-known Amazon river basin (Fig. 7.5a–p). In the Northern region where the Amazon is located, the influence of Nino 1+2, Nino 3.0, Nino 3.4, Nino 4.0, STA, ENSO, PDO, and IOD results in considerable magnitudes of over 40 mm to the amplitudes of TWS (Figs. 7.5e–p and 7.9a–b). The influence of these climate modes to the deseasonalized TWS over South America strongly corresponds to their temporal relationships with TWS. Extreme wet conditions in the region have been linked to changes in the characteristics (e.g., amplitude, propagation, etc.) of ENSO events (Boulanger et al. 2005) and such impacts are expected to modulate the distribution

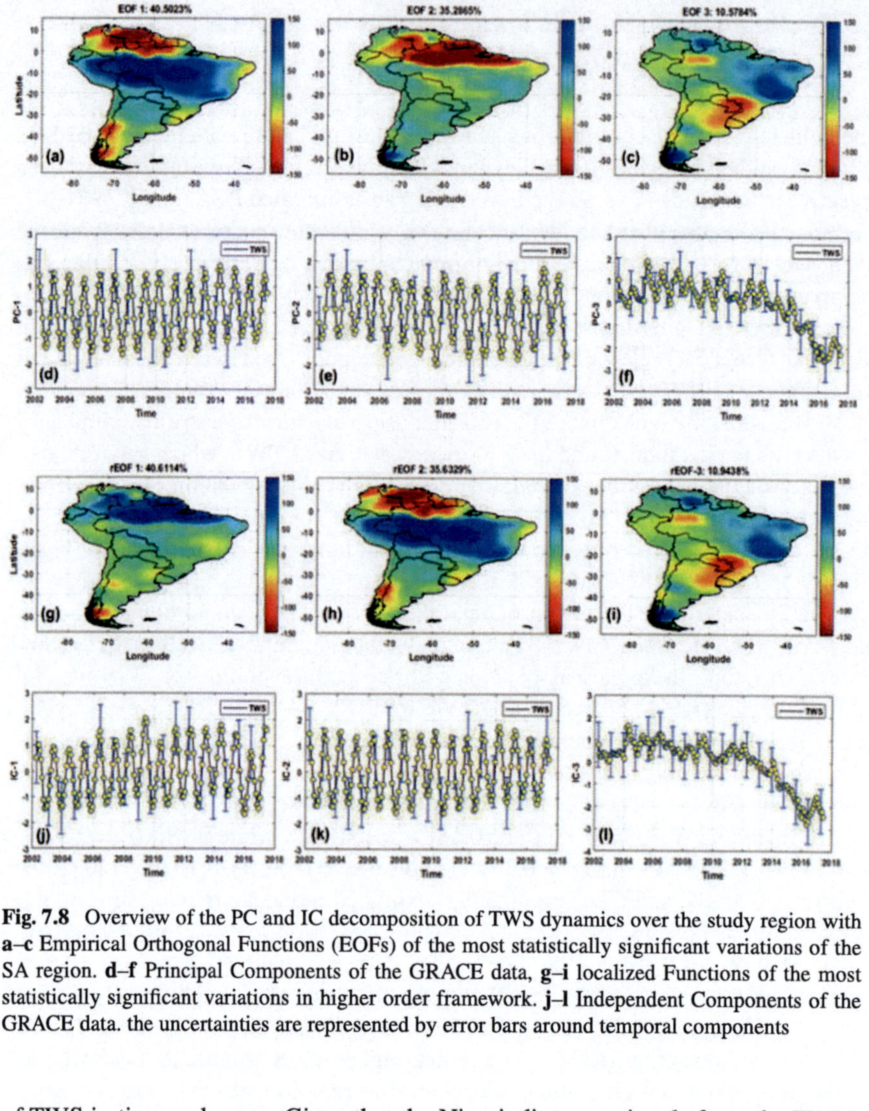

Fig. 7.8 Overview of the PC and IC decomposition of TWS dynamics over the study region with **a–c** Empirical Orthogonal Functions (EOFs) of the most statistically significant variations of the SA region. **d–f** Principal Components of the GRACE data, **g–i** localized Functions of the most statistically significant variations in higher order framework. **j–l** Independent Components of the GRACE data. the uncertainties are represented by error bars around temporal components

of TWS in time and space. Given that the Nino indices are signals from the ENSO region, which is duly characterized by the Eastern and Central Pacific, it follows, therefore, that the change in TWS over the continent is strongly driven by the ENSO phenomenon, given the strong El-Niño and La-Niña events in the region (Kayano and Capistrano 2014).

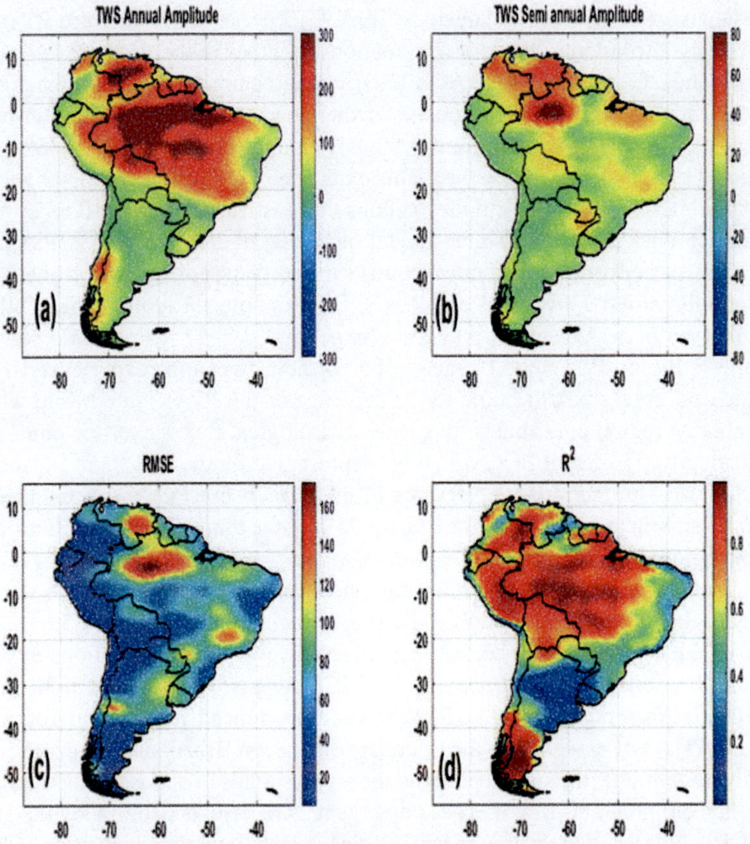

Fig. 7.9 Modeling TWS using multi-linear regression. **a** Amplitude **b** semi-annual amplitude, **c** RMSE, **d** coefficient of determination

7.5.2 Identifying Dominant Independent Patterns

In this experiment, the spatial patterns associated with the independent modes of TWS were stable. For both PCA-BPNN and ICA-BPNN models, the temporal components of the predictor datasets (i.e., all columns of $x(t)$ derived from hydrological products, SSTs, or teleconnection patterns) were correlated with the individual PCs or ICs of the predictand (i.e. TWS) respectively.

The first, second, and third independent modes of the TWS anomalies derived from GRACE for the two statistical frames are shown in Fig. 7.8. The first dominant independent PC mode, which explains 40.50% of the variance, characterizes the annual water variability over the study region from April 2002–June 2017. The spatial pattern of the first principal component (Fig. 7.8a–c) illustrates the most dominant pattern localized over the North and Central part of the study region. The strong annual dynamics of increase and decrease in TWS experienced in the third PC mode can be largely attributed to the large fluctuations in fluxes in the Amazon basin.

The second PCA modes of GRACE-TWS contain annual variability depicting a concentration of the EOF loadings on the northern part of Brazil, Guyana, Suriname and French Guiana. The trend is homogenously characterized as a net TWS loss of the depicted regions (Fig. 7.8d and e). This explains 35.29% of the total variance in South America. The third PCA mode explains the least variance with 10.58% of the total variance, characterizing an inter-annual variability of the study region. Its spatial pattern depicts a concentration of inter-annual variability over different hotspots in the continent with varied TWS gains and losses. For example, eastern Brazil, southern Venezuela, southern Argentina, and Chile experienced TWS gain, while western and southern Brazil, Paraguay, Uruguay, and northern Argentina experienced TWS loss. This can largely be attributed to the presence of the River Plate basin, whose presence in the region contributes to driving hydrological fluxes over the continent. The statistically significant PCs, which are represented by the first three PC modes (Fig. 7.8a–c) gave a cumulative variance of 86.37%. These PCs were assumed as statistically significant signals denoting over 95% of the total TWS fluctuations over the study region.

The results from the independent component analysis were similar to the PC results with the first, second, and third IC modes maintaining 40.61%, 35.63% and 10.94% of the region's variance, respectively. However, slight variations exist in the first and second EOF modes of PC and IC analysis. For the first EOF mode in the ICA, stronger amplitudes of TWS was experienced in northern Brazil as reported in PC1, but extended to the boundaries between Brazil and Venezuela. For the second EOF mode, the spatial patterns show two signals in opposite phase around Venezuela/Colombia and Brazil. These divergent EOF visuals patterns by the PCA and ICA is a pointer to their unique capabilities in statistical decomposition, which encouraged their application (PCA and ICA) in this BPNN analysis.

The cumulative variance of the IC modes is 87.19% and is assumed to consist of the most significant signals responsible for TWS fluctuations in the study region. The temporal evolutions of the IC are also characterized as annual for the first (Fig. 7.8g and j), annual variations for the second (Fig. 7.8h and k), and inter-annual signals for the third (Fig. 7.8i and l). The latter is similar to the temporal evolutions observed in the PCA approach (Fig. 7.8d–f). The PC and IC results correspond to the annual (Fig. 7.9a) and semi-annual (Fig. 7.9b) amplitude of the region's TWS, as well as the trend dynamics (Fig. 7.1d). The northern and central parts of the continent contribute mostly to the fluctuations in TWS. This was expected, as they are the biggest river basins that contribute substantially to the dynamics of hydrological fluxes in the region. The strong changes in terrestrial hydrology in this case study also warranted the usage of localized hydrological products (Fig. 7.10a) as input predictors, to maintain the stability and efficiency of model estimates. Their effect in regulating TWS dynamics of any region is well captured in the water budget equation (Chen et al. 2020), and contributes toward changes in TWS compared to other set of predictors.

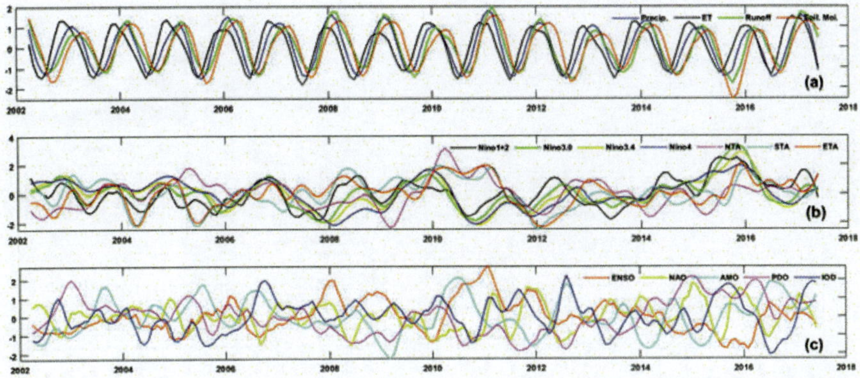

Fig. 7.10 Normalizing the original input predictors to determine the inter-relationships between them with **a** localized hydrological fluxes contributing mostly to the region's TWS, **b** the six SST products and **c** the four teleconnections patterns whose impacts affect TWS dynamics and climatic patterns in continental scales

7.5.3 Predicting Total Water Storage

To predict total water storage (TWS) using the PCA-BPNN and ICA-BPNN algorithms, 80% of the overall predictor and GRACE-TWS datasets (April 2002–May 2014) were employed for training, and the remaining 20% of the data (June 2014–June 2017) were used to test the performance of the new models. The original input parameters used for prediction were optimized prior to its injection to the deep learning framework (Sect. 7.4). For the PCA-BPNN, the predictor datasets was implemented with varying order of magnitude (Fig. 7.11a) whose eigenvectors functioned as the model input (Fig. 7.11b–c). The box plot in Fig. 7.11a represents the normalized magnitude for all predictor variables having their median values at 0.05 (precipitation), 0.12 (evapotranspiration), 0.10 (runoff), −0.05 (soil moisture), 0.00 (Nino 1+2), 0.03 (Nino 3.0), 0.02 (Nino 3.4), 0.00 (Nino 4.0), −0.03 (NTA), −0.03 (STA), 0.04 (ETA), 0.02 (ENSO), −0.02 (NAO), −0.02 (AMO), 0.00 (PDO), and 0.00 (IOD). Using the characteristic vector and values produced from the R values, it was possible to decipher the loadings used to predict the GRACE-derived TWS for the training period.

For the ICA-BPNN model, the central procedure of detecting the noise IC after the initiation of the latent ICs was adopted. With the aid of the TnA technique, it was possible to identify the IC containing noise (i.e., IC 7). This was implemented using the RHD procedure, which calculates the uncertainty between the original and reconstructed parameters. Due to the functioning of the RHD, it was possible to detect IC7 as the parameter whose value is closest to zero and eliminated it from the training to avoid compromising the test-datasets (Table 7.4). Using the optimal weight and bias parameters, the test results for both the testing and training phase were estimated (Table 7.5). To control the indefiniteness of the Hessian (H) matrix,

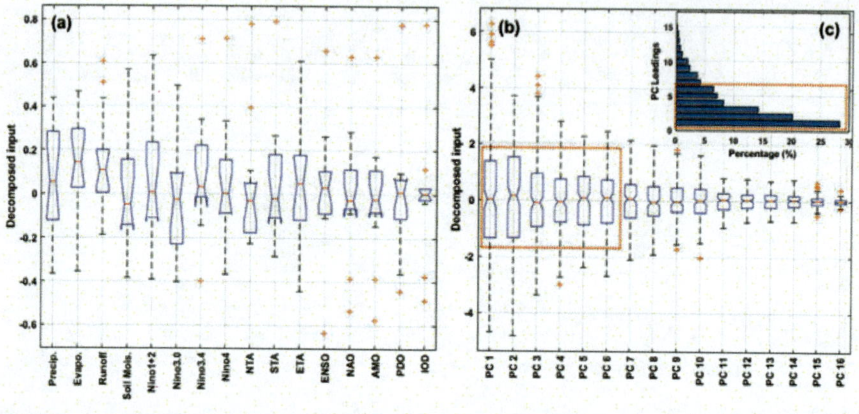

Fig. 7.11 Principal component coefficients of the input layer parameters (hydrological fluxes, SSTs, and climate teleconnections) **a** decomposed coefficients of each discrete principal component using singular value decomposition, **b** principal components of the decomposed input layers showing the loadings of all generated PCs **c** selected PCs for training

the determining parameter was set to 5.0×10^{-5}. The Marquardt adjustment parameter ranged from $1.0e^{-03}$ to $1.0e^{-07}$ to obtain the best minimum of our non-linear function over our space of input loadings. The box plot in Fig. 7.12a represents the normalized magnitude for all input loadings having their median values at −0.02 (precipitation), −0.04 (evapotranspiration), 0.01 (runoff), 0.02 (soil moisture), 0.10 (Nino 1+2), 0.02 (Nino 3.0), 0.00 (Nino 3.4), 0.01 (Nino 4.0), −0.01 (NTA), 0.01 (STA), −0.02 (ETA), −0.02 (ENSO), 0.12 (NAO), 0.02 (AMO), 0.00 (PDO), and −0.01 (IOD). Figure 7.12b on the other hand represents the independent components of the decomposed input layers showing the loadings of all generated latent ICs. To evaluate the performance of our proposed PCA-BPNN and ICA-BPNN models, the two sets of input datasets to analyze the models by comparison was selected, especially since the two of them were predicting the same quantity. Using the last 20% of the total dataset for the experiment, it was possible to test the performance of the trained BPNN models.

7.5.4 Training and Testing Phase

To make adequate estimations of the optimum PCA-BPNN and ICA-BPNN, we injected the first 146 months of each mode of GRACE-derived TWS (Fig. 7.8) and the first 146 months of all temporal modes of hydrological fluxes, SSTs, and global teleconnection patterns in Equations 7.15 and 7.16, respectively. It was pertinent to observe that the time series datasets were synchronized with the GRACE CSR-TWS temporal pattern, prior to the performance of the BPNN model. This was then followed by running the BPNN simulation step for both PC and IC modes. Given,

Table 7.5 PCA-BPNN and ICA-BPNN training and test results through two hidden layers with 6 neurons each for the study region (2002–2017). Quasi predictions of the PCA-BPNN and ICA-BPNN models are highlighted in red

		BPNN – PC Decomposition (Mode 1)					BPNN- IC Filtering (Mode 1)				
Time (T) (months)	Predictand (TWS)	Input	(initial)	(initial)	bias	Predicted	Input	(initial)	(initial)	bias	Predicted
1	1.6342	PC1 – PC6	1	1.8	1.5	1.6340	IC1 – IC6	1	1.6	1.5	1.6342
2	0.2117	PC1 – PC6	1	1.0	1.5	0.2111	IC1 – IC6	1	0.2	1.5	0.2117
3	0.2248	PC1 – PC6	1	0.4	1.5	0.2248	IC1 – IC6	1	0.5	1.5	0.2248
4	0.2397	PC1 – PC6	1	0.2	1.5	0.2397	IC1 – IC6	1	0.8	1.5	0.2397
183	0.6228	PC1 – PC6	1	0.7	1.5	0.6228	IC1 – IC6	1	0.9	1.5	0.6228
T	-	**BPNN – PC Decomposition (Mode 2)**					**BPNN- IC Filtering (Mode 2)**				
1	0.1218	PC1 – PC6	1	1.5	1.5	0.1218	IC1 – IC6	1	0.5	1.5	0.1218
2	0.4178	PC1 – PC6	1	1.8	1.5	0.4178	IC1 – IC6	1	0.6	1.5	0.4178
3	1.1015	PC1 – PC6	1	0.6	1.5	1.1015	IC1 – IC6	1	1.0	1.5	1.1015
4	1.4658	PC1 – PC6	1	1.2	1.5	1.4658	IC1 – IC6	1	1.2	1.5	1.4658
183	0.2247	PC1 – PC6	1	1.5	1.5	0.2247	IC1 – IC6	1	1.7	1.5	0.2247
T		**BPNN – PC Decomposition (Mode 3)**					**BPNN- IC Filtering (Mode 3)**				
1	–1.2474	PC1 – PC6	1	0.2	1.5	–1.2477	IC1 – IC6	1	0.9	1.5	–1.2471
2	–1.8367	PC1 – PC6	1	0.1	1.5	–1.8367	IC1 – IC6	1	1.6	1.5	–1.8361
3	–0.3384	PC1 – PC6	1	1.3	1.5	–0.3384	IC1 – IC6	1	0.1	1.5	–0.3384
4	0.4526	PC1 – PC6	1	0.4	1.5	0.4526	IC1 – IC6	1	0.3	1.5	0.4530
183	–0.2274	PC1 – PC6	1	0.8	1.5	–0.2274	IC1 – IC6	1	0.7	1.5	–0.2274

that we want to keep the prediction simple, higher orders of the PC-BPNN and IC-BPNN models for both the predictors and the predictands were not considered. The numerically simulated results for both PC1-PC6 and IC1-IC6 of GRACE-TWS suggest that the BPNN model provides the best fitting at optimal weight and seed values (Table 7.5) (Fig. 7.13).

The RMS of differences between the modeled-TWS values from the BPNN models and the PC and ICs of the GRACE-TWS was used as the fit criterion. Our simulation results indicate that the PCA-BPNN models provide a fit of 90%, 87%, and 86% for the first, second, and third independent modes of the PC-GRACE-TWS, while the ICA-BPNN model provides a fit of 94%, 91%, and 88% for the first, second, and third independent modes of the IC-GRACE-TWS for simulating the three dominant components of the GRACE-derived TWS (Figs. 7.14 and 7.15). To understand the collinearity of the optimized input predictors for the PC and IC (Fig. 7.14a–f) models, the decomposed parameters were overlaid on top of each other. The simulation fit of the BPNN corresponding to PC2 (Fig. 7.15b), PC3 (Fig. 7.15c), IC2

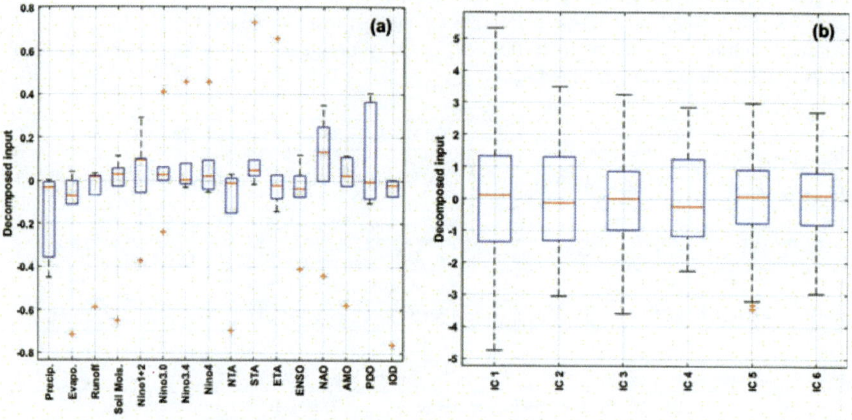

Fig. 7.12 Independent component coefficients of the input layer parameters (hydrological fluxes, SSTs, and climate teleconnections) **a** decomposed coefficients of each discrete independent component, **b** independent components of the decomposed input layers showing the loadings of all generated ICs

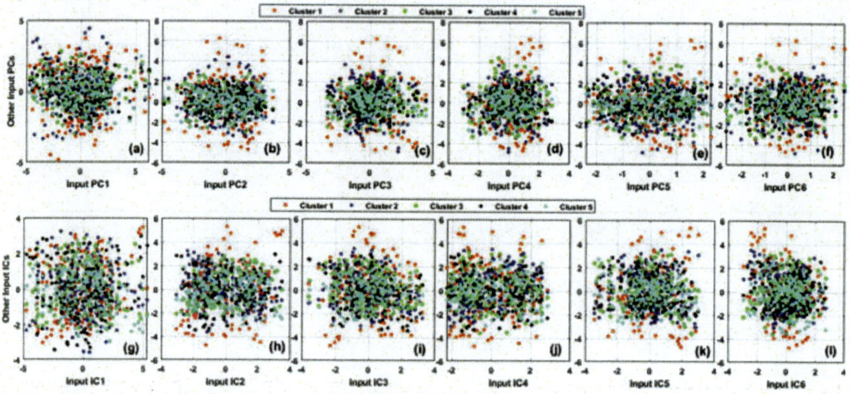

Fig. 7.13 Partitioning the statistically decomposed PCs and ICs to be used for training into clusters using the kmean algorithm as a validation procedure before injecting them into the BPNN **a** PC1-other PC's **b** PC2-other PC's **c** PC3-other PC's **d** PC4-other PC's **e** PC5-other PC's **f** PC6-other PC's **g** IC1-other IC's **h** IC2-other ICs **i** IC3-other ICs **j** IC4-other ICs **k** IC5-other ICs **l** IC6-other ICs

(Fig. 7.15e), IC3 (Fig. 7.15f) are, however, lower than that of PC1 (Fig. 7.15a) and IC1 (Fig. 7.15d). This is because they consist of temporal patterns appearing to be much more complex than the annual patterns of PC and IC1, respectively (Fig. 7.14). Therefore, they might not have been fully captured by the predictors (Fig. 7.15).

Prior to the PC- and IC- BPNN training, the decomposed input datasets were validated using the k-means clustering technique for both PC and IC features. This was warranted to warm start the position of centroids, and easily adapts the optimized

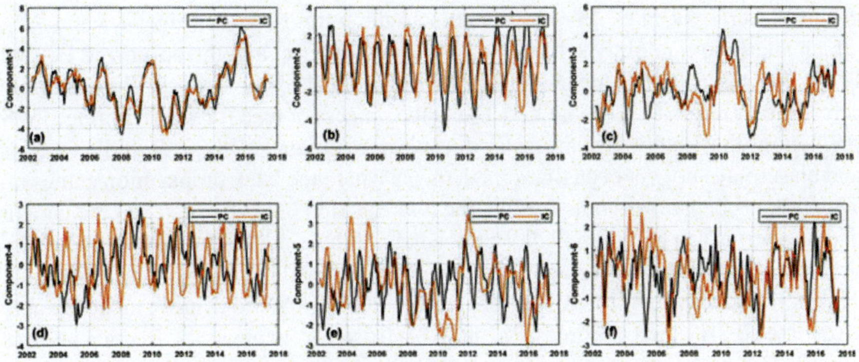

Fig. 7.14 Overlay of PCA and ICA decomposition of the input parameters used for the BPNN training

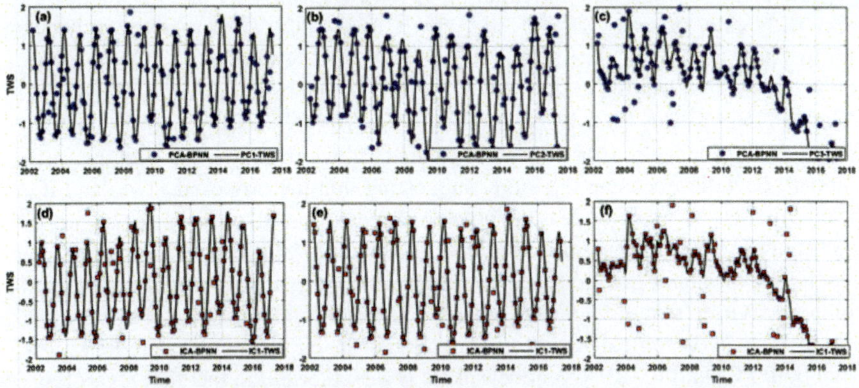

Fig. 7.15 TWS prediction using the PCA—and ICA—BPNN framework developed in this study using the fine-tuning process of back propagation to refine the output model to minimize the error rate

results to the BPNN model (Fig. 7.13). After the training procedure, the remaining 20% of the dataset was used to test the prediction step and performance of the models in predicting TWS of the SA region.

7.5.5 Prediction Step and Performance Assessment

The back propagation technique was essential in fine-tuning the predictions made by the input datasets, as the weight and bias terms of the model was minimized through the comprehensive iteration. This iteration led to a minimization of the error rate, which is also referred to as the loss function is obtained in each epoch. The goal for fine-tuning these estimates was to obtain a more reliable system model and ensure lower error rates. For the first epoch, the leftmost layer (Figs. 7.3 and 7.7), which takes the randomly selected bias and weight term, and the optimized PC and IC inputs begins the training. The four sets of hidden layers also take the randomly selected bias term of 1.5 (Table 7.5). Using the process of feed forward learning, TWS predictions that do not give the desired result were obtained (Fig. 7.7) as the weights and biases have not yet been fine-tuned, this is where the back propagation comes in. The back propagation entails feeding the loss function backwards so that the weight and bias terms can be fine-tuned. With the help of the gradient descent optimization, the weights of the networks are optimized to yield smaller loss in the next iteration. This iteration was performed 23 times for the PCA-BPNN to achieve an acceptable error rate (0.3%), and 25 times for the ICA-BPNN to achieve an acceptable error rate (0.1%).

The index of the forecast for PC-BPNN1 and IC-BPNN1 when compared to the observed GRACE-TWS values, after 5 months, achieved a fitting of 82% and 88%, respectively. The other two individual modes for each system are characterized as: PCA-BPNN2 (78%), PCA-BPNN3 (85%), ICA-BPNN-2 (77%), and ICA-BPNN3 (80%). After 10 months, the fit was reduced to PCA-BPNN1 (65%) and ICA-BPNN1 (69%) for the first leading modes, while the second and third modes were reduced to PCA-BPNN2 (60%), PCA-BPNN3 (58%), ICA-BPNN2 (64%), and ICA-BPNN3 (60%). The propagation of uncertainty index shows that after 10 months, fitting standard deviation became too large, suggesting that the proposed PCA- and ICA-BPNN models is more or less reliable for predictions of only up to 10 months. The fit of the forecast is significantly affected by multi-annual climatic oscillations, which are prominent in the SA region, regardless of the all-encompassing predictor components used to build our network (Fig. 7.15b). From the results of this analysis, the localized hydrological parameters maintained the closest fit to the predicted TWS from both PC- and IC-BPNN models. This is a pointer to the fact that regardless of the impact of global climate indices, localized hydrological signals are the aftermath of changes in global climate in climatic hotspots like SA. When the loadings and ICs of the GRACE-TWS were compared with the BPNN models, there was no apparent deviations, meaning that the fit reported for these experiments of the two systems are significantly reliable (Fig. 7.16).

For inter-annual time scale series (Fig. 7.8f and l, 7.15c and f), great care should be observed, since the simulation and prediction of the BPNN models are very sensitive to the temporal patterns of the input parameters, especially the localized products. When IC's of the predictors are not well-defined, for, e.g., due to the inclusion of the noise IC generated from the RHD parameters, the inter-annual forecast of the BPNN tend to perform very poorly. This was tested by the inclusion of IC7 in the input signals (results not shown here), and a poor fit was observed. The bias generated from the test was largely found in the inter-annual predictions of TWS-2 and TWS-3, and was attributed to the inclusion of the noise IC in the predictions. The outliers generated from Table 7.5 were used to plot a residual plot for the two BPNN models (Fig. 7.17). The fit level for the PC- and IC- BPNN models (Fig. 7.12), confusion matrix (Fig. 7.16), and residual plot (Fig. 7.17) suggest a reliable framework for TWS prediction and forecasting in the TWS-dynamic SA region. This framework is

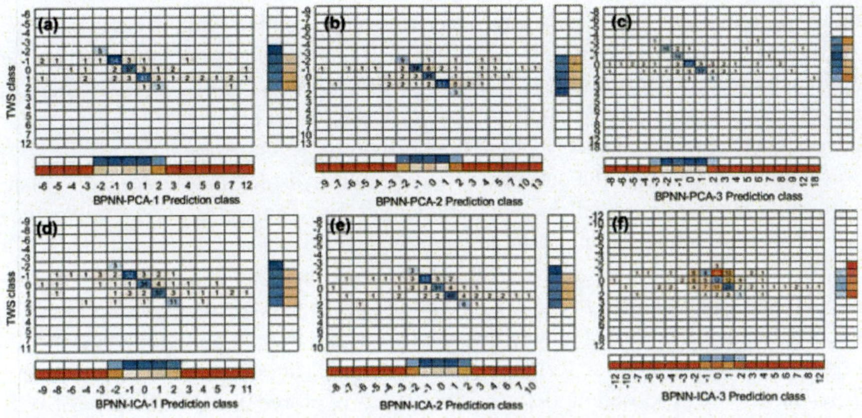

Fig. 7.16 A confusion matrix chart from the true labels (GRACE-derived TWS class) and the predicted labels (PCA—ICA-BPNN framework). The diagonals and off-diagonal cells correspond to correctly and incorrectly classified observations, respectively

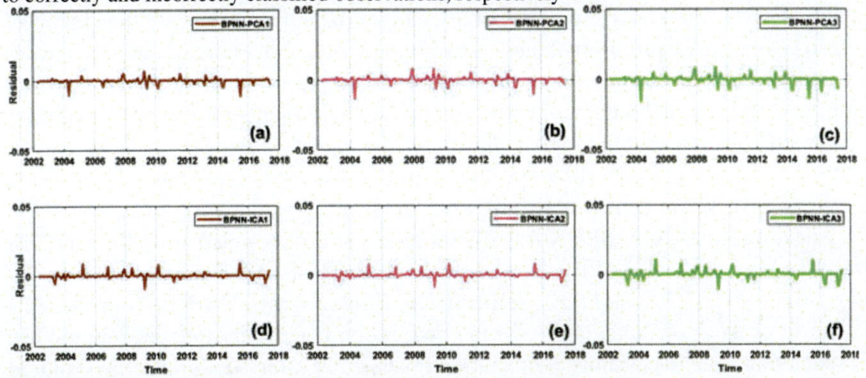

Fig. 7.17 Residual plot of the duple-based prediction algorithm developed in this experiment (axis: m)

scalable and can predict TWS in other humid regions at river basin scale, e.g., the Congo basin.

7.6 Concluding Remarks

The application of traditional machine learning algorithms in diagnosing the impacts of global climate on TWS can be constrained due to complex climate interactions in some regions. The development of convoluted techniques, which tend to accommodate these complexities in their predictive operations is warranted, thus the implementation of the deep learning technique in analyzing the footprints of global climate on continental changes in TWS. The impacts of global climate change on the availability of water resources in large sub-regions has been studied. However, the approach to identifying climatic footprints on terrestrial hydrology is still limited. In this chapter, the novel integration of principal component analysis (PCA) and independent component analysis (ICA) to the back propagation neural network (BPNN) was employed to assess the complex interaction of hydrological fluxes and climate variability indices from well-known oceanic hotspots (Indian, Atlantic, and Pacific) with freshwater changes on a continental scale.

The BPNN technique propagates the total losses back into the neural network to understand how much of the loss every node is responsible for, and subsequently updates the weights in such a way that minimizes the loss by giving the nodes with higher error rates, a lower weights and vice versa. The statistical framework and efficiency of the proposed model in this chapter is explored using a 15-year GRACE data, which served as the dependent variable against sixteen independent variables. This experiment corroborated the strength of the Atlantic and Pacific indices in regulating TWS in the SA region. This validates the claim that global climate indices in the nearby oceans have more impact on changes in land water storage than far away oceans. Apart from proximity being a key point of teleconnection influence, the roles of human-induced climate fluctuations are expected to result in an increased TWS dynamics. This is evidenced in the northern SA region where higher precipitations produced stronger amplitude and flux-impacts against TWS. This process can be attributed to the presence of large fluxes in the Amazon River basin, which is a known key driver in continental and global land water storage processes.

The developed PCA-BPNN framework in this study provided good fits for TWS with the first, second, and third orthogonal modes of the PC-GRACE-TWS showing a fit of 90%, 87%, and 86%, respectively. However, the ICA-BPNN produced a better fit of 94%, 91%, and 88% for the first, second, and third independent modes, respectively. The uncertainty propagation reveals that after 10 months, the standard deviation became too large, suggesting that the assimilated BPNN framework is more or less reliable for predictions of only up to 10 month. Given that the ICA technique explores higher order statistical methods than the PCA, it is not surprising that its system model outperformed the PCA in the prediction of TWS. Also, this experiment corroborates as earlier stated that independence is a more robust condition in statistical decomposition analysis. The application of an assimilated system model to uncover hidden patterns that underpin the prediction of TWS is largely successful, as it revealed a thorough interaction and link between multi-global climate teleconnection patterns and changes in continental freshwater in the studied region. However, further investigation is required to adequately quantify the contributions of non-idealized natural systems, such as human water management and anthropogenic influences on continental freshwater resources. This is particularly critical for regions susceptible to a wide array of climatic processes such as South America.

References

Ahmadi E, McLellan B, Mohammadi-Ivatloo B, Tezuka T (2020) The role of renewable energy resources in sustainability of water desalination as a potential fresh-water source: an updated review. Sustainability 12(13). https://doi.org/10.3390/su12135233

Ahmed M, Sultan M, Elbayoumi T, Tissot P (2019) Forecasting GRACE data over the African watersheds using artificial neural networks. Remote Sens 11(15). https://doi.org/10.3390/rs11151769

Aires F, Rossow WB, Chédin A (2002) Rotation of EOFs by the independent component analysis: toward a solution of the mixing problem in the decomposition of geophysical time series. J Atmos Sci 59(1):111–123. https://doi.org/10.1175/1520-0469(2002)059<0111:ROEBTI>2.0.CO;2

Amari S, Murata N, Muller K-R, Finke M, Yang H (1997) Asymptotic statistical theory of overtraining and cross-validation. IEEE Trans Neural Netw 8(5):985–996. https://doi.org/10.1109/72.623200

Ambrizzi T, de Souza EB, Pulwarty RS (2004) The Hadley and Walker regional circulations and associated ENSO impacts on South American seasonal rainfall. Springer Netherlands, Dordrecht, pp 203–235. https://doi.org/10.1007/978-1-4020-2944-8_8

Barnard E, Wessels L (1992) Extrapolation and interpolation in neural network classifiers. IEEE Control Syst Mag 12(5):50–53. https://doi.org/10.1109/37.158898

Bebis G, Georgiopoulos M (1994) Feed-forward neural networks. IEEE Potentials 13(4):27–31. https://doi.org/10.1109/45.329294

Berner E, Berner R (2012) Global environment: water, air, and geochemical cycles, 2nd edn. Princeton University Press, p 464. https://press.princeton.edu/books/hardcover/9780691136783/global-environment

Bond NA, Cronin MF, Freeland H, Mantua N (2015) Causes and impacts of the 2014 warm anomaly in the NE Pacific. Geophys Res Lett 42(9):3414–3420. https://doi.org/10.1002/2015GL063306

Boulanger J, Leloup J, Penalba O et al (2005) Observed precipitation in the Paraná-Plata hydrological basin: long-term trends, extreme conditions and ENSO teleconnections. Clim Dyn 24:393–413. https://doi.org/10.1007/s00382-004-0514-x

Brêda J, de Paiva R, Collischon W et al (2020) Climate change impacts on South American water balance from a continental-scale hydrological model driven by CMIP5 projections. Clim Chang 159:503–522. https://doi.org/10.1007/s10584-020-02667-9

Burke EJ, Brown SJ, Christidis N (2006) Modeling the recent evolution of global drought and projections for the twenty-first century with the Hadley Centre climate model. J Hydrometeorol 7(5):1113–1125. https://doi.org/10.1175/JHM544.1

Cai W, McPhaden M, Grimm A et al (2020) Climate impacts of the El Niño Southern Oscillation on South America. Nat Rev Earth Environ 1(4):215–231. https://doi.org/10.1038/s43017-020-0040-3

Cardoso JF, Souloumiac A (1993) Blind beamforming for non-Gaussian signals. IEE Proc 140(6):362–370

Cavole L M et al (2016) Biological impacts of the 2013–2015 warm-water anomaly in the northeast pacific: winners, losers, and the future. Oceanography 29(2):273–285. https://doi.org/10.5670/oceanog.2016.32

Cazenave A, Chen J (2010) Time-variable gravity from space and present-day mass redistribution in the earth system. Earth Planet Sci Lett 298(3):263–274. https://doi.org/10.1016/j.epsl.2010.07.035

Chen J, Cazenave A, Dahle C et al (2022) Applications and challenges of GRACE and GRACE follow-on satellite gravimetry. Surv Geophys 43:305–3455. https://doi.org/10.1007/s10712-021-09685-x

Chen J, Tapley B, Rodell M, Seo K-W, Wilson C, Scanlon BR, Pokhrel Y (2020) Basin-scale river runoff estimation from grace gravity satellites, climate models, and in situ observations: a case study in the amazon basin. Water Resour Res 56(10):e2020WR028032. https://doi.org/10.1029/2020WR028032

Chiew FHS (2006) Estimation of rainfall elasticity of streamflow in Australia. Hydrol Sci J 51(4):613–625. https://doi.org/10.1623/hysj.51.4.613

Chu H-J, Chang L-C (2009) Optimal control algorithm and neural network for dynamic groundwater management. Hydrol Process

Church JA, Gregory JM et al (2001) Changes in sea level, pp 639–694. IPCC. https://www.ipcc.ch/site/assets/uploads/2018/03/WGI_TAR_full_report.pdf on 26th August 2022

Chylek P, Klett J, Dubey M et al (2016) The role of Atlantic multi-decadal oscillation in the global mean temperature variability. Clim Dyn 47:3271–3279. https://doi.org/10.1007/s00382-016-3025-7

Collins M, Chandler R, Cox P et al (2012) Quantifying future climate change. Nat Clim Chang 2:403–409. https://doi.org/10.1038/nclimate1414

Dai H, MacBeth C (1997) Effects of learning parameters on learning procedure and performance of a BPNN. Neural Netw 10(8):1505–1521. https://doi.org/10.1016/S0893-6080(97)00014-2

de Marsily G, Abarca-del Rio R (2016) Water and food in the twenty-first century. Surv Geophys 37:503–527. https://doi.org/10.1007/s10712-015-9335-1

De Pryck K (2021) Intergovernmental expert consensus in the making: the case of the summary for policy makers of the IPCC 2014 synthesis report. Glob Environ Polit 21(1):108–129. https://doi.org/10.1162/glep_a_00574

Delpla I, Jung A-V, Baures E, Clement M, Thomas O (2009) Impacts of climate change on surface water quality in relation to drinking water production. Environ Int 35(8):1225–1233. https://doi.org/10.1016/j.envint.2009.07.001

DiNezio P, Tierney J (2013) The effect of sea level on glacial Indo-Pacific climate. Nat Geosci 44:485–491. https://doi.org/10.1038/ngeo1823

Döll P, Flörke M (2005) Global-scale estimation of diffuse groundwater recharge. Frankfurt Hydrology Paper 03, Institute of Physical Geography, Frankfurt University, Frankfurt am Main, Germany. Retrieved from chrome-extension://efaidnbmnnnibpcajpcglclefindmkaj/https://d-nb.info/1054768056/34 on 26th August 2022

Famiglietti JS, Cazenave A, Eicker A, Reager JT, Rodell M, Velicogna I (2015) Satellites provide the big picture. Science 349(6249):684–685. https://doi.org/10.1126/science.aac9238

Fausett L, Elwasif W (1994) Predicting performance from test scores using backpropagation and counter propagation. In: Proceedings of 1994 IEEE international conference on neural networks (ICNN'94), vol 5, pp 3398–3402. https://doi.org/10.1109/ICNN.1994.374782

Feng S, Kang S, Huo Z, Chen S, Mao X (2008) Neural networks to simulate regional ground water levels affected by human activities. Groundwater 46(1):80–90. https://doi.org/10.1111/j.1745-6584.2007.00366.x

Ferreira VG, Montecino HD, Ndehedehe CE, del Rio RA, Cuevas A, de Freitas SRC (2019a) Determining seasonal displacements of Earth's crust in South America using observations from space-borne geodetic sensors and surface-loading models. Earth, Planets Space 71(1):84. https://doi.org/10.1186/s40623-019-1062-2

Ferreira VG, Ndehedehe CE, Montecino HC, Yong B, Yuan P, Abdalla A, Mohammed AS (2019b) Prospects for imaging terrestrial water storage in South America using daily GPS observations. Remote Sens 11(6). https://doi.org/10.3390/rs11060679

Fletcher D, Goss E (1993) Forecasting with neural networks: an application using bankruptcy data. Inf Manag 24(3):159–167. https://doi.org/10.1016/0378-7206(93)90064-Z

Goethals P, Dedecker A, Gabriels W et al (2007) Applications of artificial neural networks predicting macroinvertebrates in freshwaters. Aquat Ecololgy 41:491–508. https://doi.org/10.1007/s10452-007-9093-3

Good SP, Noone D, Bowen G (2015) Hydrologic connectivity constrains partitioning of global terrestrial water fluxes. Science 349(6244):175–177. https://doi.org/10.1126/science.aaa5931

Gormsen E (1997) The impact of tourism on coastal areas. GeoJournal 42:39–54. https://doi.org/10.1023/A:1006840622450

Grimm AM (2003) The El Niño Impact on the summer monsoon in Brazil: regional processes versus remote influences. J Clim 16(2):263–280. https://doi.org/10.1175/1520-0442(2003)016<0263:TENIOT>2.0.CO;2

Grimm AM, Ambrizzi T (2009) Teleconnections into South America from the tropics and extratropics on interannual and intraseasonal timescales. Springer Netherlands, Dordrecht, pp 159–191. https://doi.org/10.1007/978-90-481-2672-9_7

Hayashi A, Akimoto K, Tomoda T et al (2013) Global evaluation of the effects of agriculture and water management adaptations on the water-stressed population. Mitig Adapt Strat Glob Chang 18:591–618. https://doi.org/10.1007/s11027-012-9377-3

House-Peters LA, Chang H (2011) Urban water demand modeling: review of concepts, methods, and organizing principles. Water Resour Res 47(5). https://doi.org/10.1029/2010WR009624

Huang Z, Yuan X, Liu X (2021) The key drivers for the changes in global water scarcity: water withdrawal versus water availability. J Hydrol 601:126658. https://doi.org/10.1016/j.jhydrol.2021.126658

Huffman GJ, Adler RF, Bolvin DT, Gu G, Nelkin EJ, Bowman KP, Hong Y, Stocker EF, Wolff DB (2007) The TRMM multisatellite precipitation analysis (TMPA): quasi-global, multiyear, combined-sensor precipitation estimates at fine scales. J Hydrometeorol 8:38–55. https://doi.org/10.1175/JHM560.1

Humphrey V, Rodell M, Eicker A (2023) Using satellite-based terrestrial water storage data: a review. Surv Geophys. https://doi.org/10.1007/s10712-022-09754-9

Ilin A, Valpola H, Oja E (2005) Semiblind source separation of climate data detects El Nino as the component with the highest inter annual variability. In: Proceedings of the IEEE international joint conference on neural networks, vol 3, pp 1722–1727. https://doi.org/10.1109/IJCNN.2005.1556139

Ivits E, Horion S, Fensholt R, Cherlet M (2014) Drought footprint on European ecosystems between 1999 and 2010 assessed by remotely sensed vegetation phenology and productivity. Glob Chang Biol 20(2):581–593. https://doi.org/10.1111/gcb.12393

Jury WA, Vaux HJ (2007) The emerging global water crisis: managing scarcity and conflict between water users. Advances in agronomy, vol 95, pp 1–76. https://doi.org/10.1016/S0065-2113(07)95001-4

Kalu I, Ndehedehe CE, Okwuashi O, Eyoh AE (2021) Assessing freshwater changes over Southern and Central Africa (2002–2017). Remote Sens 13(13). https://doi.org/10.3390/rs13132543

Kalu I, Ndehedehe CE, Okwuashi O, Eyoh AE, Ferreira VG (2022a) An assimilated deep learning approach to identify the influence of global climate on hydrological fluxes. J Hydrol 614:128498. https://doi.org/10.1016/j.jhydrol.2022.128498

Kalu I, Ndehedehe CE, Okwuashi O, Eyoh AE, Ferreira VG (2022b) A new modelling framework to assess changes in groundwater level. J Hydrol: Reg Stud 43:101185. https://doi.org/10.1016/j.ejrh.2022.101185

Kalu I, Ndehedehe CE, Okwuashi O, Eyoh AE, Ferreira VG (2023) Reconstructing terrestrial water storage anomalies using convolution-based support vector machine. J Hydrol: Reg Stud 46:101326. https://doi.org/10.1016/j.ejrh.2023.101326

Kanellopoulos I, Wilkinson GG (1997) Strategies and best practice for neural network image classification. Int J Remote Sens 18(4):711–725. https://doi.org/10.1080/014311697218719

Kayano MT, Capistrano VB (2014) How the Atlantic multidecadal oscillation (AMO) modifies the ENSO influence on the South American rainfall. Int J Climatol 34(1):162–178. https://doi.org/10.1002/joc.3674

Kundzewicz ZW, Mata LJ, N. W. A et al (2008) The implications of projected climate change for freshwater resources and their management. Hydrol Sci J 53(1):3–10. https://doi.org/10.1623/hysj.53.1.3

Lenard MJ, Alam P, Madey GR (1995) The application of neural networks and a qualitative response model to the auditor's going concern uncertainty decision. Decis Sci 26(2):209–227. https://doi.org/10.1111/j.1540-5915.1995.tb01426.x

Li Y, Huang C, Ding L, Li Z, Pan Y, Gao X (2019) Deep learning in bioinformatics: introduction, application, and perspective in the big data era. Methods 166:4–21. https://doi.org/10.1016/j.ymeth.2019.04.008

Liano K (1996) Robust error measure for supervised neural network learning with outliers. IEEE Trans Neural Netw 7(1):246–250. https://doi.org/10.1109/72.478411

Litjens G, Kooi T, Bejnordi BE et al (2017) A survey on deep learning in medical image analysis. Med Image Anal 42:60–88. https://doi.org/10.1016/j.media.2017.07.005

Liu H, Wang J (2011) Integrating independent component analysis and principal component analysis with neural network to predict Chinese stock market. Math Probl Eng 2011(382659):15. https://doi.org/10.1155/2011/382659

Lu R, Dong B (2005) Impact of Atlantic sea surface temperature anomalies on the summer climate in the western North Pacific during 1997–1998. J Geophys Res: Atmos 110(D16). https://doi.org/10.1029/2004JD005676

Maier HR, Dandy GC (2000) Application of artificial neural networks to forecasting of surface water quality variables: issues, applications and challenges. Artif Neural Netw Hydrol 287–309. https://doi.org/10.1007/978-94-015-9341-0_15

Maier HR, Jain A, Dandy GC, Sudheer K (2010) Methods used for the development of neural networks for the prediction of water resource variables in river systems: current status and future directions. Environ Model Softw 25(8):891–909. https://doi.org/10.1016/j.envsoft.2010.02.003

Masters T (1993) Practical neural network recipies in C++. https://doi.org/10.1016/C2009-0-22399-3

Merrifield MA, Thompson PR, Lander M (2012) Multidecadal sea level anomalies and trends in the western tropical Pacific. Geophys Res Lett 39(13). https://doi.org/10.1029/2012GL052032

Milly P, Dunne K, Vecchia A (2005) Global pattern of trends in streamflow and water availability in a changing climate. Nature 438:347–350. https://doi.org/10.1038/nature04312

Montazerolghaem M, Vervoort W, Minasny B, McBratney A (2016) Long-term variability of the leading seasonal modes of rainfall in south-eastern Australia. Weather Clim Extrem 13:1–14. https://doi.org/10.1016/j.wace.2016.04.001

Mu Q, Zhao M, Running SW (2011) Improvements to a MODIS global terrestrial evapotranspiration algorithm. Remote Sens Environ 115(8):1781–1800. https://doi.org/10.1016/j.rse.2011.02.019

Mudigonda M, Ram P, Kashinath K et al (2021) Deep learning for detecting extreme weather patterns, chapter 12. Wiley, pp 161–185

Muringai RT, Mafongoya PL, Lottering R (2021) Climate change and variability impacts on sub-Saharan African fisheries: a review. Rev Fish Sci Aquac 29(4):706–720. https://doi.org/10.1080/23308249.2020.1867057

Naghibi S, Pourghasemi H, Dixon B (2016) GIS-based groundwater potential mapping using boosted regression tree, classification and regression tree, and random forest machine learning models in Iran. Environ Monit Assess 188(44):1–27. https://doi.org/10.1007/s10661-015-5049-6

Ndehedehe C (2022a) Global freshwater systems. Springer International Publishing, Cham, pp 19–32

Ndehedehe C (2022b) Groundwater from space. Springer International Publishing, Cham, pp 211–230

Ndehedehe C (2022c) Remote sensing hydrology. Springer International Publishing, Cham, pp 3–17

Ndehedehe C (2022d) Satellite geodetic missions. Springer International Publishing, Cham, pp 53–70

Ndehedehe C, Awange J, Agutu N, Kuhn M, Heck B (2016a) Understanding changes in terrestrial water storage over West Africa between 2002 and 2014. Adv Water Resour 88:211–230. https://doi.org/10.1016/j.advwatres.2015.12.009

Ndehedehe CE (2019) The water resources of tropical West Africa: problems, progress and prospect. Acta Geophys 67(2):621–649. https://doi.org/10.1007/s11600-019-00260-y

Ndehedehe CE, Ferreira VG (2020a) Assessing land water storage dynamics over Southern America. J Hydrol 580:124339. https://doi.org/10.1016/j.jhydrol.2019.124339

Ndehedehe CE, Ferreira VG (2020b) Identifying the footprints of global climate modes in time-variable gravity hydrological signals. Clim Chang 159:481–502. https://doi.org/10.1007/s10584-019-02588-2

Ndehedehe CE, Agutu NO, Okwuashi OH, Ferreira VG (2016b) Spatio-temporal variability of droughts and terrestrial water storage over Lake Chad Basin using independent component analysis. J Hydrol 540:106–128. https://doi.org/10.1016/j.jhydrol.2016.05.068

Ndehedehe CE, Awange J, Kuhn M, Agutu N, Fukuda Y (2017) Climate teleconnections influence on West Africa's terrestrial water storage. Hydrol Process 31(18):3206–3224. https://doi.org/10.1002/hyp.11237

Ndehedehe CE, Awange JL, Agutu NO, Okwuashi O (2018) Changes in hydro-meteorological conditions over tropical West Africa (1980 – 2015) and links to global climate. Glob Planet Chang 162:321–341. https://doi.org/10.1016/j.gloplacha.2018.01.020

Ndehedehe CE, Ferreira VG, Onojeghuo AO, Agutu NO, Emengini E, Getirana A (2020) Influence of global climate on freshwater changes in Africa's largest endorheic basin using multi-scaled indicators. Sci Total Environ 737:139643. https://doi.org/10.1016/j.scitotenv.2020.139643

Ndehedehe CE, Ferreira VG, Agutu NO, Onojeghuo AO, Okwuashi O, Kassahun HT, Dewan A (2021) What if the rains do not come? J Hydrol 595:126040. https://doi.org/10.1016/j.jhydrol.2021.126040

Ndehedehe CE, Ferreira VG, Adeyeri OE, Correa FM, Usman M, Oussou FE, Kalu I, Okwuashi O, Onojeghuo AO, Getirana A, Dewan A (2023) Global assessment of drought characteristics in the Anthropocene. Resour, Environ Sustain 12:100105. https://doi.org/10.1016/j.resenv.2022.100105

Ni S, Chen J, Wilson CR, Li J, Hu X, Fu R (2017) Global terrestrial water storage changes and connections to ENSO events. Surv Geophys. https://doi.org/10.1007/s10712-017-9421-7

Nourani V, Molajou A, Najafi H, Danandeh Mehr A (2019) Emotional ANN (EANN): a new generation of neural networks for hydrological modeling in IoT. Springer International Publishing, Cham, pp 45–61. https://doi.org/10.1007/978-3-030-04110-6_3

Odorico P, Chiarelli DD, Rosa L et al (2020) The global value of water in agriculture. Proc Natl Acad Sci 117(36):21985–21993. https://doi.org/10.1073/pnas.2005835117

OECD (2015) Drying wells, rising stakes: towards sustainable agricultural groundwater use. OECD Studies on WaterOECD Publishing, Paris. https://doi.org/10.1787/9789264238701-en

Okwuashi O, Ndehedehe C (2017) Tide modelling using support vector machine regression. J Spat Sci 62(1):29–46. https://doi.org/10.1080/14498596.2016.1215272

Okwuashi O, Ndehedehe CE (2020) Deep support vector machine for hyperspectral image classification. Pattern Recognit 103:107298. https://doi.org/10.1016/j.patcog.2020.107298

Okwuashi O, Ndehedehe CE, Olayinka DN, Eyoh A, Attai H (2021) Deep support vector machine for PolSAR image classification. Int J Remote Sens 42(17):6498–6536. https://doi.org/10.1080/01431161.2021.1939910

Patuwo E, Hu MY, Hung MS (1993) Two-group classification using neural networks. Decis Sci 24(4):825–845. https://doi.org/10.1111/j.1540-5915.1993.tb00491.x

Phillips T, Nerem RS, Fox-Kemper B, Famiglietti JS, Rajagopalan B (2012) The influence of ENSO on global terrestrial water storage using GRACE. Geophys Res Lett 39(L16705):2012. https://doi.org/10.1029/2012GL052495

Piramuthu S, Shaw MJ, Gentry JA (1994) A classification approach using multi-layered neural networks. Decis Support Syst 11(5):509–525. https://doi.org/10.1016/0167-9236(94)90022-1

Ponti MA, Ribeiro LSF, Nazare TS, Bui T, Collomosse J (2017) Everything you wanted to know about deep learning for computer vision but were afraid to ask. In: 30th SIBGRAPI conference on graphics, patterns and images tutorials (SIBGRAPI-T), pp 17–41. https://doi.org/10.1109/SIBGRAPI-T.2017.12

Qadir NU (2020) Influence of principal component analysis as a data conditioning approach for training multilayer feedforward neural networks with exact form of Levenberg-Marquardt algorithm. Glob J Technol Optim 11:1–6

Reboita MS, Ambrizzi T et al (2021) Impacts of teleconnection patterns on South America climate. Ann N Y Acad Sci 1504(1):116–153. https://doi.org/10.1111/nyas.14592

Robinson B, Cohen JS, Herman JD (2020) Detecting early warning signals of long-term water supply vulnerability using machine learning. Environ Model Softw 131:104781. https://doi.org/10.1016/j.envsoft.2020.104781

Rodell M, Famiglietti JS, Wiese DN, Reager JT, Beaudoing HK, Landerer FW, Lo MH (2018a) Emerging trends in global freshwater availability. Nature 557(7707):651–659. https://doi.org/10.1038/s41586-018-0123-1

Rodell M, Famiglietti JS, Wiese DN, Reager JT, Beaudoing HK, Landerer FW, Lo M-H (2018b) Emerging trends in global freshwater availability. Nature 557:651–659. https://doi.org/10.1038/s41586-018-0123-1

Rodell M, Famiglietti J, Wiese D, Reager J, Beaudoing H, Landerer F, Lo M (2019) Emerging trends in global freshwater availability. Nature 557(7739). https://doi.org/10.1038/s41586-018-0831-6

Rogers LL, Dowla FU (1994) Optimization of groundwater remediation using artificial neural networks with parallel solute transport modeling. Water Resour Res 30(2):457–481. https://doi.org/10.1029/93WR01494

Rosenzweig CG, Casassa D, Karoly A, Imeson C et al (2007) Assessment of observed changes and responses in natural and managed systems. Climate change 2007: impacts, adaptation and vulnerability. In: Hanson (eds) Contribution of working group II to the fourth assessment report of the intergovernmental panel on climate change. Cambridge University Press, Cambridge, UK, pp 79–131. Accessed from https://www.pubs.giss.nasa.gov/abs/ro02900h.html on 26th August 2022

Sajjad H, Ghaffar, (2019) Observed, simulated and projected extreme climate indices over Pakistan in changing climate. Theor Appl Climatol 137:255–281. https://doi.org/10.1007/s00704-018-2573-7

Savitzky A, Golay MJE (1964) Soothing and differentiation of data by simplified least squares procedures. Anal Chem 36(8):1627–1639

Shah M, Javed M, Abunama T (2021) Proposed formulation of surface water quality and modelling using gene expression, machine learning, and regression techniques. Environ Sci Pollut Res 28:13202–13220. https://doi.org/10.1007/s11356-020-11490-9

Shahamiri SR (2021) Speech vision: an end-to-end deep learning-based dysarthric automatic speech recognition system. IEEE Trans Neural Syst Rehabil Eng 29:852–861. https://doi.org/10.1109/TNSRE.2021.3076778

Singh SP, Kumar A, Darbari H, Singh L, Rastogi A, Jain S (2017) Machine translation using deep learning: an overview. In: International conference on computer, communications and electronics (comptelix), pp 162–167. https://doi.org/10.1109/COMPTELIX.2017.8003957

Sun AY (2013) Predicting groundwater level changes using GRACE data. Water Resour Res 49(9):5900–5912. https://doi.org/10.1002/wrcr.20421

Swenson S, Yeh PJ-F, Wahr J, Famiglietti J (2006) A comparison of terrestrial water storage variations from GRACE with in situ measurements from Illinois. Geophys Res Lett 33(16). https://doi.org/10.1029/2006GL026962

Tapley B, Bettadpur S, Watkins M, Reigber C (2004) The gravity recovery and climate experiment: mission overview and early results. Geophys Res Lett 31:1–4. https://doi.org/10.1029/2004GL019920

Theis FJ, Gruber P, Keck IR, Meyer-bäse A, Lang EW (2005) Spatiotemporal blind source separation using double-sided approximate joint diagonalization. In: Proc, EUSIPCO

Tourian MJ, Reager JT, Sneeuw N (2018) The total drainable water storage of the Amazon River Basin: a first estimate using GRACE. Water Resour Res 54(5):3290–3312. https://doi.org/10.1029/2017WR021674

Trenberth KE (1997) The definition of El Niño. Bull Am Meteorol Soc 78(12):2771–2777

Usman M, Ndehedehe CE, Farah H, Manzanas R (2021) Impacts of climate change on the streamflow of a large river basin in the Australian tropics using optimally selected climate model outputs. J Clean Prod 315:128091. https://doi.org/10.1016/j.jclepro.2021.128091

Vergara W, Deeb A, Valencia A, Bradley R, Francou B, Zarzar A, Grünwaldt A, Haeussling S (2007) Economic impacts of rapid glacier retreat in the Andes. Eos, Trans Am Geophys Union 88(25):261–264

Walczak S, Cerpa N (1999) Heuristic principles for the design of artificial neural networks. Inf Softw Technol 41(2):107–117. https://doi.org/10.1016/S0950-5849(98)00116-5

Wang F (1994) The use of artificial neural networks in a geographical information system for agricultural land-suitability assessment. Environ Plan A: Econ Space 26(2):265–284. https://doi.org/10.1068/a260265

Weigend AS, Huberman BA, Rumelhart DE (1990) Predicting the future: a connectionist approach. Int J Neural Syst 01(03):193–209. https://doi.org/10.1142/S0129065790000102

Wiese DN, Landerer FW, Watkins MM (2016) Quantifying and reducing leakage errors in the JPL RL05M GRACE mascon solution. Water Resour Res 52(9):7490–7502. https://doi.org/10.1002/2016WR019344

Xu T, Valocchi AJ (2015) Data-driven methods to improve baseflow prediction of a regional groundwater model. Comput Geosci 85:124–136. https://doi.org/10.1016/j.cageo.2015.05.016

Young T, Hazarika D, Poria S, Cambria E (2018) Recent trends in deep learning based natural language processing [review article]. IEEE Comput Intell Mag 13(3):55–75. https://doi.org/10.1109/MCI.2018.2840738

Yu Q, Wang S, He H, Yang K, Ma L, Li J (2021) Reconstructing grace-like TWS anomalies for the Canadian landmass using deep learning and land surface model. Int J Appl Earth Obs Geoinf 102:102404. https://doi.org/10.1016/j.jag.2021.102404

Chapter 8
Deep Belief Network for Groundwater Modeling

8.1 Introductory Remarks

Subsurface water in aquifers or groundwater is a critical part of the terrestrial water cycle. Its availability depends on how it is quickly being replaced by precipitation through recharge after abstraction. The soil and geologic properties (rock materials), evapotranspiration rates, surface water, and changes in land use cover play critical roles in this recharge. Groundwater requires monitoring and forecast to underpin management as it can disappear in some areas due to prolonged drought and over extraction, thus limiting water supply to communities (Castle et al. 2014). A large proportion of available freshwater utilized by our diverse ecosystem are stored beneath the land surface, and play significant roles in global food security and helping humans adapt to climate change (Tapley et al. 2019; Green et al. 2011). This, for example, is exemplified in the expansion of irrigated agriculture into typically dry regions where precipitation and surface water are limited (Bierkens and Wada 2019). This makes its availability critical for global food security as well as the replenishment of both terrestrial and aquatic ecosystems, especially during drought.

However, increasing groundwater stress in many of the world's largest aquifers is now a concern. Large depletions of groundwater in the US, India, Middle East, and some Asian countries have been documented. This problem has been linked to both climate and anthropogenic causes, which are key drivers of aquifer depletion around the world (e.g., Kalu et al. 2022b; Ndehedehe 2022; Rateb et al. 2020; Green et al. 2011). Over the years, the increased rate of water withdrawal from aquifers is without doubt the result of unfavorable hydrological conditions caused by climatic factors (e.g., Thomas and Famiglietti 2019; Chen et al. 2014; Scanlon et al. 2012). This includes increased water demand resulting from rising human population (Okello et al. 2015), urbanization and development (Kumar et al. 2018), intensified agricultural water needs and changes in dietary requirements (Bhattarai et al. 2021; Bierkens and Wada 2019), anthropogenic influence including high rates of groundwater extraction (Thomas and Famiglietti 2019; Ashraf 2017), and climate change (Ranjan et al. 2006). This has resulted in the increased need to enhance spatio-temporal diagnosis

C. Ndehedehe, *Hydro-Climatic Extremes in the Anthropocene*, Springer Climate,
https://doi.org/10.1007/978-3-031-37727-3_8

of groundwater resources by designing optimal models and frameworks for more accurate groundwater predictions. A major limitation to this strategy is the insufficiency of groundwater wells in many regions of the world where this assessment is needed. This therefore beckons on the performance of machine learning and satellite remote sensing methods to aid in a more profound knowledge of water resource management and increased resilience to global change.

Over the past decade, the rapid progress in technology led to the emergence of several novel models and techniques aimed at improving groundwater level forecasts. For example, the application of data-driven stochastic algorithms for short-term groundwater forecasts has been severely implemented over aquifers in large sub-regions, especially due to its lucidity and intelligibility in implementation (Yin et al. 2020). Data-driven techniques including self-organized maps (Nourani et al. 2015), boosted regression trees (Sharafati et al. 2020), multivariate adaptive regression splines and M5 model trees (Rezaie-balf et al. 2017), support vector machines (lei Wei et al. 2020), Gaussian process regressions (Elbeltagi et al. 2021), and the well-known neural networks (Wu et al. 2021; Sun et al. 2019) have been extensively implemented in groundwater level forecasts and predictions. These data-driven techniques have been used in parameter estimations, calibration, and they are also efficient in determining uncertainty sources better than their physically-based counterparts (Okwuashi et al. 2020). However, unlike physically-based models, data-driven models utilize learning capabilities of making groundwater water level forecasts even without having prior knowledge of the hydrogeological parameters of the monitoring wells and aquifers.

These advanced capabilities make it the most popular and preferred deep learning framework for global groundwater estimations (e.g., Kalu et al. 2022b). Most of these estimations focus on various assessments of groundwater hydrogeology, including, groundwater fluctuation forecasts (Mohapatra et al. 2021), groundwater potential forecasts (Lee et al. 2020), aquifer nitrate evaluations (Knoll et al. 2019), susceptibility assessment of groundwater salinity (Mosavi et al. 2020), and other hydro-climatic related studies and applications (Yu et al. 2020). Data-driven techniques are increasingly becoming reliable in handling groundwater complexities predominant in groundwater estimates, including data limitations and parameter estimations over varied settings. These significant factors have been documented in past studies, especially in the areas of convoluted groundwater systems and permafrost groundwater level dynamics, and coastal aquifer assessments (e.g., Chang et al. 2015; Yoon et al. 2011). Deep learning models have been gaining considerable attention in hydrological studies (e.g., Kalu et al. 2023; Sun et al. 2020). Essentially, these models are fundamentally characterized by neural network modes with multiple layers between the input and output layers.

Various deep learning frameworks exist, however, they bear significant similitude in the aspects of neuron number and ability, synapses, weights, bias terms, as well as covariance functions. Consequently, non-linear hydrological systems, including groundwater analysis are being diagnosed using novel deep learning approaches like the Deep belief network (DBN), and which permitted the use of prior groundwater level estimates to serve as inputs for the next estimate to help improve forecast accuracy (e.g., Kalu et al. 2022b). In this chapter, the potential of DBN as a novel

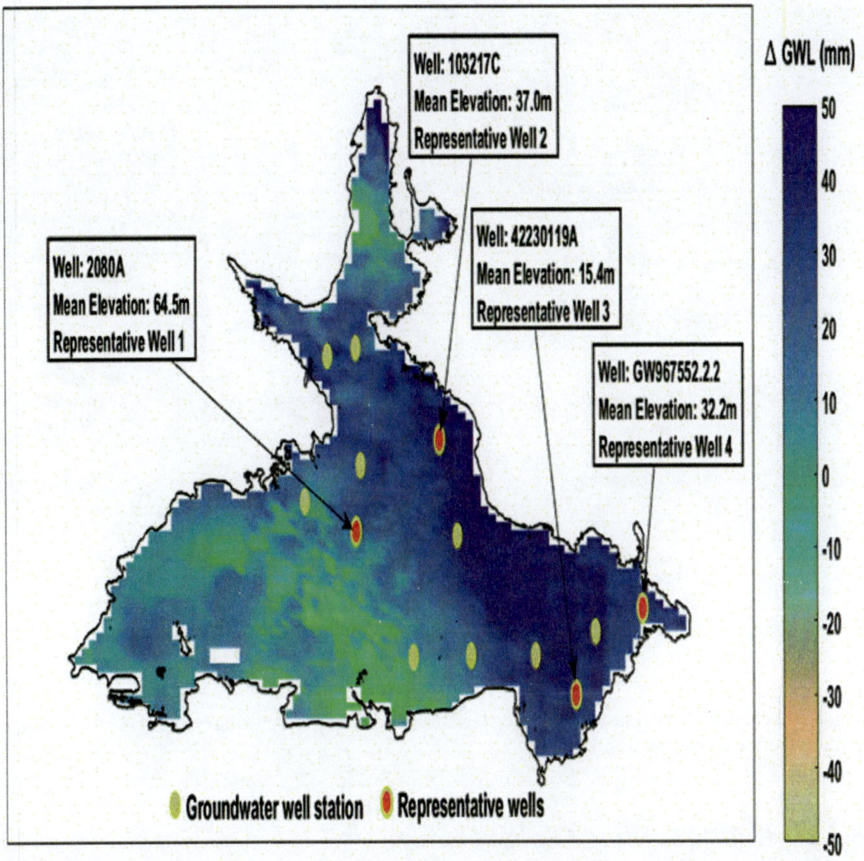

Fig. 8.1 Change in groundwater level trends showing the selected groundwater well locations and the four representative well stations used in our deep learning framework

machine learning framework in predicting hydrological stores like groundwater is explored leveraging the Great Artesian Basin (Fig. 8.1) as a tentative test bed. Following the pioneering work of Kalu et al. (2022b) over South Africa, the influence of TWS is diagnosed and the effects of localized hydrological parameters (e.g., evapotranspiration, rainfall), and the footprints of global climate teleconnection patterns (e.g., ENSO) on groundwater level fluctuations and forecasts are explored using the framework detailed in Fig. 8.2. Further in this chapter, the deep learning framework and steps to implementation, the building blocks for designing, training are discussed in detail.

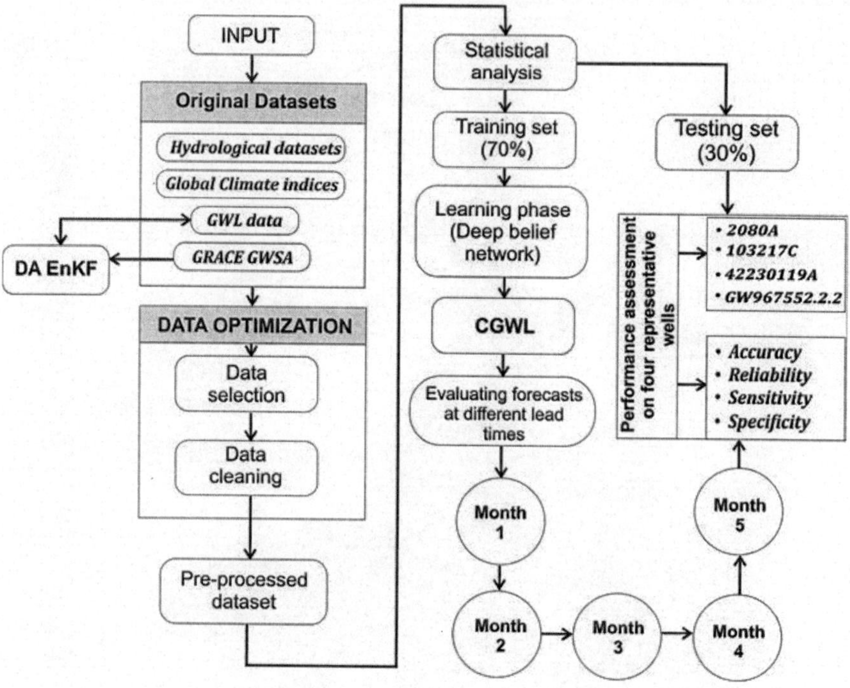

Fig. 8.2 Schematic illustration of the deep learning method for forecasting changes in groundwater level in the great artesian basin

8.1.1 Understanding Global Change

Global environmental challenges are increasing, and are evidenced in rising water scarcity, terrific groundwater and surface water losses, global mean sea level, frequent devastating floods and droughts, wildfire events, among several other ongoing transformations in the earth system (e.g., Hugonnet et al. 2021; Tellman et al. 2021; Bierkens and Wada 2019; van Dijk et al. 2013). The complexities associated with these transformations will continue to grow, affecting human well-being and constraining the health of natural ecosystems. Changes in terrestrial hydrology and the fluctuations in hydrological cycle are caused by both natural and anthropogenic factors. Other factors, which include CO2 concentration and greenhouse gases are primary regulators and contributors to global climate. For more than five decades now, atmospheric CO2 concentrations have been reported around the Pacific Ocean atop Mauna Loa, Hawaii at the National Center for Environmental Prediction (NCEP), and a steady increase in its concentration and rate of change have been reported since then (Keeling et al. 2004). The Intergovernmental Panel on Climate Change (IPCC) explained that the seasonal and multi-decadal variations appear to represent most of the residual short-term variability. Forward extrapolation of the fitted curve is not used, and future projections of greenhouse gas concentrations are based on complicated "storylines" (https://www.ipcc.ch/report/ar4/wg1/). Projected climate change reports are principally based on simulated responses to these storylines of emissions and resulting greenhouse gases.

8.1.2 Growing Concerns on the Impacts of Climate Change on Groundwater Dynamics

In recent times, studies to better understand how freshwater resources interact and respond to climate change are emerging, especially with the advent of novel machine learning approaches (Abowarda et al. 2021). However, most of these studies are limited to surface water systems due to their easy accessibility, visibility, and since the influence of climate change on surface waters can be easily tracked. Only recently have relevant institutions, government agencies, and water resource managers intensified effort in assessing the role played by groundwater aquifers in meeting the demands for agriculture, industry, and even basic human applications including domestic water use etc. Groundwater plays a key role in sustaining both terrestrial and aquatic ecosystems, and has proven useful in mitigating the impacts of climate change and coupled human activities on terrestrial water storage. The IPCC has projected significant global warming and alterations in rate and frequency of precipitation from 2000 to 2100.[1] It is therefore expected that these variations in global climate will affect the hydrological cycle by altering surface water levels and groundwater recharge, with various other associated impacts on natural ecosystems and our environment. Inasmuch as the most noticeable impacts of climate change are largely felt on surface water, subsurface water storage witnesses just as much influence through irregular patterns of hydrological processes (e.g., changes in recharge due to changes in the intensity and amplitudes of rainfall and runoff), concentration of human activities due to urbanization and population growth, as well as external forcings resulting from the actions of global teleconnection patterns and climate indices. Localized human driven impacts (e.g., irrigation demand) on groundwater fluctuations have been widely studied (e.g., Sahoo et al. 2017). However, the impacts of climate variability indices on groundwater resources are still largely understudied. Ocean currents and Sea Surface Temperature (SST) and climate variability indices such as the El-Nino Southern Oscillation (ENSO) have contributed significantly to groundwater level fluctuations. With the application of a novel deep learning framework, this chapter will attempt to identify the footprints of these climate indices on groundwater dynamics.

[1] https://www.chrome-extension://efaidnbmnnnibpcajpcglclefindmkaj/https://www.ipcc.ch/site/assets/uploads/2018/03/TAR-13.pdf.

8.1.3 Transboundary Water Resources

Climate zones traverse international political boundaries, in the similitude of surface-water and groundwater resources. Some regions have surface water rights in place depending on geographical jurisdictions. And given that changes in supply will have notable implications on surface water management and allocation, such rights and the legal aspect of water access are important. This is also the case with groundwater, as groundwater aquifers are not confined to geographical boundaries, thus making it a shared resource. Groundwater equally serves as a freshwater reservoir for prolonged periods of shortage in water resources, thus helps in circumventing drought and other climate extremes. Withdrawals from transboundary aquifers have critical political and regional management implications. Puri and Aureli (2005) provides further discussion on the issues associated with transboundary aquifers, as well as a global map identifying such transboundary groundwater resources.

8.1.4 Where Deep Learning Comes In

Deep learning technique is a new class of machine learning models, and it has only recently attracted substantial research interest. This is particularly so in fields like pattern recognition, signal processing, time series analysis, as well as complex modeling tasks like image processing (Kalu et al. 2022b; Okwuashi et al. 2021; Okwuashi and Ndehedehe 2020). Given that the technique is relatively new, its application in groundwater studies is still largely insufficient. This chapter pioneers an ensemble DBN in a comprehensive assessment of changes in groundwater level using the Great Artesian basin as a test bed. The DBN is a generative graphical model made up of multiple layers of hidden units with connections within the layers (Fig. 8.3). Given its proficiency in probabilistically reconstructing its input during training (e.g., Kalu et al. 2022b), it was adapted as a preferred choice due to the complexity and nonlinearity of groundwater level fluctuations (Fig. 8.4). Past studies have also shown the skills of the DBN technique in groundwater studies. For instance, the study undertaken by Liu et al. (2021) for the groundwater level assessment over the Jiuquan Basin using the DBN reported a satisfactory 1–3-month groundwater level forecast. However, there are widespread lags, especially in the number of running-well dataset and the limited hydrological parameters employed in the forecast. Additionally, their study did not take into account the impact of land use, surface water, and geology as significant contributing factors in groundwater level prognosis. These lags could impact on the reliability of the results obtained from predictions.

The proposal of the DBN for a sub-regional groundwater level forecast is aimed at circumventing these limitations, given that an ensemble deep learning architecture allows a comprehensive assimilation of different independent variables. For this experiment, the predictors/independent variables are grouped into,

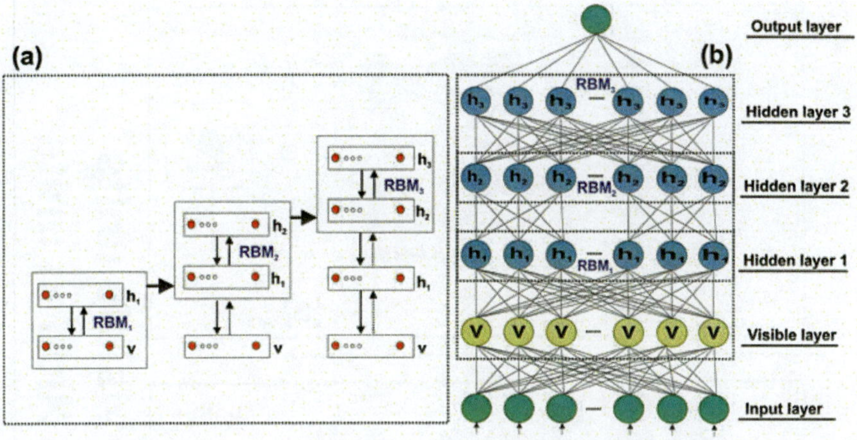

Fig. 8.3 Deep Belief Network (DBN) architecture composed by **a** stacked restricted Boltzmann machines (RBMs). Each RBM is characterized by a units in the visible layer (v) and a single hidden layer (h). RBM1 is trained using the visible layer. The hidden layers h2 and h3 are trained using the outputs of the previously trained layers h1 and h2 of the RBM1 and RBM2, respectively. The output of the h2 is the input of the next RBM3. **b** contains the trained layers h1 to h3, which form the stacked architecture. Finally, the DBN is fine-tuned with a standard back propagation algorithm

- hydrological fluxes comprising TWS, precipitation, and land surface temperature
- SSTs/global climate teleconnection indices consisting of ENSO, North Atlantic Multi-decadal Oscillation (NAO), Atlantic Multi-decadal Oscillation (AMO), Indian Ocean Dipole (IOD), and the Pacific Decadal Oscillation (PDO), and
- thirteen well-distributed groundwater well stations for in situ groundwater estimates (Fig. 8.2).

Prior to the commencement of the deep learning phase of the DBN, the original predictors are partitioned into training and testing datasets, which are made up of 70% and 30% of the total input variables, respectively. The objective of this study is to forecast the changes in groundwater levels over the Great Artesian Basin (GAB) on a 1 to 5-month time scale using extensive climatic parameters to adequately delineate the footprints of global climate on subsurface water operations.

As part of an effort to leverage machine learning to improve groundwater assessment, this chapter addresses the following key objectives;

- exploring the efficiency of the deep learning frameworks in diagnosing changes in groundwater levels at subsequent 1 to 5-month lead times. This is achieved by exploring the capabilities of the DBN,
- assessing the influence of GRACE-TWS as a forcing variable to the DBN in predictions, so as to learn the impact of terrestrial water dynamics,

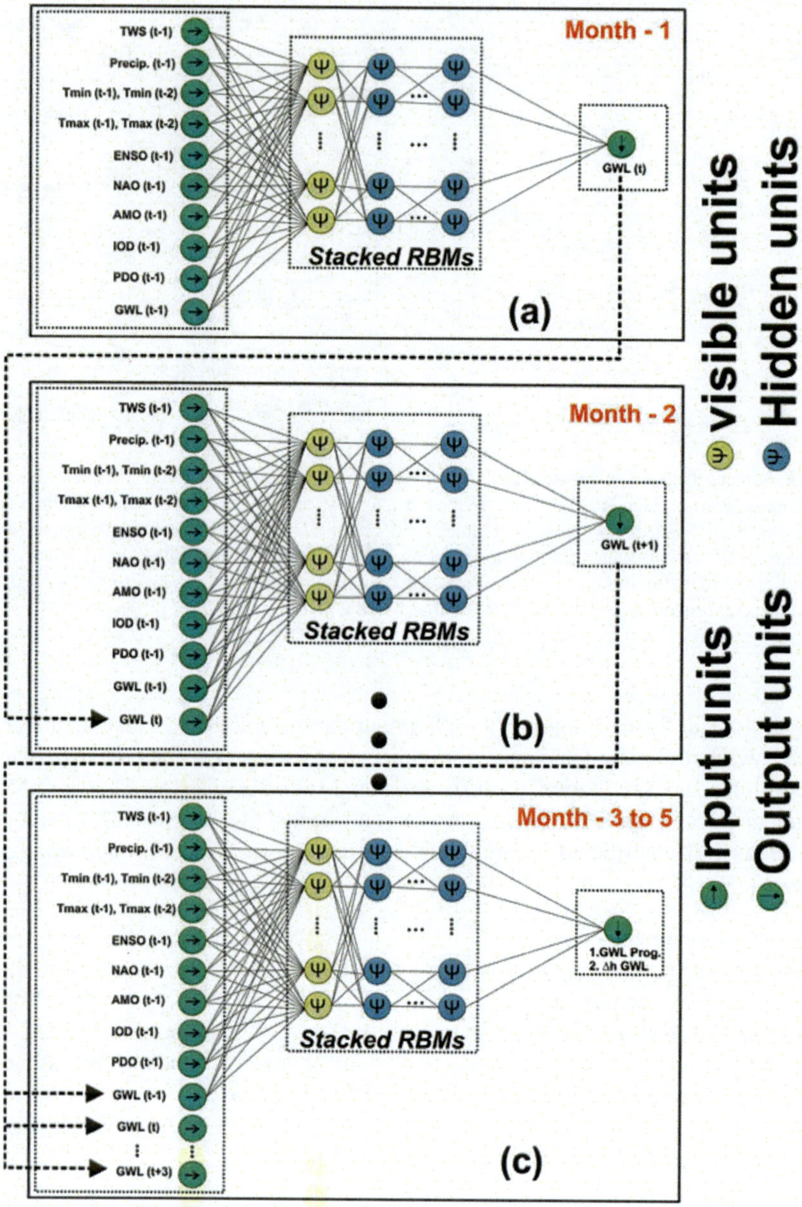

Fig. 8.4 Structure of DBN used for **a** 1 month ahead, **b** 2 months ahead, and **c** 3 to 5 months ahead forecasts. In panel **a**, the input layer consists of 10 neurons, which are antecedent TWS, precipitation, minimum and maximum temperature, climate indices (ENSO, NAO AMO, IOD, PDO), and the in situ ground water level dataset. In panel **b**, the input layer consists of 11 neurons, and in **c**, the input layer contains 15 neurons. The predictand is the groundwater level forecasts and the changes in water level; t-1, t, to t+3 indicated 1, 2 to 5 months ahead. The connection weights and bias help in fine-tuning our predictions

- understanding the interplay between global climate indices and groundwater levels by studying their influences and assimilating their footprints in subsurface water analysis.

In addition to the deep learning framework for groundwater level forecasts, this study employs a deterministic approach through the quantile regression model in addressing uncertainties, which is a key factor often left undiscussed in prediction analysis and forecast studies. Rahmati and et al. (2019) defines the quantile regression model as a powerful tool, which is characterized the ability of uncertainty detection even from different sources. This tool has been widely applied in conditional quantile estimations (Liu et al. 2021). Therefore, its adoption as a post-processing tool for the estimation of uncertainties in the entire groundwater level forecast over the GAB is warranted.

8.2 Exemplifying Deep Learning in the Great Artesian Basin

8.2.1 Hydro-Climatology and Groundwater

The Great Artesian Basin (Fig. 8.1) is the largest and deepest artesian basin in the world, stretching over 1.7 million square kilometers over three major states and four sub-regions in Australia. These states include New South Wales, South Australia, and Queensland while the sub-regions are Western Eromanga, Central Eromanga, Carpentaria, and Surat. The hydrological setting of the GAB makes it a unique region to study, as the aquifer system lies in predominantly arid and semi-arid regions, having a groundwater temperature scale ranging from 30 to 100 degrees Celsius. The process of groundwater recharge in the basin is largely contributed through precipitation, which percolates the uplifted eastern edge of the Great Dividing Range before draining into the western edge, which is regarded as a lower recharge zone, given its sparse precipitation rates. Therefore, the southeast and central parts of the GAB are regarded as the most significant groundwater recharge zones in the basin.

Groundwater discharge across the basin is a combined factor of evapotranspiration, anthropogenic influences, climate change, urbanization, and extraction activities by humans (Kaushik et al. 2021; de Rijke et al. 2016; White et al. 2016). The flow regime and direction of groundwater in the region is in the south western direction, while some of the groundwater which are composed of a gradient ascent through various fractures and faults are responsible for the creation of ecologically important springs (Rossini et al. 2018). These springs are very important sources of water for biodiversity and serve as constant replenishment of the aquatic ecosystem and habitats (Kaushik et al. 2021).

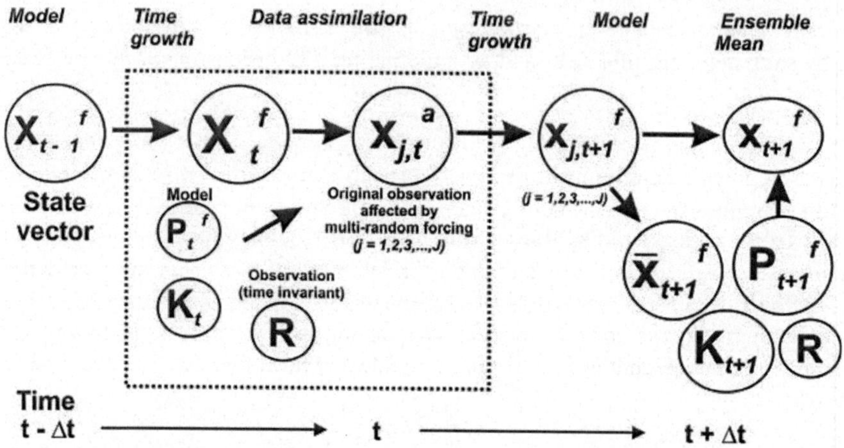

Fig. 8.5 Schematic illustration depicting the EnKF time chart for assimilating two sources of data (in situ groundwater and model groundwater datasets) to estimate possible states of the system in time so as to fill the inherent data gaps from our primary observations

8.2.2 Groundwater Challenges and Water Use

Studies have reported an increased groundwater level over the GAB since 1900 due to the activities of mining companies in the region (Chen and Randall 2013). However, groundwater extraction and stress caused by mining and increased water demand underpinned by climate change and agricultural expansion contributed to varying groundwater levels, and changes to surface discharge and aquifer recharge rates and storage over time. From 2002 to 2017, the GAB region witnessed a significant variation in the distribution of average precipitation and evapotranspiration (ET) (Kaushik et al. 2021). This makes it an interesting region and suitable testbed for demonstrating the deep learning approach in groundwater monitoring and assessment.

8.3 Hydrogeology of the Subsurface

Discussion and research into climate change impacts on subsurface hydrology and surface-subsurface hydrological interactions are increasingly becoming more frequent. The impacts of climate change and or climate variability and anthropogenic influence have the potential to affect the components of the global hydrologic cycle. These components of the global hydrological cycle include surface runoff and stream flow, soil temperature, snow covering and melting of ice and glaciers, precipitation and evapotranspiration patterns, and atmospheric water vapor content. The variations in these components have the potential of leading to changes in the subsurface component of the water cycle, within the vadose zone, the soil, as well as in large aquifers around the world. However, the potential effects of climate change on groundwater and its sustainability are yet to be fully explored.

Groundwater sustainability is defined as the sustainable use of groundwater resources such that it can be conserved for a very long time without leading to negative economic, environmental, and social impacts.[2] The interaction between climate variables and groundwater, as would be studied here using the deep learning architecture, is more complicated than with surface water (Green et al. 2011). This is because groundwater-residence times range from days to thousands of years, which provide adequate time to delay and help buffer the influence of climate change, and making it difficult to detect immediate response in groundwater changes (Chen et al. 2014).

8.3.1 Surface Water, Evapotranspiration, and Precipitation

Human activities are enhancing the Earth's natural greenhouse effects, which have considerable potential of leading to increased global warming through the emission of greenhouse gases. The capacity of the Earth's climate and atmosphere to hold water is directly proportional to temperature, and thus, global precipitation rates and intensity is expected to increase. However, spatial dynamics in projected precipitation patterns denote both positive and negative seasonal patterns. Slight changes in precipitation patterns can affect the flow, availability, and amount of freshwater resources on our planet. Warming trends, which are becoming rampant, also contribute to fluctuations in global ET patterns, and are directly linked to the sustainability of both surface and groundwater resources. Precipitation and ET are the leading hydrological fluxes that directly affect groundwater recharge and indirectly affect human groundwater withdrawal rates and lead to profound impacts. The smallest changes in precipitation patterns can result in notable changes in the freshwater recharge of some semi-arid and arid regions (Ndehedehe 2022). Similarly, perceived increase in precipitation intensity and variability will likely increase the risk of flooding in many regions, as well as trigger the risk of floods in those regions. In periods of above average precipitation intensity, it is likely that the rates of groundwater recharge will increase, and water demand decreases due to the lowering of temperature, solar radiation, and higher humidity during the period. In contrast, the proportion of land surface in extreme drought is expected to rise under future climate change.

[2] https://www.chrome-extension://efaidnbmnnnibpcajpcglclefindmkaj/https://pubs.usgs.gov/circ/circ1186/pdf/circ1186.pdf.

8.3.2 Saturated Groundwater

Groundwater in the saturated zone, which is located beneath the vadose zone or aquifers (comprises porous materials saturated with water), is an essential part of the global water balance. Groundwater aquifers consist of an estimated 30% of the Earth's freshwater resources and 96% of liquid freshwater (Chap. 5). Groundwater provides an avenue for drought mitigation, especially as a result of given the large storage capacity of most aquifers. These aquifers serve as a freshwater reservoir, and their relatively lower sensitivity to climate change compared to surface water resources, make them critical in cushioning the impacts of mitigating drought. Surface water relies on groundwater discharge to maintain base flow conditions. However, the ability to utilize groundwater storage to protect against the impact of rainfall deficits that influence surface water resources is constrained by the need to protect groundwater-dependent ecosystems. Groundwater resources and systems have witnessed significant changes as a result of non-climatic forcings, especially since the 1950s. These include contamination, reduction in recharge, loss of storage due to groundwater mining, and lowering of the water table. Climate-related fluctuations in groundwater systems have been relatively small compared with non-climatic drivers (Kundzewicz et al. 2008). Also, groundwater systems respond slower to climate change, and have more substantial temporal lag as compared to surface water systems. Severe and persistent dry periods change the hydraulic properties of aquifers, such as the transmissivity of a regional karst aquifer in Europe (Larocque et al. 1998).

Factors such as precipitation variability in semi-arid and arid regions drive current vulnerabilities in water resources. The regions which exhibit this phenomenon are vulnerable to the impacts of climate change, especially in the absence of groundwater reservoirs and or limited storage in aquifer systems. Even with the presence of groundwater systems, communities would be vulnerable to climate change impact if the ratio of stored groundwater volumes to recharge is smaller and if there exist no other local water resources. This could mean overtly strong reliance of all biodiversity components, agriculture, industry, and other human water use on groundwater systems. The IPCC previously reported that groundwater levels correlate more strongly with precipitation intensity than with temperature, however, temperature is a significant hydrological flux for existing shallow aquifers (Kundzewicz et al. 2008). The complexity is worsened because predictions of global precipitation patterns are less certain than temperature predictions as a result of climate-induced changes in atmosphere and processes. This has made the study of evidenced-based assessments of climate related impacts on groundwater systems unpopular. However, an important aspect of this study is to apply novel deep learning techniques to highlight footprints of global climate on groundwater systems, and its availability. Details about climate change impacts on groundwater systems, such as recharge, discharge, flow and storage, groundwater quality, and surface–subsurface hydrological interactions, are summarized in the sections below.

8.3.2.1 Recharge

Understanding the processes and interactions influencing groundwater recharge over time is critical to the assessment of subsurface water amount as well as quality. The application of machine learning processes provides a learning platform for the use of climatic variables (e.g., rainfall, evapotranspiration, etc.) to help determine their influence and effects on this freshwater reservoir. Aquifer recharge is mostly considered to be vertical, and assumes a volumetric flux pattern of water across the water table. However, it also involves the inter-aquifer flow from adjacent hydro-geologic formations. Recharge occurs through two generic pathways depending on various environmental impacts. These are divergent recharge to the water table and convergent recharge. The latter occurs at locations where surface-water flow is assembled at land surface, which includes topographic depressions, lakes, stream channels, irrigated agricultural lands, and other flow pathways (See Chap. 5). Therefore, the mechanism of groundwater recharge is a function of climate, hydrological fluxes, topography, local geology and soil, vegetation, surface-water hydrology, land use activities, crop rotation, woodland establishment, urbanization, and irrigation practices. However, our understanding of the knowledge and mechanism of recharge rates is still insufficient. Excess precipitation rates and runoff that are not stored in reservoirs become part of the soil, flow back into the ocean or percolate to groundwater systems. Recharge will be largely influenced by variations in precipitation pattern, and intensity. Groundwater resources have been categorized into four for the purpose of understanding climate change effects on recharge and groundwater resources (Sherif and Singh 1999). These classifications are detailed below.

- Confined aquifers having upper impermeable layers, which accommodate recharge only from precipitation as the water bearing formations outcrop at land surface.
- Unconfined aquifers in wet regions where an indirect proportion exist between rainfall and ET rates, through respectively varying highs and lows. These are highly renewable aquifers because precipitation exceeds ET throughout much of the year and are not expected to witness significant threats to climate change.
- Unconfined aquifers in semi-arid and arid regions that are expected to entail shifting annual balances between precipitation and evapotranspiration rates and a generic drying trend under most climate change forecasts.
- Coastal aquifers extremely susceptible to rising sea level and salt-water intrusion. The world's largest population live in coastal areas and excessive groundwater extraction increases the occurrence of saltwater intrusion.

8.3.2.2 Discharge

Groundwater discharge involves the loss of water from an aquifer into the atmosphere, a surface water body, or abstraction for human utilization. Groundwater depletion is said to have occurred when rates of groundwater discharge are greater than rates of recharge. Over the past decade, groundwater depletion from direct or indirect channels of climate change and anthropogenic influences, including groundwater pumping for irrigated agriculture has increased tremendously, from being a local problem to one that affects several regions around the world. Changes in global groundwater discharge have seen marked contributions to sea-level rise during the past decades. Sea level rise, notwithstanding, could have been significantly higher if substantial quantities of freshwater had not been stored in land-surface reservoirs or conducted into aquifers by irrigation return flows. Several groundwater infrastructures could be significantly influenced by climate change impacts, regardless of the present groundwater pumping rates. For example, the Edward aquifer in Texas, USA (Loáiciga et al. 2000), and the Chalk aquifer in eastern England (Yusoff et al. 2002) exhibited these similarities in their hydro-geological assessments.

The direct or indirect effects of climate change on groundwater discharge include changes in water demand, soil degradation, and changes in irrigation practices. The significant rise in groundwater depletion beginning in the mid-1900s coincides with the rising human population in many regions of the world and the development of advanced well pumps that are utilized in the wide support of agricultural industries and public/private drinking water supplies. Submarine groundwater discharge also referred to as the net groundwater discharge that occurs beneath the ocean is a substantial element of the global hydrologic cycle, which accounts for as much as 12,000 km cube per year, and may otherwise be responsible for freshwater source for human needs (Burnett et al. 2006). Quantifying submarine groundwater discharge and the biogeochemical implications it has on ocean bodies provides critical knowledge toward the understanding of climate change effects on oceanic processes. For instance, high dissolved nitrogen-phosphorus ratios in submarine groundwater discharge flows relative to surface water may result in phosphorus limitation over coastal oceans within the coming decade, which can ultimately lead to change in the present nitrogen limited coastal primary productions (Taniguchi et al. 2008).

8.3.2.3 Groundwater Depletion and Surface Movements

The variations in groundwater storage and agricultural-related groundwater pumping in active semi-arid basins are substantial. However, they are yet to be fully understood as a notable component of water balance. As drought impact escalates, groundwater use escalates as surface water storage becomes increasingly more limited due to severe and prolonged drought events. Prior to development in the industrial age, water storage in most aquifers worldwide was based on local-climate conditions, interactions with surface water, and ecological demands. However, loss of storage worldwide coupled with ecological demands during the second half of the twentieth century became consistent with the development of high capacity well pumps, warming climate, and aquifer development for human use. Even though regions like parts of Russia may have sufficiently dependable groundwater systems and resources under future climate change the rate of global groundwater depletion during the second half of the twentieth century however, increased substantially across several regions of the world (Condon et al. 2021).

Arguably, if groundwater losses continue into the future, the number of people globally that will live in water stressed regions will increase substantially. This problem will be complicated by future global-population growth, which relates to higher groundwater pumping rates that further jeopardize the sustainability of groundwater reservoirs and aquifers at a global scale. Other problems that result from population growth and associated increase in the demand for freshwater resources through excessive groundwater pumping and temporary loss of groundwater storage is substantial land-subsidence for water stressed region. Some of these regions include Mexico, California, the Middle East, and some parts of Asia like Japan where land surface movements associated with climate and human activities like groundwater pumping are common. For instance, in response to land subsidence problems, regulations, and groundwater management in Osaka and Tokyo, Japan has mandated a reduction in groundwater pumping rates aimed to mitigate subsidence problems (Taniguchi et al. 2008). However, other problems have emerged, which include damage to underground systems caused by rising groundwater levels.

8.3.2.4 Flow and Storage

Changes to flow and infiltration rates also influence groundwater storage systems and tend to increase groundwater storage if infiltration rates are higher than normal. In the United States, for example, the water-table experienced a significant decline, and the loss of groundwater storage was recorded in the 1940s, when aquifer development became widespread, until around the early 1980s when rates of water table drawdown reduced (Rosenberg et al. 1999). Other factors that could have led to the worsening of the water table decline includes a total acreage of irrigated agriculture between the early 1980s and mid-1990s, as well as a number of technological and economic factors, but mostly due to the fact that precipitation rates in the High Plains between 1980 and 1999 were well-above normal. This means that the responsiveness of the High Plains aquifer and other similar aquifers indicate that both natural and anthropogenic variations in precipitation, temperature, humidity, albedo, etc. have significant effect on the availability and future viability of groundwater systems. Another important contributor to the above normal precipitation witnessed across the High Plains aquifer region is teleconnection patterns resulting from natural variations in sea-surface temperatures and atmospheric pressures over the nearby oceans (Atlantic and Pacific oceans).

8.3.2.5 Climate Change and Water Extraction Impact on Subsurface Storage

The complex nature of climate dynamics on multiple time scales tend to confuse the role of anthropogenic activities and impacts on groundwater availability associated with climate change. The impact of human activities on aquifers, including groundwater pumping is often on the same time scale as some natural-climate dynamics. This makes it complex to distinguish between the two as the impacts of human water extraction on groundwater during drought could be rather convoluted in some cases. When these natural variations in climate are combined, they can have intense effects on the surface hydrologic cycle largely due to the magnitude of changes in extreme climatic events, such as flooding, heat waves, drought, and low stream flows. As a result, recent studies have attempted to outline the inter-relationship between subsurface hydrologic and geochemical response to climate variability on different time scales such as inter-annual and multi-decadal. This is because variability on these time scales have the most quantifiable implications on water resource management. The effects of climatic forcings on these timescales, which are identified by the impacts of the PDO, AMO, and ENSO, etc., have been recognized as having significant control on recharge and water-table dynamics of the High Plains aquifer and other aquifer systems in different regions of the world.

8.3.2.6 Groundwater Quality

Studies that discussed the influence of climate change and fluctuations on groundwater systems have focused on the processes that affect discharge, recharge, storage dynamics, as well as the associated physical processes that oversee subsurface-water flow. Only few studies, however, focused on the processes that will affect groundwater quality. Groundwater quality is a key function of the physical, chemical, and biological features of the resource. Therefore, groundwater quality can be presumed to respond to changes in climate and linked anthropogenic influences due to the impact of discharge, recharge, and land use on groundwater systems. Groundwater quality is considered as a value-specific concept because the quality of water is correlated to specific water-use standards, and this may vary in different climes. The enhancement and protection of groundwater quality has been a high-priority environmental concern due to direct implications for drinking water and health standards. Also, if groundwater becomes too saline due to rising sea levels, then water quality will be a limiting factor or problem for other groundwater uses, which include irrigated agriculture, industry, or biodiversity needs. Therefore, the sustainability of water supplies under future climate change and dynamics is not only reliant on the quality and quantity of groundwater resources, but also on the physical hydro-geologic features of the aquifers, regulations, laws, and even socioeconomic factors that influence the control, distribution, ultimate demand, and use of groundwater resources.

Other factors that contribute to the decline of groundwater quality are highlighted below.

- Climate change: This can affect groundwater quality by causing a decline of fresh groundwater through reduced recharge or increased pumping. This has the tendency of disrupting the current balance of freshwater/saline water boundary, thereby leading to saline water intrusion in both inland aquifers and saline water intrusion systems.
- Urbanization and population growth: Coastal regions contain approximately one-quarter of the global population, but contain less than 10% of the global renewable water supply and are experiencing rapid population growth.

The IPCC has a high confidence level that sea-level rise (which is majorly a function of climate change and human activities), fluctuations in precipitation and evapotranspiration patterns, (which may affect recharge, and increase groundwater pumping) will lead to more groundwater salinization in many coastal regions. With the compound effect of population growth contributing to the problem, the quality of groundwater systems in the nearest future is not guaranteed unless strategic steps are taken to curtail these occurrences.

8.4 Deep Learning Datasets

8.4.1 Groundwater Level Data

The groundwater level data used in this study were collected from the Australian Bureau of Meteorology over the Great Artesian Basin (GAB). They consist of Northern territory, Queensland, South Australia, and New South Wales. The Australian Bureau of Meteorology is a federal agency of the Australian government responsible for providing and administering weather services over Australia and its environs. It was established in 1906 under the Meteorology Act and brought together the meteorology services that existed before its inception. This Bureau has helped in the improvement of groundwater assessment, monitoring, and information management at a continental scale. The antecedent in situ groundwater formed part of the predictors in our deep learning architecture and framework. The EnkF-DA technique was applied to estimate the monthly lags from the original groundwater level datasets (Fig. 8.5). The predicted changes in groundwater level for each month were integrated into the testing parameters for the next month as summarized in the DBN framework (Fig. 8.4).

8.4.2 Terrestrial Water Storage

Terrestrial water storage (TWS) is the sum of stored water in the soil, underground, over the land surface in lakes and rivers, canopy and even as snow (Chap. 4). For this experiment, the monthly mass concentration (mascon) GRACE (Gravity Recovery and Climate Experiment) product (RL-05) (November 2007–October 2012) available at 0.5-degrees resolution obtained from the Centre for Space Research (CSR) was used. The use of GRACE in global assessments of changes in TWS or land water storage has been fully detailed in Chap. 4 of this book and is only briefly described here. The CSR mascon approach applied in this study uses regularization matrix (Save et al. 2016) to derive the mascon solution. This allows the composition of a time-variable regularization based only on the GRACE data.

8.4.3 Precipitation

The monthly generated Tropical Rainfall Measuring Mission (TRMM) v7 3B43 precipitation estimates (November 2007–October 2012) from the National Aerospace and Space Administration (NASA) Goddard Space Flight Center (GSFC) assimilated with multiple precipitation datasets such as (i) integrated Multi-satellitE Retrievals for Global Precipitation Measurements (IMERG), (ii) Special Sensor Microwave Imager (SSM/I), (iii) TRMM Microwave Imager (TMI), (iv) Visible and Infrared Scanner (VIRS), and (v) rain gauge data were applied in this study. The parameters are obtained by averaging the TRMM 3B42 V6 precipitation estimates. They contain total monthly precipitation from 1998 at a spatial resolution of 0.25 arc-degree. This dataset was downloaded from NOAA GSFC (NOAA, 2022). It was necessary to rescale the TRMM/IMERG rainfall dataset to preserve and maintain a consistent spatial resolution with the other input parameters used in the deep training technique.

8.4.4 Temperature

The temperature data used in this experiment was obtained from the European center for medium-range weather forecasts Re-Analysis (ERA Interim) project. The global ERA interim assimilates the 2 m air temperature using various observational springs, which include data from stations, satellites, and radiosonde at a spatial and temporal resolution of 0.75 arc-degree and 6h, respectively (Dee and Uppala 2011).

8.4.5 Climate Indices

8.4.5.1 Multivariate ENSO Index

The El-Nino Southern Oscillation (ENSO) is a recurring climatic pattern that involves changes in the temperatures over the central and eastern tropical Pacific Ocean. It describes the presence of abnormally warm (El-Nino) and cold (La-Nina) Sea Surface Temperature (SST) anomalies in the eastern pacific (Ndehedehe et al. 2017). The Multivariate ENSO Index (MEI) is the first principal component of the assimilated regularized fields of total cloudiness fractions, meridional and zonal wind components, surface air pressure, as well as the sea level pressure. The units of MEI are standardized, thus a score of 1 denotes a complete standard deviation departure of the principal component for the corresponding period. Given the association of the MEI (derived from http://www.esrl.noaa.gov/psd/enso/mei/) with inter-annual variability of water availability as well as its integration of six other variables and anomalies describing the Pacific, it was preferred in this analysis as a significant contributor to climate change. The El-Nino and La-Nina are parts of the natural cycle collectively called ENSO and are further described in what follows.

1. **El-Nino:** This describes a warming of the ocean surface, or above-average SST (+0.5° Celsius above normal), over the eastern and central tropical Pacific Ocean. It occurs naturally, and on average it occurs every two to seven years, with each episode typically lasting between 9 and 12 months. Over Indonesia, precipitation reduces considerably while strong increases over the central and eastern tropical Pacific Ocean and some areas of Eastern Africa are seen. The low-level surface winds from east to west along the equator (often referred to as easterly winds) weaken instead, and in some cases, start blowing in a reverse direction (i.e., from west to east, often referred to as westerly winds). Generally, the warmer the ocean temperature anomalies, the stronger the El-Nino phenomenon and vice versa.
2. **La-Nina:** A cooling of the ocean surface, or below-average SST, over the central and eastern tropical Pacific Ocean. The Australia Bureau of Meteorology describes its occurrence as *when equatorial trade winds become stronger, changing ocean surface currents and drawing cooler deep water up from below. This results in a cooling of the central and eastern tropical Pacific Ocean.* Over Indonesia, precipitation patterns tend to increase while precipitation patterns decrease over the central and eastern tropical Pacific ocean. The easterly winds along the equator become even stronger. In Australia, La-Nina typically means higher rainfall amount across much of the country among several other impacts. Generally, the cooler the ocean temperature anomalies, the stronger the La-Nina and vice versa.
3. **Neutral:** This is neither El-Nino nor La-Nina. Usually, tropical Pacific SSTs are closer to average. The different types of the MEI capture several oceanic patterns and variability in global climate, and is used as a predictor in our deep learning framework.

8.4.5.2 North Atlantic Oscillation

The North Atlantic Oscillation (NAO) is a weather phenomenon over the North Atlantic Ocean indicating fluctuations in the difference of sea level atmospheric pressure between the Azores High and the Icelandic Low. Even though the strength of the Icelandic Low and the Azores High is characterized by fluctuation patterns, they control the direction and strength of westerly winds and location of storm tracks across the North Atlantic. The NAO was uncovered through several studies in the early twentieth century. However, unlike the ENSO phenomenon in the Pacific, the NAO is largely characterized by atmospheric mode. It is one of the most significant climate signals in the North Atlantic and surrounding humid climates, thus making it important in this deep learning assessments. This is particularly so given that its proximity and influence can be easily identified in freshwater systems over the GAB. The NAO index is composed of a north-south dipole of anomalies. The first is located around Greenland, while the second having an opposite sign span through the central latitudes of the North Atlantic region (i.e., 350N and 400N). These opposing phases of the NAO have been identified as critical drivers for basin-wide moisture dynamics in some regions (Ndehedehe et al. 2019; Sheffield and Wood 2008). This results in the high-level variations of global precipitation and temperature patterns which is regarded as an indirect contributor to groundwater systems, given its impact on sub-regional climate change (Ndehedehe 2019). The NAO dataset for the period of November 2007–October 2012 was downloaded from NOAA data portal (http://www.cpc.ncep.noaa.gov/products/precip/CWlink/pna/nao.shtml), and employed in this analysis

8.4.5.3 Atlantic Multi-decadal Oscillation

The Atlantic Multi-decadal Oscillation (AMO) is a coherent variability pattern in the North Atlantic having duration of 720–960 months and usually computed from the mean SST anomalies of the Northern Atlantic (Ndehedehe et al. 2017; Trenberth and Shea 2006). Recent studies have correlated multi-decadal changes in the Sahel summer rainfall to the activities of the AMO, as well as identify its contribution to global warming (Ndehedehe et al. 2017; Trenberth and Shea 2006). Its indirect influence in regulating hydrological fluxes like evapotranspiration, precipitation patterns, and latent heat transfer makes it a potential indirect driver of groundwater level changes over large freshwater basins and subsurface storage (Ndehedehe et al. 2017). For this study, the unsmoothed AMO index from the Kaplan SST V2 was adopted as part of the deep learning input predictors over the climatic analysis of the GAB. The data can be downloaded from NOAA's repository http://www.esrl.noaa.gov/psd/data/timeseries/AMO/.

8.4.5.4 Indian Ocean Dipole

The Indian Ocean Dipole (IOD) is defined by the difference in SSTs between two poles or areas—an eastern pole in the eastern Indian Ocean south of Indonesia, and a western pole in the Arabian Sea (western Indian Ocean). The IOD influences the climate of Australia and other regions that surround the Indian Ocean Basin. It has also been reported to serve as a significant contributor to precipitation variability over the region (Ashok et al. 2003). This warranted its inclusion as part of the indices used as predictors in the deep learning methods developed in this work. The IOD bears a similarity with ENSO because the variations in temperature gradients across the Indian Ocean often lead to variations in regions of rising and descending moisture and air. The IOD is largely characterized by irregular oscillation of SSTs, which allows the alternation of warm and cold temperatures between the western and eastern part of the Indian Ocean. It is primarily described by the anomalous SST gradient between the southeastern equatorial Indian Ocean and the western equatorial Indian Ocean. The latter is popularly known as the Dipole Mode Index (DMI). This study explored the monthly correlation between this dataset and groundwater level changes for the period of November 2007–October 2012. This data was downloaded from the Japanese Marine-Earth Science and Technology (JAMSTECH) platform.[3]

8.4.5.5 Pacific Decadal Oscillation

The Pacific Decadal Oscillation (PDO) is a recurring ocean-atmosphere climate variability indices at the center of the mid-latitude Pacific ocean. It is identified as either warm or cool surface waters in the Pacific Ocean, north of 200N. The PDO can be likened to a long-lived El-Nino like pattern of Pacific climate dynamics having extremes symbolized by widespread oscillations in the North American climate and the Pacific Basin. Considering its oscillating patterns, which are irregular and characterized by low frequency dynamics, it serves as a potential tool for diagnosing the impacts of future climate on hydrology, especially as it contributes to regulating climate patterns on continental scales (Ndehedehe 2019). The PDO dataset for the period of November 2007–October 2012 was downloaded from NOAA's data portal (https://www.ncdc.noaa.gov/teleconnections/pdo/), and served as an input variable making up the predictor datasets used in our deep learning system.

8.5 Methodology

8.5.1 Deep Learning Framework

This section details the deep learning frameworks, including the building blocks for designing, training (Fig. 8.6), and validating the deep belief method in our groundwater level forecast (Fig. 8.2).

(i) **Input/Predictor datasets:** This involves the hydrological fluxes (Section 8.4) and global climate indices (Section 8.4.5) that served as the independent variable parameters against the groundwater level estimates over the study period and region.

[3] http://www.jamstec.go.jp/frcgc/research/d1/iod/HTML/Dipole%20Mode%20Index.html.

(ii) **Ensemble Kalman filtering for data assimilation:** The Ensemble Kalman filter (Enkf) data assimilation technique provided a platform that allowed the integration of different sources of data to estimate possible states of a system as it evolves in time. It assimilates observation data and the underlying dynamical theories controlling the system to estimate the system state. For this experiment, the observations were the in situ groundwater levels for the thirteen wells, while the model input was the isolated variable of the groundwater estimates from the GRACE-TWS (Fig. 8.5).

(iii) **Data Optimization:** This involves a comprehensive collection of the relevant climate datasets and managing them efficiently to maximize the speed and effectiveness of analyzing, extracting, and utilizing critical information from them.

(iv) **Data Selection:** This is a pre-processing phase entailing the search for the best datasets that contribute mostly to changes in groundwater levels over our study region. Upon downloading these datasets, they underwent the process of data cleaning.

(v) **Data Cleaning:** Data cleaning is the process of fixing or removing incorrect, corrupted, incorrectly formatted, duplicate, or incomplete data within a dataset. When combining multiple data sources, there is a high tendency for data to be mislabeled or duplicated.

(vi) **Pre-processed dataset:** This involves the comprehensive arrangement of the datasets that had undergone the previous steps highlighted above.

(vii) **Statistical analysis:** This entails the learning and validation methods based on statistical and evaluation metrics.

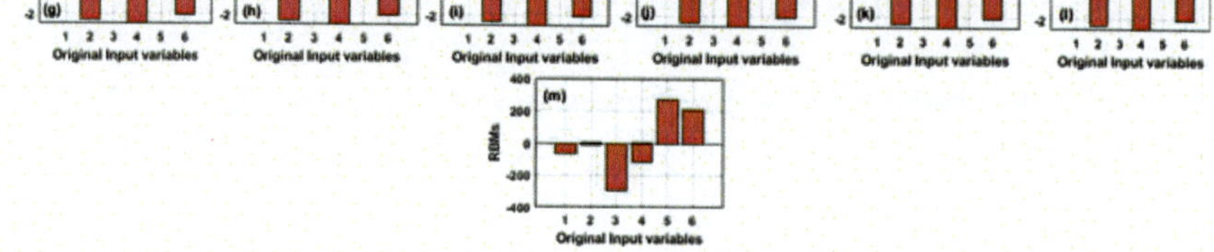

Fig. 8.6 A 9-6-6-1 architecture for the DBM training depicting the dynamics of the RBM from the input to visible layer **a–f**, visible to hidden layer **g–l**, and the output layer **h** for system models of type M1 (No EnkF). The 9 represents the original input parameters, the first 6 represents the visible RBM modes, the second 6 represents the hidden RBM nodes, and 1 is the output which represents the change in groundwater levels. The RMSE of this operation is 0.1442

(viii) **Testing and training samples:** Seventy percent of the entire model datasets were used for training, while the remaining thirty percent were excluded from training. This was warranted to provide satisfactory validation to our deep learning methods and approach, as well as improve the reliability of our estimates to aid water management.

(ix) **Learning phase:** The DBN is composed of two popular techniques which are (a) the layer-by-layer training and (b) the fine tuning training. Further details on these approaches are provided in the next section below.

(x) **Changes in groundwater levels (CGWL) estimates:** This points to the monthly forecasts of groundwater levels in the representative wells made possible by our deep learning methods. The pattern of groundwater fluctuations was studied and learned using the DBN. The original predictor inputs helped in strengthening the robustness of the input model, while providing answers to the influence of TWS and global climate on groundwater level fluctuations.

(xi) **Forecasts at different lead times:** These experiments made forecast for up to 5 months for each representative well. These estimates can be relied upon for making decisions that affect the availability of freshwater to subsurface water users. These estimates can also be relied upon for subsidence monitoring in alternate studies.

(xii) **Performance assessment on the four representative wells:** The use of quantile regression among other statistical methods was applied in the validation of the accuracy, reliability, sensitivity, and specificity of the developed model for accurate groundwater level forecasts.

8.6 Deep Belief Network

The Deep Belief Network (DBN) is an emerging method in the field of neural networks. Recently, it has shown finer accuracy than many prominent and conventional deep learning techniques, especially in image recognition, speech/handwriting recognition, and other classification problems (e.g., Gencoglu et al. 2014). One significant feature of the DBN that distinguishes it from other shallow learning networks is its capability to refine the identification of deep neural patterns, facilitate reasoning abilities, and the possibility of encapsulating the deep difference between normal and faulty datasets. The DBN is a class of deep neural networks that invents training datasets by exploiting the maximum probability in multiple sequentially stacked Restricted Boltzmann Machines (RBM) targeted at training the weights between neurons in its network (Fig. 8.6). The RBM is regarded as the primary unit of the DBN and is composed consists of two layers of neurons (i.e., the visible and hidden layers) used for training input and feature detection tasks, respectively.

Even though a distinction between the adjacent layers of the DBN network exists, there are no associations and links among the entire RBM layers. Proposed by Geoffrey Hinton in 2006, the DBN is popular for deciphering unsupervised learning tasks to minimize feature dimensionality. It can also be utilized in solving supervised learning tasks to hypothesis classification or even regression models. The DBN is composed of two popular techniques which are

- the layer-by-layer training and
- the fine tuning training.

In the former (layer-by-layer training), an unsupervised training technique of each RBM is utilized to learn all the top-down systems and most important generative weights in the network. This is intended to enable the hidden layer inferred from the data vector to serve as that of the succeeding layer. Conversely, the latter implements the error back-propagation technique to fine-tune the DBN parameters after the unsupervised training is completed. For this experiment, the dual learning process of the DBN was assimilated to enhance the accuracy and precision of the forecasts (Fig. 8.2). For the unsupervised pre-training process, no target variable was assumed after the training (Kalu et al. 2022b). In this phase, after training one of the RBMs, the learning parameters are sustained. Simultaneously, the outputs of the hidden layers are utilized as input for the next round of RBM training (Figs. 8.3 and 8.4). Following this procedure, the RBM training experiences a progressive training order till all the RBMs are finally trained. This phase is critical to achieving an optimum solution of the initial weights and bias updates of the network, and the obtained parameters can be easily applied in phase two, which is the supervised fine tuning phase. Phase two helps to improve the model structure and successive supervised learning labeled dataset (Fig. 8.7).

The modeled DBN in this study is characterized by various RBMs stacked on top of one another. The propagative RBMs employed in the deep learning consists of a layer of visible units (v) and a layer of binary hidden units (h). The assimilated configuration of the visible and hidden units produces a total energy function given by

$$\Xi(v,h) = -\Sigma_{i,j}\mathbf{v}_i\mathbf{h}_j W_{i,j} - \Sigma_i \mathbf{v}_i a_i - \Sigma_j \mathbf{h}_j b_j \tag{8.1}$$

where the indices of the visible and hidden layers are represented as i and j, respectively; $W_{i,j}$ is the connection of the weights between the *ith* visible and the *jth* hidden unit. The states of the *ith* and *jth* hidden unit are characterized by v_i and h_j, while a_i and b_j represent the biases of the visible and the hidden layers. The first term $\Sigma_{i,j}\mathbf{v}_i\mathbf{h}_j W_{i,j}$ depicts the energy between the hidden and visible units with their connected weights. The second term $\Sigma_i \mathbf{v}_i a_i$ represents the energy in the visible layer, while the third term $\Sigma_j \mathbf{h}_j b_j$ represents the energy in the hidden layer. The joint probability assigned to the visible and hidden layer by the RBM is given by

$$p(v,h) = \frac{e^{-E(v,h)}}{\mathbf{Z}}, \tag{8.2}$$

where $\mathbf{z}$ given as $\mathbf{z} = \Sigma_{v,h}$ is the partition function, derived by summing the energy of all possible visible and hidden layer configurations (Kalu et al. 2022b). If the probability of a data point represented by a state v is to be determined, the marginal probability is implemented, eradicating the state of the hidden layer or vice versa,

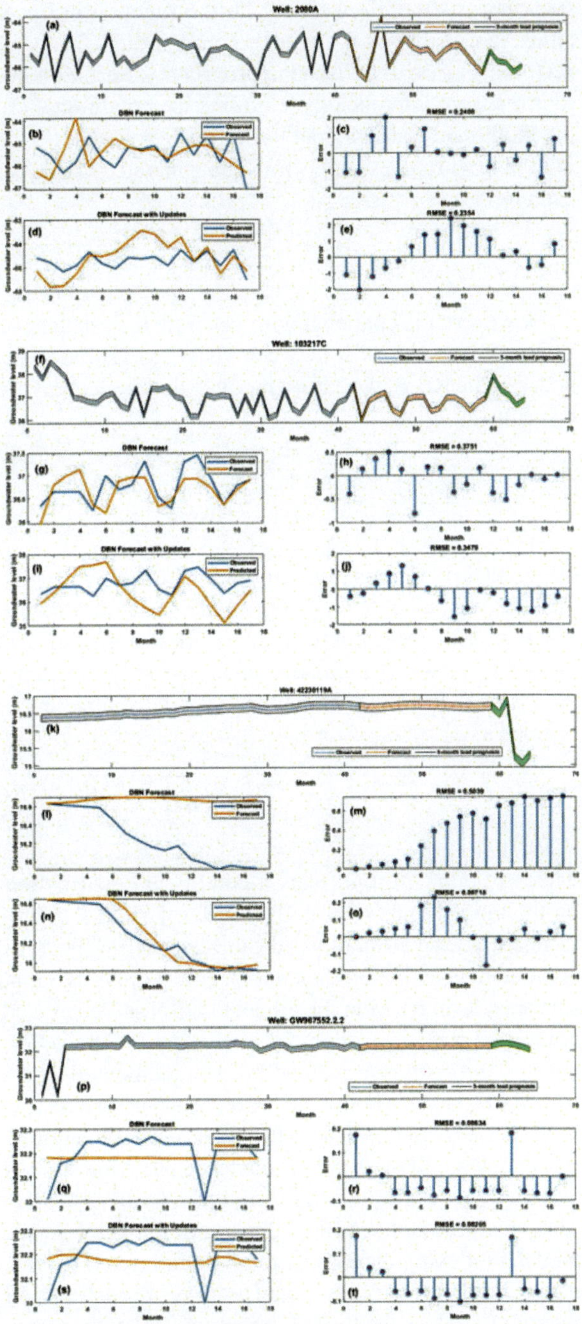

Fig. 8.7 The time series of monthly groundwater level fluctuations over four representative wells: (i) 2080A; (ii) 103217C; (iii) 42230119A; (iv) GW969552.2.2. In the respective panels, **a** depicts the DBN training of the groundwater levels in this station. The gray bound represents 70% of the observed data used for training (42 data points), the red bound represents 30% of the observed data used for testing (18 data points), while the green bound represents the DBN forecasts of change in groundwater level of 2080A for a 5-month time lead, **b, g, l, q** represents a direct DBN forecast of the testing period without the reintegration of precedent groundwater levels for future predictions while **c, h, m, f** represents the corresponding error estimate, **d, i, n, s** represents the DBN forecast of the training period using the precedent action for future prediction as explained in Fig. 8.2. **e, j, o, t** represents the corresponding error estimate. The unit of the y-axis is in meters

such that $p(v) = \Sigma_h p(\mathbf{v.h})$. With the function in Eq. 8.2 above, the probability of either the visible or the hidden configuration for any given input parameter to be activated can be established appropriately. Furthermore, the parameters can be used to investigate inference to determine the conditional probabilities in the model. For the likelihood of the model to be fully exploited, the gradient of the log-likelihood with respect to the weights must be generated. The gradient of the first term can be captured as

$$\frac{\delta \log(\Sigma_h \exp(-E(v, h)))}{\delta W_{i,j}} = \mathbf{v_i}.p(\mathbf{h_j} = 1|\mathbf{v}). \tag{8.3}$$

However, it is challenging to generate the gradient of the second term. Inasmuch as no connections exist between visible to visible or between the hidden-to-hidden nodes in the RBM, we can extrapolate the connection between the visible and hidden layers using,

$$p(\mathbf{h_j} = 1|\mathbf{v}) = \sigma(b_j + \Sigma_i \mathbf{v_i} W_{ji}) \tag{8.4}$$

$$p(\mathbf{v_i} = 1|\mathbf{h}) = \sigma(a_i + \Sigma_i \mathbf{h}_i W_{ji}) \tag{8.5}$$

where $\sigma(.)$ represents the sigmoid function.

Generally, a DBN is composed of an arbitrary number of stacked RBMs serving as a stochastic recurrent neural network on top of each other. This generates an integration of partially-undirected and a partially-directed graphical model. Therefore, the assimilated dissemination between the visible layer $\boldsymbol{v}$ and the l hidden layers h^T is given as

$$p(v, h^l, ..., h^T) = \prod_{k=0}^{l-2} P(h^T|h^{T+1}) P(h^{l-1}, h^l) \tag{8.6}$$

where $p(h^k|h^{k+1})$ represents a conditional distribution for the visible units based on the hidden units of the RBM at level T, and $p(h^{(l-1)}, h^l)$ shows the integrated visible-hidden distribution in the top-level RBM. For the purpose of this experiment, a six-hidden layer (i.e., $h_1(j)$, $h_2(j)$, and $h_3(j)$) DBN model was simulated (Fig. 8.6). The apex level RBM in the modeled DBN served as a complimentary prior from the bottom-level directed sigmoid likelihood function. For the first training phase, the "greedy unsupervised method" was implemented by training separately each RBM from the model in a bottom to top style and utilizing the derived hidden layer from the prior phase as an input layer for the next RBM (Figs. 8.3, 8.4, and 8.6). In addition, the modeled DBN was used to envisage our prior estimates acquired from both in situ and satellite remote sensing techniques to another state space using binary values. This was accomplished by connecting the prior states in the bottom layer of the model, and extrapolating the top hidden layer from them. Each RBM translates its data distribution into a summative posterior distribution over its hidden units. This, therefore, divides the task of modeling its data into two;

- in the first task, it learns generative weights that can decipher the summed posterior distribution over the hidden units back to the data distribution, and

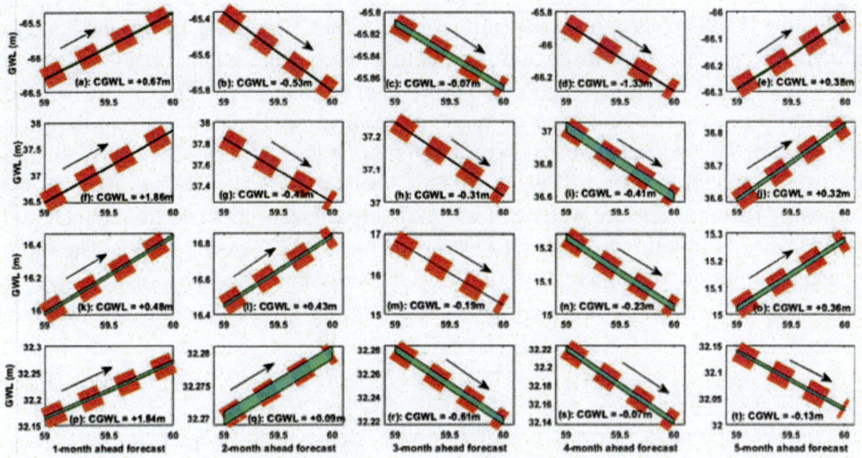

Fig. 8.8 Depiction of the DBN training for change in groundwater levels of the four representative stations from 1 to 5 month lead times. The shaded green line represents the groundwater change index for each month, the block red on the shaded green line represents the time step of the CGWL of each month, while the black arrow represents the CGWL gradient flow for each month

- in the second phase, it learns to model the summed posterior distribution over hidden units.

However, the second task is less complex for the next RBM than it was for the original input datasets because the summed posterior distribution bears a reasonable proximity to a distribution that an RBM can model perfectly. Finally, it was possible to assimilate the top hidden layer parameters directly into the Q-learning model of the DBN for further examination, to aid learning and prediction of CGWL (changes in groundwater level) lead times (Fig. 8.8).

8.6.1 *Data Assimilation*

As a result of the large data gaps in the in situ groundwater data, which is usually common in groundwater wells, the data assimilation (DA) technique was adopted to make the adequate estimation of those gaps. The goal here is to achieve a more comprehensive result with an adequate number of input parameters. DA allows the combination of different sources of data to estimate possible states of a system as it evolves in time. It assimilates observation data and the underlying dynamical theories controlling the system to estimate the system state, which is better than could be generated with just the discrete data or model alone. Also, its use in the optimal estimation of hydrological variables by integrating in situ data or remote sensing data into models has been considerably studied and widely applied over the past years (Liu et al. 2016; Montzka et al. 2011). For this experiment, the ensemble Kalman filtering technique (EnkF) was applied for estimating the data gaps in the in situ data due to its efficiency in merging observation data into a dynamical system, obeying Kalman filter theory and the Monte Carlo method (Fan et al. 2017). The DA model was fed with an optimal in situ groundwater information and model groundwater information was obtained by isolating the groundwater component from the GRACE-TWS using the water budget approach (Fig. 8.5). This was accomplished by subtracting (or removing) discrete surface water and soil moisture datasets obtained from the Global Land Data Assimilation System (GLDAS) from the GRACE-derived TWS. The snow and canopy water component of the TWS was overlooked since they are not relevant (or part of water storage changes) to the study region. This was achieved through

$$GW = TWS_{GRACE} - SM_{NOAH} \tag{8.7}$$

where GW represents the obtained groundwater data and SM depicts the soil moisture information obtained from the NOAH-LSM model from NASA's global land data assimilation system. The surface water component (SW), which is an integral component of the TWS was masked from the GLDAS data. In the study area used to demonstrate the DBN, SW change is typically limited and negligible given that much of the region is semi-arid and dry. From the GRACE-hydrological signal perspective, reduced SW change or its absence implies gravity fluctuations associated with surface water component as observed by GRACE is limited-that is, in relation to other hydrological stores like soil moisture and groundwater (Ndehedehe 2022; Kaushik et al. 2021).

The NOAH-LSM used in this experiment was retrieved from the GLDAS version 2.1 available from NASA's earth data site (http://earthdata.nasa.gov). Note, the difference between TWS_{GRACE} and TWS_{NOAH} is used to quantify GRACE-informed groundwater anomalies. Also considering the error covariance and the bias based on the inverse of squared observation and model error covariance, we introduce the iteration procedure of our DA algorithm. Regardless of the large time consumption of our DA model, we employed the sequential updating of the Kalman gain with an ensemble mean of the observation data and model error covariance to achieve a finely determined predictive model. In the conventional EnKF scheme, the model error covariance is quantified from the ensemble forecasts in the routine of time growth of the model state. The model error covariance estimate accounts for the most significant source of intensification in the computational time elapsed for completing a DA cycle. To obtain more accurate model error covariance estimates, larger ensemble sizes are enforced on the system state to achieve more computation time.

The time growth in the system at time t was accomplished through the integrated results from the model, which was however slightly affected by several random errors (Table 8.1). Through an ensemble of the predicted results, the mean model state and model error covariance at the next time steps are estimated for 60-month in situ data employed in this experiment using various architectural frames (Fig. 8.10), and trend analysis (Fig. 8.9). The observation equation at time t is represented as

$$y_t = \mathbb{E} + n_t \tag{8.8}$$

Table 8.1 Geophysical features of the monitoring wells used in our experiment depicting the EnKF approach utilized in assimilating in situ and groundwater model datasets for missing months for each well. Each well data was forced to generate groundwater level data to complete a range from 1 November 2007–31 October 2012 to serve as part of our antecedent deep learning input parameters. The four well cells in bold are the representative wells used to test our model. N/B: In the model update column, M1 represents "no assimilation" while M2 represents "assimilation" in the model

Well ID	Lat (deg)	Lon (deg)	Average surface elevation	In situ to model input ratio	% EnKF DA	% Error covariance	Model update
87A	–22.2648	141.2889	65.27	7:3	18.33%	0.24%	M2
2080A	**–26.9224**	**144.5376**	**–64.33**	**1:9**	**95.00%**	**0.31%**	**M2**
11346A	–24.5319	145.3664	–14.28	1:9	95.00%	0.28%	M2
14333A	–19.376	141.2923	–12.01	1:9	95.00%	0.25%	M2
15587A	–19.4188	140.7853	–7.81	1:9	95.00%	0.28%	M2
93079A	–21.5967	141.016	26.59	2:8	86.67%	0.15%	M2
103217C	**–23.6024**	**146.1294**	**36.55**	**3:7**	**80.00%**	**0.18%**	**M2**
41620017A	–28.6714	149.2163	4.34	7:3	16.67%	0.19%	M2
42230119A	**–27.4227**	**151.2176**	**16.24**	**10:0**	**0.00%**	**0.00%**	**M1**
GW021266.3.4	–30.1636	149.6003	45.28	10:0	0.00%	0.00%	M1
GW036909.1.1	–29.8374	148.2732	15.04	10:0	0.00%	0.00%	M1
GW036958.1.1	–28.6917	149.4064	11.39	10:0	0.00%	0.00%	M1
GW967552.2.2	**–29.4365**	**150.1336**	**32.18**	**6:4**	**35.28%**	**0.11%**	**M2**

where the model state at the time, t is depicted by x_t ; y_t represents the system state and observation data at time t, E is the time-invariant transform matrix, while n represents the observation error at time t. The system state is updated at every time following the execution of DA to

$$x_t^a = x_t^f + K_t(y_t - \mathbb{E}x_t^f) \tag{8.9}$$

where the assimilated model state at every time, t is represented by x_t^a, the forecast model state is depicted by x_t^f, while K_t is the Kalman gain, the innovation vector is represented by $(y_t - \mathbb{E}x_t^f)$. It is important to adopt the Kalman gain computational efficiency as it attempts to weigh and balance the uncertainties between the estimate and observation before proceeding to the analysis step, as well as curtails the covariance threshold $(x_t^a - x_t^f)(x_t^a - x_t^f)^T$.

$$K_t = P_t^f \mathbb{E}^T (\mathbb{E}P_t^f \mathbb{E}^T + R)^{-1} \tag{8.10}$$

where P_t^f depicts the model error covariance at time t and R depicts the observation error covariance with time-invariant parameters. The time growth of the model state is achieved using several initial conditions of J pieces using an ocean circulation

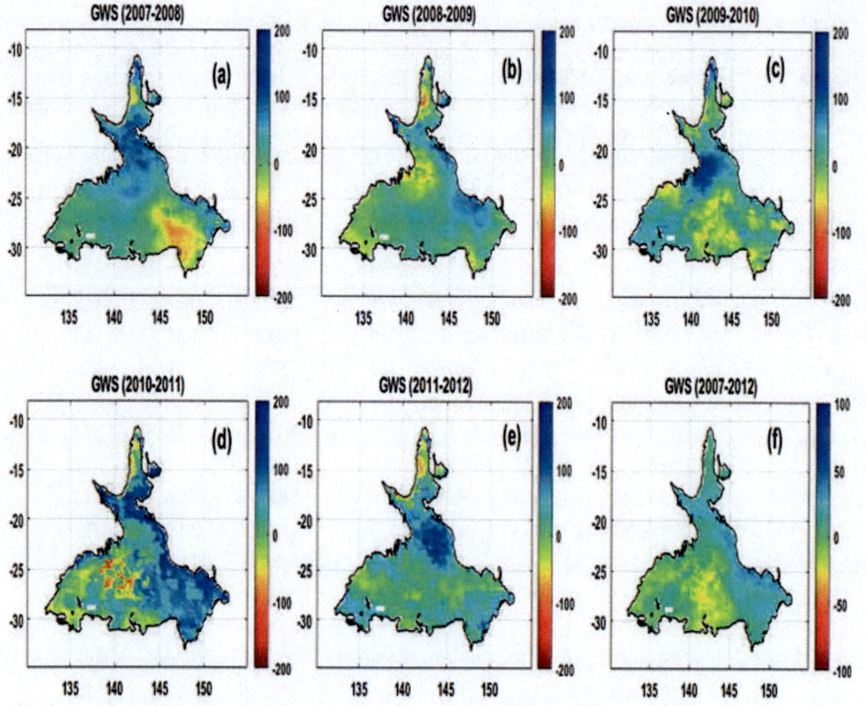

Fig. 8.9 Spatial distribution in GWS extracted from the GRACE dataset using the GLDAS model as depicted in equation (7) over the GAB for **a** 2007–2008, **b** 2008–2009, **c** 2009–2010, **d** 2010–2011, **e** 2011–2012, and **f** 2007–2012 (long term GWS representation)

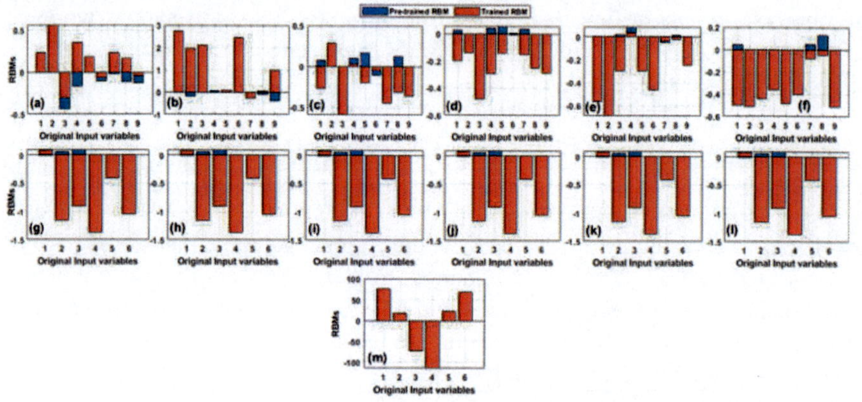

Fig. 8.10 A 9-6-6-1 architecture for the initial DBM training (Table 8.2) depicting the dynamics of the RBM from the input to visible layer **a**–**f**, visible to hidden layer **g**–**l** and the output layer **h** for system models of type M2 (EnkF). The 9 represents the original input parameters, the first 6 represents the visible RBM modes, the second 6 represents the hidden RBM nodes, and 1 is the output which represents the change in groundwater levels. The RMSE of this operation is 0.1874

Table 8.2 An experimental framework for the DBN modeling procedure for CGWL prognosis at each time step (i.e., month 1–month 5)

	Cases	Input	Output	Nodes	Model type	Batch size	Verbose	Step ratio	Drop-out rate	Object	Predict	RMSE
t	Case 1	X1a	Y1	9-6-6-1	BBDBN	15	True	0.1	0.5	Squares	F1	0.117
	Case 2	X1b	Y1	10-6-6-1	BBDBN	15	True	0.1	0.5	Squares	F1	0.148
	Case 3	X1c	Y1	11-6-6-1	BBDBN	15	True	0.1	0.5	Cross entropy	F1	0.223
	Case 4	X1d	Y1	12-6-6-1	BBDBN	15	True	0.1	0.5	Cross entropy	F1	0.210
	Case 5	X1e	Y1	13-6-6-1	GBDBN	15	True	0.1	0.5	Cross entropy	F2	0.183
t+1	Case 1	X2a	Y2	9-6-6-1	BBDBN	15	True	0.1	0.5	Squares	F1	0.127
	Case 2	X2b	Y2	10-6-6-1	BBDBN	15	True	0.1	0.5	Squares	F1	0.135
	Case 3	X2c	Y2	11-6-6-1	BBDBN	15	True	0.1	0.5	Cross entropy	F1	0.108
	Case 4	X2d	Y2	12-6-6-1	BBDBN	15	True	0.1	0.5	Cross entropy	F1	0.250
	Case 5	X2e	Y2	13-6-6-1	GBDBN	15	True	0.1	0.5	Cross entropy	F2	0.112
t+2	Case 1	X3a	Y3	9-6-6-1	BBDBN	15	True	0.1	0.5	Squares	F1	0.027
	Case 2	X3b	Y3	10-6-6-1	BBDBN	15	True	0.1	0.5	Squares	F1	0.032
	Case 3	X3c	Y3	11-6-6-1	BBDBN	15	True	0.1	0.5	Cross entropy	F1	0.055
	Case 4	X3d	Y3	12-6-6-1	BBDBN	15	True	0.1	0.5	Cross entropy	F1	0.054
	Case 5	X3e	Y3	13-6-6-1	GBDBN	15	True	0.1	0.5	Cross entropy	F2	0.158
t+3	Case 1	X4a	Y4	9-6-6-1	BBDBN	15	True	0.1	0.5	Squares	F1	0.056
	Case 2	X4b	Y4	10-6-6-1	BBDBN	15	True	0.1	0.5	Squares	F1	0.193

(continued)

Table 8.2 (continued)

	Cases	Input	Output	Nodes	Model type	Batch size	Verbose	Step ratio	Drop-out rate	Object	Predict	RMSE
	Case 3	X4c	Y4	11-6-6-1	BBDBN	15	True	0.1	0.5	Cross entropy	F1	0.014
	Case 4	X4d	Y4	12-6-6-1	BBDBN	15	true	0.1	0.5	Cross entropy	F1	0.127
	Case 5	X4e	Y4	13-6-6-1	GBDBN	15	True	0.1	0.5	Cross entropy	F2	0.037
t+4	Case 1	X5a	Y5	9-6-6-1	BBDBN	15	True	0.1	0.5	Squares	F1	0.066
	Case 2	X5b	Y5	10-6-6-1	BBDBN	15	True	0.1	0.5	Squares	F1	0.145
	Case 3	X5c	Y5	11-6-6-1	BBDBN	15	True	0.1	0.5	Cross entropy	F1	0.062
	Case 4	X5d	Y5	12-6-6-1	BBDBN	15	True	0.1	0.5	Cross entropy	F1	0.101
	Case 5	X5e	Y5	13-6-6-1	GBDBN	15	True	0.1	0.5	Cross entropy	F2	0.019

Baseline

X1a—TWS(t), Precip.(t), Temp.(t-1, t-2), ENSO(t), NAO(t), AMO(t), IOD(t), PDO(t), GWL(t-1); **X1b**—CGWL(t)[X1a], TWS(t), Precip.(t), Temp.(t-1, t-2), ENSO(t), NAO(t), AMO(t), IOD(t), PDO(t), GWL(t-1); **X1c**–CGWL(t)[X1a], CGWL(t+1)[X1b], TWS(t), Precip.(t), Temp.(t-1, t-2), ENSO(t), NAO(t), AMO(t), IOD(t), PDO(t), GWL(t-1); **X1d**–CGWL(t)[X1a], CGWL(t+1)[X1b], CGWL(t+2)[X1c], TWS(t), Precip.(t), Temp.(t-1, t-2), ENSO(t), NAO(t), AMO(t), IOD(t), PDO(t), GWL(t-1); **X1e**–CGWL(t)[X1a], CGWL(t+1)[X1b], CGWL(t+2)[X1c], CGWL(t+3), TWS(t), Precip.(t), Temp.(t-1, t-2), ENSO(t), NAO(t), AMO(t), IOD(t), PDO(t), GWL(t-1)

Y1 (CGWL at t step); **Y2** (CGWL at t+1 step); **Y3** (CGWL at t+2 step); **Y4** (CGWL at t+3 step); **Y5** (CGWL at t+4 step)

F1: No model update at subsequent prediction period, **F2**: Updated predicting model randomly with the predicted CGWL as input in the training process

BBDBN: all input RBMs in this phase are the Bernoulli-Bernoulli RBMs, GBDBN: the input RBM here is the Gaussian-Bernoulli RBM

Training and Testing ensembles

Study period: Monthly data from 1 Nov 2007–31 Oct 2012 (total 60 months), with 70% of the original datasets serving training samples, and the remaining 30% as testing samples (Figs. 8.11, 8.12 and 8.13)

model

$$x^f_{j,t+1} = F(x^a_{j,t}, w_{j,t} + q_t) \; for \; j = 1, 2, , , J \tag{8.11}$$

where $w_{j,t}$ represents the external forcing, generated from a series of unsystematic perturbed open-boundary conditions with zero mean and a given covariance (Gaussian distribution) and q_t represents the model error. The ensemble mean, however, is generated from several forecast model states,

$$\tilde{x}^f_{t+1} = \frac{1}{J}\sum_{j=1}^{J} x^f_{t+1}. \tag{8.12}$$

The standard deviation of the ensemble mean is derived from

$$S^f_{t+1} = \sqrt{\frac{\sum_{j=1}^{J}(x^f_{t+1} - \tilde{x}^f_{t+1})^2}{j-1}}. \tag{8.13}$$

Model error covariance is determined by

$$P^f_{t+1} = S^f_{t+1}(S^f_{t+1})^T, \tag{8.14}$$

where $S^f_{t+1} = [S^f_{t+1,1}, S^f_{t+1,2},...,S^f_{t+1,N}]$ and N represents the model grid. The vectors $S^f_{t+1,k} \in R^N$ are the *kth* column of matrix $S^f_{t+1} \in R^{N\times J}$.

The observation error covariance (R) is generated from the covariance of travel time divergence for each transmission line generated in a high-frequency band. The observation error covariance abates a diagonal matrix, considering that the observed travel time uncertainties are uncorrelated between individual transmission lines.

$$R = \begin{bmatrix} e_1^2 . . .0 \\ . . . . \\ . . . . \\ . . . . \\ 0. . . .e_M^2 \end{bmatrix},$$

The variation of travel time errors for the *ith* ray is given by e_i^2 $(i = 1, 2, M)$. Time growth to the next time $t + 1$ is derived by the model using the integrated results;

$$x^f_{t+1} = F(x^a_t). \tag{8.15}$$

This recursive looping continues until the system model convergence.

8.6.2 *Quantile Regression: An Uncertainty Evaluation Approach*

For this experiment, the predictive uncertainty of the developed DBN models was determined using the quantile regression (QR) algorithm. Koenker and Hallock (2001) defined the QR algorithm as a linear method that can be used to estimate the quantiles of a response variable even in the absence of a prescribed hypothesis. The QR technique is a popular method used in evaluating the strength of model and has been applied in several fields for over five decades (Kasraei et al. 2021). The robustness and efficiency of applying the QR in estimating forecast errors with respect to the forecasted water levels are important features that have been identified in the literature as it provides highly accurate results in the entire quantitative modeling process (Weerts et al. 2011). In this study, the relationship between the observed (y) and predicted (YPredicted) for each quantile is expressed as follows:

$$y = Y_{Predicted} + b \tag{8.16}$$

where the slope and intercept of Eq. (8.16) above is given by and b respectively, and is estimated by minimizing the sum or residuals such that

$$\min \sum_{i=1}^{N} (y_i - (Y_{predicted(i)})) \tag{8.17}$$

where y_i and $Y_{predicted(i)}$ are i-th samples from the dataset and QR function of the $-th$ quantile is which is given by

$$\rho(\varepsilon_i) = \begin{cases} \{(\tau - 1) \times \varepsilon_i; & \varepsilon_i < 0 \\ \tau \times \varepsilon_i; & \varepsilon_i \geq 0. \end{cases} \tag{8.18}$$

The deep learning methods generated different uncertainty values, which were estimated as a set of forecasted quantiles for every desired quantile level (i.e., = 0.90 and 0.95).

8.7 Results and Discussion

8.7.1 *Model Performance Evaluation*

Given that four representative wells were set aside for the validation of 1-5 month-groundwater forecasts against our modeled DBN models, several statistical models were implemented. This includes the mean absolute error (MAE), correlation coefficient (R), Nash-Sutcliffe efficiency coefficient (NSE), and root mean square error

(RMSE) to ascertain the precision and robustness of our model as a relevant tool for hydrological projections. These statistical tools are well known and have been extensively applied in the performance evaluation of several hydrological models (Kalu et al. 2021; Ndehedehe and Ferreira 2020).

8.7.2 Parameter Selection

Prior to commencing training of the original parameters, a preliminary evaluation was undertaken to enable the understanding of their performance based on the correlation analysis of the input variables (Fig. 8.2). This is a prerequisite for the training of deep learning algorithms, which require coarse correlation fit for their original input variables. This is because a good correlation between them signifies that they are comprise more repeated information, and as such may lead to a reduction in the prediction precision of the model, while increasing the computational complexity of the entire process (Kalu et al. 2022a; Liu and Wang 2011). Hydrological variables (TWS, Precipitation, Temperature), climate variability indices (ENSO, NAO, AMO, IOD, PDO), and the in situ groundwater levels at different time lags are used as predictors, while the change in groundwater levels served as predictand (Fig. 8.11). The choice of input variables covers a wide range of factors influencing groundwater fluctuations in the GAB. Some of the factors taken into consideration in this process are

- temperature change,
- decline in TWS,
- rainfall fluctuation,
- global climate indices in regulating hydrological fluxes and patterns in large sub-regional scales.

All these processes are either directly or indirectly associated with groundwater fluctuations in the respective sub-regions and time scales (Fig. 8.9). As such, they will provide relevant information in understanding the dynamics of groundwater levels, thus their utilization in our experiment. Given that the study area is relatively large, the strength of correlation between the Changes in groundwater levels (CGWL) and the selected climate variables tend to be affected by the region's dynamics in the landscape and climatic patterns. Also, high population density in regions like New South Wales and Queensland are equally direct contributors to groundwater stress in those regions, and as well as affect the correlation strength between the CGWL and parameters such as TWS and rainfall over these regions. Generally, the correlation coefficient between the CGWL and hydrological-related variables is positive and significant at most of the well locations. On the other hand, the correlation between CGWL and climate teleconnection indices variables is dominated negatively at most stations with a magnitude below -0.6. This in turn resulted in more negative changes in groundwater fluctuation levels for our representative wells (Fig. 8.8b-d, g-i, m-n, r-t).

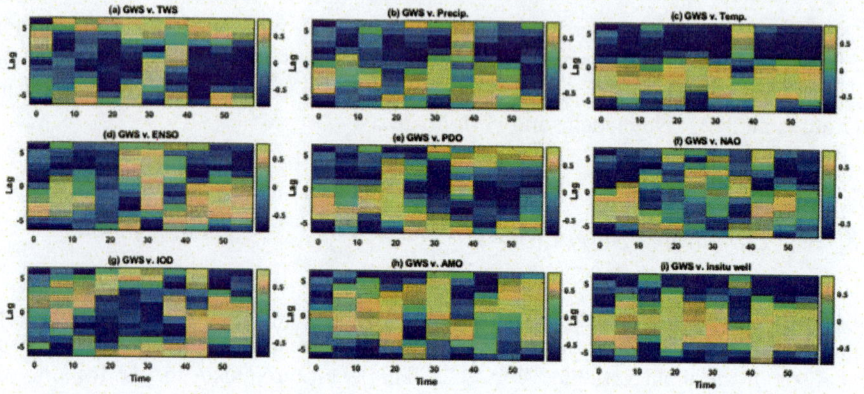

Fig. 8.11 Influence of hydrologic parameters **a–c**, teleconnection patterns **d–h** and in situ groundwater levels (i) on changes in groundwater forecasts using a windowed cross correlation approach

This is a pointer to the indirect influence of teleconnection patterns on groundwater fluctuations. This shows that those precedent hydrological parameters are likely to significantly influence CGWL of the river basin. From the foregoing analysis, TWS, rainfall, temperature, climate indices, and in situ groundwater well information are selected as potential input variables in our DBN network.

8.7.3 Implementation of the DBN Model

In this experiment, the DBN was implemented to predict the CGWL at 1 to 5-month time scales. The flow chart of the DBN model is provided in (Fig. 8.2). The prior input parameters are picked from our original data (i.e., rainfall, temperature, TWS, climate teleconnection indices, and in situ groundwater well levels) and are used in the generation of our DBN model. The DBN constructed an initial learning model using a training set of nine input variables, six visible and hidden nodes, and one output data, which represented the CGWL (Figs. 8.6 and 8.10). In general, the DBN consists of two steps;

- the first phase is constructing the RBM network using unsupervised learning and
- the second phase where the RBM network learns the back-propagation algorithm of supervised learning.

The RBM setting of epoch, batch size and momentum for the construction and back-propagation phases are duly represented in Table 8.2. RBMs hold two sets of random parameters, and is here referred to as neurons; one layer where the inputs go into which is referred to as the visible nodes to represent observable data, and another layer of hidden variables to encompass dependencies of visible variables by approximating the probability distribution of the attributes (Figs. 8.6 and 8.10).

However, the task of deciding the number of nodes to use in our training for both the visible and hidden layers is a vital aspect in the regulation of the accuracy index of our model. Few hidden neurons provide faster training but result in scarce feature detectors. In contrast, more hidden neurons result in a longer training period. For instance, if two identical networks having similar errors on the training datasets with dissimilar hidden layers are examined, the one with few hidden units is likely to produce a more accurate prediction on new cases. However, the more complex one, while diminishing the model's ability to generalize on new datasets has an increased chance of over fitting on the training dataset. Also, an increased number of hidden layer parameters allow a generative model to refine its closeness of fit, while a smaller number boosts the smoothness or extrapolation proficiency of the network.

A generative model with one visible and one hidden layer can be assumed to perform linearly and can estimate any function as long as the visible and hidden layers consist of sufficient functions. However, some studies that implemented generative models in their predictive analysis suggest initiating with at least two visible and hidden layers (Flood and Kartam 1994). For this study, the trial-and-error method was implemented to determine the optimal number of visible and hidden layers and their respective discrete nodes, which was based on the experience of our dataset. This strategy was implemented because of the highly problematic dependence of many optimal generative model geometries. Five cases were evaluated to predict the CGWL at subsequent one-to-five-month-time scales using precedent datasets of TWS, precipitation, temperature, global indices, and in situ groundwater data as predictors.

8.7.4 Analysis of the Representative Stations

Four representative wells were chosen to establish the goodness of fit of the DBN model with the in situ groundwater wells (Fig. 8.12); they are represented in red with yellow rings in Fig. 8.1. The first two (2080A and 103217C) had the least variance and the last two stations (42230119A and GW967552.2.2) had the largest variance in the entire range of well datasets. The correlation coefficient for the large-variance wells (42230119A and GW967552.2.2) between CGWLs and hydrological-related variables (TWS, rainfall, and temperature) is positive and relatively weak (correlation coefficient less than 0.5) at current and lagged 1 to 4 months. On the other hand, the correlation is relatively strong for the least variance wells at current and lagged up to five months. However, the correlation coefficient between the other input variables such as the ENSO, NAO, AMO, IOD, PDO, and the corresponding changes in groundwater measurements varied among the selected four stations (Fig. 8.11).

Due to the non-uniformity of groundwater wells, many stations show weak signals at current and lagged for up to five months. The magnitude, however, is usually higher at the lagged 1 to 3 months for groundwater anomalies. In general, the heterogeneity of the CGWL pattern against hydrological fluxes and global climate indices shows the distinct association between CGWL and climatic parameters at different time scales. The goodness of fit between the predicted and observed CGWL at 1-5 months

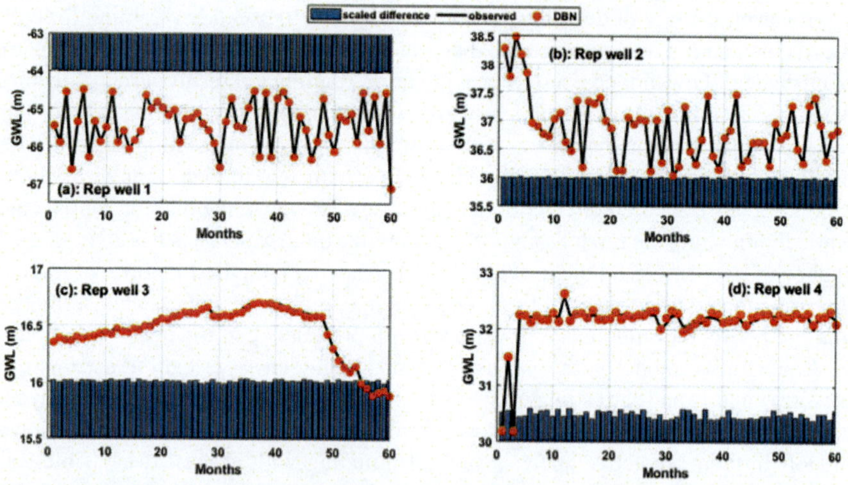

Fig. 8.12 GWL forecasts using the DBN model after applying the fine-tuned weights and bias, **a** first representative well (2080A) **b** second representative well (103217C), **c** third representative well (42230119A), **d** fourth representative well (GW969552.2.2); the difference between the observed and DBN GWL estimates have been scaled to have its origin at the max of the abscissa of the **a** axis, and the min. of the abscissa of the other axis **b–d**, unit of GWL is in meters

are provided in Figs. 8.7, 8.8, and 8.12. For the CGWL forecasts in subsequent one-month, the DBN performs better at stations 103217C and 42230119A than at stations 2080A and GW967552.2.2 (Figs. 8.12 and 8.13) according to the quantile regression validation process. The addition of precedent GWL outputs to Case 2 and Case 5 improves the performance of DBN at most selected stations according to the validation statistical metrics.

8.7.5 *Validation of the Change in Groundwater Level Estimates*

Typically, higher CGWL prediction accuracy means low discrepancies and smoother fluctuations, which is a critical point in the use and planning of groundwater resources. The assessment of forecast validation in this experiment was carried out on four representative wells. Consistent with the goodness of fit index, the time series of the predicted CGWL at the subsequent 1 to 5 months coincide well with the observed CGWL at representative well 1 (2080A) and well 3 (42230119A) using the DBN model (Fig. 8.7a, c). Following the increase in the lead time, the DBN performance for the representative wells slightly degraded between the third- and fourth-month lead times. However, a slight improvement of the DBN performance is noticed in the fifth case as a result of constantly re-injecting the output parameters, especially at

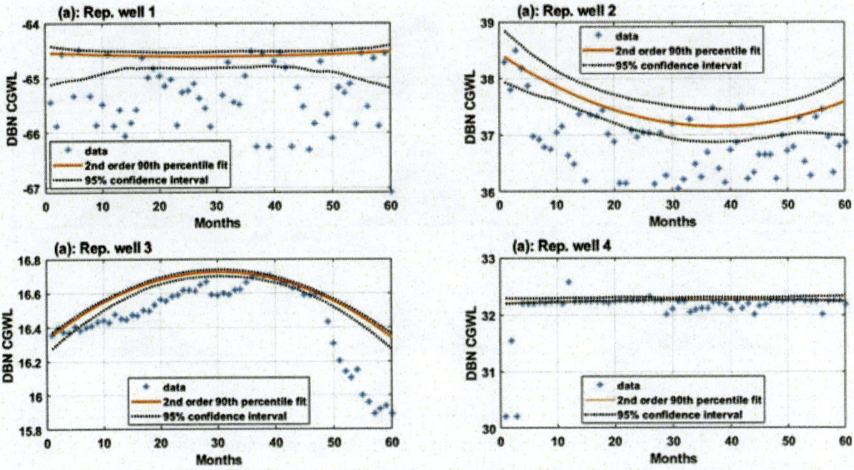

Fig. 8.13 Quantile regression estimates for the DBN predicted CGWL for the four respective wells. Unit of the y-axis which represents the DBN CGWL is in meters

representative wells 2 (103217C) and 3 (42230119A). The impact of the re-injected in situ groundwater data in improving the performance of the predictions (Fig. 8.7d-e, i-j, n-o, and s-t), which showed better RMSE values than the un-updated DBN testing phases, except for representative well 2 (103217C), is obvious (Fig. 8.7b-c, g-h, l-m, and q-r). By comparison, among the four selected representative stations, the DBN performance in Cases 1-4 at stations 1 and 4 is comparatively weaker at representative wells 2 and 3, according to its higher bias and poor goodness of fit metrics. The higher bias at representative wells 1 and 4 can be attributed to the significant abnormality in the CGWL trend, which is hard to be captured even by the DBN using the limited input parameters. Also, the lower correlation between the CGWLs and the input parameters are represented in wells 3 and 4 depicts such poor performance.

Using the quantile regression estimate to determine the level of uncertainties in the DBN predictions at 90% and 95% confidence levels (degree 2), this experiment showed that representative wells 2 (Fig. 8.13b) and 3 (Fig. 8.13c) maintained the least uncertainty in our experiment, whereas wells 1 (Fig. 8.13a) and 4 (Fig. 8.13d) had the highest uncertainly level in the predictions. This shows that the level of non-linearity in groundwater level can also influence the prediction mode of any deep learning model. As analyzed above, the DBN model can adequately predict the CGWL up to a 5-month lead time even with limited input parameters at most of the wells. Besides this, hydrological fluxes and teleconnection patterns among other factors can also affect groundwater level fluctuations, thus the uncertainties in our results. The addition of GWLs as input variables slightly fine-tuned the DBN performance index as represented in the Bayesian regression plot for the histogram of inferred parameters against CGWLs (Fig. 8.14).

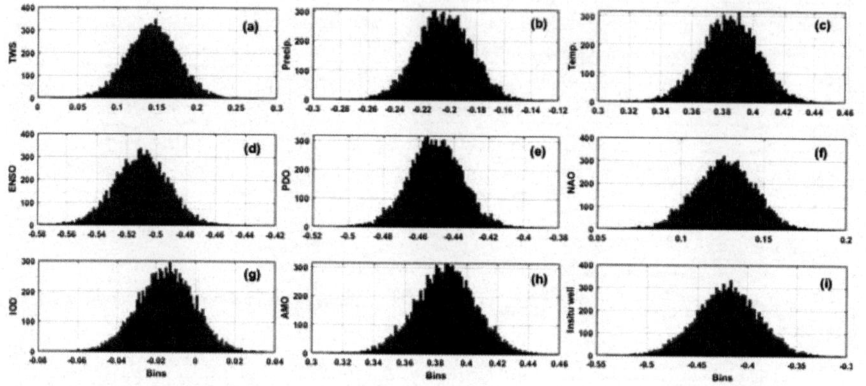

Fig. 8.14 Bayesian regression plot for the histograms of inferred repressors/parameters against changes in groundwater level of the study region, **a** TWS, **b** Precipitation, **c** Temperature, **d** ENSO, **e** PDO, **f** NAO, **g** IOD, **h** AMO, **i** in situ well. The data points are grouped and rendered based on a bin-value of 100 (x-axis) for each of the predictors. The entire range of data values is divided into a series of non-overlapping intervals

8.8 Concluding Remarks

This experiment showcased the application of a DBN as a tool that underpins groundwater forecast (at one to five-month lead times) using the Great Artesian basin as a tentative test bed. The influence of global climate and in situ groundwater withdrawal rate on groundwater level fluctuation was further evaluated as part of an overall assessment of key predictors of groundwater using deep learning methods. The modeling phase of this project covered thirteen (13) independently monitored well locations over the GAB. Results from this study indicated that the DBN technique has capabilities of predicting changes in groundwater levels even up to 5-month lead time with reasonable accuracy. This is particularly so due to the re-injection process adopted, where the prior predicted CGWLs are forced back into the system model as part of the input predictors, thus strengthening the predictor network, and fine-tuning the learning phase of the DBN. The DBN is a robust learning method that utilizes the maximum probability in its adoption of a multiple sequential stacking of RBMs.

In applying the DBN for the construction of a CGWL prediction model, this study employed, layer to layer training procedure, which provided a framework to perform unsupervised training on each RBM to learn all the top-down approach and most important generative weights in the system. This was aimed at liberating the hidden layer parameters inferred from the data vector to serve as the succeeding layer. Additionally, the fine tuning process applies the error back-propagation technique to refine the obtained DBN parameters after the unsupervised training process. This pseudo-back propagation procedure exhibited by the DBN for modeling non-linear systems also accomplishes optimum solution of the initial weights of the network, thereby, resulting in refined outputs and forecasts following the enhancement of the model structure and successive supervised learning label data by supervised fine-tuning. Key outcomes from this study are summarized as follows.

- The inclusion of a robust and comprehensive set of predictor class datasets improved the performance of the DBN in the CGWL predictions. However, the performance and accuracy of the prediction results were noticed to reduce as the lead time increased from one to three months. Given that a re-injection format was adopted, which allowed prior CGWL datasets to become inputs for the next estimate, the fourth and fifth lead time recorded significant accuracy after the training.
- The non-linear relationship existing between global climate indices and groundwater level fluctuations is often undermined, given that there exists no direct relationship between them. However, these experiments show that their extents in driving groundwater levels are exacerbated by the role they play in land use change, which affects withdrawal rate, and compounds anthropogenic influences over the GAB. This experiment observed that the DBN performance is highly susceptible to local climate variables and global climate indices such as ENSO and NAO. They drive climate change, resulting in considerable changes in groundwater level. The modeling framework developed in this study can help forecast local groundwater resources ahead of time to support management strategies that aim to mitigate risk.
- The strength of the forecast model developed in this study was also energized due to contributions from the data assimilation process. It was very essential in balancing the numerous data gaps obtained from the original in situ well datasets. Stations having very low in situ to model ratios had a higher percentage error rate than stations with average and high in situ to model ratios. It is observed that the EnKF enhanced the prediction process and provided sufficient understanding of the considered system estimate. Deep learning frameworks provide a comprehensive and systematic assessment approach for the proper evaluation of non-linear processes in complex groundwater terrains. This study assessed the influence of global climate in subsurface water analysis, which bears a significant non-collinearity, but has a notable effect on groundwater fluctuations. The DBN provided insight to the robustness and capabilities of the deep learning frameworks in making informed decisions to (a) better understand groundwater dynamics and (b) help in guiding water managements to mitigate the impacts of climate extremes through reliable lead time forecasts.

References

Abowarda AS, L. B. et al (2021) Generating surface soil moisture at 30 m spatial resolution using both data fusion and machine learning toward better water resources management at the field scale. Remote Sens Environ 255:112301. https://doi.org/10.1016/j.rse.2021.112301

Ashok K, Guan Z, Yamagata T (2003) Influence of the Indian ocean dipole on the Australian winter rainfall. Geophys Res Lett 30(15). https://doi.org/10.1029/2003GL017926

Ashraf B, A. A. A. A et al (2017) Quantifying anthropogenic stress on groundwater resources. Scientific Report, p 12910. https://doi.org/10.1038/s41598-017-12877-4

Bhattarai N, Pollack A, Lobell DB, Fishman R, Singh B, Dar A, Jain M (2021) The impact of groundwater depletion on agricultural production in india. Environ Res Lett 16(8):085003. https://doi.org/10.1088/1748-9326/ac10de

Bierkens MFP, Wada Y (2019) Non-renewable groundwater use and groundwater depletion: a review. Environ Res Lett 14(6):063002. https://doi.org/10.1088/1748-9326/ab1a5f

Burnett W, Aggarwal P, A. A. et al (2006) Quantifying submarine groundwater discharge in the coastal zone via multiple methods. Sci Total Environ 367(2):498–543. https://doi.org/10.1016/j.scitotenv.2006.05.009

Castle SL, Thomas BF, Reager JT, Rodell M, Swenson SC, Famiglietti JS (2014) Groundwater depletion during drought threatens future water security of the Colorado River basin. Geophys Res Lett 41(16):5904–5911. https://doi.org/10.1002/2014GL061055

Chang J, Wang G, Mao T (2015) Simulation and prediction of suprapermafrost groundwater level variation in response to climate change using a neural network model. J Hydrol 529:1211–1220. https://doi.org/10.1016/j.jhydrol.2015.09.038

Chen C, Randall A (2013) The economic contest between coal seam gas mining and agriculture on prime farmland: it may be closer than we thought. J Econ Soc Policy 15(3):87–118

Chen J, Li J, Zhang Z, Ni S (2014) Long-term groundwater variations in Northwest India from satellite gravity measurements. Global Planet Change 116:130–138. https://doi.org/10.1016/j.gloplacha.2014.02.007

Condon LE, Kollet S, Bierkens MFP et al (2021) Global groundwater modeling and monitoring: Opportunities and challenges. Water Resour Res 57(12):e2020WR029500. https://doi.org/10.1029/2020WR029500

de Rijke K, Munro P, de Lourdes Melo Zurita, M, (2016) The great artesian basin: A contested resource environment of subterranean water and coal seam gas in Australia. Soc Nat Resour 29(6):696–710. https://doi.org/10.1080/08941920.2015.1122133

Dee DP, Uppala SM et al (2011) The era-interim reanalysis: configuration and performance of the data assimilation system. Quart J Royal Meteorol Soc 137(656):553–597. https://doi.org/10.1002/qj.828

Elbeltagi A, Azad N, A. A. et al (2021) Applications of gaussian process regression for predicting blue water footprint: Case study in Ad Daqahliyah. Egypt. Agric Water Manag 255:107052. https://doi.org/10.1016/j.agwat.2021.107052

Fan Y, Huang G, Baetz B, Li Y, Huang K, Chen X, Gao M (2017) Development of integrated approaches for hydrological data assimilation through combination of ensemble kalman filter and particle filter methods. J Hydrol 550:412–426. https://doi.org/10.1016/j.jhydrol.2017.05.010

Flood I, Kartam N (1994) Neural networks in civil engineering. i: Principles and understanding. J Comput Civil Eng 8(2):131–148. https://doi.org/10.1061/(ASCE)0887-3801(1994)8:2(131)

Gencoglu O, Virtanen T, Huttunen H (2014) Recognition of acoustic events using deep neural networks. In: 2014 22nd European signal processing conference (EUSIPCO), pp 506–510

Green TR, Taniguchi M, Kooi H, Gurdak JJ, Allen DM, Hiscock KM, Treidel H, Aureli A (2011) Beneath the surface of global change: Impacts of climate change on groundwater. J Hydrol 405(3):532–560. https://doi.org/10.1016/j.jhydrol.2011.05.002

Hugonnet R, McNabb R, Berthier E et al (2021) Accelerated global glacier mass loss in the early twenty-first century. Nature 592:726–731. https://doi.org/10.1038/s41586-021-03436-z

Kalu I, Ndehedehe CE, Okwuashi O, Eyoh AE (2021) Assessing freshwater changes over Southern and Central Africa (2002–2017). Remote Sens 13(13). https://doi.org/10.3390/rs13132543

Kalu I, Ndehedehe CE, Okwuashi O, Eyoh AE (2022) Estimating the seven transformational parameters between two geodetic datums using the steepest descent algorithm of machine learning. Appl Comput Geosci 14:100086. https://doi.org/10.1016/j.acags.2022.100086

Kalu I, Ndehedehe CE, Okwuashi O, Eyoh AE, Ferreira VG (2022) A new modelling framework to assess changes in groundwater level. J Hydrol: Reg Stud 43:101185. https://doi.org/10.1016/j.ejrh.2022.101185

Kalu I, Ndehedehe CE, Okwuashi O, Eyoh AE, Ferreira VG (2023) Reconstructing terrestrial water storage anomalies using convolution-based support vector machine. J Hydrol: Reg Stud 46:101326. https://doi.org/10.1016/j.ejrh.2023.101326

Kasraei B, Heung B, Saurette DD, Schmidt MG, Bulmer CE, Bethel W (2021) Quantile regression as a generic approach for estimating uncertainty of digital soil maps produced from machine-learning. Environ Model Soft 144:105139. https://doi.org/10.1016/j.envsoft.2021.105139

Kaushik PR, Ndehedehe CE, Burrows RM, Noll MR, Kennard MJ (2021) Assessing changes in terrestrial water storage components over the great artesian basin using satellite observations. Remote Sens 13(21). https://doi.org/10.3390/rs13214458

Keeling CD, Brix H, Gruber N (2004) Seasonal and long-term dynamics of the upper ocean carbon cycle at station ALOHA near Hawaii. Global Biogeochem Cycles 18(4). https://doi.org/10.1029/2004GB002227

Knoll L, Breuer L, Bach M (2019) Large scale prediction of groundwater nitrate concentrations from spatial data using machine learning. Sci Total Environ 668:1317–1327. https://doi.org/10.1016/j.scitotenv.2019.03.045

Koenker R, Hallock KF (2001) Quantile regression. J Econ Perspect 15(4):143–156. https://doi.org/10.1257/jep.15.4.143

Kumar P, Masago Y, Mishra BK, Fukushi K (2018) Evaluating future stress due to combined effect of climate change and rapid urbanization for pasig-marikina river, manila. Groundwater Sustain Develop 6:227–234. https://doi.org/10.1016/j.gsd.2018.01.004

Kundzewicz ZW, Mata LJ, N. W. A et al (2008) The implications of projected climate change for freshwater resources and their management. Hydrol Sci J 53(1):3–10. https://doi.org/10.1623/hysj.53.1.3

Larocque M, Mangin A, Razack M, Banton O (1998) Contribution of correlation and spectral analyses to the regional study of a large karst aquifer (Charente, France). J Hydrol 205(3):217–231. https://doi.org/10.1016/S0022-1694(97)00155-8

Lee S, Hyun Y, Lee S, Lee M.-J (2020) Groundwater potential mapping using remote sensing and gis-based machine learning techniques. Remote Sens 12(7). https://doi.org/10.3390/rs12071200

Lei WZ, Fei WD, Yue SH, Yan X (2020) Comparison of a physical model and phenomenological model to forecast groundwater levels in a rainfall-induced deep-seated landslide. J Hydrol 586:124894. https://doi.org/10.1016/j.jhydrol.2020.124894

Liu D, Mishra AK, Yu Z (2016) Evaluating uncertainties in multi-layer soil moisture estimation with support vector machines and ensemble kalman filtering. J Hydrol 538:243–255. https://doi.org/10.1016/j.jhydrol.2016.04.021

Liu H, Wang J (2011) Integrating independent component analysis and principal component analysis with neural network to predict chinese stock market. Math Prob Eng 382659:15. https://doi.org/10.1155/2011/382659

Liu W, Yu H, Yang L, Yin Z, Zhu M, Wen X (2021) Deep learning-based predictive framework for groundwater level forecast in arid irrigated areas. Water 13(18). https://doi.org/10.3390/w13182558

Loáiciga H, Maidment D, Valdes J (2000) Climate-change impacts in a regional karst aquifer, texas, usa. J Hydrol 227(1):173–194. https://doi.org/10.1016/S0022-1694(99)00179-1

Mohapatra JB, Jha P, Jha MK, Biswal S (2021) Efficacy of machine learning techniques in predicting groundwater fluctuations in agro-ecological zones of India. Sci Total Environ 785:147319. https://doi.org/10.1016/j.scitotenv.2021.147319

Montzka C, Moradkhani H, Weihermüller L, Franssen H-JH, Canty M, Vereecken H (2011) Hydraulic parameter estimation by remotely-sensed top soil moisture observations with the particle filter. J Hydrol 399(3):410–421. https://doi.org/10.1016/j.jhydrol.2011.01.020

Mosavi A, Hosseini FS, Choubin B, Goodarzi M, Dineva AA (2020) Groundwater salinity susceptibility mapping using classifier ensemble and bayesian machine learning models. IEEE Access 8:145564–145576. https://doi.org/10.1109/ACCESS.2020.3014908

Ndehedehe C (2022) Groundwater from space. Springer International Publishing, Cham, pp 211–230 https://doi.org/10.1007/978-3-030-99577-5_9

Ndehedehe CE (2019) The water resources of tropical West Africa: propblems, progress and prospect. Acta Geophys 67(2):621–649. https://doi.org/10.1007/s11600-019-00260-y

Ndehedehe CE, Anyah RO, Alsdorf D, Agutu NO, Ferreira VG (2019) Modelling the impacts of global multi-scale climatic drivers on hydro-climatic extremes (1901–2014) over the Congo basin. Sci Total Environ 651:1569–1587. https://doi.org/10.1016/j.scitotenv.2018.09.203

Ndehedehe CE, Awange J, Kuhn M, Agutu N, Fukuda Y (2017) Climate teleconnections influence on West Africa's terrestrial water storage. Hydrol Proces 31(18):3206–3224. https://doi.org/10.1002/hyp.11237

Ndehedehe CE, Ferreira VG (2020) Assessing land water storage dynamics over Southern America. J Hydrol 580:124339. https://doi.org/10.1016/j.jhydrol.2019.124339

Nourani V, Alami MT, Vousoughi FD (2015) Wavelet-entropy data pre-processing approach for ann-based groundwater level modeling. J Hydrol 524:255–269. https://doi.org/10.1016/j.jhydrol.2015.02.048

Okello C, Tomasello B, Greggio N, Wambiji N, Antonellini M (2015) Impact of population growth and climate change on the freshwater resources of lamu island, kenya. Water 7(3):1264–1290. https://doi.org/10.3390/w7031264

Okwuashi O, Ndehedehe CE (2020) Deep support vector machine for hyperspectral image classification. Pattern Recogn 103:107298. https://doi.org/10.1016/j.patcog.2020.107298

Okwuashi O, Ndehedehe CE, Attai H (2020) Tide modeling using partial least squares regression. Ocean Dyn 70:1089–1101. https://doi.org/10.1007/s10236-020-01385-1

Okwuashi O, Ndehedehe CE, Olayinka DN, Eyoh A, Attai H (2021) Deep support vector machine for PolSAR image classification. Int J Remote Sens 42(17):6498–6536. https://doi.org/10.1080/01431161.2021.1939910

Puri S, Aureli A (2005) Transboundary aquifers: A global program to assess, evaluate, and develop policy. Groundwater 43(5):661–668. https://doi.org/10.1111/j.1745-6584.2005.00100.x

Rahmati O, B. C et al (2019) Predicting uncertainty of machine learning models for modelling nitrate pollution of groundwater using quantile regression and UNEEC methods. Sci Total Environ 688:855–866. https://doi.org/10.1016/j.scitotenv.2019.06.320

Ranjan P, Kazama S, Sawamoto M (2006) Effects of climate change on coastal fresh groundwater resources. Glob Environ Change 16(4):388–399. https://doi.org/10.1016/j.gloenvcha.2006.03.006

Rateb A, Scanlon BR, Pool DR, Sun A et al (2020) Comparison of groundwater storage changes from grace satellites with monitoring and modeling of major u.s. aquifers. Water Resour Res 56(12):e2020WR027556. https://doi.org/10.1029/2020WR027556

Rezaie-balf M, Naganna SR, Ghaemi A, Deka PC (2017) Wavelet coupled MARS and M5 model tree approaches for groundwater level forecasting. J Hydrol 553:356–373. https://doi.org/10.1016/j.jhydrol.2017.08.006

Rosenberg N, Epstein D, Wang D et al (1999) Possible impacts of global warming on the hydrology of the Ogallala Aquifer Region. Clim Change 42(42):677–692. https://doi.org/10.1023/A:1005424003553

Rossini RA, Fensham RJ, Stewart-Koster B, Gotch T, Kennard MJ (2018) Biogeographical patterns of endemic diversity and its conservation in australia's artesian desert springs. Divers Distrib 24(9):1199–1216. https://doi.org/10.1111/ddi.12757

Sahoo S, Russo TA, Elliott J, Foster I (2017) Machine learning algorithms for modeling groundwater level changes in agricultural regions of the u.s. Water Resour Res 53(5):3878–3895. https://doi.org/10.1002/2016WR019933

Save H, Bettadpur S, Tapley BD (2016) High-resolution CSR GRACE RL05 mascons. J Geophys Res: Solid Earth 121(10):7547–7569. https://doi.org/10.1002/2016JB013007

Scanlon BR, Faunt CC, Longuevergne L, Reedy RC et al (2012) Groundwater depletion and sustainability of irrigation in the us high plains and central valley. Proceed Nat Acad Sci 109(24):9320–9325. https://doi.org/10.1073/pnas.1200311109

Sharafati A, Asadollah SBHS, Neshat A (2020) A new artificial intelligence strategy for predicting the groundwater level over the rafsanjan aquifer in Iran. J Hydrol 591:125468. https://doi.org/10.1016/j.jhydrol.2020.125468

Sheffield J, Wood EF (2008) Global trends and variability in soil moisture and drought characteristics, 1950–2000, from observation-driven simulations of the terrestrial hydrologic cycle. J Clim 21(3):432–458. https://doi.org/10.1175/2007JCLI1822.1

Sherif MM, Singh VP (1999) Effect of climate change on sea water intrusion in coastal aquifers. Hydrol Process 13(8):1277–1287. https://doi.org/10.1002/(SICI)1099-1085(19990615)13:8<1277::AID-HYP765>3.0.CO;2-W

Sun AY, Scanlon BR, Zhang Z, Walling D, Bhanja SN, Mukherjee A, Zhong Z (2019) Combining physically based modeling and deep learning for fusing GRACE satellite data: Can we learn from mismatch? Water Resour Res 55(2):1179–1195. https://doi.org/10.1029/2018WR023333

Sun Z, Long D, Yang W, Li X, Pan Y (2020) Reconstruction of GRACE data on changes in total water storage over the global land surface and 60 basins. Water Resour Res 56(4):e2019WR026250. https://doi.org/10.1029/2019WR026250

Taniguchi M, Burnett WC, Ness GD (2008) Integrated research on subsurface environments in Asian urban areas. Sci Total Environ 404(2):377–392. https://doi.org/10.1016/j.scitotenv.2008.04.033

Tapley B, Watkins M, Flechtner F, Reigber C, Bettadpur S, Rodell M, Sasgen I, Famiglietti J, Landerer F, Chambers D, Reager J (2019) Contributions of grace to understanding climate change. Nat Clim Change 9(5):358–369. https://doi.org/10.1038/s41558-019-0456-2

Tellman B, Sullivan J, Kuhn C, Kettner A, Doyle C, Brakenridge G, Erickson T, Slayback D (2021) Satellite imaging reveals increased proportion of population exposed to floods. Nature 596:80–86. https://doi.org/10.1038/s41586-021-03695-w

Thomas B, Famiglietti J (2019) Identifying climate-induced groundwater depletion in GRACE. Sci Rep 9:4124. https://doi.org/10.1038/s41598-019-40155-y

Trenberth KE, Shea DJ (2006) Atlantic hurricanes and natural variability in 2005. Geophys Res Lett 33(12):L12704. https://doi.org/10.1029/2006GL026894

van Dijk AIJM, Beck HE, Crosbie RS et al (2013) The millennium drought in southeast Australia (2001–2009): Natural and human causes and implications for water resources, ecosystems, economy, and society. Water Resour Res 49(2):1040–1057. https://doi.org/10.1002/wrcr.20123

Weerts AH, Winsemius HC, Verkade JS (2011) Estimation of predictive hydrological uncertainty using quantile regression: examples from the national flood forecasting system (england and wales). Hydrol Earth Syst Sci 15(1):255–265

White DC, Lewis MM, Green G, Gotch TB (2016) A generalizable ndvi-based wetland delineation indicator for remote monitoring of groundwater flows in the Australian Great Artesian Basin. Ecol Ind 60:1309–1320. https://doi.org/10.1016/j.ecolind.2015.01.032

Wu C, Zhang X, W. W. et al (2021) Groundwater level modeling framework by combining the wavelet transform with a long short-term memory data-driven model. Sci Total Environ 783:146948. https://doi.org/10.1016/j.scitotenv.2021.146948

Yin W, Li T, Zheng W et al (2020) Improving regional groundwater storage estimates from GRACE and global hydrological models over Tasmania, Australia. Hydrogeol J 28:1809–1825. https://doi.org/10.1007/s10040-020-02157-3

Yoon H, Jun S-C, Hyun Y, Bae G-O, Lee K-K (2011) A comparative study of artificial neural networks and support vector machines for predicting groundwater levels in a coastal aquifer. J Hydrol 396(1):128–138. https://doi.org/10.1016/j.jhydrol.2010.11.002

Yu J, Li F, Wang Y, Lin Y, Peng Z, Cheng K (2020) Spatiotemporal evolution of tropical forest degradation and its impact on ecological sensitivity: A case study in jinghong, xishuangbanna, china. Sci Total Environ 727:138678. https://doi.org/10.1016/j.scitotenv.2020.138678

Yusoff I, Hiscock KM, Conway D (2002) Simulation of the impacts of climate change on groundwater resources in eastern England. Geol Soc London Special Publ 193(1):325–344. https://doi.org/10.1144/GSL.SP.2002.193.01.24

Chapter 9
Integrated Machine Learning in Satellite Hydrology

9.1 Introductory Remark

The application of the Gravity Recovery and Climate Experiment (GRACE, Tapley et al. 2004) satellite data in assessing freshwater status and changes in terrestrial stored water, including drought monitoring, flood potential assessments, global sea level changes, groundwater storage estimates, and ocean mass analysis has increased over the last decade (see, e.g., Chen et al. 2016; Ferreira et al. 2020; Ndehedehe et al. 2016b; Quinn and Ponte 2010; Thomas and Famiglietti 2019). However, the corrections and bias-update processes involved in the processing chain of the datasets introduce errors in the original data, thus leading to a significant limitation or constraint in its application in hydro-climatic studies. The problem of data gaps (or months with missing observations) in the GRACE data is another key challenge that limits its use in climate change studies and data-driven frameworks attempting to decipher the variabilities in terrestrial water storage (TWS) in many regions of the world. These gaps provide an opportunity to reconstruct the TWS time series captured from the GRACE mission or extend it backward to favor climate change studies. To address these limitations, the use of statistical methods and machine learning models to reconstruct GRACE data or extend its records is gradually emerging (see, e.g., Gyawali et al. 2022; Humphrey and Gudmundsson 2019; Jing et al. 2020; Kalu et al. 2023; Sun and Scanlon 2019; Sun et al. 2020; Yang et al. 2018). The litany of recent studies on this subject confirms the importance of accurately filling these data gaps to enhance freshwater-related studies that utilize GRACE-datasets in analysis and predictions of terrestrial water storage. This chapter attempts to bring a balance between various data types from satellite, ground, and model platforms using an integrated data-driven algorithm with a promise of faster convergence rates, finer predictions, and more efficient reconstructive properties for non-linear systems. An important aspect of this chapter is developing an integrated machine learning process to exemplify the reconstruction of TWS that will underpin the use of GRACE data to improve understanding of climate influence on terrestrial hydrology.

C. Ndehedehe, *Hydro-Climatic Extremes in the Anthropocene*, Springer Climate,
https://doi.org/10.1007/978-3-031-37727-3_9

9.1.1 Traditional Machine Learning Techniques

Traditional machine learning algorithms are recently being substituted for deep or integrated learning techniques in predictions and pattern recognitions (Okwuashi and Ndehedehe 2020). The benefits of traditional methods include innovative applications that can enhance the processing of computational tasks. However, regardless of its numerous advantages, the traditional technique is still laden with some challenges, especially with the emergence and application of "big data" in contemporary geospatial analysis. Some of these challenges, are increasingly constraining modern-day studies and they include:

- complexities surrounding error diagnosis and corrections,
- the traditional systems have time constraint in learning, given that their predictive accuracy is largely based on the bigger and more exposed they are to historic datasets,
- it is difficult to prove with complete certainty that the predictions made by traditional machine learning systems is suitable for all scenarios leading to problems with verification, and
- the traditional systems are still prone to hidden and unintentional bias, thereby requiring significant human intervention for their output evaluations, thus limiting their prediction proficiency.

These updated (integrated) algorithms are useful in small and large datasets, as they attempt to explore higher dimensions and uncover hidden patterns even in severely non-liner datasets, which translates to a more reliable and stable prediction. These integrated machine learning procedures that are designed for extensive inspection and learning between severely non-linear datasets of varying quality have undergone substantial refinements over the past decade. They have been applied in diverse disciplines due to their promise of fine-tuned results. For example, their application in bioinformatics, speech recognition, image classification, engineering, and medicine is a significant pointer to their importance across multi-disciplinary fields and in modeling (El Ayadi et al. 2011; Guo et al. 2021; Okwuashi and Ndehedehe 2020).

9.1.2 The Emergence of Integrated Machine Learning Approach

Integrated machine learning approach is gaining attention in several applications in commerce, science, and technology. Machine learning is a set of algorithms that can be trained to mimick reality to improve the forecast of non-linear systems. Given sample of training datasets $(x_1, y_1), (x_n, y_n)$ from $R^d R$, a predictor can be learned, such that $h_n : R^d R$ and is applied in the prediction of the y label of a new data point of x, which are used as testing samples (Fig. 9.1). The selection of the predictor h_n is usually from function class, bearing a significant relationship with the predictand

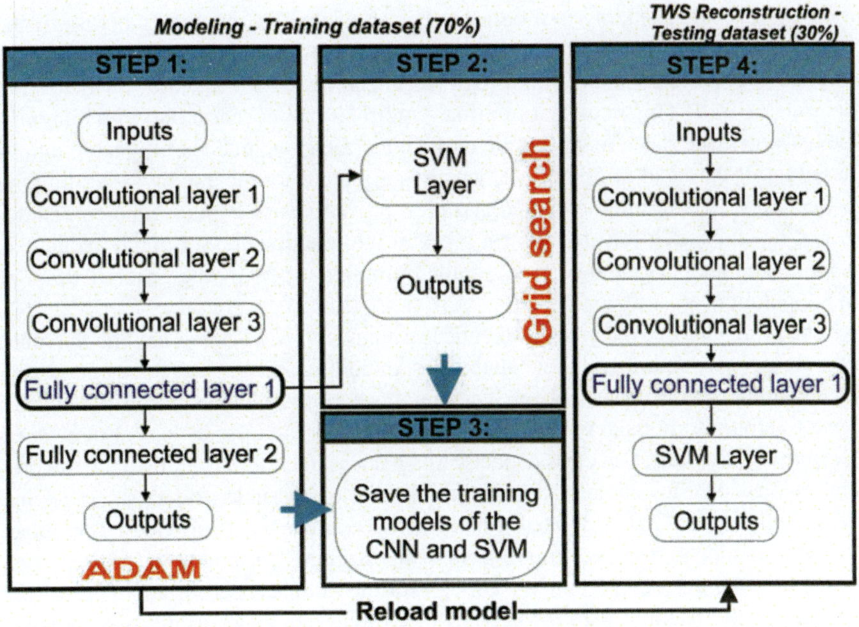

Fig. 9.1 Framework and computational routine of the integrated machine learning process (C-SVM) proposed in this study

(for e.g., in the prediction of terrestrial water storage, rainfall, evapotranspiration, and runoff can serve as predictor datasets). This is chosen within the confinements of the machine learning architecture having the empirical risk minimization (ERM) and its variants. In ERM, the predictor functions play a major role in the minimization of empirical (or training) risk, by attempting to abate the loss function for regression problems, or the $0-1$ loss for classification problems.

The major target of machine learning operators is to determine the predictor datasets (h_n) that provide very good fits for the set of testing datasets, which are not included in the training process. To understand the performance of previously learned algorithms on new datasets (otherwise regarded as generalization), it is assumed that the training samples are randomly sampled from a probability distribution P over $R^d R$ and then the h_n is evaluated on a new test sample (x, y) drawn randomly and independently from P. One of the challenges of traditional machine learning techniques is that there is a potential mismatch between the aim of minimizing the empirical risk (i.e., the explicit target of ERM algorithms/optimization) and minimizing the true risk. This emanates from poor collinearity between the learned algorithms and the independently selected testing datasets (i.e., the aim of machine learning).

In most cases, the peril of machine learning procedures is underfitting or overfitting of the response variables. This suggests the management of the capacity of the function class based on the bias-variance trade off. If the function class is too small, all the predictors may underfit the training data, thus constituting a larger empirical risk, and probably leading to a poor prediction of the test datasets, whereas, if the function class is too large, there is a tendency of the empirical risk minimizer overfitting spurious patterns in the training data, thus leading to poor accuracy on the testing datasets as well. This depicts a minimal empirical risk scenario but with a significantly larger true risk. An advantage of the integrated machine learning technique is based on its classical thinking rule, which is principally concerned with finding the "sweet spot". Using implicit (regularization procedure) or explicit (choice of integrated machine architecture) control types of the function class, suitable balance is achieved on the performance of h_n on the training datasets by the generalization of the population, P.

The proficiency of integrated machine learning approach extends beyond the function class management, as other capabilities, including number of hidden and visible layers, number of neurons, entropy and weights contribute to the machines having a low or zero training risk even for highly non-linear predictors, such as climate data. Despite the high function class capacity and near-perfect fit to training data, the predictors provide considerable accurate predictions on the independent set of testing data (Belkin et al. 2019). Indeed, this integration process has provided a template for best practice in deep learning for choosing machine learning architectures, such that the network is comprehensive enough for the seamless permission of zero-loss training of the training data. Moreover, in a direct encounter to the bias-variance trade-off philosophy, contemporary empirical evidence shows that kernel machines in the integrated networks trained to interpolate the training datasets during learning obtain close-optimal results, regardless of the severe non-linearity of the training data constituted by high level noise and uncertainties (Zhang et al. 2021). Hydro-climatic data, such as terrestrial water storage is a good example of this data, and remains considerably useful in understanding the changes in hydro-climatic patterns over any specific region.

The application of several machine learning regression learners in TWS predictions has metamorphosed from physically-based algorithms to data-driven algorithms. Traditionally, physically-based algorithms rely on data derived from meteorological and satellite observations for the TWS learning process. Therefore, in comparison to data-driven algorithms, they maintain a significant difference in simulation results due to the large amounts of uncertainties from too many input variables and bias terms inherent in the model structures (Yao et al. 2017). Data assimilation models are popularly known for the integration of satellite-based parameters into biophysical or land surface models to enhance system state estimates. However, data assimilation technique is still limited. This restriction is that the model accuracy of the data assimilation systems is severely affected by the accuracy of the satellite-based input parameters (Yao et al. 2017) by exhibiting a strong dominance over the model variables and their corresponding influences. As a result, this leads to impedance of the robustness and efficiency of traditional TWS reconstruction algorithms. These traditional empirical/semi-empirical algorithms integrate ground-based measurements, primary meteorological parameters, and satellite-based datasets in the learning process of TWS patterns. Their modeling procedures include a sound theoretical process which entails the establishment of strong chain of interrelationships between the input (i.e., predictor) and output (i.e., predictand) variables. However, their over-reliance on large number of training datasets for comprehensive system behavior impedes their robustness in modern-day learning operations. Also, their large data requirement is not feasible in data-deficient regions like Africa and some parts of Asia, and this also contributes to the reduction in accuracy and computational efficiency of their predictions (Kalu et al. 2021). An integrated machine learning framework (Fig. 9.1) comes handy to help address some of these constraints, which are detailed in subsequent sections of this chapter.

9.1.3 Satellite-Derived Terrestrial Water Storage Changes

Since March 2002, the GRACE mission has provided first hand estimates of land water storage changes by a global assessment of time-variable components of the Earth's gravity field (Tapley et al. 2004). The observed changes in Earth's gravity fields are often used to infer spatio-temporal variations in terrestrial water storage. Terrestrial water storage (TWS) refers to all the forms of water stored above, on or beneath the Earth surface. It is a basic component of the global and terrestrial hydrologic cycle, thus contributing significantly to the dynamics of the Earth's climate system. Surface water influences freshwater rates and nutrient and sediment transport, and plays a key role in greenhouse gas emission to the atmosphere. In the calculation of current rates of global mean level rise, Pokhrel et al. (2012) defines TWS as a key unknown in the equation, however, having a strong impact on the variation in Earth's rotation, thus affecting parameters such as length of day. Even as an integrated measure of surface and groundwater availability, TWS has substantial applications in water resource management and administration.

Despite the manifold importance of TWS, it was not until recently that its application in global assessment became rampant. This was largely due to the lack of a comprehensive global network for routine TWS monitoring. While observations such as soil moisture contents and surface water levels can be achieved by ground and satellite-based methods (Alsdorf and Lettenmaier 2003; Alsdorf et al. 2003; Njoku et al. 2003), there was no direct observation of TWS until 2002 when the GRACE mission came onboard. After the launch of the GRACE mission in 2002, the dearth of direct observations of large-scale TWS estimates was resolved as it provided a remarkable opportunity of quantifying changes in continental water storage across the globe. Even though the mission's primary objective was to capture changes in the Earth's gravity fields at approximately 30 days interval, it has demonstrated remarkable capability in assessing water mass dynamics on a global scale that underpins the hydrologists' and water resource managers' ability in freshwater reporting (Chen et al. 2016; Kalu et al. 2021; Rodell et al. 2018; Syed et al. 2008).

To date, most GRACE hydrological studies have attempted to compare the derived terrestrial water storage anomalies (i.e., this refers to TWS deviations from the mean rather than monthly estimates) to the limited available estimates from models. Others have attempted to compare the methods of data processing and error analysis, as well as their emerging applications in the assessment of TWS components and flux patterns (Li et al. 2012). The capability of the GRACE mission in a thorough assessment of continental scale variations and changes in monthly water storage is now well documented. However, it is important that more studies address the fundamental problems ranging from the characterization of its space-time changes and its function in global hydro-climatology to the distribution of terrestrial water storage changes among terrestrial subsurface and surface stores and how the fluxes of the water budget equation act to either enhance or dissipate the storages. The limitation in this aspect is obvious, given the missing months in TWS estimates as observed by the GRACE satellite, thus drawing attention to a reconstruction process that can further enhance the study and provide reliable estimates for future hydrological processes and monitoring.

9.1.4 GRACE Data Gaps and Implications

The TWS datasets observed by the GRACE missions (including the GRACE-FO) have proved essential in the comprehensive assessment of land water storage changes for almost two decades. However, the nearly one year gap existing within the GRACE missions (i.e., GRACE and GRACE-FO) breaks the continuity of the observations. This break or gap (Fig. 9.2) affects the study on short-term TWS variations and has the tendency of introducing bias and uncertainties in GRACE-based data analysis as well as in analysis and interpretation of GRACE results. Also, these gaps in GRACE-TWS data (Fig. 9.2) introduce errors in the phase, amplitude of the annual cycle and secular trends. While residual variabilities can also be affected, these observational gaps may increase the level of uncertainty in the spectral modeling of this data. There is a 20 month data gap in the GRACE observations (163 monthly solutions out of 183 potential months from April 2002 to June 2017), two-month gap in the GRACE-FO observations (i.e., August and September, 2018, starting from June 2018 to July 2019), and 11-month gap between the transition period of the GRACE and GRACE-FO missions (i.e., from July 2017 to May 2018) (Fig. 9.2). These long gaps also have the tendency of obscuring the temporal patterns of TWS datasets and consequently distorting the outcomes of any statistical analysis (Fig. 9.2). They affect the implementation of multivariate methods like the principal component analysis to assess dominant patterns in TWS is difficult (Ndehedehe et al. 2016a). Moreover, it is difficult to track the complete directly-observed TWS estimates over the Murray–Darling Basin from Jan, 2003 to Dec, 2012 due to the several missing months existing within this window. Therefore, this raises the question of how to reconstruct the continuous

GRACE - L2	Jan	Feb	Mar	April	May	Jun	July	Aug	Sep	Oct	Nov	Dec
2002				1	2			3	4	5	6	7
2003	8	9	10	11	12		13	14	15	16	17	18
2004	19	20	21	22	23	24	25	26	27	28	29	30
2005	31	32	33	34	35	36	37	38	39	40	41	42
2006	43	44	45	46	47	48	49	50	51	52	53	54
2007	55	56	57	58	59	60	61	62	63	64	65	66
2008	67	68	69	70	71	72	73	74	75	76	77	78
2009	79	80	81	82	83	84	85	86	87	88	89	90
2010	91	92	93	94	95	96	97	98	99	100	101	102
2011		103	104	105	106		107	108	109	110	111	112
2012	113	114	115	116		117	118	119	120		121	122
2013	123	124		125	126	127	128			129	130	131
2014	132		133	134	135	136		137	138	139	140	
2015	141	142	143	144	145		146	147	148			149
2016	150	151	152		153	154	155	156			157	158
2017	159		160	161	162	163						

GRACE Level 2 Products

No GRACE Level 2 Products available (i.e. missing months)

Fig. 9.2 Illustration of missing monthly observations in the Gravity Recovery and Climate Experiment data record (2002–2017)

observations of these GRACE-TWS estimates to aid freshwater assessment and track impacts of climate variability.

9.1.5 Reconstruction of GRACE Satellite Data

Past studies have attempted to reconstruct the GRACE-TWS using variety of datasets and techniques at both grid and basin scales (e.g., Eicker et al. 2014; Humphrey and Gudmundsson 2019; Jing et al. 2020; Yang et al. 2018). The study undertaken by Gyawali et al. (2022) is one of the latest in this category, exploring three deep learning techniques (e.g., generalized linear model and gradient boosting machine) underpinned by other data to fill these gaps through reconstruction. The grid-scale reconstruction of GRACE-derived TWS is possible and reconstructed data capture the intrinsic variabilities amplitudes of TWS caused by climate variability (Fig. 9.3).

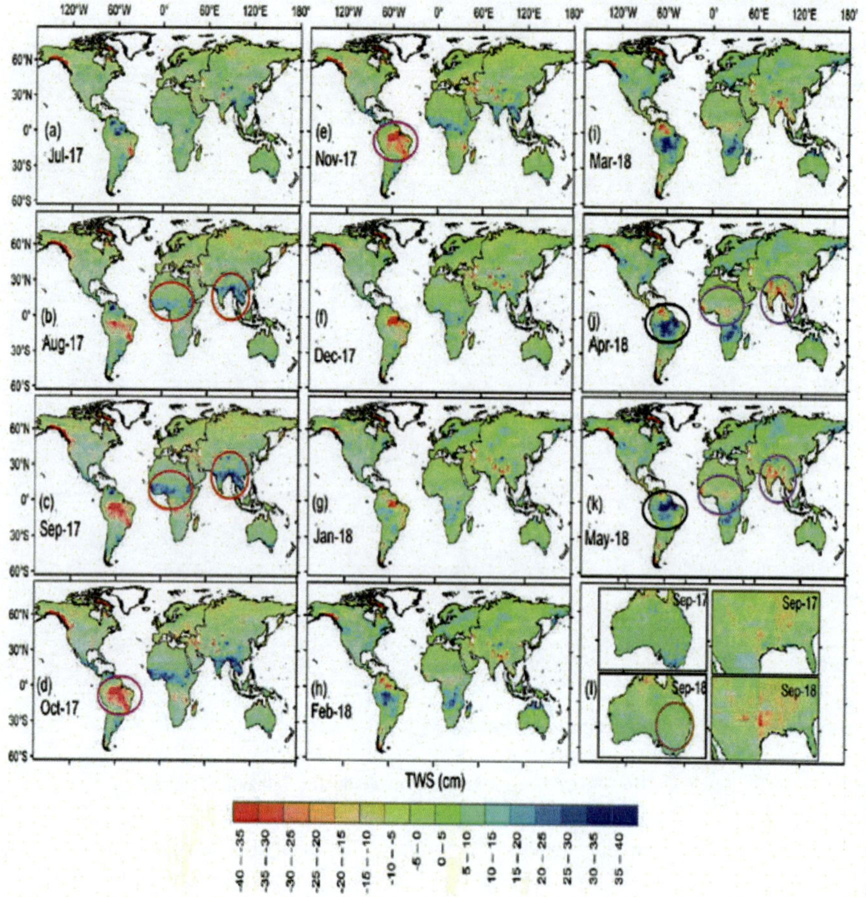

Fig. 9.3 Reconstructed grid-scale GRACE-TWS for the period between the GRACE and GRACE-Follow-On missions. *Source* Gyawali et al. (2022)

For instance, the impact of the 2017 hurricane season is captured in the reconstructed data (Fig. 9.3l) among several other notable signals (Gyawali et al. 2022). The recent work of Kalu et al. (2023) builds on this by introducing a convolution-based variant of the support vector machine technique to fill data gaps in TWS in humid African sub-regions where in situ data is lacking. The current experiment showcased in this chapter is developed after the similitude of their work, and the robustness of their approach can be explored further by varying the independent variables provided to the machine learning framework and adjusting their hyper-parameters to prioritize a better fitting index.

Some of the datasets that have been applied in the GRACE-TWS reconstruction process include but are not limited to measured and modeled hydro-climatic and gravimetric data. The hydro-climatic variables were utilized in the reconstruction process by correlating the existing spatio-temporal variability in these variables with the existing spatio-temporal variability in the GRACE-TWS. Some of these climate variables are precipitation, temperature, sea surface temperatures, and global climate indices, while hydrological variables are runoff, water level, soil moisture, evapotranspiration, etc. (e.g., Becker et al. 2011; Yi and Sneeuw 2021). Gravimetric data from different low Earth orbiting satellite missions have been utilized in reconstructing and filling the missing months of the GRACE-TWS. For example, past literatures have highlighted the application of the geodetic satellite laser ranging data that provides temporal variations of the Earth's gravity field at the lowest degrees of the spherical harmonic spectrum in filling these data gaps in GRACE-TWS (Loomis et al. 2019). Low-resolution gravity models obtained from the European Space Agency (ESA)'s swarm satellite have been widely applied in the recovery of temporal variations of the Earth's gravity field and in the reconstruction of GRACE-TWS datasets (Friis-Christensen et al. 2006). However, a major limitation in the application of these products in reconstructions lie in their coarse spatial resolutions (approximately 1500 km). This spatial resolution results in spatial heterogeneity between the two platforms, thus constraining a smooth reconstruction process of the GRACE-TWS datasets using ESA's swarm platforms.

9.1.6 Some Gap Filling Techniques in Satellite Hydrology

A variety of techniques have also been applied in filling these data gaps in the GRACE-TWS using a wide array of hydro-climatic and hydrological variables. Some of them include, signal decomposition methods, interpolation, water balance equations, land surface models (LSM), data assimilation, and statistical/data mining algorithms (e.g., Becker et al. 2011; Gyawali et al. 2022; Yi and Sneeuw 2021). One of the most popular techniques is the interpolation of the neighboring months, which have been widely used to fill the gaps in the GRACE-TWS estimates (Ndehedehe et al. 2016a; Rodell et al. 2018). However, in mapping severely non-linear systems such as the GRACE-TWS, linear interpolation techniques tend to be less accurate in making estimations, especially in a case where two or more successive monthly values are missing. Also, other higher order interpolants, including the spline are less accurate in mapping extreme values such as those associated with extreme flood and drought episodes, and they usually suffer from overfitting as a result of the usual smoothing effects (Wang 2011).

9.1.7 Singular Spectrum Analysis

The singular spectrum analysis techniques have been getting considerable attention due to their proficiency in estimating these missing months. They seem efficient for extracting substantial information from noisy and short time series datasets. Through their statistical decomposition capability, multivariate data can be decomposed into trends and seasonal signals, and noise even without prior physical knowledge of the time series information. However, some studies have reported distorted reconstruction results in the estimates of the singular spectrum analysis (Yi and Sneeuw 2021). The LSM-derived TWS outputs have also been utilized as proxies for the estimation of missing GRACE-TWS datasets. This method depends on the degree to which these models can simulate natural episodic occurrences, which are largely dependent on the structure and physics of these LSMs, and on the coverage, resolution, and accuracy of these meteorological forcing datasets. Over the African continent, Ahmed et al. (2016) showed that some of the LSMs resulted in the underestimation of summer TWS values and the overestimation of winter TWS values when compared to GRACE-TWS estimates. As with their use in quantifying and assessing land water storage changes, the performance (in terms of uncertainties) of soil moisture products from a range of land surface models in much of Africa suggests a key limitation in the use of some of these products (Agutu et al. 2021). Globally, LSMs underestimate GRACE-TWS trends and are incapable of capturing the seasonal TWS amplitudes. Such models have the potential of simulating GRACE-TWS variabilities emanating from natural phenomena. However, they also less proficient at simulating variabilities resulting from anthropogenic influences because these forcings are classically not captured in the LSM inputs (Oleson et al. 2008).

9.1.8 The Water Balance Approach

The study undertaken by Mueller et al. (2011) attempted to reconstruct GRACE-TWS using basin-scale water balance equations, combining precipitation, evapotranspiration, and runoff. Their results showed consistency between reconstructed product and the original data. In another related study, Pan and et al. (2012) used a similar approach to reconstruct the GRACE-TWS over thirty two global river basins using remote sensing measurements, ground-based observations and LSM simulations. However, the effectiveness of this approach is usually hampered by the uncertainties in precipitation and evapotranspiration estimates, as well as the unavailability of in situ-runoff data in many regions of interest for validation and calibration. Also, this technique does not take advantage of information about the natural variabilites and anthropogenic influences captured by the GRACE-TWS estimates. The use of data assimilation procedures for TWS reconstructions with higher spatio-temporal resolutions has been gaining considerable attention in recent years. For model enhancement, assimilation methods integrate GRACE-TWS estimates with LSMs for improved performance.

9.1.9 Data Assimilation

Data assimilation provides the opportunity to merge different sources of observational data to estimate possible states of a system as it evolves in time (e.g., Kalu et al. 2022). An ensemble-based Kalman filter approach has been used to assimilate GRACE-TWS data into the water global analysis and prognosis (WaterGAP) LSM for the prediction of grid-based TWS over the United States (Eicker et al. 2014). It should be noted that the purpose of assimilating GRACE-TWS into several LSMs is to explore the possibility of improving model's output, especially the TWS component. Another benefit of the assimilation technique is its proficiency in horizontally downscaling coarse resolution and vertically disaggregating GRACE-TWS estimates. However, given the heterogeneity in both spatial and temporal resolutions between the GRACE-TWS and LSMs, as well as the complexities existing due to the error properties of the GRACE-TWS estimates (Fig. 9.4a–d), the assimilation process requires complex algorithms that can be computationally encompassing. Moreover, the work undertaken by Kalu et al. (2022) developed a novel deep belief network to predict the changes in monthly groundwater levels for 27 groundwater wells in sub Sahara Africa. Given the large data gaps in these in situ groundwater data over the region, they used a data assimilation technique to fill those gaps to improve upon the robustness of the results.

9.1.10 Statistical and Data Mining Techniques

The statistical and data mining techniques correlate changes in GRACE-TWS with variations in various hydro-climatic and hydrologic units, including temperature, precipitation, evapotranspiration, vegetation indices, soil moisture, etc. These approaches are often more efficient compared to other methods because they simplify the process of GRACE-TWS reconstructions, either at grid or basin scales. The former generates relevant variables for each pixel (e.g., 0.5 degree by 0.5 degree), while the latter approach averages relevant variables over a specific basin (Humphrey and Gudmundsson 2019). A few of such techniques include, multi-linear regression, support vector machines (Yang et al. 2018), random forest algorithms (Jing et al. 2020), generalized linear model (Gyawali et al. 2022; Yang et al. 2018), deep neural network (Sun et al. 2020), convolution neural network (Sun and Scanlon 2019), artificial neural network (Li et al. 2021), general regression neural networks, seasonal auto regressive integrated moving average with exogenous variables, linear statistical models (e.g., Gyawali et al. 2022; Ndehedehe et al. 2020; Sun et al. 2020), autoregressive exogenous, and non-linear autoregressive with exogenous, automated machine learning, and extreme gradient boosting (Li et al. 2020; Sun et al. 2021). Each of these techniques have their own shortcomings including:

- suitability and effectiveness of some of them only in large-scale basins, exceeding areas of 200,000 km^2,

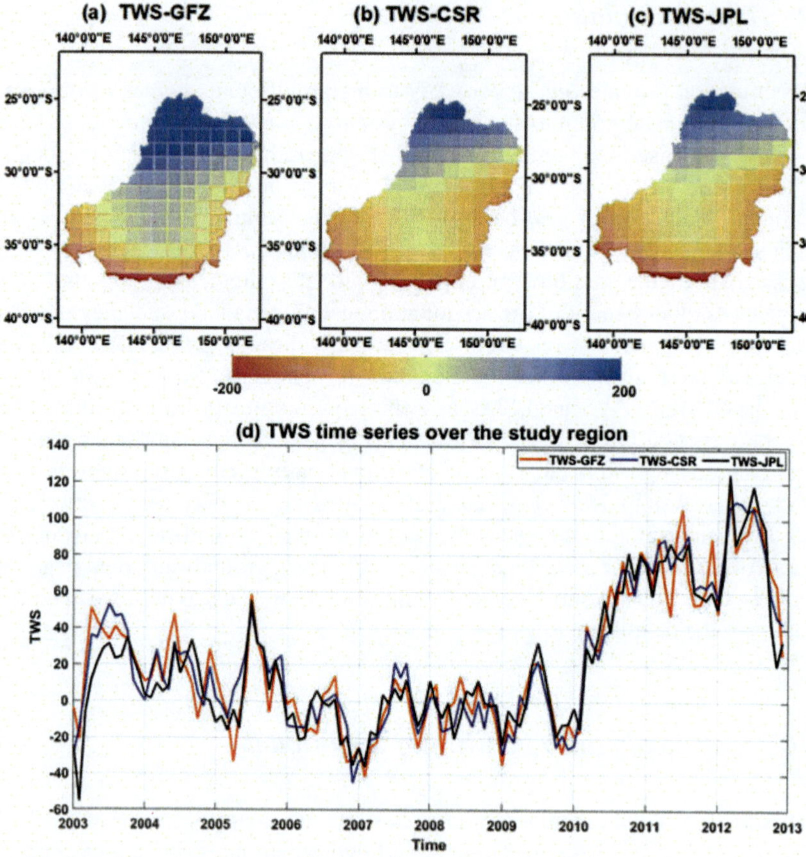

Fig. 9.4 Spatial (**a–c**) and temporal (**d**) depictions of the three variants of GRACE level 2 products (Jan 2003–Dec 2012) that served as our predictands

- requiring knowledge of basin characteristics and hydrogeology, which may not be available for many basins,
- being restricted to specific regions like those experiencing significant ocean-land-atmospheric interactions,
- less effective in reconstructing the trends and changes of GRACE-TWS data, especially over hydrogeologic systems largely influenced by anthropogenic activities (e.g., Sun et al. 2020).

Moreover, most of these approaches either:

- perform comparative tasks of multiple models without identifying the leader model for each grid cell or basin extent even on global scales,

- choose the leading model without providing an all-inclusive comparison of other conventional models' performance either at grid cell or basin extent levels on a global scale, or
- determine the leading model without investigating model inputs that substantially control the variabilities of GRACE-TWS at each grid or basin.

Other limitations of these past methods also include their inability to withstand large gaps in the GRACE-TWS time series, such as the gap between the GRACE and the GRACE-FO missions globally on both grid and basin scales. The integrated machine learning approach proposed in this chapter attempts to address these limitations by offering innovative solutions to overcome the majority of these limitations with the combined proficiency of two well-known machine learning algorithms, i.e., the convolution neural network and the support vector machine.

9.2 The Murray–Darling Basin

The Murray–Darling Basin (MDB) is popularly regarded as Australia's food bowl and is home to various iconic freshwater bodies that are ecologically and culturally important. The recent and most intense Millennium drought over the continent, which lasted from mid-1990s to 2009 was the most severe hydrological drought ever witnessed since the arrival of the European migrants to Australia. The severe climatic impact on the basin during the Millennium drought is a clarion call to the need for the continued assessments of freshwater availability. To cushion impacts of the ongoing declines in water resources, and to protect the region from future impacts of climate change, the Australian government is currently attempting to introduce a novel, comprehensive integrated technique toward intense monitoring and management of the basin's water resources. This makes it a good reason for choosing this basin as a potential test bed for TWS reconstruction processes using the proposed integrated machine learning methods (Fig. 9.4).

The MDB covers a land mass of 1.06×10^6 km^2 which translates to about one-seventh (14%) of the entire Australian continent. The topography of the basin is composed of vast plains, bounded by the east and south by the great dividing range characterized by undulating hills reaching an elevation of up to 2228 m asl (Mount Kosciuszko). Under normal conditions, the Murray river which is about 2530 km long drains into the southern ocean through the lower lakes system and consists of three main tributaries, the Murrumbidgee (app. 1700 km), Darling (app. 2750 km), and the Goulburn Broken (app. 300 km), which under normal conditions supply 12%, 14%, and 17% of the total Murray river inflows, respectively (Leblanc et al. 2012). Available surface water resources in the basin represent only 6% of total precipitation, whereas 42% of surface water are harvested from water courses and farm dams, and are mostly used for irrigation farming. Additionally, the substantial hydrological drought which resulted in unprecedented losses of freshwater resources whose sources cannot be attributed using conventional estimation techniques occurred largely in the lowland river reaches, which represents a combination of evapotranspiration losses and the replenishment of wetlands and floodplains (Leblanc et al. 2012). Under current development conditions, it is projected that over longer seasons, approximately 17% of river flows would reach the Murray mouth. This represents 1% of total precipitation; the remainder is lost by evaporation. More than 900 nationally important wetlands exist in Australia and over 100 of such wetlands[1] are within the basin and this include, the Morella Watercourse/Boobera Lagoon/Pungbougal Lagoon, Warrego River Waterholes, Great Cumbung Swamp, and Lower Goulburn River floodplain among several others. Much of the several wetlands in the basin are located on private parcels but are identified as valuable for breeding and feeding of waterbirds and native fish, thereby sustaining the aquatic ecosystem. The basins' river system also maintain a significant number of local and internationally notable plant and animal species, which includes the 95 species recently listed as threatened (Leblanc et al. 2012).

The five Australian territories that share the vast drainage system of MDB includes, the Australian Capital Territory, South Australia, Victoria, New South Wales, and Queensland. These have contributed significantly to the development of the basin's water resources since the second half of the twentieth century. The MDB has agriculture as the most dominant economic activity covering approximately 80% of the basin's area. Ninety percent of the cotton and almost all the Australian rice are produced in the northern and southern MDB respectively. The basin also serves as a freshwater resort to over 2 million people living in the region, and also contributes to the water supply of Adelaide. The task of maintaining healthy wetlands and rivers has long been a challenge in the basin due to the complicated history of water regulations, salinisation, increasing demands from irrigation areas, and the challenges of water allocation among the five states, all governed by different polices and legislations. This warrants the need for a more robust method of assessing the basin, which suggests the inclusion of emerging techniques that leverage data from remote sensing platforms. TWS is a crucial component of the water cycle, and a well-known indicator to assess flooding, drought, groundwater depletion, surface water shortages, and other natural hazards which the MDB could be vulnerable to. Therefore, with the advent of GRACE, monitoring TWS variations over the basin can underpin efforts in assessing both hydro-climatic and anthropogenic influence on freshwater availability in the region.

Given that government and water experts agree that urgent actions are required to save the water system of the MDB for its cultural, ecological, and economic values, a comprehensive assessment of the basing using satellite methods is therefore is important. However, the data gap in the GRACE-TWS data record affects continuity of observations, as well as introduce uncertainties and bias to data analysis. Therefore, the task of filling data gaps in missing GRACE observations (monthly data) over the

[1] https://www.mdba.gov.au/importance-murray-darling-basin/environment/significant-sites.

MDB is necessary for maintaining data continuity and facilitating data analysis to improve understanding of changes in long-term TWS.

9.3 Data and Methodology

9.3.1 Independent Parameters

Independent parameters in machine learning are variables through which the value or effect of an output (dependent parameter) variable is explained. This happens through a chain relationship (cause-and-effect) between the independent (cause) and dependent (effect) variables. Mathematically, this assumes the general formula, $y = f(x)$, where x and y are the independent and dependent variables, respectively. This means that changes in the magnitude and direction of x will affect the value of y. However, it is important to note that this is just an association and not causality. Let us assume that we need to predict flood occurrence in a particular land parcel or location. It is important to analyze the various potential parameters that influence flooding and reduce runoff in the parcel. Some possible variables or parameters which may influence flooding and reduce runoff are:

- soil type
- precipitation rate,
- available drainage systems,
- presence of vegetation.

In Fig. 9.5, we can assume that if the soil is porous enough, with low precipitation rate in the area, good drainage systems and presence of vegetation in the area, it is highly unlikely that the parcel will be flooded. On the contrary, the parcel will be flooded if the independent variables are opposite of what they are now. It is clear that the inter-relationship among these set of parameters, such as soil type, precipitation rate, available drainage systems, and presence of vegetation can have a determining effect on flooding in the parcel. Therefore, the variables that aid in determining the outcome (Fig. 9.5) are independent, while the outcome, (which in this case is flooding) is a dependent variable.

9.3.1.1 Hydrological Variables

These constitute part of the independent variables that contribute to the fluxes in the hydrological cycle between the oceans, the atmosphere, the biosphere, surface water bodies (i.e., rivers, streams and lakes) and groundwater in subsurface aquifers. They are well- known drivers of TWS as evidenced in the water budget equation (Ferreira et al. 2020). We selected the following products as variables representing hydrological fluxes over our test bed:

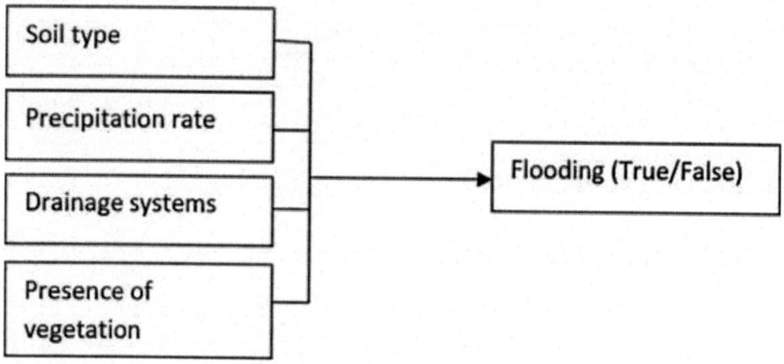

Fig. 9.5 Illustration of some independent or potential parameters that can influence flood occurrence

- TRMM-Precipitation,
- Climate water deficit,
- Evapotranspiration,
- Runoff,
- Wind-speed at 10 m,
- Albedo.

The dataset in item 1 is a monthly TRMMv7 3B43 satellite-derived rainfall parametric product with a temporal period of Jan, 2003 to Dec, 2012. It was obtained from the National Aerospace and Space Administration (NASA) Goddard Space Flight Center (GSFC) merged with the integrated Multi-satellitE Retrievals for Global Precipitation Measurements (IMERG) through the link, https://disc.gsfc.nasa.gov/datasets/TRMM_3B43_7/summary. Items 2–6 in the above list were extracted from the Global Land Data Assimilation Model (GLDAS). This model generates and provides high quality fields of land surface states and fluxes by integrating satellite- and ground-based observational data products, with the use of advanced land surface modeling and data assimilation techniques. A link summarizing these datasets can be accessed from NASA repository.[2]

9.3.1.2 Climate Indices and Sea Surface Temperatures

A climate index can be defined as the estimated values that can be used to describe the state and variations of the climate system. The climate at a defined region is regarded as the averaged weather pattern of a region over a considerably long period usually thirty years or could be less. Climate change is much slower than weather change, which can vary substantially with days. The first classical climate teleconnection indices of the atmosphere have been defined already about a hundred years ago. For

[2] https://disc.gsfc.nasa.gov/datasets/GLDAS_NOAH025_3H_2.1/summary.

example, the first teleconnection pattern is the North Atlantic Oscillation (NAO). The NAO index or the NAO pattern can be determined through data acquired from monitoring stations or identified by means of Empirical Orthogonal Functional (EOF) analysis. Climate indices are considerably important climatic products that allow the assessment of dependent hydro-climatic variables, such as terrestrial water storage and how they change in time and space, especially due to the effects of climate change. They provide the means to carry out analysis and comparison of time series, extremes and trends. Their influence in driving TWS warranted their inclusion as independent predictors of the integrated machine learning system presented in this chapter. The following indices were used for the assessment:

- Multivariate ENSO Index (MEI),
- Atlantic multi-decadal oscillation (AMO),
- North Atlantic Oscillation Index (NAO),
- Pacific Decadal Oscillation (PDO),
- Indian Ocean Dipole (IOD).

These climate indices were previously described in detailed in Chaps. 7 and 8 of this book. For a full and comprehensive assessment of climatic impacts on the MDB's terrestrial water cycle, sea surface temperature (SST) products were also included to the predictor datasets. The goal here is to strengthen the predicted capability of the proposed integrated machine learning. SST products are essential climatic variables that propel our understanding of the climate system and provide a basis for the quantification of the ongoing climate change. Long-term climate data records of SST have been developed from satellite datasets. The averaged SST products applied in this analysis include, Ninos 1+2 (0-10S, 90W-80W), 3.0 (5N-5S, 150W-90W), 3.4 (5N-5S, 170W-120W), and 4.0 (5N-5S, 160E-150W). Other SST products such as the NTA, ETA, and STA from the Atlantic Ocean were also used in this experiment.

9.3.2 Overview of the Convolution Neural Networks and Support Vector Machines

Convolution neural networks (CNN) are known as regularized versions of multilayer perceptrons. Multilayer perceptrons usually mean fully connected networks, which allow each neuron in one layer to connect to all other neurons in the next layer (e.g., Kalu et al. 2023). A CNN is largely composed of an input, hidden, and output layer. In any feed-forward neural network, the middle layer is usually referred to as the "hidden layer". This is because, the middle layer serves as a bridge between the input and output nodes, which are masked by the activation function and final convolution. In a convolution neural network, the hidden layer includes layers that perform convolutions. Distinctively, this includes a layer that performs a dot product of the convolution kernel with matrices of the input layer. The product is usually the

Frobenius inner product and usually makes use of the ReLU activation function. As the convolution kernel slides along the input matrix for the layer, the convolution operation creates a feature map, which in turn interposes on the input of the next layer. This is followed by other layers which includes pooling layers, fully connected layers, and the normalization layers.

The support vector machine is a popular machine learning algorithms which falls into the supervised learning category and is usually associated with learning algorithms that analyze data for both classification and regression tasks (see, e.g., Kalu et al. 2023; Okwuashi and Ndehedehe 2020; Okwuashi et al. 2021). Given a set of training datasets, the support vector machine training algorithm designs a model that assigns new samples to one category or the other, thereby making it a non-probabilistic binary linear classifier. The machines map training sample to points in space to maximize the gap width between the two categories. New examples are then mapped into the same space and predicted to a class or category due to the side of the gap they fall into. Furthermore, their proficiency in performing regression tasks, which is the major concern in this study is well founded. As a supervised learning regressor, they are used to predict discrete values, using the same principles as the classifiers. The fundamental idea behind the regression performance is finding the line of best fit, which is the hyperplane that has the maximum number of points (Yao et al. 2017).

Contrary to other regression models that attempt the minimization of the bias and uncertainty factors between the real and predicted values, the support vector regressor attempts to fit the best line within a threshold value. This threshold value is the distance between the hyperplane and the boundary line. The support vector regressor has a fit time complexity of both first and second degrees which must be homogenous with the number of samples. This means that datasets having thousands of samples will be very difficult to be scaled using the support vector regressor, as they are not suitable for very large datasets. However, in a case where large datasets must be used, regressors such as the linear or SGD regressors can suffice. The linear support vector regressor provides a faster implementation than the basic support vector regressor, but is limited to the operation of the linear kernel alone. The basic function of the support vector machine kernel is to take data as input and transform them into the required and needed form. Various support vector algorithms make use of different types of kernel functions (e.g., Ndehedehe et al. 2020). These functions can be of different types, such as the linear, non-linear, sigmoid, radial basis function, and polynomial (e.g., Okwuashi and Ndehedehe 2017). Also, the kernel functions are very proficient in returning the inner product between two points in a suitable feature space. Its robustness and efficiency in support vector regression stems from its idea of defining a notion of similarity between datasets with little computational cost even in very high-dimensional spaces. It is important to note that the model produced by the support vector regressors depend only on a subset of the entire training datasets, because the cost function pays little or no attention to samples whose prediction is close to their targets.

To build a robust and more efficient model for TWS reconstruction, this chapter attempts to integrate these two machine learning algorithms. The justification for this integration is based on the capabilities of the individual machine learning frameworks. The CNN is a deep learning techniques based on neural networks and is very robust for mining big data but less effective in smaller datasets. On the other hand, support vector machines are extremely robust in mining smaller datasets but less effective in big data mining. Therefore, the purpose of integrating them for the creation of a convoluted support vector machine (C-SVM) is aimed at harnessing the advantage of both algorithms and deep learning capabilities for diverse datasets to maximize prediction capabilities and tests for the reconstruction of GRACE-TWS estimates.

9.3.3 *Integrating CNN with SVM—a Convolution Based Support Vector Machine Framework*

The modeled C-SVM utilized a convolution product for its operation by assimilating the CNN and support vector machine methods. CNNs are very proficient in the creation of good filters and extricating the in-depth features of their training datasets efficiently. The SVM on the other hand is largely based on the structural risk minimization approach (e.g., Ndehedehe et al. 2020; Okwuashi and Ndehedehe 2017), and consists of a proficient non-linear mapping performance having the capabilities of efficiently mapping data inputs into their respective high-dimensional feature space and performing a linear regression in its feature space (Mi et al. 2019). For the SVM integration, the linear, polynomial, and radial basis function kernels were used (Fig. 9.6). The computational steps of the C-SVM are summarized as follows.

- **Step 1:** The original input training datasets which are the independent variables are used to build and train the CNN model. Each CNN model is characterized by three convolution layers and two fully connected layers. The convolution layers occupy channels 4, 8, and 16, respectively. For the first hidden layer, the "Adam" and "RELU"algorithms were used as the activation and optimization functions, and for the second hidden layer, the sigmoid algorithm was used as the activation function. Also, the generalization performance of the CNN was enhanced through the capabilities of the "dropout" and "Lasso regularization" techniques in the multilayered architecture (Ambach and Schmid 2017). A dropout ratio of $p = 0.7$ was used between the convolution layer 3 and fully connected layer 1, while the lasso regularization maintained a proportion of 0.01.
- **Step 2:** The extracts of the fully connected layer 1 of the CNN are now used to inject into the SVM as optimized input training datasets, after which the modeling and training of the SVM framework starts. To search for ε and C, the "grid search" optimizer was used, given that it performs a random search order, but with a uniform sampling mode even without replacement from the grid. This was important to improve the robustness and generalization of the integrated machine learning model.

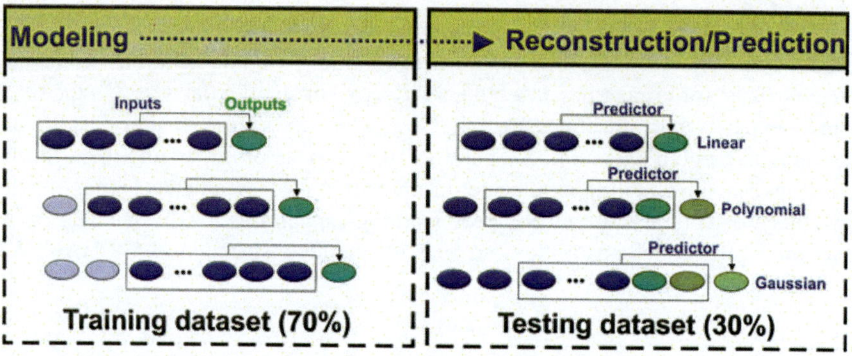

Fig. 9.6 Various Kernel reconstruction steps of the convolution-based support vector machine

- **Step 3:** The training models of the convolutional neural networks and the support vector machines are saved (Fig. 9.1).
- **Step 4:** The optimized training model is reloaded into the system, while the testing model is rebuilt. The rebuilding of the testing model is accomplished by directly interfacing the fully connected layer 1 with the SVM layer. Additionally, the optimized training and testing model parameters are segmented for TWS regressions using hydrological flux parameters, climate indices, and sea surface temperature components.

The aim of this procedure is to estimate the influence of each individual set of hydrological variables on the TWS of the MDB. In the convolution-based SVM framework (C-SVM), the relationship among the activation function, kernel functions, prediction function, and non-linear mapping functions are detailed as:

- Six activation function phases are utilized in the C-SVM frame; with the RELU algorithm controlling the first input transfer functions to the convolutional layer 1. The second transfer function which is controlled by the sigmoid algorithm is translated to the convolution layer 2, and the third function translates the convolution layer 2 to the convolution layer 3. The fourth function translates the convolution layer 3 to the fully connected layer 2, and the fifth function translates the fully connected layer 1 to the fully connected layer 2, while the sixth function translates the fully connected layer 2 to the outputs. Note, the fully connected layer 2 maintains the attributes and characteristic parameters of all the other prior connections, which improve clarity on the output predictions.
- The non-linear mapping function can communicate the fully connected layer 1 to the SVM layer, while the kernel function represents the transformation of the non-linear mapping function (Fig. 9.6).
- The output reconstruction model is built on the prediction function, $f(x, a_i, a_i^*) = {}_{i=1}^{n}(a_i - a_i^*)K_m(x, x_i) + b(m = 1, 2, 3)$.

In this experiment, we revised all values in filters of the proposed convolution products using the simplified swarm optimization model(SSOM). Yeh (2013) defined the SSOM as a classic machine learning algorithm due to its update mechanism. Originally introduced by Yeh (2013), the simplicity and efficiency of the SSOM have been corroborated by past studies to be very efficient in optimization and even data mining problems (Yeh 2013). The basic idea of the SSOM is that each variable, for example, the *j*th variable in the *i*th solution, $x_{i,j}$ undergoes a stepwise update routine as suggested by Yeh (2013):

$$x_{i,j} = \begin{cases} g_i \; if \; \rho[0.1] \epsilon (0, C_g) \\ P_{i,j} \; if \; \rho[0, 1] \epsilon (C_g, C_p) \\ x_{i,j} \; if \; \rho[0, 1] \epsilon (C_p, C_w) \\ x, \; otherwise \end{cases} \tag{9.1}$$

The value [0.1] ϵ [0, 1] is randomly produced by the system, while the parameters, C_g, $C_p - C_g$, $C_w - C_p$, and $1 - C_w$ all have a range of [0,1] which equally represent the probabilities of the existing parameters that are obtained from the best of all solutions. That is, the optimal *i*th solution, the existing solution, and a randomly generated feasible value, respectively. Using the no free lunch theorem, we customized and reduced the four items in Eq. 9.1 to three items, to improve efficiency. The C_g, C_p and C_w parameters are self-adapted. Special equations are selected to substitute g_j, $P_{i,j}$, $x_{i,j}$, and x, which means that only a limited number of parameters are selected to be updated. Despite this, the SSOM update mechanism is largely based on the stepwise function.

9.3.4 One Filter One Variable Greedy Update Mechanism

In the C-SVM implemented here, all values in filters are parameters that need to be authenticated to carry out the convolution products. As a result of the duration of variable update required by the system, which is aimed at deepening the SVM regressors, especially without help from the graphics processing unit (GPU), it was pertinent to explore an alternative update mode. Hence, instead of utilizing the traditional algorithms, including the SSOM, the particle swarm optimization, *n*, and genetic algorithm, of which all solutions need to be updated, we select only one random solution for updating each generation of our proposed SSOM update mechanism. Let's select a solution *sol* to be updated based on the given model:

$$sol = \begin{cases} g_{Best} \; if \; \rho[0.1] \epsilon (0, C_{g,1}) \\ P_{[1,Nsol]} \; if \; \rho[0, 1] \epsilon (C_{g,1}, C_{p,1}) \\ sol \; if \; \rho[0, 1] \epsilon (C_{p,1}, C_{w,1}) \\ 0, \; otherwise \end{cases} \tag{9.2}$$

Where, [0.1] is a random floating-point generated from the interval [0,1], and $P_{[1,Nsol]}$, is an element of a real number ranging from 1 to Nsol, and signifies the index of the randomly selected solution. g_{Best} is the index of the best solution found, while we assign 0 as a new randomly generated solution. The ensuing processes enable a decision to be made as to if we will discard or replace the newly updated solution X_{sol} with the prior X_{sol} value. From Eq. 9.2, we select sol as the solution to be updated. In the proposed new SSOM, we select randomly, only one filter, *fil* in solution sol. Given that, *fil* $= 1, 2, pFilter[fil] = j$. Moreover, one additional parameter, for instance, $x_{sol,fil,k}$, where k is given by $1, 2, N_{var}$, in filter $x_{(sol,fil)}$ is equally selected at random to undergo the update process. Details on the update process are further outlined in the work of Yeh et al. (2021). However, for this machine learning, the values -0.03 and 0.09 were used as the lower and upper bounds of the randomly generated numbers, $[-0.03, 0.09]$. Additionally, a step size of 0.045 was used in the local search to ensure an adequate optimal solution is achieved for the integrated system.

9.4 Results and Discussions

9.4.1 Fitting the Proposed Algorithm to the Three Level-2 GRACE Products

The integrated machine learning technique proposed in this chapter was designed to mimic the operation of the convolution neural network. The distinctive form of using SVMs as regression models applies the same principle as the classification-based SVM processes. The training phase, which is composed of 70% of the input datasets tried to find the best fit line between the independent and dependent parameters. The best line of fit is the hyper plane, which contains the maximum number of points that the SVM attempts to fit within a threshold range. This mechanism is different from the traditional regression model, which attempts to minimize the error between the actual and predicted/reconstructed value (e.g., Okwuashi and Ndehedehe 2017). The threshold value is the distance between the hyper plane and the boundary line. The SVM maintains a fit time complexity of above quadratic normal with the number of samples making it difficult to scale large sample datasets.

In Fig. 9.7, the comparison of the resulting output of the first order training models is indicated. These models are the linear, polynomial (order 2), and RBF (gamma 0.05) on the three solutions (variants) of GRACE-TWS, which served as the target output. From the fits observed in the models, we observe that SVM-RBF 1 made the best fit for TWS-GFZ compared to the other variants (Fig. 9.7g), SVM-linear marinated the best fit for TWS-CSR when compared to other variants (Fig. 9.7b) while the SVM-RBF 1 maintained the best fit for TWS-JPL when compared to other variants (Fig. 9.7i). Generally, it is observed that the reconstruction fit with the minimal bias was between the TWS-CSR and the SVM-RBF 1. The CSR-TWS

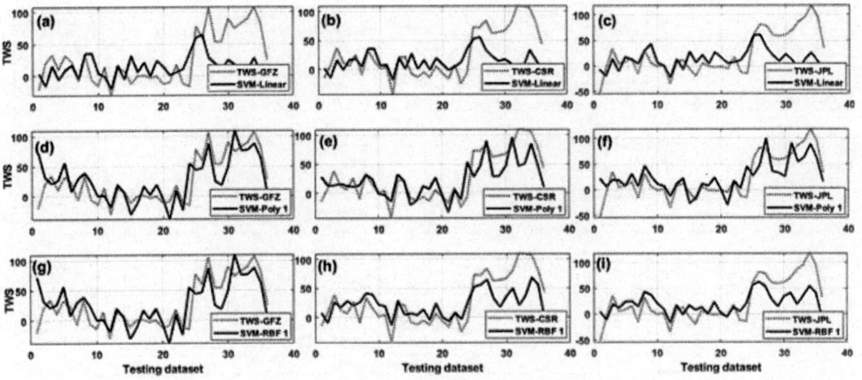

Fig. 9.7 First order reconstruction procedure for the three variant (TWS-GFZ, TWS-CSR, and TWS-JPL) level 2 GRACE products using the integrated machine learning framework (i.e., C-SVM)

product also showed minimal bias with the other first order models better than the other GRACE variants explored in the experiment (Fig. 9.7b, e, and h). A large contributor to this could be the C30 replacement solutions from satellite laser ranging from the GRACE solutions, and the degree 1 coefficient corrections implemented in the solutions using the estimates in TN-13a, which inadvertently results in a better fitting index for our SVM-linear mode. Another plausible explanation could be the state-of-the-art mascon gridded GRACE-TWS products which are undergo glacial isostatic adjustments and the monthly averages of the modeled non-tidal ocean bottom pressure deviation corrections, which make them more suited for hydrological analysis.

The reconstruction procedure undertaken in both the first (Fig. 9.7) and higher order (Fig. 9.8) SVM levels used a total of 22 predictors, which consist of the HFs, CIs, and the SSTs. The regularization parameter was used to prepare the model to enable the balance of weights and biases in the SVM-training phase. The weights, which are depicted as extracts from the SVM-hyper plane are decided by obtaining the coordinates of the corresponding vectors which are orthogonal to the hyper plane. Given that the directions provide us with relevant details about the TWS reconstruction exercise, we take the dot product of any point with the vector to examine if it belongs to the positive or negative class.

For the higher order training models (Fig. 9.8), which are composed of the polynomial (order 3), RBF (gamma 0.02), and RBF (gamma 0.03) kernels, it was observed that the fitting index of the actual (3-variant GRACE-TWS products) and predicted (reconstructed TWS) information was better than the first order models. For the higher order model fits, we note that for TWS-GFZ, SVM-RBF 3 made the best fit when compared to other variants (Fig. 9.8g). But for TWS-CSR product, SVM-RBF 2 maintained the best fit compared to other variants (Fig. 9.8e), while for TWS-JPL product, SVM-RBF 3 maintained the best fit compared to other variants (Fig. 9.8i). It was observed from both first and higher order reconstruction phases that

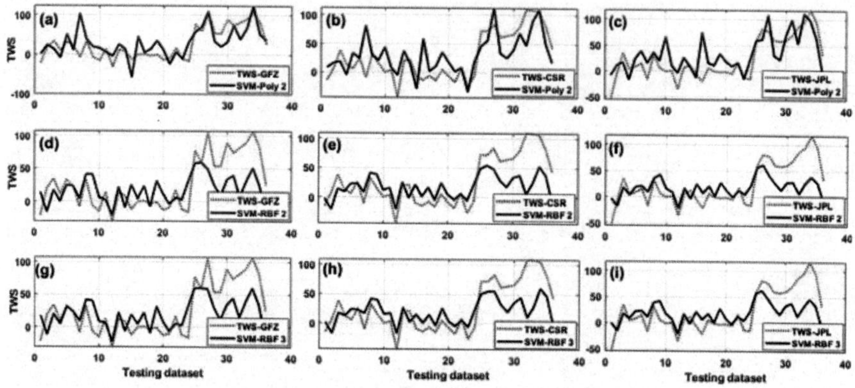

Fig. 9.8 Higher order reconstruction procedure for the three variant (TWS-GFZ, TWS-CSR, and TWS-JPL) level 2 GRACE products for the integrated machine learning framework (i.e., C-SVM)

the RBF reconstructed and predicted TWS better than other kernels. This corroborates the superior performance of the RBF in regression tasks as has been reported in previous studies (e.g., Yao et al. 2017). The popularity of the RBF kernels stems from their close similarity to K-nearest neighborhood (KNN) algorithm. It equally bears the advantages of the KNN, and overcomes the space complexity problems in the model, given that the support vectors are stored during training and not the entire dataset.

9.4.2 Estimating the Impacts of HF's, CI's, and SSTs in Regulating TWS

Part of the objectives of this chapter was to assess the influence of the three sets of climate variables (H.Fs, C.Is, and SSTs) extracted from the predictor classes in modeling TWS discretely. This was warranted as past studies have attempted to explore the impacts of hydrological fluxes, climate indices, and sea surface temperatures in tropical humid regions (Kalu et al. 2021; Ndehedehe et al. 2017). Extreme hydro-meteorological conditions are erratic outcomes of variations in sub-regional and global climate modes, which in turn impact on changes in TWS. For instance, global precipitation records display an average increase of approximately 9 mm of precipitation over land areas during the twentieth century. However, studies have shown an increased frequency of wet days in several sub-regions of the world and an intensifying proportion of total precipitation in many of the world's largest freshwater reservoirs (Knapp et al. 2008). This also contributes significantly in the regulation of TWS, as expressed in the water budget equation,

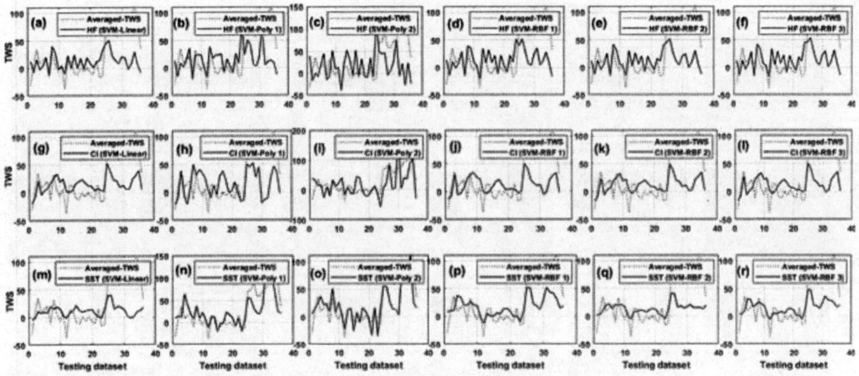

Fig. 9.9 Modeling the relationship between the averaged GRACE-TWS and discrete hydrological fluxes (**a**–**f**), climate indices (**g**–**l**), and sea surface temperatures (**m**–**r**) datasets, and observing their performance against the three SVM kernels used in this experiment

$$TWS = P - R - ET, \tag{9.3}$$

where P, R, and ET represent precipitation, runoff, and evapotranspiration, respectively.

Also, we observed that one hydrological product has a key influence over the behavioral patterns of other hydrological products. For instance, intensification in latent heat transfer leads to increase in land surface temperature, increase in soil temperature, increase in evapotranspiration rates, and a substantial decrease in soil moisture content. This interplay between hydrological variables provides us with clear information as to their role in the regional TWS dynamics. Figure 9.8 shows that RBF 3 provided the best fit between the hydrological parameters and the averaged TWS. However, a significant fitting range from month 5 to month 10 was observed in the training samples (Fig. 9.9a–f). Months 4 to 14 represent wet months extracted to form part of the independent (training) dataset, and this is yet another significant pointer to precipitation as a leading driver of changes in TWS, irrespective of the influence of other hydrological variables employed during the modeling exercise. Another important thing to note is that months 4 to 14 are dominated by annual signals, which also show that in complex hydrological modeling, the convolution-based SVMs developed in this experiment performs poorly in multi-annual time series. This is usually seen in large freshwater basins where the dynamics of the hydrological process are significantly strong, such as the Congo River basin. Therefore, performing a C-SVM-modeling procedure for such regions may not give optimal results. This is particularly so given the strong land-atmosphere dynamics and interaction, which drive extreme changes in the hydro-climatology of such regions.

Evidence of rapid changes in the global atmosphere and hydrological cycle due to the influence of climate variability has been reported in past literatures (e.g.,

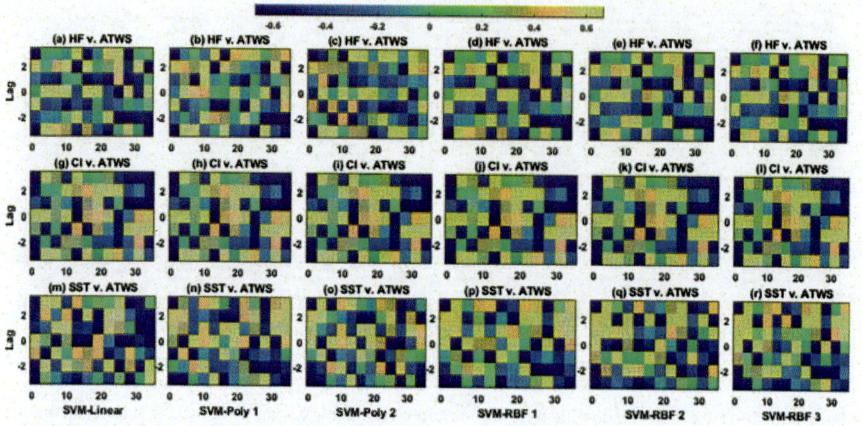

Fig. 9.10 Windowed cross-correlation estimating the influence of **a–f** HFs on TWS, **g–l** CIs on TWS, and **m–r** SSTs on TWS for all six kernel variants explored in this experiment

Ndehedehe et al. 2017). Such changes are affect water resources causing incessant and extreme hydro-meteorological conditions. For instance, in Africa and South America, the dominant impact of AMO, ENSO, and IOD indices on changes in GRACE-derived TWS was identified in the studies undertaken by Ndehedehe et al. (2017), Phillips et al. (2012), and Ndehedehe and Ferreira (2020). The integrated C-SVM adopted in this chapter attempts to relate these datasets with an averaged GRACE-TWS by exploring both first and higher order kernels. Results from this analysis show that the SVM-poly 1 (Figs. 9.9h and 9.10h) made the best attempt in relating these two products. However, when compared to the performance of the hydrological products in estimating TWS, we observe that the CIs did not perform well on their own in this modeling phase. This however is not surprising, as their influence on TWS is usually augmented by other hydrological products, which did not form part of the training at this stage. Also, the influence of these indices is usually localized, while only a few of them, like ENSO have a more dominant widespread influence. SST datasets performed better than climate indices in this experiment. This is due to:

- spatially distributed measurements of the SST better represent oceanic dynamics than the single indices produced by the teleconnection patterns,
- the gridded monthly SST datasets are composed of severally averaged SST components which provided the convolution-based SVM a more comprehensive training dataset than the climate indices,
- SST predicts precipitation better than the pressure anomaly usually applied in the development of climate indices. Given that precipitation plays a significant role in regulating TWS, therefore SST properties outperform climate indices in TWS estimations.

Results from this study show that of the three kernels used in the estimation, the SVM-poly 1 (Figs. 9.9n and 9.10n) produced the best fit between SSTs and aerial averaged TWS. The polynomial kernel performs well in modeling predictands whose predictors do not maintain sufficient relationships between them even more than the RBF kernels. This also suggests that the assertion made by H. Fizazi Izabatene and Ghardaoui (2010), which postulates that the increasing polynomial order increases accuracy in regression analysis does not apply in all cases, especially when non-linear predictors and predictands form part of the experiment's architectural framework.

9.4.3 Performance of the Different Kernel Variants Applied in the Proposed System

Integrated machine learning frameworks are emerging as considerably efficient techniques in pattern recognition analysis and have garnered significant attention as explained earlier in this chapter. Here, we attempt to evaluate the skills of the linear, polynomial, and RBF kernels in hydrological modeling. The linear kernel is regarded as the most basic of all the other kernels. It does not need to project data into a higher dimension but only functions with the inner product of its latent variables while maintaining an optional constant in some scenarios. It is important to point out that the incredible benefit of the linear kernel lies on its simplicity of use, application, and operation. The linear kernel is applied on datasets having a substantial amount of correlation, and thus increments in the dimensionality of these datasets does not necessary translate to increase in regression accuracy. In this experiment, the SVM-linear demonstrated a substantial regression fit for the three variants of TWS using the joint inputs from the hydrological flux parameters, climate indices, and sea surface temperature products as predictor datasets.

A random selection of three testing samples was undertaken to perceive the reconstruction efficiency of all the kernels explored in our experiment. For the linear kernel, a bias of TWS-GFZ (0.011 mm, 0.127 mm, and 0.335 mm), TWS-CSR (0.241 mm, 0.115 mm, and 0.068 mm), and TWS-JPL (0.144 mm, 0.153 mm, and 0.213 mm) was observed. For the same random points, we equally measured the regression strengths of the discrete HFs, CIs, and SSTs on TWS. Therefore, our results shows a percentage bias of Av. TWS–HF's (24%, 30%, and 25%), Av. TWS–CI's (60%, 57%, and 58%), and Av. TWS–SSTs (44%, 39%, and 37%), respectively. The least bias was noticed in the regressions where the x values maintain a significant relationship with their corresponding xi values. This clearly shows that the linear kernel performs better when the predictors are substantially correlated with the predictand, as earlier stated. This clearly shows that results from the polynomial kernel proved to be more significant than the linear kernel in many cases. The poly 1 kernel as used in this study depicts the polynomial kernel of order 2. From the same test points for TWS reconstruction, we achieved a bias of TWS-GFZ (0.023 mm, 0.416 mm, and 0.444 mm), TWS-CSR (0.023 mm, 0.051 mm, and 0.100 mm), and TWS-JPL (0.042 mm,

0.010 mm, and 0.072 mm), respectively. For the regression strengths of the discrete components, we recorded Av. TWS–HF (27%, 25%, 21%), Av. TWS–CI (51%, 55%, 60%), and Av. TWS–SST (37%, 34%, 33%).

The poly 2 kernel, on the other hand, represents polynomial kernel of order 3. Over the same test points, a bias of TWS-GFZ (0.132 mm, 0.071 mm, and 0.044 mm), TWS-CSR (0.374 mm, 0.034 mm, and 0.011 mm), and TWS-JPL (0.041 mm, 0.011 mm, and 0.009 mm) was recorded. The regression strengths for the discrete components read Av. TWS–HF (19%, 22%, 26%), Av. TWS–CI (52%, 49%, 51%), and Av. TWS–SSTs (40%, 29%, 34%). We determined a similarity between the results from the polynomial order 2 and 3 kernels and the linear kernel in terms of trends, however, they differed in fine-tuned accuracy index, especially in cases where the predictors are significantly correlated to the predictands. This is because the polynomial kernel unlike the linear kernel takes the inner products from a higher dimensional space to aid regression fits. This is why the SVM-poly 2 fit, which assessed order three, outperformed the SVM-poly 1, which only assessed order two in most cases. Therefore, it is safe to assume that the higher order we access, the more fitted our predictors gets to the predictand. Notwithstanding, this is not usually the best case as it commonly results in over fitting. Therefore, we avoided making use of polynomial orders higher than 3 to avoid over fitting and corruption of our result. The robustness of the RBF kernels was also explored. These kernels tend to have better fitting capabilities in comparison to other SVM kernels detailed in past literatures (e.g., Khalil et al. 2006). The RBF kernel is composed of a free parameter that scales the amount of effect two points have on each other. It is basically represented as gamma. As opposed to the polynomial kernel, which looks at the additional dimensions that can be explored, the RBF expands into an infinite number of dimensions. This expansion helps to enhance the reconstructive capabilities of the kernel and minimize the regression loss significantly.

In this experiment, the SVM-RBF kernel was divided into three to explore the capabilities of three different gamma values. SVM-RBF 1 has gamma value of 0.05, SVM-RBF 2 has gamma value of 0.02, and SVM-RBF 3 has gamma of 0.03. For RBF 1, the results over randomly selected test points show a bias of TWS-GFZ (0.009 mm, 0.007 mm, and 0.011 mm), TWS-CSR (0.009 mm, 0.012 mm, and 0.009 mm), TWS-JPL (0.007 mm, 0.002 mm, and 0.001 mm), respectively. For the regression strengths of the discrete components, we recorded a % bias of Av. TWS–HF (18%, 20%, 21%), Av. TWS–CI (46%, 49%, 48%), Av. TWS–SST (38%, 36%, 35%), respectively. For RBF 2, our results show that over the randomly selected test points, a bias of TWS-GFZ (0.008 mm, 0.002 mm, and 0.011 mm), TWS-CSR (0.001 mm, 0.005 mm, and 0.002 mm), TWS-JPL (0.001 mm, 0.008 mm, and 0.010 mm), respectively. For the segmented predictors that measured percentage impact, we recorded a % bias of Av. TWS–HF (25%, 16%, 12%), Av. TWS–CI (48%, 50%, 38%), Av. TWS–SST (30%, 28%, 31%), respectively. For RBF 3, our results from the randomly selected test points show a bias of TWS-GFZ (0.004 mm, 0.009 mm, and 0.010 mm), TWS-CSR (0.002 mm, 0.001 mm, and 0.005 mm), TWS-JPL (0.035 mm, 0.002 mm, and 0.001 mm), respectively. For the segmented predictors that measured percentage impact, we observed a % bias of Av. TWS–HF (19%, 16%, 14%), Av. TWS–CI

(44%, 47%, 40%), Av. TWS–SST (26%, 32%, 23%), respectively. The RBF kernel outperformed the linear kernel in both cases. This was to be expected given that the kernel is equivalent to the inner product of the two data points having an infinite number of dimensions. The optimal gamma and epsilon values were generated using the grid search method. It generated the least bias between predictors and predictand, even in the case of the measured percentage impact to estimate the influence of HFs, CIs, and SSTS on the averaged TWS.

9.4.4 Comparison Between the C-SVM and Other Conventional Models in TWS Regressions

It was important to compare the efficiency and robustness of the C-SVM model with other conventional regression models. This is to help determine their fit index using the same datasets, and the prevalence of integrating machine learning models for twenty-first century hydrological modeling and operations. In using the multi-linear regression (MLR) to model the TWS-GFZ, it was observed that the MLR performed better in fitting test months 9 to 14 than the remaining months. This was the case in both the CSR and JPL-derived GRACE products. However, there was a stronger fitting index for the TWS-JPL variant that the other TWS products. The bias in the CSR predictions shows the unpredictability of the MLR in reconstructing several variants of the TWS, especially due to its significant non-linearity. For the partial least squares regression model (PLSR), some studies have shown their robust and efficient applications in machine learning operations (e.g., Okwuashi and Ndehedehe 2020). In this experiment, the PLSR model performed well in reconstructing the TWS-CSR and TWS-JPL better than the TWS-GFZ. Its robustness is seen in the fitting index. However, significant regression errors were noticed in the fitting of the TWS, especially from the 14th to 36th month (Fig. 9.11d–f). The C-SVM outperforms the PLSR in the fitting index, given that a standard deviation of TWS-GFZ (± 0.455), TWS-CSR (± 0.317), and TWS-JPL (± 0.373) was recorded for the PLSR, whereas the polynomial kernel of order 2 (poly 2) showed a standard deviation of ± 0.147 (TWS-GFZ), ± 0.111 (TWS-CSR), and ± 0.124 (TWS-JPL). This suggests that the PLSR is more affected in cases of disturbed predictor signals, low concentration impacts, and significant non-linearity in the correlation estimates of predictors and predictands more than the C-SVM.

The back propagation neural networks (BPNN) are historically known for performing a series of linear integrations combined with non-linear activation functions across several layers. Results from this analysis show that the BPNN modeled TWS better than the SVM-linear and SVM-poly 1 kernels. However, the RBF kernels outperformed them in all cases. Despite this, the BPNN arguably, has the potential of outperforming the C-SVM in a case of very high-dimensional datasets, given that this experiment only made use of 22 original input predictors. The Gaussian process regression (GPR) is a conventional non-parametric Bayesian approach to regression

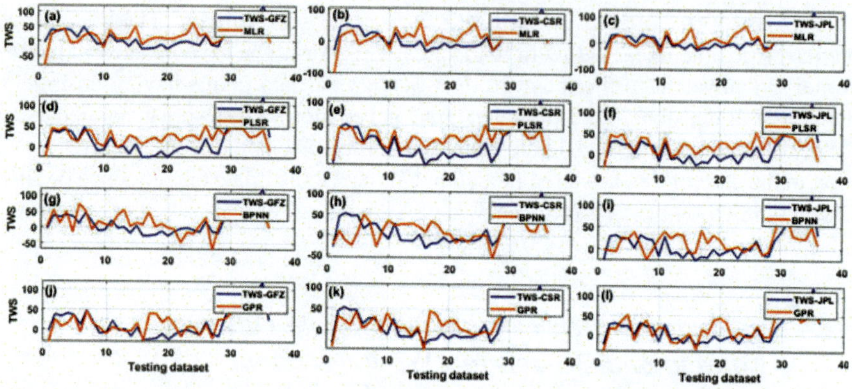

Fig. 9.11 Testing the robustness of other well-known regression models (MLR, PLSR, BPNN, and GPR) in reconstructing the three variant GRACE-TWS which served as our predictands

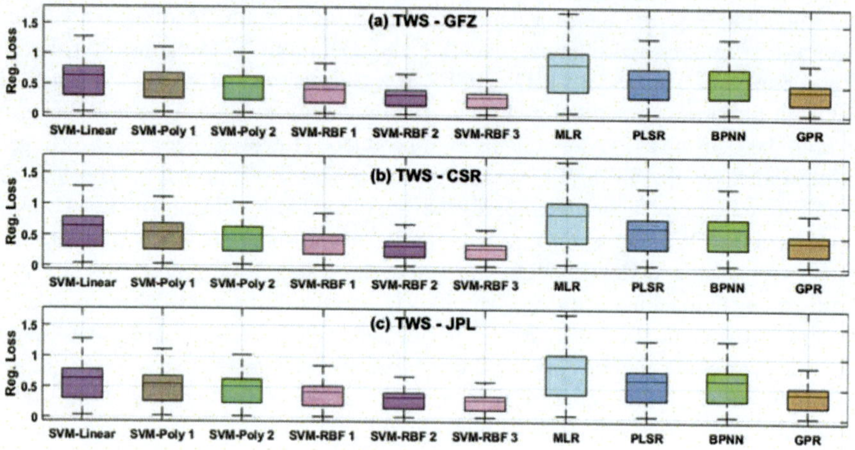

Fig. 9.12 Regression loss for the convolution-based SVM kernel variants and other conventional regression models used for this experiment

and reconstruction tasks and is making significant waves in machine learning operations and applications. It showed significant skill in modeling TWS, especially the TWS-GFZ and TWS-JPL (Fig. 9.11j, l). This is to be expected as the strength of the GPR is mostly dominant in cases of small to medium input datasets (Fig. 9.12). Also, given that the prediction is probabilistic, it allows the system to predict the confidence intervals and make decisions based on those that should go for re-fitting the prediction. Nevertheless, the robustness of the exponentiated quadratic kernel of the GPR, which was applied in this regression outperformed all the SVM kernels used in this study except the SVM-RBF 3 kernel (Fig. 9.12). This shows that in cases of kernel sampling of the GPR, it is possible to achieve better results from the GPR than our C-SVM.

9.5 Concluding Remarks

The emergence of integrated machine learning techniques in data science problems is gaining considerable attention, especially due to their proficiency in regression, fitting, and reconstructing the corresponding predictand data. In this chapter, the capabilities of the C-SVM in performing highly non-linear climatic operations in TWS reconstructions, as well as measuring the impacts of HFs, CI's, and SSTs on averaged TWS over the Murray–Darling basin was explored. The motivation toward integrating the CNN with the SVM is because the CNN is known for its proficiency in generating optimal filters that enable the extraction of in-depth features of their consistent data. However, the SVM has the ability to efficiently map big data inputs into their corresponding high-dimensional feature space to aid the performance of regression analysis in its feature space using the conventional structural risk minimization approach. Therefore, the underlying argument here is that the idea of integrating these two robust algorithms is in line with the recent assimilated and deep learning methods that are taking the lead in machine learning and artificial intelligence fields. For this experiment, twenty two sets of data comprising HF–CI and SST data were employed. With these original input predictors, which served as the independent variables, the CNN model was built and trained. The CNN model was characterized by three convolution layers and two fully connected layers. The activation and optimization functions of the first hidden layers were achieved with the RELU and Adam algorithms, while the sigmoid algorithm was used as the activation function of the second hidden layer. The support vectors from the SVM were then fed with the input features from the fully connected layer 1 of the CNN model, after which the SVM operation started. To enhance the robustness and generalization of the C-SVM model, the SVM used the grid search optimizer to investigate the necessary latent parameter index, such as the and C variables. After saving the CNN and SVM models, the training samples were reloaded into the system. Also, the SSOM was applied in updating all the filter values of our proposed convolution products, to enable a better fit for the reconstruction of the three TWS products explored in these experiments.

Another key objective of this chapter was to estimate the impact on and contribution of hydrological fluxes, climate indices, and sea surface temperatures to the localized data, which formed our predictand for the integrated machine learning build up. To achieve this, the training and testing models were segmented to make predictions of an averaged TWS using HF's, CI's, and SSTs. This was warranted to estimate the extent and impact of these sets of climatic observations in regulating region-based TWS. Results from this analysis showed that the hydrological data provided the most significant impacts and contributions to TWS, whereas the climate indices showed the least impact on TWS. This was to be expected as the climate variability indices are globalized data (capture variability in specific oceanic hotspots), while changes in TWS are localized. The SST data on the other hand provided a minimal impact on TWS, but was stronger than that of climate indices. Changes in SST have a far-reaching impact on the global climate, and the relatively stronger impact exhibited by

SST over climate teleconnections further stresses the importance of the nearby ocean on changes in GRACE-TWS. Also, the outcome from this experiment confirms that the integrated machine learning method (C-SVM) provides better results in cases of more robust and comprehensive input parameters when compared to other traditional machine learning techniques.

References

Agutu N, Ndehedehe C, Awange J, Kirimi F, Mwaniki M (2021) Understanding uncertainty of model-reanalysis soil moisture within Greater Horn of Africa (1982–2014). J Hydrol 603:127169. https://doi.org/10.1016/j.jhydrol.2021.127169

Ahmed M, Sultan M, Yan E e a (2016) Assessing and improving land surface model outputs over Africa using GRACE, field, and remote sensing data. Surv Geophys 37:529–556. https://doi.org/10.1007/s10712-016-9360-8

Alsdorf DE, Lettenmaier DP (2003) Tracking fresh water from space. Science 301(5639):1491–1494. https://doi.org/10.1126/science.1089802

Alsdorf D, Lettenmaier D, Vörösmarty C (2003) The need for global, satellite-based observations of terrestrial surface waters. Eos, Trans Am Geophys Union 84(29):269–276. https://doi.org/10.1029/2003EO290001

Ambach D, Schmid W (2017) A new high-dimensional time series approach for wind speed, wind direction and air pressure forecasting. Energy 135:833–850. https://doi.org/10.1016/j.energy.2017.06.137

Becker M, Meyssignac B, Xavier L, Cazenave A, Alkama R, Decharme B (2011) Past terrestrial water storage (1980–2008) in the amazon basin reconstructed from GRACE and in situ river gauging data. Hydrol Earth Syst Sci 15(2):533–546. https://doi.org/10.5194/hess-15-533-2011

Belkin M, Hsu D, Ma S, Mandal S (2019) Reconciling modern machine-learning practice and the classical bias-variance trade-off. Proc Natl Acad Sci 116(32):15849–15854. https://doi.org/10.1073/pnas.1903070116

Chen J, Wilson C, Tapley B, Scanlon B, Güntner A (2016) Long-term groundwater storage change in Victoria, Australia from satellite gravity and in situ observations. Glob Planet Chang 139:56–65. https://doi.org/10.1016/j.gloplacha.2016.01.002

Eicker A, Schumacher M, Kusche J, Döll P, Schmied HM (2014) Calibration/data assimilation approach for integrating GRACE data into the waterGAP global hydrology model (WGHM) using an ensemble Kalman filter: first results. Surv Geophys 35(6):1285–1309. https://doi.org/10.1007/s10712-014-9309-8

El Ayadi M, Kamel MS, Karray F (2011) Survey on speech emotion recognition: features, classification schemes, and databases. Pattern Recognit 44(3):572–587. https://doi.org/10.1016/j.patcog.2010.09.020

Ferreira V, Yong B, Tourian M, Ndehedehe C, Shen Z, Seitz K, Dannouf R (2020) Characterization of the hydro-geological regime of yangtze river basin using remotely-sensed and modeled products. Sci Total Environ 718:137354. https://doi.org/10.1016/j.scitotenv.2020.137354

Friis-Christensen E, Lühr H, Hulot G (2006) Swarm: a constellation to study the earth's magnetic field. Earth Planet Space 58:351–358. https://doi.org/10.1186/BF03351933

Guo Y, Zhang Z, Tang F (2021) Feature selection with kernelized multi-class support vector machine. Pattern Recognit 117:107988. https://doi.org/10.1016/j.patcog.2021.107988

Gyawali B, Ahmed M, Murgulet D, Wiese DN (2022) Filling temporal gaps within and between GRACE and GRACE-FO terrestrial water storage records: an innovative approach. Remote Sens 14(7). https://doi.org/10.3390/rs14071565

Humphrey V, Gudmundsson L (2019) GRACE-REC: a reconstruction of climate-driven water storage changes over the last century. Earth Syst Sci Data 11(3):1153–1170. https://doi.org/10.5194/essd-11-1153-2019

Izabatene HF, Benhabib W, Ghardaoui S (2010) Contribution of kernels on the SVM performance. J Appl Sci 58(10):831–836. https://doi.org/10.3923/jas.2010.831.836

Jing W, Zhang P, Zhao X et al (2020) Extending GRACE terrestrial water storage anomalies by combining the random forest regression and a spatially moving window structure. J Hydrol 590:125239. https://doi.org/10.1016/j.jhydrol.2020.125239

Kalu I, Ndehedehe CE, Okwuashi O, Eyoh AE (2021) Assessing freshwater changes over Southern and Central Africa (2002–2017). Remote Sens 13(13). https://doi.org/10.3390/rs13132543

Kalu I, Ndehedehe CE, Okwuashi O, Eyoh AE, Ferreira VG (2022) A new modelling framework to assess changes in groundwater level. J Hydrol: Reg Stud 43:101185. https://doi.org/10.1016/j.ejrh.2022.101185

Kalu I, Ndehedehe CE, Okwuashi O, Eyoh AE, Ferreira VG (2023) Reconstructing terrestrial water storage anomalies using convolution-based support vector machine. J Hydrol: Reg Stud 46:101326. https://doi.org/10.1016/j.ejrh.2023.101326

Khalil AF, McKee M, Kemblowski M, Asefa T, Bastidas L (2006) Multiobjective analysis of chaotic dynamic systems with sparse learning machines. Adv Water Resour 29(1):72–88. https://doi.org/10.1016/j.advwatres.2005.05.011

Knapp AK, Beier C, Briske DD, e a (2008) Consequences of more extreme precipitation regimes for terrestrial ecosystems. BioScience 58(9):811–821. https://doi.org/10.1641/B580908

Leblanc M, Tweed S, Van Dijk A, Timbal B (2012) A review of historic and future hydrological changes in the Murray-Darling Basin. Glob Planet Chang 80–81:226–246. https://doi.org/10.1016/j.gloplacha.2011.10.012

Li B, Rodell M, Zaitchik B, Reichle R, Koster R, van Dam T (2012) Assimilation of GRACE terrestrial water storage into a land surface model: evaluation and potential value for drought monitoring in western and central Europe. J Hydrol 446–447:103–115. https://doi.org/10.1016/j.jhydrol.2012.04.035

Li F, Kusche J, Rietbroek R e a (2020) Comparison of data-driven techniques to reconstruct (1992–2002) and predict (2017–2018) GRACE-like gridded total water storage changes using climate inputs. Water Resour Res 56(5):e2019WR026551. https://doi.org/10.1029/2019WR026551

Li F, Kusche J, Chao N, Wang Z, Löcher A (2021) Long-term (1979-present) total water storage anomalies over the global land derived by reconstructing GRACE data. Geophys Res Lett

Loomis BD, Rachlin KE, Luthcke SB (2019) Improved earth oblateness rate reveals increased ice sheet losses and mass-driven sea level rise. Geophys Res Lett 46(12):6910–6917. https://doi.org/10.1029/2019GL082929

Mi X, Liu H, Li Y (2019) Wind speed prediction model using singular spectrum analysis, empirical mode decomposition and convolutional support vector machine. Energy Convers Manag 180:196–205. https://doi.org/10.1016/j.enconman.2018.11.006

Mueller B, Hirschi M, Seneviratne SI (2011) New diagnostic estimates of variations in terrestrial water storage based on era-interim data. Hydrol Process

Ndehedehe C, Awange J, Agutu N, Kuhn M, Heck B (2016a) Understanding changes in terrestrial water storage over West Africa between 2002 and 2014. Adv Water Resour 88:211–230. https://doi.org/10.1016/j.advwatres.2015.12.009

Ndehedehe CE, Awange JL, Corner R, Kuhn M, Okwuashi O (2016b) On the potentials of multiple climate variables in assessing the spatio-temporal characteristics of hydrological droughts over the Volta Basin. Sci Total Environ 557–558:819–837. https://doi.org/10.1016/j.scitotenv.2016.03.004

Ndehedehe CE, Awange J, Kuhn M, Agutu N, Fukuda Y (2017) Climate teleconnections influence on West Africa's terrestrial water storage. Hydrol Process 31(18):3206–3224. https://doi.org/10.1002/hyp.11237

Ndehedehe CE, Ferreira VG (2020) Identifying the footprints of global climate modes in time-variable gravity hydrological signals. Clim Chang 159:481–502. https://doi.org/10.1007/s10584-019-02588-2

Ndehedehe CE, Haile GG, Agutu NO, Ferreira VG, Getirana A, Okwuashi O (2020) Hydrological hotspots of climatic influence in Brazil: a two-step regularization approach. Atmos Res 246:105116. https://doi.org/10.1016/j.atmosres.2020.105116

Njoku E, Jackson T, Lakshmi V, Chan T, Nghiem S (2003) Soil moisture retrieval from AMSR-E. IEEE Trans Geosci Remote Sens 41(2):215–229. https://doi.org/10.1109/TGRS.2002.808243

Okwuashi O, Ndehedehe C (2017) Tide modelling using support vector machine regression. J Spat Sci 62(1):29–46. https://doi.org/10.1080/14498596.2016.1215272

Okwuashi O, Ndehedehe CE (2020) Deep support vector machine for hyperspectral image classification. Pattern Recognit 103:107298. https://doi.org/10.1016/j.patcog.2020.107298

Okwuashi O, Ndehedehe CE, Olayinka DN, Eyoh A, Attai H (2021) Deep support vector machine for PolSAR image classification. Int J Remote Sens 42(17):6498–6536. https://doi.org/10.1080/01431161.2021.1939910

Oleson KW, Niu G-Y, Yang Z-L e a (2008) Improvements to the community land model and their impact on the hydrological cycle. J Geophys Res: Biogeosciences 113(G1). https://doi.org/10.1029/2007JG000563

Pan M, Sahoo AK, et al (2012) Multisource estimation of long-term terrestrial water budget for major global river basins. J Clim 25(9):3191–3206. https://doi.org/10.1175/JCLI-D-11-00300.1

Phillips T, Nerem RS, Fox-Kemper B, Famiglietti JS, Rajagopalan B (2012) The influence of ENSO on global terrestrial water storage using GRACE. Geophys Res Lett 39(L16705):2012. https://doi.org/10.1029/2012GL052495

Pokhrel YN, Hanasaki N, Yeh PJ, Yamada TJ, Kanae S, Oki T (2012) Model estimates of sea-level change due to anthropogenic impacts on terrestrial water storage. Nat Geosci 5(6):389–392. https://doi.org/10.1038/ngeo1476

Quinn KJ, Ponte RM (2010) Uncertainty in ocean mass trends from GRACE. Geophys J Int 181(2):762–768. https://doi.org/10.1111/j.1365-246X.2010.04508.x

Rodell M, Famiglietti JS, Wiese DN, Reager JT, Beaudoing HK, Landerer FW, Lo M-H (2018) Emerging trends in global freshwater availability. Nature 557:651–659. https://doi.org/10.1038/s41586-018-0123-1

Sun AY, Scanlon BR et al (2019) Combining physically based modeling and deep learning for fusing GRACE satellite data: can we learn from mismatch? Water Resour Res 55(2):1179–1195. https://doi.org/10.1029/2018WR023333

Sun AY, Scanlon BR, Save H, Rateb A (2021) Reconstruction of GRACE total water storage through automated machine learning. Water Resour Res 57(2):e2020WR028666. https://doi.org/10.1029/2020WR028666

Sun Z, Long D, Yang W, Li X, Pan Y (2020) Reconstruction of GRACE data on changes in total water storage over the global land surface and 60 basins. Water Resour Res 56(4):e2019WR026250. https://doi.org/10.1029/2019WR026250

Syed TH, Famiglietti JS, Rodell M, Chen J, Wilson CR (2008) Analysis of terrestrial water storage changes from GRACE and GLDAS. Water Resour Res 44(2). https://doi.org/10.1029/2006WR005779

Tapley B, Bettadpur S, Watkins M, Reigber C (2004) The gravity recovery and climate experiment: mission overview and early results. Geophys Res Lett 31:1–4. https://doi.org/10.1029/2004GL019920

Thomas B, Famiglietti J (2019) Identifying climate-induced groundwater depletion in GRACE. Sci Rep 9:4124. https://doi.org/10.1038/s41598-019-40155-y

Wang Y (2011) Smoothing splines: methods and applications. CRC Press, p 384

Yang P, Xia J, Zhan C, Wang T (2018) Reconstruction of terrestrial water storage anomalies in Northwest China during 1948–2002 using GRACE and GLDAS products. Hydrol Res 49(5):1594–1607. https://doi.org/10.2166/nh.2018.074

Yao Y, Liang S et al (2017) Improving global terrestrial evapotranspiration estimation using support vector machine by integrating three process-based algorithms. Agric For Meteorol 242:55–74. https://doi.org/10.1016/j.agrformet.2017.04.011

Yeh W-C (2013) New parameter-free simplified swarm optimization for artificial neural network training and its application in the prediction of time series. IEEE Trans Neural Netw Learn Syst 24(4):661–665. https://doi.org/10.1109/TNNLS.2012.2232678

Yeh W-C, Jiang Y, Tan S-Y, Yeh C-Y (2021) A new support vector machine based on convolution product. Complexity 9932292. https://doi.org/10.1155/2021/9932292

Yi S, Sneeuw N (2021) Filling the data gaps within grace missions using singular spectrum analysis. J Geophys Res: Solid Earth 126(5):e2020JB021227. https://doi.org/10.1029/2020JB021227

Zhang C, Bengio S, Hardt M, Recht B, Vinyals O (2021) Understanding deep learning (still) requires rethinking generalization. Commun ACM 64(3):107–115. https://doi.org/10.1145/3446776

Chapter 10
Regularized Neural Network for Tide Modeling

10.1 Introductory Remarks

Tide is an expression used to describe the periodic variations in water level, typically characterized by the vertical rise and fall of sea levels induced by the gravitational forces of the moon, sun, and rotation of the earth (e.g., de Jong et al. 2010; Tebaldi et al. 2012; Zaron and Ray 2017). Basically, the three types of tide observed in different parts of the globe are diurnal, semi-diurnal, and mixed tides. Diurnal tide is characterized by one high and one low tide each day, the semi-diurnal tide has two nearly equal high and low tides each day, while the mixed tide has two uneven tides a day (or in other words is a combination of diurnal and semi-diurnal tides) (e.g., Okwuashi and Ndehedehe 2017; Okwuashi and Olayinka 2017). Whereas the marine industries rely on tide data for several of their activities, accurate tide prediction is vital for construction activities in coastal and offshore areas, installation of platforms, navigation, among others (Xu et al. 2017).

Tide prediction[1] typically involves the use of observed consecutive time series measurements to forecast future values (e.g., Field et al. 2017; Merrifield et al. 2013). The pioneering work of Newton in 1687 formed the basis for recent development of tides as natural phenomena due to the gravitational forces of the sun and moon acting on a rotating earth (Shu 2003). Apart from these aforementioned tide studies, conventional tide modeling experiments based on the method of Least Squares (LS) have also been reported (e.g., Okenwa 1978; Shu 2003). Apparently, the application of the Standard Neural Network (SNN) or Artificial Neural Network (ANN) to tide prediction has been widely studied (see, e.g., Chang and Lin 2006; Lee and Jeng 2002; Lee 2004). In fact, due to its ability to develop complex non-linear models, the SNN has evolved over the years as a reliable modeling and prediction tool with a wide range of environmental applications, be it hydro-climatic science, water resources, or the studies of physical processes at various catchment scales (see, e.g., Campolo

[1] This is an invited chapter based on the experiment undertaken by Dr. Onuwa Okwuashi, University of Uyo, Nigeria and Dr. Christopher Ndehedehe, Griffith University, Queensland, Australia.

C. Ndehedehe, *Hydro-Climatic Extremes in the Anthropocene*, Springer Climate,
https://doi.org/10.1007/978-3-031-37727-3_10

et al. 1999; Chen et al. 2012; Ndehedehe et al. 2018; Sertel et al. 2008; Sudha Rani et al. 2017; Yin et al. 2015). In the next section, some examples of tide modeling using machine learning are briefly discussed.

10.1.1 Machine Learning and Statistical Models for Tide Forecast

Even though the SNN nonparametric model has proven to be substantially more robust than the parametric models, e.g., the LS, etc. (e.g., Darwish et al. 2014; Pombeiro et al. 2017), some weakness associated with the SNN approach necessitated the need to further compare the efficiency of several neural network algorithms. For instance, Sertel et al. (2008) studied the daily mean sea level heights using the Multiple Linear Regression (MLR), and three ANN algorithms namely Feed Forward Back Propagation (FFBP), Radial Basis Function (RBF), and Generalized Regression Neural Network (GRNN) algorithms. Their numerical results showed that the FFBP, RBF, and MLR algorithms produced significantly better results than the GRNN method, and the best performance was obtained using the FFBP algorithm. To correct poor calculations with a two-dimensional hydrodynamic model in predicting storm surge height during typhoon events, Chen et al. (2012), explored the Back Propagation Neural Network (BPNN) and Adaptive Neuro-Fuzzy Inference System (ANFIS) algorithms. Their results demonstrated the utility of the ANFIS technique in predicting water levels along the east coast of Taiwan during typhoon events.

Apart from the relatively higher uncertainties of neural networks, e.g., in predicting groundwater level fluctuations (Yoon et al. 2011), SNNs, similar to other machine learning techniques (e.g., kernel regression) have over fitting and over-parameterization problems that make resulting predictions exceed the range of the training sample (e.g., El-Shafie and Noureldin 2011; Okut 2016). As a result, Okut (2016) argues that the use of regularized neural networks attenuate the bias of parameter estimates considerably, thereby improving the generalization ability of the model. For instance, a study on the inflow forecasting at Aswan High Dam revealed that two Generalized Neural Network (i.e., Regularized Neural Network-RNN and Ensemble Neural Network) can be used to address the over-fitting of training data in SNN or classical ANN models (El-Shafie and Noureldin 2011). As a way of improving modeling performance, the joint use of Bayesian regularization with other generalization approaches was recommended by Anctil and Lauzon (2004). Although similar concept was also echoed in Khashei and Bijari (2010), they went further to show that integrating different ANN models under a hybrid formation is an effective approach to improving the forecasting accuracy and predictive performance of neural networks in general. However, despite the improved accuracy of regularized neural networks indicated in some of the aforementioned studies, that is, when compared to SNN or classical ANN models, the application of RNN in tide modeling is largely unreported.

Tide levels can be predicted using various techniques, including traditional parametric methods such as the least squares model and nonparametric support vector machine regression (SVMR) models. Unlike least squares, SVMR models aim to construct a hyperplane that lies as close as possible to the data points by selecting a hyperplane with a small norm that minimizes the sum of the distances from the data points to the hyperplane. To achieve generalised performance, SVMR attempts to minimize the generalization error bound rather than just the observed training error. The idea behind SVMR is to compute a linear regression function in a high-dimensional feature space, where the input data are mapped through a non-linear function. In previous studies (Okwuashi and Ndehedehe 2017), researchers have explored various SVMR kernels, including linear, polynomial, RBF and neural kernels, and compared their performance to that of the traditional least squares technique. Moreover, to mitigate the effect of multicollinearity, the partial least squares regression (PLSR) technique is also emerging and helps to reduce the number of regressors by decreasing the number of components and then discarding any redundant ones (e.g., Ndehedehe and Ferreira 2020). The most effective components are then selected. Researchers have found that the PLSR model is a viable alternative to the traditional model for predicting tide levels (Okwuashi et al. 2020).

10.1.2 Why Tide Forecast?

Sea level changes are caused by two main factors: thermal expansion of the oceans and a reduction in terrestrial ice storage (Okwuashi and Ndehedehe 2017). Tidal phenomena, which describe the periodic movement of water resulting from differential gravitational forces, also contribute to sea level variations (Shu 2003). This can be observed as the rise and fall of the sea level. To continuously monitor sea level, a tide gauge device is conventionally used to record tide levels relative to a datum (Okwuashi and Ndehedehe 2017; Tebaldi et al. 2012). Accurately predicting tides is essential for construction activities in coastal and offshore areas, as well as for installation of platforms and navigation. The movement of ice sheets near their marine margins experiences high-frequency variability caused by ocean tides. Past studies (e.g., Chang and Lin 2006; Okwuashi et al. 2020; Xu et al. 2017) have confirmed the importance of having precise tidal measurements and forecasts. To support the accurate prediction of tide for several marine and construction activities in coastal regions among other applications, this chapter evaluates the potential and efficiency of RNN in modeling tide levels. Notably, the RNN implemented in this study is developed by a systematic optimisation of the SNN. To assess the performance of the proposed RNN, the numerical results of the RNN are compared with those of the SNN and conventional least squares technique using the Jarque–Bera test statistics (Jarque and Bera 1980) and other prediction skill metrics or evaluation statistics. More details regarding the mathematical formulation and modeling framework are provided in the next section.

10.2 Data and Methods

The modeling was based on tide gauge data obtained at Lagos, Nigeria (WOCE 2006). The tide gauge was located at latitude: 6.423300oN and longitude: 3.411700oE. The data were extracted from consecutive hourly tidal observations for a period of one year. One thousand consecutive hourly tidal observations were extracted from the field data and split into two as shown Fig. 10.1. The first 900 datasets served as the training data. It should be noted that the neural network system evaluates each run of the experiment using regression, by splitting the first 900 datasets internally into training, validation, and testing components. Meanwhile, the second part of the data (i.e., the remaining 100 datasets) was used for testing (or forward prediction) based on the hold-out method of cross-validation (e.g., Haley 2017). The experiment was implemented with the first seven tidal constituents M_2, S_2, N_2, K_2, K_1, O_1, and P_1 (de Jong et al. 2010; Foreman 1996; Quaresma and Pichon 2013). The tidal constituents' frequencies were $M_2 = 28.9841042$o/hr, $S_2 = 30.0000000$o/hr, $N_2 = 28.4397295$o/hr, $K_2 = 30.0821373$o/hr, $K_1 = 15.0410686$ o/hr, $O_1 = 13.9430356$ o/hr and $P1 = 14.9589314$ o/hr. M_2 and S_2 denote principal lunar semi-diurnal constituents. N_2 indicates larger lunar elliptic semi-diurnal constituent. K_2 denotes the lunisolar semi-diurnal constituent. K_1 and O_1 indicate lunar diurnal constituents, and P_1 denotes the solar diurnal constituent. The basic equation for computing the

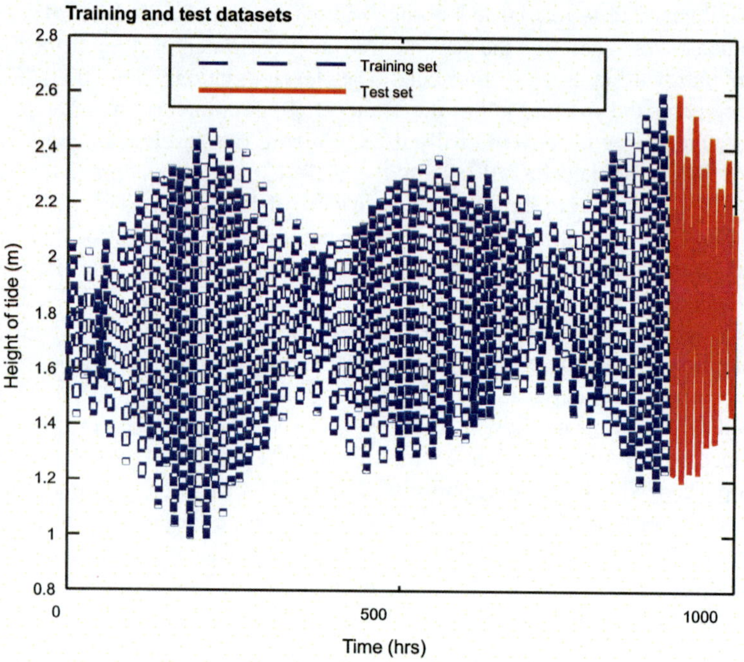

Fig. 10.1 Training and test datasets extracted from consecutive hourly observations

height of tide at any time can be expressed as the sum of harmonic terms (e.g., Foreman 1996; Najibi et al. 2013; Okenwa 1978),

$$h(t) = s_0 + \sum_{k=1}^{\propto} H_k \cos(\sigma_k t + \phi_k), \tag{10.1}$$

where s_0 is the height of mean water level above the datum in use, H_k is the amplitude of the tidal constituent, k, σ_k is the constituent frequency, and ϕ_k is the phase of the constituent. "The phase of the constituent ϕ_k depends on the varying mean longitude of the moon's perigee and sun's perigee with periods of approximately 8.61 and 2100 years respectively and the ecliptic longitude of the moon's ascending node with a period of 18.61 tropical years (Okenwa 1978)". To take these effects into account the f_k, v_k, and u_k were introduced into Eq. 10.1,

$$h(t) = s_0 + \sum_{k=1}^{n} f_k H_k \cos(\sigma_k t + (v_k + u_k) - \phi_k), \tag{10.2}$$

where n is the number of harmonic constituents, v_k is the phase angle at time zero, u_k is the nodal angle, and f_k is the nodal factor. Equation 10.2 can be expanded by using trigonometric identity as

$$h(t) = s_0 + \sum_{k=1}^{n} f_k H_k \cos[\sigma_k t + (v_k + u_k)] \cos_{\phi_k} + \sum_{k=1}^{n} f_k H_k \sin[\sigma_k t + (v_k + u_k)] \sin \phi_k. \tag{10.3}$$

The parameters H_k and ϕ_k are the harmonic constants (Foreman 1996). The parameters f_k, v_k, u_k, $h(t)$, σ_k, and t are the known parameters while s_0, H_k, and ϕ_k are the unknown parameters. Let

$$a_k = f_k H_k \cos[(v_k + u_k) - \phi_k], \tag{10.4}$$

and

$$b_k = f_k H_k \sin[(v_k + u_k) - \phi_k]. \tag{10.5}$$

We divide Eq. 10.5 by Eq. 10.4 to obtain

$$\tan[(v_k + u_k) - \phi_k] = b_k / a_k. \tag{10.6}$$

Introducing Eqs. 10.4 and 10.5 into Eq. 10.3 yields,

$$h(t) = s_0 + \sum_{k=1}^{n} a_k \cos(\phi_k.t) + \sum_{k=1}^{n} b_k \sin(\phi_k.t) \tag{10.7}$$

where a_k and b_k are the constituents'constants. For the formulation of the LS, the adjusted observations $\hat{Y}^a$ can be expressed as a function of the adjusted parameters $\hat{X}^a$,

$$\hat{Y}^a = f(\hat{X}^a). \tag{10.8}$$

Equation 10.8 can be expanded as

$$\hat{Y}^b + \hat{R} = f(\hat{X}^0 + \hat{X}), \tag{10.9}$$

where $\hat{Y}^b$ denote the original observations, $\hat{R}$ denotes the residuals, $\hat{X}^0$ denotes the approximate values of the unknown parameters, and $\hat{X}$ denotes the estimate of the unknown parameters. Linearising Eq. 10.9 using Taylor's series and truncating at the first order, Eq. 10.9 becomes

$$\hat{Y}^b + \hat{R} = f(\hat{X}^0) + \frac{\partial f(\hat{X}^0)}{\partial \hat{X}^a}\hat{X} \mid_{\hat{X}^a=\hat{X}^0} + \ldots. \tag{10.10}$$

Assuming,

$$C = \frac{\partial f(\hat{X}^0)}{\partial \hat{X}^a} \mid_{\hat{X}^a=\hat{X}^0} + \ldots, \tag{10.11}$$

and

$$\hat{Y} = (f(\hat{X}^0) - \hat{Y}^b). \tag{10.12}$$

Therefore Eq. 10.10 can be rewritten as

$$\hat{R} = (C\hat{X}^0 - \hat{Y}). \tag{10.13}$$

Using Lagrange multiplier $\hat{K}^T$ to minimize the sum of the squares of the weighted residuals $\hat{R}^T W \hat{R}$, where W denotes the weight of observations, therefore,

$$\xi = \hat{R}^T W \hat{R} - \hat{K}^T (C\hat{X} + \hat{Y} - \hat{R}) \; is \; minimum. \tag{10.14}$$

Differentiating Eq. 10.14 partially as $\frac{\partial \xi}{\partial \hat{R}}, \frac{\partial \xi}{\partial \hat{K}^T}, \frac{\partial \xi}{\partial \hat{X}}$, and solving the normal equations $\frac{\partial \xi}{\partial \hat{R}} = 0, \frac{\partial \xi}{\partial \hat{K}^T} = 0, \frac{\partial \xi}{\partial \hat{X}} = 0$ simultaneously yields the solution of $\hat{X}$,

$$\hat{X} = (C^T W C)^{-1} C^T W \hat{Y}^b. \tag{10.15}$$

Using Eq. 10.15 the constituents constants s_0, a_k, b_k, in Eq. ?? can be determined as

$$C = \begin{bmatrix} 1 & \cos\sigma_1 t_1 & \sin\sigma_1 t_1 & . \; . \; . & \cos\sigma_n t_1 & \sin\sigma_n t_1 \\ 1 & \cos\sigma_1 t_2 & \sin\sigma_1 t_2 & 4 \; 5 \; 6 & \cos\sigma_n t_2 & \sin\sigma_n t_2 \\ . & . & . & & . & . \\ . & . & . & & . & . \\ . & . & . & & . & . \\ 1 & \cos\sigma_1 t_m & \sin\sigma_1 t_1 & . \; . \; . & \cos\sigma_n t_m & \sin\sigma_n t_m \end{bmatrix},$$

$$Y = |h(t_1), h(t_2), h(t_3), \ldots, h(t_m)|^T, \tag{10.16}$$

$$\hat{X} = |s_0, a_1, b_1, \ldots, a_n, b_n|^T. \tag{10.17}$$

The corresponding weight is

$$W = \sum_{Y}^{-1} = diag|\frac{1}{\sigma_{Y_1}^2}, \frac{1}{\sigma_{Y_2}^2}, \frac{1}{\sigma_{Y_3}^2}, \ldots, \frac{1}{\sigma_{Y_n}^2}| \tag{10.18}$$

where $\sigma_{Y_1}^2, \sigma_{Y_2}^2, \sigma_{Y_3}^2, \ldots, \sigma_{Y_n}^2$ are errors in the observations. Recall $\hat{R}$ denotes the residuals that represent the errors between the actual tidal data and the predicted tidal data. From Eq. 10.13 $\hat{R}$ can be computed by substituting Eq. 10.12 into Eq. 10.13, and setting the function $f(\hat{X}^0)$ to zero, since the experiment is a linear experiment. Therefore Eq. 10.13 becomes

$$\hat{R} = C\hat{X} - \hat{Y}^b. \tag{10.19}$$

The multi-layer feed-forward neural network was used in this experiment. A multi-layer feed-forward neural network typically consists of neurons that are ordered into layers. The first layer is the input layer, the middle layer is the hidden layer, while the last layer is called the output layer (Karri and Sahu 2018). Each neuron in a layer is connected with all the neurons in the next layer (Fig. 10.2). The formulation of a multi-layer feed forward neural for output y based on inputs x and weight w can be expressed as

$$y(x, \omega) = \sum_{i=1}^{n} w_i^1 . f(w_i^0 x + b_i) + c, \tag{10.20}$$

where n denotes the number of hidden neurons, w_i^1 denotes the weight of the connection between ith hidden neurons and the output neuron, w_i^0 denotes the weight of the connection between the input neuron and the ith hidden neurons, b_i denotes the corresponding bias, and c denotes the bias of the output neuron. The activation function is the sigmoid function expressed as (Ayoobi and Tangestani 2017),

$$y(x, \omega) = \frac{1}{1 + e^{-(x, \omega)}}. \tag{10.21}$$

The training error in the network is computed as the sum of the errors of all the neurons in the output layer (e.g., D'Addona et al. 2017),

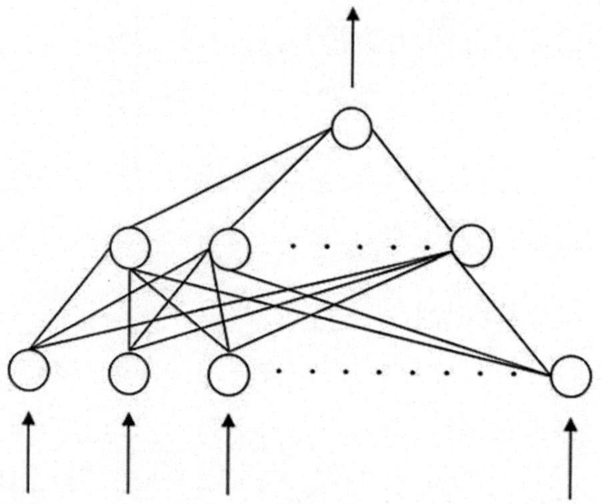

Fig. 10.2 An illustration of a multi-layer feed forward neural network architecture

$$e(x, \omega, y) = \frac{1}{n}\sum_{i=1}^{n}(f_i(x, \omega) - y_i)^2. \tag{10.22}$$

The training process of the network is basically executed through a series of patterns. The learning process involves the adjustment of interconnection weights within the input and output value (Rajasekaran et al. 2008). The BPNN learning model was used in this experiment. The procedure of the BPNN is implemented by adjusting the weights of the connections in the network in order to minimize the error (the difference between actual output vector of the network and the desired output vector). The back propagation procedure calculates the weight of the network and adjusts the weight of interconnections to minimize the output error. The learning process varies the threshold coefficient Θ_i and weight coefficient ω_{ij} to minimize the sum of the squared differences between the computed and the actual output values. In back propagation algorithm the steepest-descent minimisation method is used for the adjustment of the weight and threshold coefficients, it therefore holds that

$$\omega_{ij}^{p+1} = \omega_{ij}^{p} - \eta(\frac{\partial e}{\partial \omega_{ij}})^p \tag{10.23}$$

and

$$\Theta_i^{p+1} = \Theta_i^{p} - \eta(\frac{\partial e}{\partial \Theta_{ij}})^p, \tag{10.24}$$

where η is the rate of learning $\eta > 0$. The main problem is the calculation of the derivatives $\partial e/\partial \omega_{ij}$ and $\partial e/\partial \Theta_{ij}$. Calculation goes through the following steps: First step,

$$e = \frac{1}{2}(x_0 - \hat{x}_0)^2 = \frac{1}{2}\sum_p h_p^2 \tag{10.25}$$

where $h_p = (x_p - \hat{x}_p)$ for $p \in$ output layer, $h_p = 0$ for $p \ni$ output layer. Second step,

$$\frac{\partial e}{\partial w_{ij}} = \frac{\partial e}{\partial x_i}\frac{\partial x_i}{\partial w_{ij}} = \frac{\partial e}{\partial x_i}\frac{\partial f(x_i, \omega_i)}{\partial \omega_{ij}} \tag{10.26}$$

$$= \frac{\partial e}{\partial x_i}\frac{\partial f(x_i, \omega_i)}{\partial (x_i, \omega_i)}\frac{\partial (x_i, \omega_i)}{\partial \omega_{ij}}, \tag{10.27}$$

$$= \frac{\partial e}{\partial x_i} f'(x_i, \omega_i)\frac{\partial (\sum_{j\in\Omega_i^{-1}} \omega_{ij}x_j + \Theta)}{\partial \omega_{ij}}. \tag{10.28}$$

The third step, given that output layer, therefore, $\frac{\partial e}{\partial x_i} = h_i$ For $i \in$ hidden layer,

$$\frac{\partial e}{\partial x_i} = \sum_{r=\Omega_i}\frac{\partial e}{\partial x_r}\frac{\partial x_r}{\partial x_i} = \sum_{r=\Omega_i}\frac{\partial e}{\partial x_r}\frac{\partial f(x_r, \omega_r)}{\partial x_i}, \tag{10.29}$$

and

$$= \sum_{r=\Omega_i}\frac{\partial e}{\partial x_i}\frac{\partial f(x_r, \omega_r)}{\partial (x_r, \omega_r)}\frac{\partial (x_r, \omega_r)}{\partial x_i} = \sum_{i=\Omega_i}\frac{\partial e}{\partial x_r} f'(x_r, \omega_r)\omega_{ii} = \sum_{r=\Omega_i}\frac{\partial e}{\partial \Theta_r}\omega_{ii}. \tag{10.30}$$

Using this procedure the derivatives of the objective function for the output layer and then for the hidden layers can be recurrently calculated. This BPNN is so called because the output error propagates from the output layer through the hidden layers to the input layer (Jiang et al. 2018). For the neural network modeling, the matrix of independent variables is represented as

$$X_{m\times n} = \begin{bmatrix} \cos\sigma_1 t_1 & \sin\sigma_1 t_1 & . \; . \; . & \cos\sigma_n t_1 & \sin\sigma_n t_1 \\ \cos\sigma_1 t_2 & \sin\sigma_1 t_2 & 4\;5\;6 & \cos\sigma_n t_2 & \sin\sigma_n t_2 \\ . & . & & . & . \\ . & . & & . & . \\ . & . & & . & . \\ \cos\sigma_1 t_m & \sin\sigma_1 t_1 & . \; . \; . & \cos\sigma_n t_m & \sin\sigma_n t_m \end{bmatrix}.$$

The vector of the dependent variable is represented as

$$Y_{m\times 1} = |h(t_1), h(t_2), h(t_3), \ldots, h(t_m)|^T. \tag{10.31}$$

The potential of RNN in tide modeling was evaluated by comparing its numerical results with those of SNN and LS using the Jarque–Bera test statistic which is computed as

$$JB = \frac{n}{6}[s^2 + \frac{(k-3)^2}{4}], \tag{10.32}$$

where JB denotes Jarque–Bera statistic, n is the sample size, s is the sample skewness, and k is the sample kurtosis. The normality of the LS residual and histogram were also examined with the Jarque–Bera test (Jarque and Bera 1980; Kim 2016). The hypotheses for the normality test can be stated as follows: Null hypothesis—H_0: the residuals are normally distributed at the 95% confidence level; alternative hypothesis—H1: the residuals are not normally distributed (reject H_0 if p-value <0.05). Furthermore, to evaluate the errors in the predictions, the Root Mean Square Error (RMSE) was computed as

$$RMSE = \sqrt{\frac{1}{n}\sum_{j=1}^{n}(y_j - \hat{y}_j)^2}, \tag{10.33}$$

where y_j is the observed tide, $\hat{y}_j$ is the predicted tide, while n represents the number of observations (e.g., Rajasekaran et al. 2008).

10.3 Results

Nine hundred hourly tidal observations were used for training the neural network. By default the neural network splits the data into three: training, validation, and testing. The goal of this experiment is to train the neural network to minimize the training error to achieve the least training error through adjusting the weights of the network, by optimally selecting the hidden neurons using the process of cross-validation. Part of the training data was used to perform the validation of the network parameters. The basic equations for the SNN are the same as those of the RNN. Essentially, the two free parameters in the neural network are the random seed number and the number of neurons. The optimal random seed number is set to ensure that the outputs remain the same for the same values of the number of neurons. If not the results will continue to vary using the same number of neurons. The optimal number of neurons represents the setting of the required number of neurons in the hidden layer.

The neural network minimizes the training error between the observed and the predicted data. For the RNN, the training error is further minimized by introducing three additional parameters (the epoch parameter, the goal parameter, and the performance regularization parameter). The epoch parameter determines the period for the convergence of the training of the network. The goal parameter determines the sensitivity of the network; that is, how the uncertainty in the output can be distributed to different sources of uncertainty in its inputs. The performance regularization parameter determines the generalization performance of the network. By finding the optimal parameters for these three additional parameters, unlike the SNN

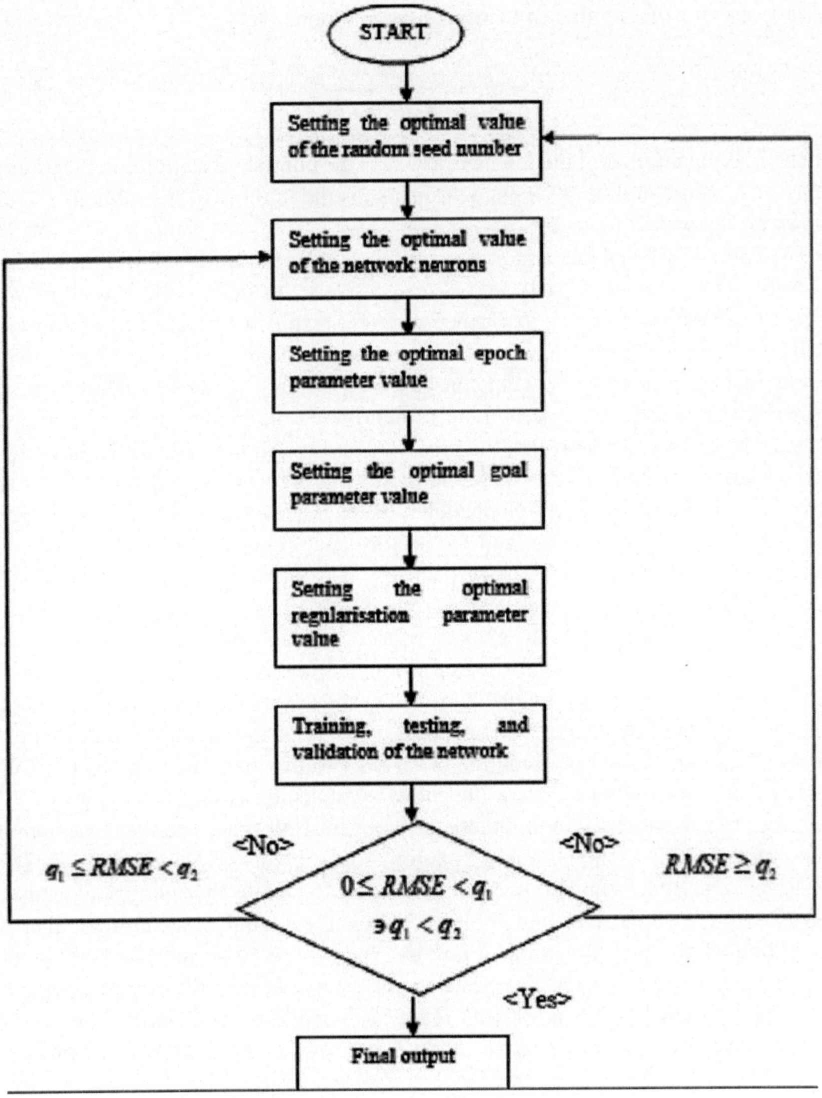

Fig. 10.3 A concise flowchart of the regularize neural network implementation

the RNN further minimizes the training error and enhances the accuracy of the prediction. The schematic flow, which shows the procedures for implementing the RNN is indicated in Fig. 10.3. In order to improve the performance of the neural network (or to further minimize the training error in the network), the network was regularized by employing a cross-validation procedure. The conventional method of selecting the optimal parameter values is through cross-validation (Anders and Korn 1999; Aras and İpek Deveci Kocakoç 2016). The following formula was used to obtain an initial guess for the number of neurons in the hidden layer,

$$n_h = \frac{n_s}{k(n_i + n_O)} \tag{10.34}$$

where n_h is the number of hidden neurons, n_s is the number of samples in the training datasets, n_i is the number of input neurons, n_O is the number of output neurons, and k is an arbitrary scaling factor.

The optimal value of the random seed number and the optimal value of the network neurons were determined as 10 and 27 respectively. By default the neural network is initialised with initial weights; hence different results were obtained in every run of the network. To ensure that the results stay the same in every run of the network the random seed number was kept constant. The training of the neural network was implemented by adjusting the number of neurons in the hidden layer. The training error is the discrepancy between the predicted and the actual value. The adjustment of the number of neurons was sustained until the highest generalization accuracy was obtained (Okwuashi and Ndehedehe 2015). The optimal values of the epoch parameter, the goal parameter, and the performance regularization parameter were determined as 20, 10e-5, and 10e-6 respectively. A Root Mean Squared Error (RMSE) threshold was used to regulate the experiment. The process was repeated when the computed RMSE was above the error threshold value. Figure 10.4a and b and Table 10.1 show the computed mean squared error; the computed gradient, mu, and val fail; and the computed training, validation, test and all regression plots, respectively. Figure 10.5a and b indicates the temporal patterns of the observed and predicted tides and the correlation between the observed and predicted tides based on RNN, respectively. The computed r^2 for the above relationship was 0.9768.

A second experiment was implemented using the SNN based on the same optimal random seed number and number of neurons found by cross-validation, as those of the RNN (recall, the random seed number was 10, while the number of neurons was 27). Figure 10.6a and b and Table 10.1 show the computed mean squared error; the computed gradient, mu, and val fail; and the computed training, validation, test and all regression plots respectively. The temporal patterns of the observed and predicted tides based on SNN is shown in Fig. 10.7a while the correlation between the observed and predicted tides based on SNN is shown in Fig. 10.7b. The computed r^2 was 0.9659. A third experiment was implemented using the LS. The temporal series of the observed and predicted tides and the correlation between the observed and predicted tides based on LS are depicted in Fig. 10.8a and b, respectively. The computed r^2 was 0.9599. The computed constituent constants were presented in Table 10.2. The computed phases using equation 6 were $M_2 = 216.5546o$, $S_2 = 47.6735o$, $N_2 = 93.5035o$, $K_2 = 93.3133o$, $K_1 = 72.6062o$, $O_1 = 13.5932o$, and $P_1 = 166.0451^0$. The computed amplitudes using either Eqs. 4 or 5 were: $M_2 = 0.4020$ m, $S_2 = 0.1295$ m, $N_2 = 0.1005$ m, $K_2 = 0.0364$ m, $K_1 = 0.0951$ m, $O_1 = 0.0201$ m and $P_1 = 0.0237$ m. The composite result of all the experiments using

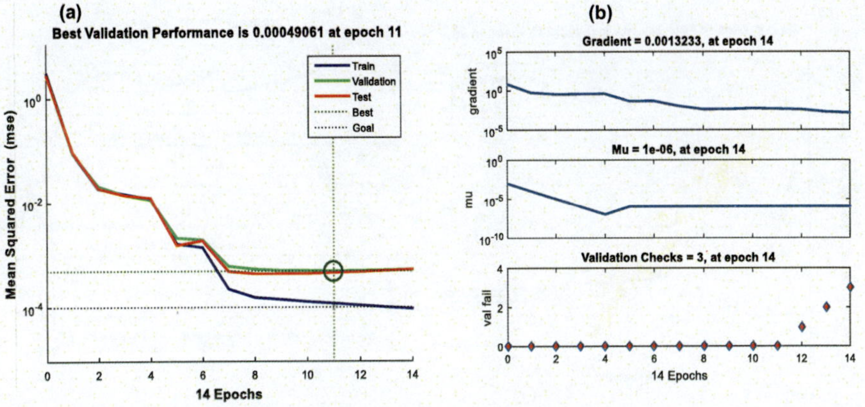

Fig. 10.4 **a** Graph showing the computed mean squared error for RNN and **b** graphs showing the computed gradient, mu, and val fail for RNN

Table 10.1 Comparing the neural network parameters and data components (training, validation, test, and all samples) for the RNN and SNN models based on observed linear correlations (r) from the regression outputs

Model	Training	Validation	Test	All samples
RNN	0.99947	0.99784	0.99808	0.9990
SNN	0.98840	0.98557	0.98695	0.98757

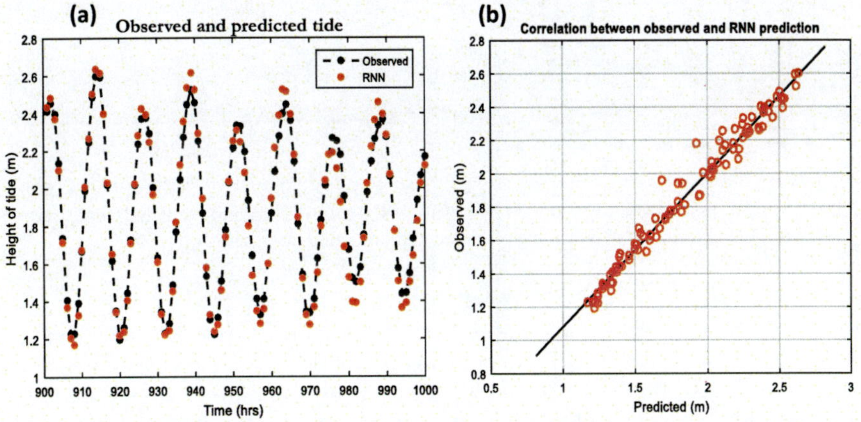

Fig. 10.5 Skill assessment of predicted tide using the proposed RNN method. **a** Observed and predicted tide based on RNN **b** the correlation between the observed and predicted tides based on RNN

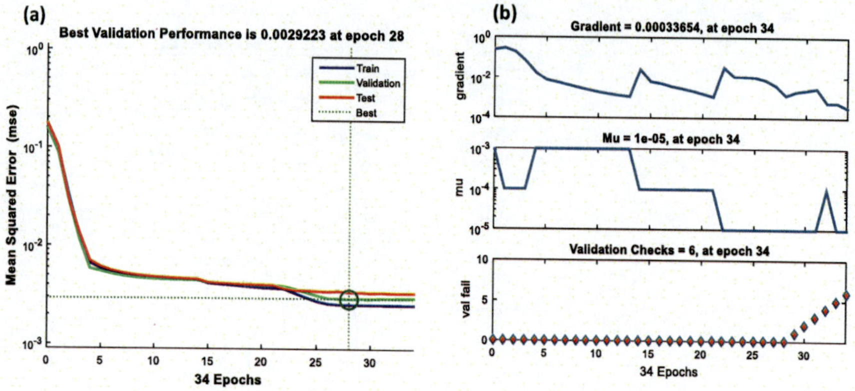

Fig. 10.6 **a** Computed mean squared error for SNN and **b** the computed gradient, mu, and val fail for SNN

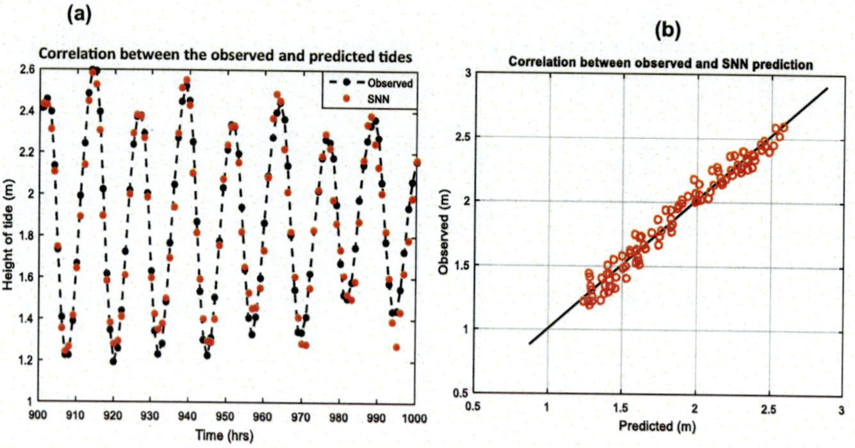

Fig. 10.7 Skill assessment of predicted tide using SNN. **a** Observed and predicted tide based on SNN and **b** the correlation between the observed and predicted tides based on SNN

the LS, SNN and RNN is indicated in Fig. 10.9. Figure 10.10 shows the computed residuals for RNN, SNN, and LS computed by subtracting the predicted tide from the observed tide. Figure 10.10 also shows the histogram plot of the LS. Only the histogram plot of the LS residuals was plotted while those of the RNN and SNN were not necessary since the computed residuals of nonparametric models, e.g., RNN and SNN do not require fulfilling the normality statistically criterion (Yadav and Chandel 2017). The distribution of the computed residuals shown in Fig. 10.10 indicated that the computed residuals for RNN, SNN, and LS were fairly normality distributed. It should be noted that the normality of the histogram plots indicated in Fig. 10.10 was also examined using the Jarque–Bera test. From the result of the Jarque–Bera

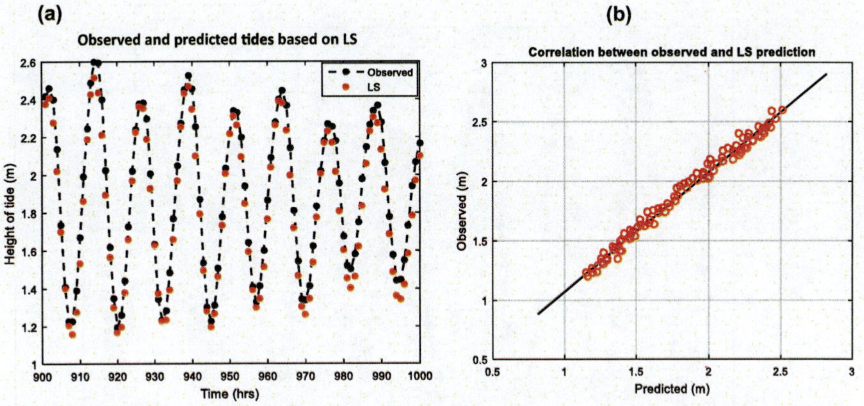

Fig. 10.8 Skill assessment of predicted tide using the LS approach. **a** Observed and predicted tide based on LS and **b** the correlation between the observed and predicted tides based on LS

Table 10.2 Least squares results for computed constituent constants. Values with asterisks (∗) are significant at $\pm t > 1.96$ or $p < 0.05$ at the 95% confidence level

Constituent constant	Coefficient	t stat	Standard error	p value
s_0	1.8075	805.8921	0.0022	0.0000*
a_1	−0.3229	−98.5366	0.0033	0.0000*
b_1	−0.2394	−73.0686	0.0033	0.0000*
a_2	0.0872	9.7905	0.0089	1.47e-21
b_2	0.0957	10.8025	0.0089	1.23e-25*
a_3	−0.0061	−1.8785	0.0033	0.0606
b_3	0.1003	30.6873	0.0033	2.02e-141*
a_4	−0.0021	−0.2374	0.0089	0.8124
b_4	0.0363	4.0959	0.0089	4.59e-05*
a_5	0.0284	3.2352	0.0088	0.0013
b_5	0.0908	10.3139	0.0088	1.25e-23
a_6	0.0196	6.0833	0.0032	1.75e-09
b_6	0.0047	1.4753	0.0032	0.1405
a_7	−0.023	−2.6035	0.0088	0.0094*
b_7	0.0057	0.6488	0.0088	0.5167

normality test for the LS residuals ($JB = 2.8524$; critical value $= 5.4314$; p value $=$ 0.1579), the computed p value was greater than 0.05, we therefore accept H_0 at the 95% confidence level that the computed residual was normally distributed. In essence if the residuals are statistically proven to be normally distributed, then the results of the experiments can be trusted. One of the methods of ascertaining the reliability of the experiment is by assessing the nature of the computed tide. Nigerian coastal waters

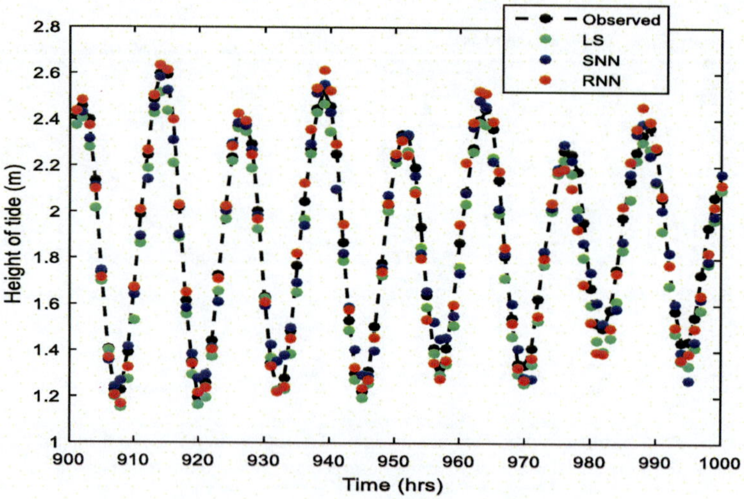

Fig. 10.9 Time series of observed and predicted tide based on LS, SNN, and RNN

are characteristically semi-diurnal in nature. This can be examined by computing the tidal form factor. According to Foreman (1996), the characterization of tide may be obtained by calculating the ratio of the amplitudes of the harmonic constituents. This value is called the "number" or "idal form factor" (Dietrich 1963; Pugh 2004). The classification of the observed tide can be determined using the formula (Foreman 1996),

$$F = \frac{K_1 + O_1}{M_2 + S_2}. \tag{10.35}$$

The tide is then said to be

1. semi-diurnal if $0 \leq F \leq 0.25$,
2. mixed if $0.25 < F \leq 3.00$,
3. diurnal if $F > 3.00$.

Recall that the computed amplitudes for LS for K_1, O_1, M_2, and S_2 were 0.0951, 0.0201, 0.4020, and 0.1295 respectively. Therefore F was computed as $F = \frac{0.0951+0.0201}{0.4020+0.1295} = 0.2167$. The computed F fell between 0 and 0.25. Figure 10.11a depicted the box-plot of the residuals for RNN, SNN, and LS respectively. For RNN, the computed minimum, median, and maximum errors were −0.1278, 0.0079, and 0.2653 respectively. For SNN, the computed minimum, median, and maximum errors were −0.1251, 0.0158, and 0.2016 respectively. For LS, the computed minimum, median, and maximum errors were −0.0293, 0.0694, and 0.1859 respectively. Figure 10.11b depicted the graph of the computed RMSEs for RNN, SNN, and LS respectively. The computed RMSEs for RNN, SNN, and LS were 0.0712 m, 0.0776 m, and 0.0869 m respectively.

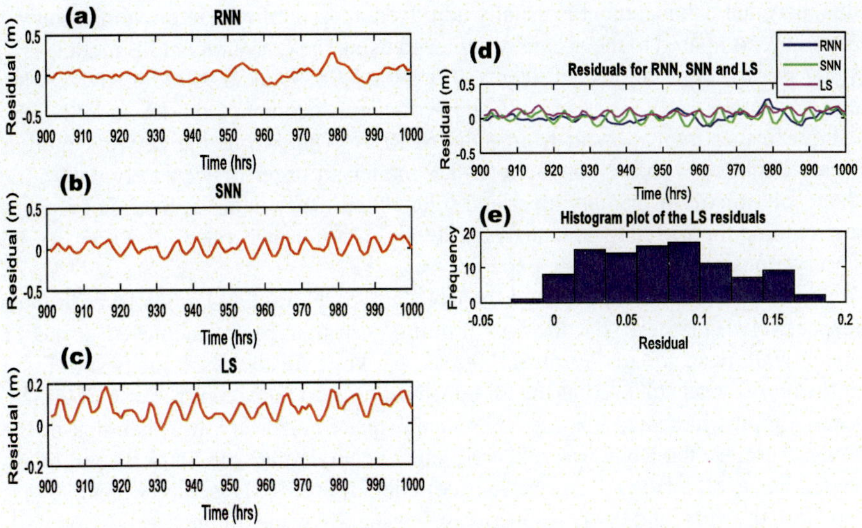

Fig. 10.10 **a**, **b**, **c** The computed residuals for RNN, SNN, and LS respectively, **d** composite plot for RNN, SNN, and LS, **e** histogram plot of LS residuals

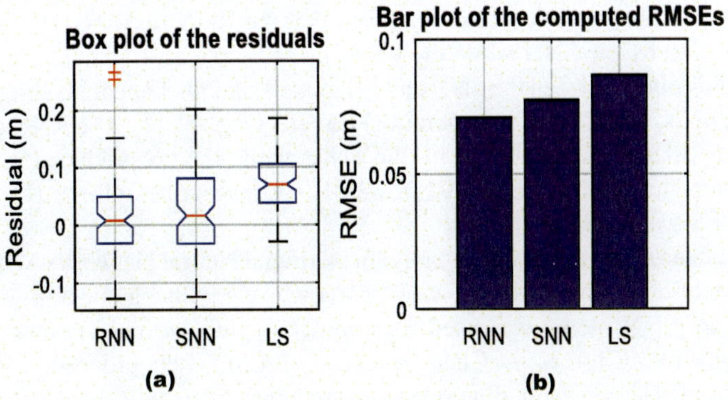

Fig. 10.11 **a** Box plot of the residuals **b** graph of the computed root mean square error (RMSE)

10.4 Discussion

The essence of aggressive research in algorithm refinement is the continuous quest for enhancement in algorithm performance. For example, research has shown that the SNN is more robust and accurate than the conventional LS (e.g., Darwish et al. 2014; Pombeiro et al. 2017). Research has also shown that the RNN is more accurate than the SNN (e.g., Adeli and Wu 1998; Jiang et al. 2018; Larsen et al. 1998; Loone and Irwin 2001; Sulistyo et al. 2017). The result of these researches showed that these new models usually improve the accuracy of the experiment marginally. Scientifically, this marginal enhancement is usually adjudged as a significant improvement on the accuracy, compared to the accuracy achieved using the conventional LS model. This study showed correlation coefficients of 0.9768, 0.9659, and 0.9599 for RNN, SNN, and LS, respectively. These results showed an enhancement from LS to SNN, and even a further improvement from SNN to RNN. The two neural network models were compared with the conventional LS model. In order to accurately determine the resolution of the primary lunar and solar constituents, at least 28 days of records are required for a reliable tidal analysis. Hence one thousand consecutive hourly tidal observations were used for the experiment.

Figures 10.4 and 10.6 depicted the RNN and SNN training hidden layer outputs respectively. For the RNN, the best training performance was achieved at the 11 epoch with mean squared error of 0.00049061 while for the SNN, the best training performance was achieved at the 28 epoch with mean squared error of 0.0029223. These outputs indicated that the RNN mean squared error is lower than that of the SNN; therefore the RNN was relatively more accurate than the SNN. The training, validation, test, and overall correlation coefficients of the RNN method were 0.99947, 0.99784, 0.99808, and 0.999 respectively. For the SNN, the training, validation, test, and overall correlation coefficients were 0.9884, 0.98557, 0.98695, and 0.98757 respectively. The output correlation coefficients indicated that the training results of the RNN were better than those of the SNN. The comparison of the predicted tide against known observed tide values showed that the RNN furnished the highest r^2 value followed by the SNN while the LS furnished the least r^2 value. This implied that the RNN outperformed the SNN and LS models. Another apparent assessment of the RNN, SNN, and LS was the computed RMSEs (Fig. 10.11). The RNN produced the lowest RMSE value followed by the SNN; while the LS produced the highest RMSE value, this implied that the RNN was the most accurate, followed by the SNN while the LS was the least accurate.

The Standard Neural Network (SNN) or Artificial Neural Network (ANN) application to tide prediction has been widely studied (e.g., Chang and Lin 2006). For example, an ANN model was used for simulating tides at multi-points considering tide-generating forces by Chang and Lin (2006). Their proposed model was found to be efficient in estimating tide at a single point. Notably, the RNN and SNN are nonparametric models. As a result, it was impossible to use them to compute the coefficients of the constituents's constants (Albert et al. 2017; Humphrey et al. 2017). The LS arguably does not have this disadvantage being a parametric model (Pino-Mejías et al. 2017). The LS result in this study (Table 10.2) indicated that nearly all the constituent constants were significant at the 95% confidence level. Another method of validating the experiment is the test of the normality of the computed residuals. Unlike the LS model, the computed residuals of the RNN and SNN are not subject to fulfilling statistical normality since the RNN and SNN are nonparametric models (Islam 2017; Pandya et al. 2017). A Jarque–Bera normality test done with the LS residuals indicated that the computed residuals were normally distributed. This normality condition must be fulfilled before the LS result can be accepted. The final validation of this experiment was to test for the nature of the tide. The Nigerian coastal waters are semi-diurnal in nature. The computed tidal form factor F fulfilled

$0 \leq F \leq 0.25$ which implied a semi-diurnal tide. As a result, the experiment can be adjudged valid.

10.5 Concluding Remarks

To support the accurate prediction of tide for several marine and construction activities in coastal regions, this study evaluated the potential and efficiency of Regularized Neural Network (RNN) in modeling tide levels. In order to examine the performance of the proposed RNN, the numerical results of the RNN were compared with those of the Standard Neural Network (SNN) and conventional Least Squares (LS) technique using prediction skill metrics and the Jarque–Bera statistics. The RNN furnished the highest r^2 value followed by the SNN while the LS furnished the least r^2 value. Further assessments based on RMSE also confirmed the RNN performed better with the lowest RMSE value as opposed to the other two approaches which showed relatively higher RMSEs. This implied that the RNN outperformed the SNN and LS models.

References

Adeli H, Wu M (1998) Regularization neural network for construction cost estimation. J Constr Eng Manag 124(1):18–24

Albert LP, Keenan TF, Burns SP, Huxman TE, Monson RK (2017) Climate controls over ecosystem metabolism: insights from a fifteen-year inductive artificial neural network synthesis for a subalpine forest. Oecologia 184(1):25–41. https://doi.org/10.1007/s00442-017-3853-0

Anctil F, Lauzon N (2004) Generalisation for neural networks through data sampling and training procedures, with applications to streamflow predictions. Hydrol Earth Syst Sci 8(5):940–958. https://doi.org/10.5194/hess-8-940-2004

Anders U, Korn O (1999) Model selection in neural networks. Neural Netw 12(2):309–323. https://doi.org/10.1016/S0893-6080(98)00117-8

Aras S, Kocakoç İpek Deveci (2016) A new model selection strategy in time series forecasting with artificial neural networks: IHTS. Neurocomputing 174:974–987. https://doi.org/10.1016/j.neucom.2015.10.036

Ayoobi I, Tangestani MH (2017) The effect of minimum noise fraction data input on success of artificial neural network in lithological mapping of a magmatic terrain with aster data; a case study from SE Iran. Remote Sens Appl: Soc Environ 7:21–26. https://doi.org/10.1016/j.rsase.2017.06.001

Campolo M, Andreuss P, Soldati A (1999) River flood forecasting with a neural network model. Water Resour Res 35(4):1191–1197. https://doi.org/10.1029/1998WR900086

Chang H-K, Lin L-C (2006) Multi-point tidal prediction using artificial neural network with tide-generating forces. Coast Eng 53(10):857–864. https://doi.org/10.1016/j.coastaleng.2006.05.001

Chen W-B, Liu W-C, Hsu M-H (2012) Comparison of ANN approach with 2D and 3D hydrodynamic models for simulating estuary water stage. Adv Eng Softw 45(1):69–79. https://doi.org/10.1016/j.advengsoft.2011.09.018

D'Addona DM, Ullah AMMS, Matarazzo D (2017) Tool-wear prediction and pattern-recognition using artificial neural network and DNA-based computing. J Intell Manuf 28(6):1285–1301. https://doi.org/10.1007/s10845-015-1155-0

Darwish HW, Hassan SA, Salem MY, El-Zeany BA (2014) Different approaches in partial least squares and artificial neural network models applied for the analysis of a ternary mixture of amlodipine, valsartan and hydrochlorothiazide. Spectrochim Acta Part A: Mol Biomol Spectrosc 122:744–750. https://doi.org/10.1016/j.saa.2013.11.045

de Jong C, Lachapelle G, Skone S, Elema I (2010) Hydrography. Delft University Press, Delft

Dietrich G (1963) General oceanography. Interscience Publishers, New York

El-Shafie A, Noureldin A (2011) Generalized versus non-generalized neural network model for multi-lead inflow forecasting at Aswan High Dam. Hydrol Earth Syst Sci 15(3):841–858. https://doi.org/10.5194/hess-15-841-2011

Field CR, Bayard TS, Gjerdrum C, Hill JM, Meiman S, Elphick CS (2017) High-resolution tide projections reveal extinction threshold in response to sea-level rise. Glob Chang Biol 23(5):2058–2070. https://doi.org/10.1111/gcb.13519

Foreman M (1996) Manual for tidal heights analysis and prediction. Pacific Marine Science Report, pp 77–10. http://www.omg.unb.ca/GGE/5013_LABS/heights.pdf on 24th May, 2018

Haley MR (2017) K-fold cross validation performance comparisons of six Naive portfolio selection rules: how Naive can you be and still have successful out-of-sample portfolio performance? Ann Financ 13(3):341–353. https://doi.org/10.1007/s10436-017-0301-4

Humphrey GB, Maier HR, Wu W, Mount NJ, Dandy GC, Abrahart RJ, Dawson CW (2017) Improved validation framework and R-package for artificial neural network models. Environ Model Softw 92:82–106. https://doi.org/10.1016/j.envsoft.2017.01.023

Islam MQ (2017) Estimation and hypothesis testing in multivariate linear regression models under non normality. Commun Stat-Theory Methods 46(17):8521–8543. https://doi.org/10.1080/03610926.2016.1183789

Jarque CM, Bera AK (1980) Efficient tests for normality, homoscedasticity and serial independence of regression residuals. Econ Lett 6(3):255–259. https://doi.org/10.1016/0165-1765(80)90024-5

Jiang YG, Wu Z, Wang J, Xue X, Chang SF (2018) Exploiting feature and class relationships in video categorization with regularized deep neural networks. IEEE Trans Pattern Anal Mach Intell 40(2):352–364. https://doi.org/10.1109/TPAMI.2017.2670560

Karri RR, Sahu J (2018) Modeling and optimization by particle swarm embedded neural network for adsorption of zinc (ii) by palm kernel shell based activated carbon from aqueous environment. J Environ Manag 206:178–191. https://doi.org/10.1016/j.jenvman.2017.10.026

Khashei M, Bijari M (2010) An artificial neural network (p, d, q) model for timeseries forecasting. Expert Syst Appl 37(1):479–489. https://doi.org/10.1016/j.eswa.2009.05.044

Kim N (2016) A robustified Jarque-Bera test for multivariate normality. Econ Lett 140:48–52. https://doi.org/10.1016/j.econlet.2016.01.007

Larsen J, Svarer C, Andersen L, Hansen L (1998) Adaptive regularization in neural network modeling. In: Neural networks: tricks of the trade (pp 113–132). Springer, Berlin, Heidelberg, pp 18–24

Lee T, Jeng D (2002) Application of artificial neural networks in tide-forecasting. Ocean Eng 29(9):1003–1022. https://doi.org/10.1016/S0029-8018(01)00068-3

Lee T-L (2004) Back-propagation neural network for long-term tidal predictions. Ocean Eng 31(2):225–238. https://doi.org/10.1016/S0029-8018(03)00115-X

Loone SM, Irwin G (2001) Improving neural network training solutions using regularisation. Neurocomputing 37(1):71–90. https://doi.org/10.1016/S0925-2312(00)00314-3

Merrifield MA, Genz AS, Kontoes CP, Marra JJ (2013) Annual maximum water levels from tide gauges: contributing factors and geographic patterns. J Geophys Res: Ocean 118(5):2535–2546. https://doi.org/10.1002/jgrc.20173

Najibi N, Abedini A, Sheibani R (2013) Harmonic decomposition tidal analysis and prediction based on astronomical arguments and nodal corrections in Persian Gulf, Iran. Res J Environ Earth Sci 7:381–392

Ndehedehe CE, Agutu NO, Okwuashi O (2018) Is terrestrial water storage a useful indicator in assessing the impacts of climate variability on crop yield in semi-arid ecosystems? Ecol Indic 88C:51–62. https://doi.org/10.1016/j.ecolind.2018.01.026

Ndehedehe CE, Ferreira VG (2020) Assessing land water storage dynamics over Southern America. J Hydrol 580:124339. https://doi.org/10.1016/j.jhydrol.2019.124339

Okenwa E (1978) Automated tidal reduction of soundings. Technical report of the Department of Surveying Engineering, University of New Brunswick, Canada. http://www2.unb.ca/gge/Pubs/TR55.pdf on 23rd May 2018

Okut H (2016) Bayesian regularized neural networks for small n big p data. Artif Neural Netw. https://doi.org/10.5772/63256

Okwuashi O, Ndehedehe C (2015) Digital terrain model height estimation using support vector machine regression. S Afr J Sci 111(9/10):5. https://doi.org/10.17159/sajs.2015/20140153

Okwuashi O, Ndehedehe C (2017) Tide modelling using support vector machine regression. J Spat Sci 62(1):29–46. https://doi.org/10.1080/14498596.2016.1215272

Okwuashi O, Ndehedehe CE, Attai H (2020) Tide modeling using partial least squares regression. Ocean Dyn 70:1089–1101. https://doi.org/10.1007/s10236-020-01385-1

Okwuashi O, Olayinka DN (2017) Tide modelling using the Kalman filter. J Spat Sci 62(2):353–365. https://doi.org/10.1080/14498596.2016.1245162

Pandya D, Dennis B, Russell R (2017) A computational fluid dynamics based artificial neural network model to predict solid particle erosion. Wear 378–379:198–210. https://doi.org/10.1016/j.wear.2017.02.028

Pino-Mejías R, Pérez-Fargallo A, Rubio-Bellido C, Pulido-Arcas JA (2017) Comparison of linear regression and artificial neural networks models to predict heating and cooling energy demand, energy consumption and co_2 emissions. Energy 118:24–36. https://doi.org/10.1016/j.energy.2016.12.022

Pombeiro H, Santos R, Carreira P, Silva C, Sousa JM (2017) Comparative assessment of low-complexity models to predict electricity consumption in an institutional building: Linear regression vs. fuzzy modeling vs. neural networks. Energy Build 146:141–151. https://doi.org/10.1016/j.enbuild.2017.04.032

Pugh D (2004) Changing sea levels: effects of tides, weather and climate. Cambridge University Press, Cambridge

Quaresma LS, Pichon A (2013) Modelling the barotropic tide along the West-Iberian margin. J Mar Syst 109–110:S3–S25. https://doi.org/10.1016/j.jmarsys.2011.09.016

Rajasekaran S, Gayathri S, Lee T-L (2008) Support vector regression methodology for storm surge predictions. Ocean Eng 35(16):1578–1587. https://doi.org/10.1016/j.oceaneng.2008.08.004

Sertel E, Cigizoglu HK, Sanli DU (2008) Estimating daily mean sea level heights using artificial neural networks. J Coast Res 727–734. https://doi.org/10.2112/06-742.1

Shu J (2003) Prediction and analysis of tides and tidal currents. Int Hydrogr Rev 4(2):24–29

Sudha Rani NNV, Satyanarayana ANV, Bhaskaran PK (2017) Assessment of climatological trends of sea level over the Indian coast using artificial neural network and wavelet techniques. Pure Appl Geophys 174(4):1527–1546. https://doi.org/10.1007/s00024-017-1501-6

Sulistyo SB, Woo WL, Dlay SS (2017) Regularized neural networks fusion and genetic algorithm based on-field nitrogen status estimation of wheat plants. IEEE Trans Ind Inform 13(1):103–114. https://doi.org/10.1109/TII.2016.2628439

Tebaldi C, Strauss BH, Zervas CE (2012) Modelling sea level rise impacts on storm surges along US coasts. Environ Res Lett 7(1):014032. http://stacks.iop.org/1748-9326/7/i=1/a=014032

WOCE SL (2006) Hourly sea level from coastal tide gauge station Lagos-C in 1993(research quality database). Sea Level Center, University of Hawaii. https://doi.org/10.1594/PANGAEA.435683

Xu J, Bao J-Y, Zhang C-Y, Zhou X-H (2017) Tide model CST1 of China and its application for the water level reducer of bathymetric data. Mar Geod 40(2–3):74–86. https://doi.org/10.1080/01490419.2017.1308896

Yadav AK, Chandel S (2017) Identification of relevant input variables for prediction of 1-minute time-step photovoltaic module power using artificial neural network and multiple linear regression models. Renew Sustain Energy Rev 77:955– 969. https://doi.org/10.1016/j.rser.2016.12.029

Yin J-C, Wang N-N, Hu J-Q (2015) A hybrid real-time tidal prediction mechanism based on harmonic method and variable structure neural network. Eng Appl Artif Intell 41:223–231. https://doi.org/10.1016/j.engappai.2015.03.002

Yoon H, Jun S-C, Hyun Y, Bae G-O, Lee K-K (2011) A comparative study of artificial neural networks and support vector machines for predicting groundwater levels in a coastal aquifer. J Hydrol 396(1):128–138. https://doi.org/10.1016/j.jhydrol.2010.11.002

Zaron ED, Ray RD (2017) Using an altimeter-derived internal tide model to remove tides from in situ data. Geophys Res Lett 44(9):4241–4245. https://doi.org/10.1002/2017GL072950

Part VI
Impacts of Extreme Events

Chapter 11
Drought Characteristics and Impacts in the Anthropocene

11.1 Introductory Remarks

The continued perturbation of the Earth system may lead to more deleterious impacts on humanity as well as natural ecosystems in the future. This is because future warming scenarios are expected to amplify these impacts and drought risk through changes in drought magnitudes and severity (e.g., Adeyeri et al. 2023; Naumann et al. 2018). Additionally, anthropogenic forcings will drive substantial changes in the characteristics (e.g., frequency, intensity, etc.) of drought events (Chiang et al. 2020). Yet, contemporary understanding of the impacts of global climate on changes in global drought characteristics (intensities, extents, duration) is still limited and not well understood. This knowledge is critical given that projected changes in climate will have implications on future water availability as well as influence decisions on how water resources are allocated (Samantaray et al. 2022; Usman et al. 2021).

Efforts to cushion drought risk are now global priority and is evidenced in the prioritization and effective allocation of revenues (Hagenlocher et al. 2019). Such efforts nonetheless will be underpinned by our capability to keep track of the changes in drought characteristics across the globe. Even though the surrounding oceans and indices of oceanic variability like the El Niño-Southern Oscillation (ENSO) and Pacific Decadal Oscillations are well-established important drivers of drought and moisture availability across the globe (Ndehedehe et al. 2021; Kiem et al. 2016; Dai 2013; Sheffield and Wood 2008), the implications of changes in key drought metrics on global hydrology are poorly understood. For instance, how will future changes in drought characteristics (e.g., duration) impact on variations in global land water storage? Furthermore, drought stress is an established constraint on crop production and the negative impacts of extreme events like drought on agricultural revenues (or national income) and manufacturing industry have recently being stressed (e.g., Furrer et al. 2022; Ndehedehe 2022a; Ahluwalia et al. 2021). The far-reaching impacts of drought on water availability and gross domestic product warrants knowledge of its potential future implications on global hydrology. In this regard, the antic-

C. Ndehedehe, *Hydro-Climatic Extremes in the Anthropocene*, Springer Climate,
https://doi.org/10.1007/978-3-031-37727-3_11

ipated increase in drought occurrence and severity across several regions of the world makes this understanding urgently crucial.

Emerging drought assessments on global scales (e.g., Xu et al. 2019; Naumann et al. 2018; Spinoni et al. 2014; Dai 2013; Sheffield and Wood 2008) are beginning to diagnose the complexities of drought, in terms of drivers, propagation, termination, frequency of dry spells, physical processes, and the role of human actions in regulating drought processes. However, in terms of future climate outlook, there is still limited knowledge on changes in drought properties (e.g., duration) and impact (e.g., extent) on a global scale, especially with the expected increase in carbon emissions that will result from social economic changes. This understanding is critical and underpins our effort in drought preparedness and ability to increase resilience through improved forecast, managing water supplies and allocation under drought. The recent work undertaken by Ndehedehe et al. (2023) combined the Coupled Model Intercomparison Project Phase 6 climate models, which incorporate the Shared Socioeconomic Pathways (2040–2070 and 2070–2100) with multi-scale earth observation data (1980–2020) to assess future changes in the characteristics of drought events. This chapter provides more details on how such changes may impact groundwater hydrology and the availability of water resources in general. Additionally, it articulates how global climate teleconnection patterns are important predictors of these changes and their role on drought intensity.

11.1.1 Drought Impact

Drought impact here is restricted to mean the areas under drought (that is, areas affected by drought event) or drought extent as highlighted in Sect. 11.3.6. In principle, drought impact encompasses the social, economic, and environmental effects of drought conditions, and it is essential to understand its various dimensions to develop effective strategies to mitigate its negative effects. Section 11.3.6 details how this can be quantified using climate data like the SPEI. Drought impact is the negative effects resulting from drought conditions on various aspects of human life and the natural environment, including impacts on hydrological systems and freshwater availability (e.g., Ndehedehe et al. 2021). This can manifest in different forms, including reduced crop yields, livestock losses, water shortages, increased risk of wildfires, reduced water quality, and economic losses, among others (Ndehedehe et al. 2023). In the context of food security and human well-being, drought impact refers to the adverse effects of drought on the availability, accessibility, and utilization of food, as well as on the physical, social, and economic well-being of affected populations (e.g., Furrer et al. 2022; Rippey 2015). These impacts can include hunger, malnutrition, reduced national income, increased food prices, as well as limited access to clean water and sanitation, among several others (e.g., Ndehedehe et al. 2020; Agutu et al. 2017). The impact of drought on food security and human well-being could be acute in developing countries where agriculture is largely rainfed and below-normal rainfall can lead to widespread hunger and malnutrition, particularly among children and vulnerable populations as frequently seen Eastern African countries, including Kenya (e.g., Agutu et al. 2017, 2019; Ndehedehe et al. 2020). Drought impact can also have long-term consequences on the natural environment, such as reduced productivity of floodplains, loss of biodiversity, and increased erosion, among others. These impacts can further exacerbate the effects of drought, leading to a vicious cycle of environmental degradation and human suffering.

11.1.2 Drought Duration

One of the main goals of this chapter is to demonstrate how climate models (e.g., Coupled Model Intercomparison Project Phase 6) (O'Neill et al. 2016) and satellite observations can be combined to improve contemporary understanding of changes in drought characteristics (e.g., duration) following the work of Ndehedehe et al. (2023). So, one of the drought properties quantified and discussed in this chapter is drought duration. This is defined as the time length (i.e., the number of months) between the drought onset and termination (e.g., Spinoni et al. 2014). To calculate this, the sum of drought durations for all drought events was calculated and then divided by the number of drought events (e.g., Diaz et al. 2019; Xu et al. 2019) as

$$DD = \frac{\sum_{i=1}^{n} d_i}{n}, \tag{11.1}$$

where DD is the drought duration (months), d_i, duration of ith drought event, and n the total number of drought events. In this study, we have classified drought into various intensities (extreme, severe, and moderate) and their occurrence is based on drought thresholds described in McKee et al. (1993). Drought classification based on these thresholds assumes a drought condition when the drought indicator is consistently negative and reaches a value of –1, which is characterized as moderate drought. A drought event becomes severe when the value ranges from –1.5 to –1.99 and extreme when it is –2.0 or less. A drought event terminates when the indicator becomes positive. There is a duration for every drought event and the drought characteristics (duration and extent) as well as their intensities studied here are based on the definition of these drought thresholds.

11.1.2.1 Groundwater Drought

Apart from drought impacts on the availability of surface water, changes in groundwater or its availability can also be affected by drought events (Ndehedehe 2022a). When drought is severe and prolonged, it can lead to decline in groundwater levels because of limited recharge, low well yields. This is groundwater drought and it has also been linked to increasing long-term human water extraction caused by extreme drought conditions (Ndehedehe 2022a). Groundwater droughts are driven primarily by climate change and could be exacerbated by extensive groundwater withdrawals, making it more difficult to characterize and assess. For this reason, indicators to assess various characteristics of groundwater drought, including, onset, duration, magnitude, and variability are still emerging (Ndehedehe 2022a; Uddameri et al. 2019; Thomas et al. 2017). The concept of groundwater drought and how it can be measured and characterized were extensively discussed in Chap. 5.

11.1.2.2 Anthropogenic Drought

Anthropogenic drought is a concept that reflects the various processes, e.g., land-atmosphere interactions, water, and energy balance that involve climate change and human actions that contribute to water-deficit conditions (Ndehedehe 2022a). This type of drought is a representation of human actions and climate change impacts on human-water systems, as well as how they interact to drive water bankruptcy (e.g., AghaKouchak et al. 2021; Madani and Shafiee-Jood 2020). Prolonged drought is usually the main trigger of anthropogenic drought and extensive human water extraction drive water bankruptcy following the drought event. The permanent surface water loss in the Lakes Chad and Urmia in West Africa and Iran, respectively, are good examples of anthropogenic drought. These cases were elaborately discussed earlier on in Chap. 2 of this book.

11.2 Changes in Global Drought Characteristics

As mentioned earlier in Chaps. 1 and 2 of this book, the Anthropocene is characterized by rapid changes in the environment caused by the impact of increasing human activities and climate change (e.g., drought and floods). These impacts have resulted in a broad range of transformations and complexities, including terrific groundwater loss, and deforestation, disappearance of large surface water bodies, sea-level rise. Even though the frequency and magnitude of droughts and floods have intensified in the last century (Ndehedehe 2023), contemporary understanding of the impacts of climate change on global drought characteristics (e.g., intensities, extents) is still limited and not well understood. The main aim of this section is to improve contemporary understanding of drought characteristics (extents and duration) in the Anthropocene using multi-scale earth observation data (1980–2020) and global climate models (2040–2070 and 2070–2100) following the work undertaken by Ndehedehe et al. (2023). These climate models, that is, the Coupled Model Intercomparison Project Phase 6 (CMIP6) (O'Neill et al. 2016) incorporates the Shared Socioeconomic Pathways (SSPs), that assume considerable socio-economic changes. The latter is expected to emanate from rapid growth in future global economy. Even though the SSPs are hypothesized and developed based on different socioeconomic and carbon emission scenarios, this study employed the ensemble mean of eight CMIP6 Global Climate Models (GCMs)s under the SSP5 emission scenario (Fig. 11.1) and combine

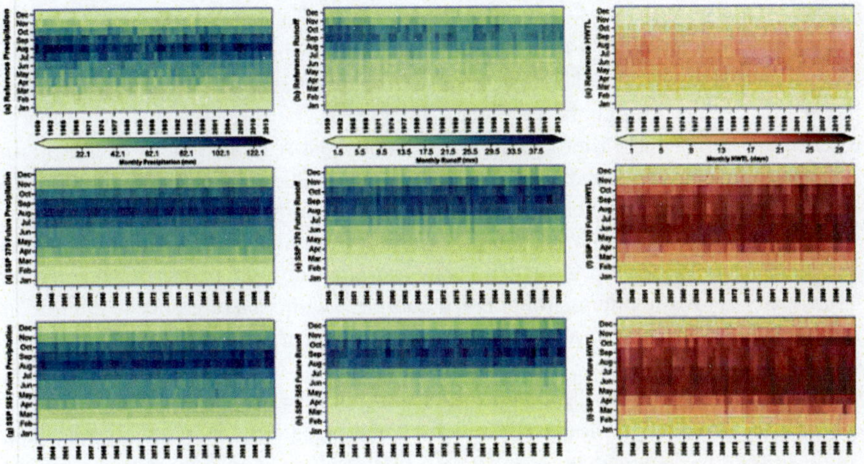

Fig. 11.1 Changes in temporal distribution of hydrological units (precipitation, runoff) and Heat Wave Total Length (HWTL) for Africa's tropical north domain based on current and future periods. The multivariate-bias-corrected CMIP6 multi-model ensemble mean is used. *Source* Adeyeri et al. (2023)

carbon emissions that integrates future social, economic, technical, and environmental changes.

11.2.1 Future Droughts and Hydrological Systems

In the preceding Chapters of this book, we mentioned that droughts affect hydrological systems. Climate is changing and extreme rainfall anomalies are triggering changes in hydrological stores (e.g., groundwater, surface water, and soil moisture) and water availability. When extreme drought conditions persevere, limited surface water, especially in typically dry regions of the world results in increased use of groundwater. In the long term, the combined impacts of prolonged drought and human water extraction can induce groundwater drought (Sect. 5.3, Chap. 5). Improving knowledge on the potential impacts of future droughts on hydrological systems is thus critical to guide future expectations about droughts and to help develop risk-management solutions to improve adaptation and resilience (Ndehedehe et al. 2021). Some of the changes in the characteristics of global drought events reported by Adeyeri et al. (2023) suggest the transition of meteorological to hydrological drought will happen slowly over the peak winter and spring months in places like Europe. Their study which was based on multi-model drought propagation and projection also predicted that most continents are expected to experience severe drought, suggesting a future with increasing vulnerability to the impacts of climate change (Fig. 11.1). Considerably humid regions like the Congo and Amazon basins, which regulate and contribute to global climatology, have long experience the impacts of changes in drought characteristics that led to large changes on surface flow and huge losses in forest ecosystems (e.g., Ndehedehe et al. 2019; Zhou et al. 2014). The impacts of the nearby oceans through the synergistic interactions between sea surface temperatures in the Indian and eastern Atlantic Ocean on west African Sahel still prevail on freshwater in the Lake Chad today (Ndehedehe et al. 2020; Giannini et al. 2003).

To date, the Lake Chad only exists as a small pool or body of water that changes slightly in extent when rainfall amounts increase in some wet seasons. This was discussed in detail in Chap. 2 (Sect. 2.3.4). Australia's drought history, especially the notoriously dry Millennium Drought, is incomplete without a reference to the role of large-scale climatic signals and global sea surface temperature and sea level pressure (White et al. 2004). Notably, the changes in wet and dry conditions across important hydrological hotspots like the Congo and Amazon basins may have consequences on other regions whose hydrology and local precipitation patterns are remotely regulated by these regions (Ndehedehe et al. 2019). Future changes in droughts in terms of their increased severity and duration may result in increased acceleration in the global water cycle and large changes in freshwater.

11.2.2 Understanding Future Drought Evolutions is Important

Throughout the latter half of the twentieth century, there has been a global increase in drought frequency, duration, and severity (Adeyeri et al. 2023; Ndehedehe 2023; Spinoni et al. 2014). This trend is projected to continue, as evidenced by the expected rise in drought frequency and intensity in future climate projections according to IPCC-AR4 (van der Molen et al. 2011). Recent studies of droughts in various regions of the world (see, e.g., Ndehedehe 2023; Chiang et al. 2020; Ferreira et al. 2018; Rippey 2015; Dai 2013) reflect our contemporary understanding of the escalating vulnerabilities of societies to the impacts of extreme climatic conditions. The effects of droughts on agricultural production systems pose multiple risks to food security and nutrition, particularly in regions reliant on rainfed agriculture and with shared water resources (e.g., Ndehedehe et al. 2021; Agutu et al. 2017). Such impacts have far-reaching consequences, even affecting distant populations through price volatility and trade disruptions, thereby increasing the vulnerability of rural agro-communities to poverty and other drought-related risks (Ndehedehe et al. 2019; FAO 2016). Despite the inadequate articulation of the economic and political basis for investing in drought risk reduction (Cenacchi 2014), efforts to address susceptibility to dangerous climatic thresholds have led to some investments in hydraulic and water infrastructures, risk-based decisions, mitigation strategies, and adaptive approaches. Tracking and predicting ocean-atmosphere interactions and processes associated with drought episodes can significantly enhance our knowledge of drought evolution and variability. Firstly, it can ensure the continued development of infrastructural and economic instruments to manage and share risks. Secondly, it could inform government and institutional efforts in raising the level of drought preparedness and promote region-specific drought risk-management policy solutions. Such solutions may include mitigation strategies and complementary interventions, such as insurance programs, that could help cushion the effects of drought risk.

11.3 Framework to Assess Global Changes in Drought Characteristics

In the coming century, these GCM outputs under SSP585 emissions (Fig. 11.1) have sufficient capacity to deliver the 8.5 W/m^2 forcing given the anticipated developments in fossil-fuel and human capital. Comparing current and future changes in drought properties will enable us understand impacts of climate change on potential future outlook on water availability that underpins increase drought preparedness, support adaptation, and management of water resources. The specific aims of this chapter are to (i) quantify changes in drought characteristics using long-term multi-scale climate data, the Standardized Evapotranspiration Index (SPEI) and outputs from GCMs, (ii) assess the response of groundwater changes to drought events on multiple scales using SPEI on 3, 6, and 12 months cumulations, and (iii) identify climatic hotspots where global climate may potentially drive changes in drought characteristics. The main goal is to improve our understanding of changes in drought characteristics (extents and duration) in the Anthropocene and the implications of such changes on future water availability.

11.3.1 Standardized Precipitation Evapotranspiration Index

The standardized precipitation evapotranspiration index (SPEI Vicente-Serrano et al. 2010a, b) was used in this study as a drought indicator. The data was retrieved from https://spei.csic.es/database.html for the period 1980–2018. The SPEI is based on a water balance approach, measuring the difference between precipitation (P) and PET (potential evapotranspiration) (Vicente-Serrano et al. 2010b). The SPEI is multi-scale drought index, incorporating the effects of variations in temperature as well as rainfall deficits on drought assessment. This makes it an important drought metric to assess the impacts of drought under a global warming scenario. In addition to the multi-temporal nature of the SPEI, it is sensitive to changes in evaporation demand, which is caused by temperature fluctuations. The multi-scalar nature of SPEI enhances its potential in identifying several drought episodes (Vicente-Serrano et al. 2010b). The SPEI database provides historical, ready-made, already computed drought conditions on global scale at 0.5 degrees spatial resolution and is available on monthly time scale. Its accumulation timescales are available for 1–48 months and

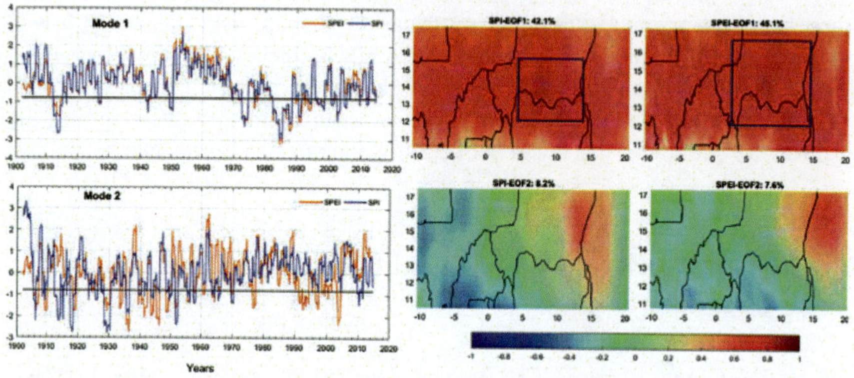

Fig. 11.2 Spatial and temporal patterns of SPI and SPEI over the Sahel region during the 1901–2014 period using the principal component analysis (PCA). The spatial patterns of SPI/SPEI (right) and the corresponding principal components (PCs), which are temporal variations (left) are the outcome of statistical decomposition using the PCA technique. The actual SPEI and SPI values representing drought or wetness conditions are jointly derived from their spatial and temporal patterns. The black solid line is drought threshold based on the classification in McKee et al. (1993). The insert on the spatial maps shows locations in Niger with the highest SPEI and SPI loadings during the entire period. *Source* Ndehedehe et al. (2020)

cover the period between January 1901 and December 2020. This data is freely and publicly available online (http://digital.csic.es/handle/10261/153475). The classification and categorization of drought in this study are based on the thresholds defined by McKee et al. (1993). Even though the multi-scalar nature of SPEI enhances its potential in identifying several drought episodes (Vicente-Serrano et al. 2010b), this index also agrees substantially with SPI (Fig. 11.2) in several regions across the world (Ndehedehe 2022a).

11.3.2 Standardized Precipitation Index

The standardized precipitation index (SPI, McKee et al. 1993) is a well-known drought indicator that is computed from long-term precipitation record (Fig. 11.2). The assumption is that drought is primarily caused by prolonged below-normal precipitation or rainfall deficits (e.g., Spinoni et al. 2014). A probability distribution is usually fitted to this long-term precipitation record, which is then transformed to a normal distribution so that the mean SPI (Fig. 11.3) for a given location and desired period is zero (Ndehedehe 2022a; Mishra and Singh 2010; McKee et al. 1993). The SPI is simple to compute as it is based on precipitation and usually standardized so that wetter and drier climates can be represented in a similar way and can be calculated for other hydrological quantities or units such as snowpack, reservoir, streamflow, soil moisture, and groundwater (McKee et al. 1993). Participants of the

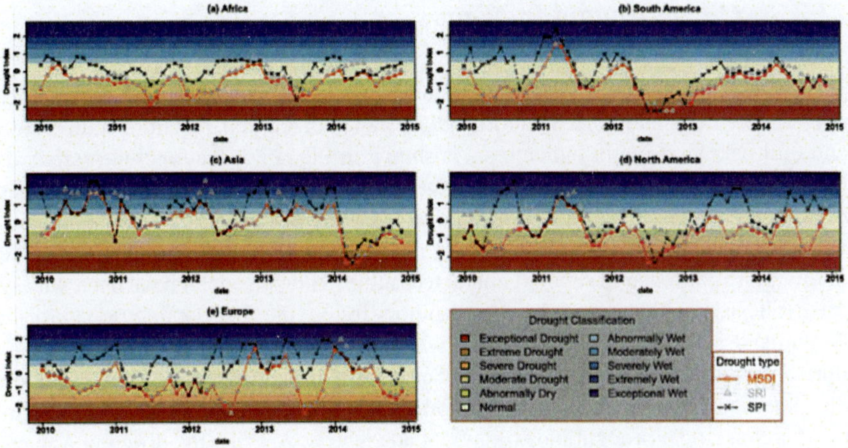

Fig. 11.3 Illustration of univariate and multivariate drought indicators for continent-scale drought assessment (2010–2014) based on reference data. The background colors show drought classifications for each category. *Source* Adeyeri et al. (2023)

Lincoln Declaration on Drought Indices unanimously adopted SPI and recommended it to World Meteorological Organization as a standard index to track meteorological drought events (Hayes et al. 2011). Furthermore, because the SPI can be calculated for different time scales (e.g., 3, 6 12, 24 months) similar to SPEI, they can be used to monitor short (meteorological), medium (soil), and long (groundwater, rivers, and reservoirs) term changes in water availability (Fig. 11.3). SPI is suitable in most regions, showing consistency with SPEI (Ndehedehe 2022a).

The statistical decomposition of SPEI and SPI using the principal component analysis show that the spatial and temporal patterns of these drought indicators over the West African Sahel rather similar looking at the two leading orthogonal modes of variability for the 1901–2014 period (Fig. 11.2). For instance, this consistency is evidenced in the fact that both SPEI and SPI confirm that the wetness in the 1950s remains considerably unique compared to other periods and despite the recovery and change in rainfall conditions after the droughts of the 1980s, the 1950 threshold has still not being reached (Fig. 11.2). The Pearson correlation coefficients between SPEI and SPI time series (Fig. 11.2) for the first and second modes of variability were 0.93 and 0.52, respectively. The moderate correlation between SPEI and SPI in the second modes of variability could suggest impacts resulting from the changing land surface conditions in and the need to incorporate biophysical indicators that underpin our understanding of drought impacts and its cascading effects (Ndehedehe et al. 2020). Even though its inability to account for outgoing water fluxes like evapotranspiration makes it somewhat limited in some areas (Ndehedehe et al. 2020), this was not an issue in this current study, which relied on future precipitation scenarios to assess changes in drought characteristics.

11.3.3 Other Drought Indicators

The purpose of this section is to briefly highlight on the existence and use of other univariate (SRI-standardized runoff index; standardized soil moisture index, etc.) and multivariate drought indicators. As shown in Fig. 11.3, they include standardized runoff index and multivariate drought index and can be used to quantify various drought intensities, including meteorological and hydrological droughts. The work of AghaKouchak (2015) and Hao and AghaKouchak (2014) laid a foundation for the multivariate approach to drought characterizations and since then have been widely used. These drought metrics by these authors follow the non-parametric approach Farahmand and AghaKouchak (2015) where instead of gamma distribution function, an empirical approach is used to derive a marginal probability that relies on the univariate form of the Gringorten plotting position (Gringorten 1963). The argument is that single drought index may not adequately describe the different aspects of drought properties such as onset, duration and termination (see Farahmand and AghaKouchak 2015; Hao and AghaKouchak 2013, 2014), and combining different hydro-climatic data (e.g., rainfall and river discharge) will lead to a multivariate index that captures these characteristics. As recently demonstrated in the work of Adeyeri et al. (2023), these indicators are efficient in continent-scale drought characterization. For instance, hydrological drought events during the 2011–2012 were observed in Europe based on these drought metrics.

11.3.4 Ensemble Climate Models

The Coupled Model Intercomparison Project Phase 6 (CMIP6) includes various scenario experiments for the future (O'Neill et al. 2016). One component of these scenarios is the Shared Socioeconomic Pathways (SSPs), a collection of various social development options. The SSPs illustrate potential future demographic, economic, technical, social, governance, and environmental changes. The SSPs are structured around five narratives that depict diverse socioeconomic and carbon emission scenarios. However, this study focused on the ensemble mean of eight CMIP6 Global Climate Models (GCM) under the SSP585 emission scenario for two future timeslices (2040–2070; 2070–2100). SSP585 is the only SSP with emissions sufficient to deliver the 8.5 W/m^2 forcing in 2100 because of fossil-fuelled development, rapid technological progress, and human capital development.

The GCM outputs were calibrated using the WATCH Forcing Data methodology applied ERA5 reanalysis data over land merged with ERA5 over the ocean (W5E5) (Cucchi et al. 2020; Weedon et al. 2014). This dataset was created as part of Phase 3b of the Inter-Sectoral Impact Model Intercomparison Project, which is focused on detecting and attributing observable impacts using the Intergovernmental Panel on Climate Change Working Group II definition. It has also been utilized as input for the hydrological modeling of several ungauged basins and for the bias correction of climate models for impact assessments (Iturbide et al. 2022). Similar to the SPEI, the ensemble GCM outputs based on future scenarios (2040–2070; 2070–2100) were used to construct a drought indicator and drought characteristics were generated from these metrics.

11.3.5 Global Groundwater From Space

Global groundwater was estimated using the Gravity Recovery and Climate Experiment (GRACE) (GRACE Tapley et al. 2004) satellite data. The GRACE data provides monthly changes in terrestrial water storage (TWS, the sum of soil moisture, groundwater, surface water, snow water, and canopy storage, etc.) across the globe. GRACE products exist in various forms, e.g., geopotential spherical harmonic Stokes coefficient, level 3 mass grids, and mass concentration solutions popularly known as mascons (Ndehedehe 2022b). But the global mascon solutions (Save et al. 2016) Release 06 (2002–2020) available at 0.25 arc-degree resolution were used in this study. The data can be retrieved from the Center for Space Research data portal (http://www2.csr.utexas.edu/grace/RL06_mascons.html).

The water budget approached used to separate groundwater storage changes from the GRACE (2002–2020) water column typically relies on auxiliary data (e.g., soil moisture, snow/ice, surface water) from models and is well known and detailed in the literature (e.g., Thomas and Famiglietti 2019; Chen et al. 2016). Groundwater was estimated using this approach by subtracting model-derived soil moisture data from GRACE-derived TWS prior to masking areas with substantial surface water variation. Furthermore, using polygon extents of globally recognized glaciers and ice caps, sub-regions (e.g., Patagonia, Alaska, Himalayas, Swiss Alps, etc.) with potentially large variations in residual ice/snow cover were masked during the GWS estimation process from GRACE to minimize uncertainties. As a result, in the spatial correlations between SPEI (Sect. 11.3.6) and GWS, similar number of grids for both data are mandatory and were maintained to enable this assessment.

11.3.6 Drought Extent

Areas under drought (in percentage) were aggregated based on the drought indicators for each time scale and step (Ndehedehe et al. 2021; Diaz et al. 2019),

$$PDA(t) = 100 \cdot \frac{\sum_{c=1}^{N} D_c(t) \cdot A_c}{\sum_{c=1}^{N} A_c}, \tag{11.2}$$

where A is the area of the cell c within the considered region, t is time step, D is the drought state in binary representation (0 and 1). That is, Eq. 11.2 gives the percentage of drought area (PDA), which quantifies the impact of drought in terms of its extent or land areas affected (in percentage).

11.3.7 Statistical Analyses

The spatial and temporal patterns of SPEI were analysed using independent component analysis by rotating the pre-orthogonalized SPEI data matrices toward statistical independence (e.g., Cardoso and Souloumiac 1993). This method has recently been established as part of a novel framework that underpins drought diagnoses and the assessment of the impacts of climate variability indices as well as groundwater to support irrigated agriculture (Ndehedehe 2022a; Agutu et al. 2019). To identify climatic hotspots where global climate may potentially drive changes in drought characteristics, the localized temporal patterns for each aggregation scale were thereafter correlated with time series of climate teleconnection using a correlation matrix plot. These climate modes are ENSO (El-Niñno Southern Oscillation), Pacific Decadal Oscillation (PDO), Indian Ocean Dipole (IOD), Atlantic Multi-decadalOscillation (AMO), Atlantic Meridional Mode (AMM), Quasi-Biennial Oscillation (QBO), and Maiden Julian Oscillation (MJO) and are available in NOAA's global data repository https://psl.noaa.gov/data/climateindices/list/. Twelve independent patterns (temporal and spatial), which represent hydrological hotspots with significant variability during the 1980–2020 period, were selected for correlation with these climate modes.

11.4 Understanding Changes in Global Drought Characteristics

The impacts of climate change on global drought characteristics, such as intensities and extents, are not well understood despite contemporary efforts. This knowledge is crucial in predicting future water availability and resource allocation. Therefore, the primary objective of this chapter is to enhance the understanding of drought characteristics, including extents and durations, in the Anthropocene era, which is marked by the combined influence of human activities and climate change. To assess these characteristics and identify climatic hotspots, this chapter employed multi-scale earth observation data (1980–2020) and the Coupled Model Intercomparison Project Phase 6 climate models, which consider the Shared Socioeconomic Pathways (2040–2070 and 2070–2100). The results are detailed in what follows.

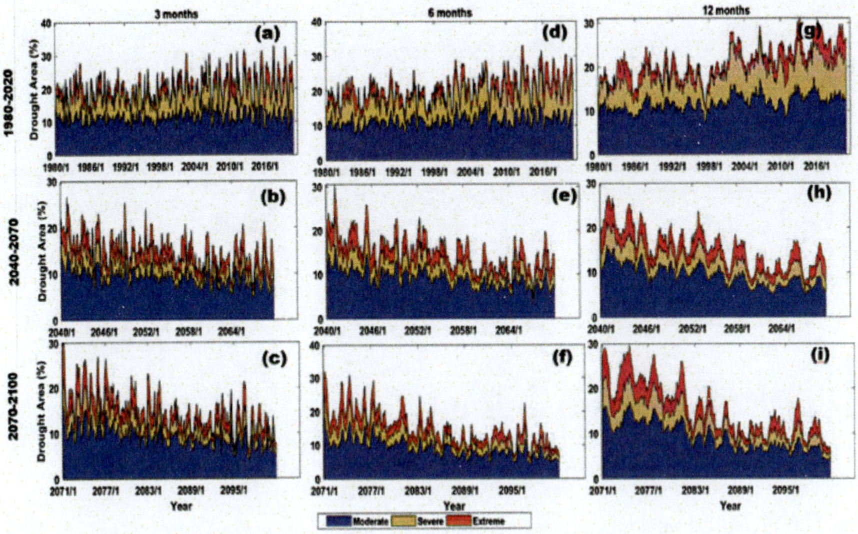

Fig. 11.4 Global aerial averaged temporal variations of different drought intensities. Land areas under drought (in percentage) are indicated for current (1980–2020) and future epochs (2040–2070 and 2070–2100). *Source* Ndehedehe et al. (2023)

11.4.1 Drought Events: Extents and Duration

The impacts of droughts (percentage of global land areas under drought) and its characteristics (duration) are studied based on three levels of aggregation (3, 6, and 12 month) (Fig. 11.4). These aggregation capture the impacts of short (meteorological), medium (agricultural or soil moisture) and long term changes (hydrological, respectively, including groundwater and surface water) in rainfall deficit, which can lead to any category of droughts (meteorological-3 month, agricultural-6 month, hydrological, socio-economic, and groundwater-12 month), including anthropogenic drought, a new drought concept that reflect the various processes, such as land–atmosphere interactions, water, and energy balance that exacerbates the impacts of climate change and human actions on water-deficit conditions (Ndehedehe 2022a). Results summarizing the duration and aerial extents of droughts for current (1980–2020) and future (2040–2070; and 2070–2100) climatological periods are presented in what follows.

11.4.1.1 Land Areas Affected by Droughts

The three levels of aggregation (3, 6, and 12 month) used to capture various drought intensities (moderate, severe, and extreme) for the 1980–2020 period were also used to generate drought characteristics for future climate scenarios (Fig. 11.4). The areas under drought for each of these intensities provide holistic view of drought impacts

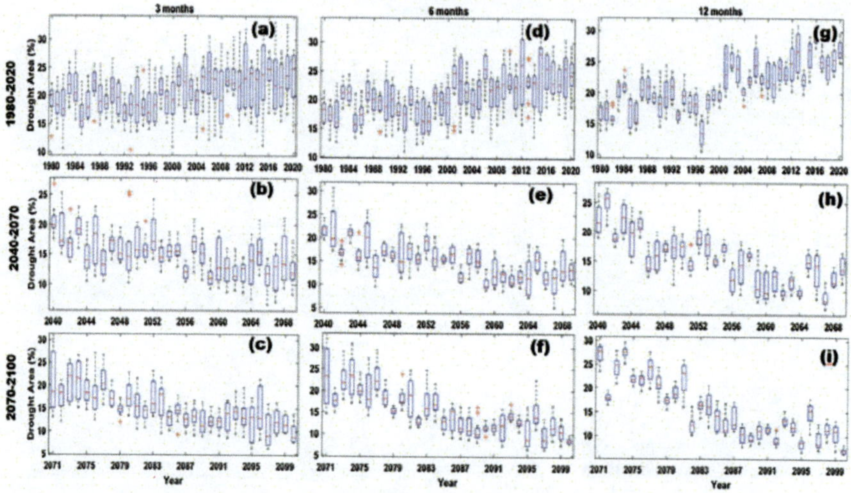

Fig. 11.5 Temporal variations of Land areas under drought (in percentage) for current (1980–2020) and future epochs (2040–2070 and 2070–2100). *Source* Ndehedehe et al. (2023)

and are shown as percentage of land areas under drought (Fig. 11.4a–i). The evolutionary drought patterns at the 3-month scale show no clearly noticeable change in the extent of various drought intensities looking into the future unlike the 6 and 12 month aggregation (Fig. 1.1a–i). While drought extents show obvious increase since the 1980s on the 12-month scale (Figs. 11.4g–i), future scenarios indicate a considerable fall in the proportion of land areas that will be affected by moderate, severe, and extreme drought conditions (Fig. 11.4g). The extent of extreme drought event based on 12 month scale show relatively higher amplitude at the start of each climatological periods in 2040 and 2070 and fluctuates between 30% and about 15%. Overall, global land areas that have experienced moderate, severe and extreme drought and those that will be affected by similar drought intensities in the future will not exceed 30%.

Away from the extent of drought intensities, the aerial averaged temporal variations of land areas under drought indicated in Fig. 11.5a–i provide vivid summary of the earth system in terms of drought impact. Since 1980, short (3-month), medium (6 month) and long-term changes (12 month) in rainfall deficit have shown obvious increase (Fig. 11.5a, d, g). This is in contrast to the declines observed in future emission scenarios (Fig. 11.5b–c, e–f, and h–i), which are even stronger on a 12-month scale. The ongoing increase in the areas under drought may continue into the future and beyond 2030, given that the start of the 2040–2070 and 2070–2100 periods show similar proportion (about 30%) of land areas under drought (Fig. 11.5h–i).

But other periods (e.g., 2050–2070 and 2075–2100) in the future emission scenarios show declines in the proportion of areas affected by drought (Fig. 11.5h–i). The color coded patterns of areas under drought (Fig. 11.6) enable the visualization of areas affected by drought, which are distributed across each month of the year.

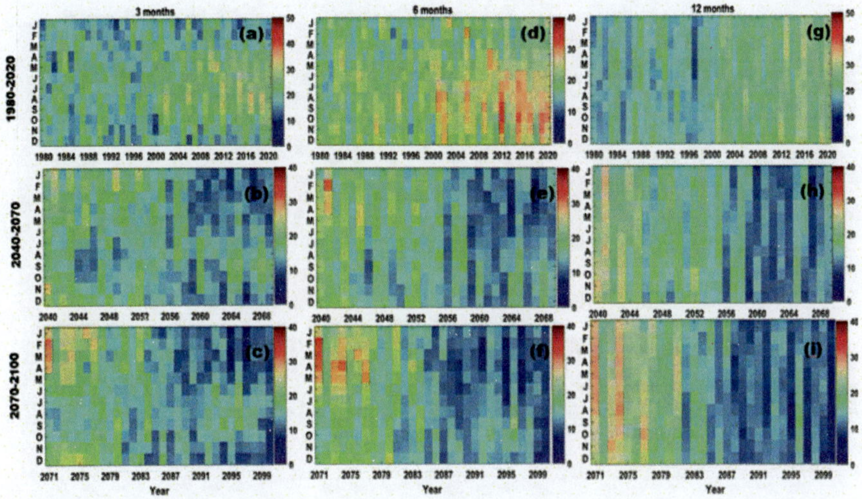

Fig. 11.6 Averaged areas under drought for annual (in percentage) during the current (1980–2020) and future epochs (2040–2070 and 2070–2100). The y-axis is the month (January to December), which are represented using alphabets. The x-axis is the year of occurrence. *Source* Ndehedehe et al. (2023)

The rows represent the month (January to December), which are indicated as alphabets and the columns (x-axis) are the corresponding year of drought occurrence, but this is shown as percentage of drought areas (Fig. 11.6a–i). For instance, in the 6-month drought aggregation, areas under drought between 2000 and 2020 ranged from 30% to 40% and mostly occurred in the months between July and December (Fig. 11.6d).

Overall, on a 6 and 12 months drought aggregation, a shift in drought seasonal cycles (January to December) is expected looking into the future (Fig. 11.6). For instance, the highest drought extent in 2012 (about 35%) for the 6 month aggregation occurred between September and November (Fig. 11.6d). Future occurrence of drought showing similar impact (i.e., 35% land areas affected) is noticeable between February and April (Fig. 11.6f). Moreover, future scenarios (Fig. 11.6h, i) are indicating that impacts of drought events based on the 12 months drought aggregation will be similar to current period (Fig. 11.6g), with about 25–30% of global areas under drought affected all through January to December.

11.4.1.2 Duration of Drought Events

Drought duration is another key drought metric that this study focused on. As with drought extent, drought duration was computed for different aggregation scales (3, 6, and 12 month) and they have been averaged on each of these scales (Fig. 11.7a–i). For the current period, higher drought duration (6 to 12 months duration) is noticeably widely spread on a 6 month and 12 cumulation along Amazon (or northern

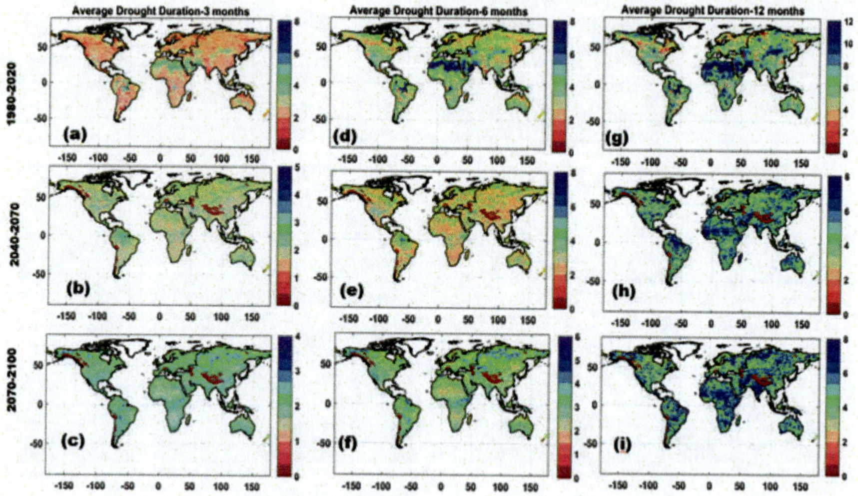

Fig. 11.7 Drought duration (months) during the current (1980–2020) and future epochs (2040–2070 and 2070–2100). Areas showing zero duration have no data as they were masked out. *Source* Ndehedehe et al. (2023)

South America), Northern and Central Africa, Middle East, and Asia (Fig. 11.7d, g). Compared to other regions, these same locations stand out in the 3-month scale mean drought duration but with lower drought duration (mostly 3 to 4 months) as would be expected compared to the 6 to 12 months duration (Fig. 11.7a, d, g). Some of these regions are typically dry or semi-arid (e.g., Northern Africa, Middle East, and Asia) and prolonged rainfall deficit in the future is expected to increase drought duration in some regions (Fig. 11.7e–f and h–i). On higher cumulation scale (12 month), there is evidence drought duration will vary under future emission scenarios, thus leading to increase in most regions of the world and decrease in other areas. On the one hand, regions like Australia, West Africa, South and North America, East Africa and much of Middle East and Asia will experience a rise in drought duration (Fig. 11.7g, h, i). On the other hand, the arid North Africa will see a fall in drought duration (Fig. 11.7g, h, i).

11.4.2 Drought Impacts on Groundwater

The spatial correlation between SPEI and changes in GWS was undertaken to help identify hotspots or locations where climate variability will potentially affect water availability or variations in GWS. The correlation between climate data (i.e., SPEI) based on three levels of aggregation scales (3, 6, and 12 month) and GWS show spatially distinct relationships. The Spearman's Rho correlation coefficients are positive in much of Africa, South America, south Asia and negative toward the north pole,

Europe, Australia (west and east coast) and eastern US and South Africa (Fig. 11.8a, b, c). Compared with the SPEI-3 shorter integration scale relationship with GWS (Fig. 11.8a), the Spearman's Rho correlation coefficients (0.5 and greater) for SPEI-6 versus GWS tend to show relatively moderately strong values widely distributed across the globe. Specific hotspots with these correlations include North-east Brazil, South Asia, East/Central Africa, central US and northern Australia, among others (Fig. 11.8b). However, Spearman's Rho correlation coefficients show considerably stronger positive values (0.8 and greater) at the 12-month SPEI aggregation in central US and some locations in South America and East/central Africa (Fig. 11.8c). Unlike SPEI-6-GWS relationship (Fig. 11.8b), the distribution of moderately strong Spearman's Rho correlation values is much higher in the SPEI-12-GWS relationship (Fig. 11.8c). Additionally, considerably stronger negative correlation coefficients are also obvious in South Africa, Western Australia, and south-east Asia. Except for western Europe (around the UK) and some areas in the high-latitude regions, the magnitude of these negative correlations increases with SPEI accumulation scales (Fig. 11.8a, b, c).

11.4.3 Spatial and Temporal Drought Variability

The localized spatial and temporal variations of SPEI across all aggregation scales were undertaken by statistically rotating them toward independence using the independent component analysis (ICA Cardoso and Souloumiac 1993). The twelve independent modes of variability (spatial and temporal) were used to establish links between climatic hotspots where climate teleconnection patterns may drive changes in drought characteristics and occurrence. An illustration of the ICA decomposition of SPEI-12 into spatial and temporal patterns is captured in Figs. 11.9 and 11.10. In the spatial patterns of SPEI-12 (Fig. 11.9), the signals or data are localized over Middle East (SPEI-R1), Australia (SPEI-R2), Europe (SPEI-R3), Russia (SPEI-R4), Asia (SPEI-R5), West Africa (SPEI-R6), South America (SPEI-R7), etc. Drought occurrence in time in a specific location (Fig. 11.9) can be deduced by jointly interpreting the spatial patterns with the temporal time series in Fig. 11.10. For instance, between the year 2000 and 2020 around the Middle East (SPEI-R1), there have been three periods with prolonged droughts (Figs. 11.9 and 11.10). The temporal pattern of SPEI in this region also suggests that since 1999, the Middle East seems to have been dryer compared to the period between 1980 and pre-2000 (Fig. 11.10).

11.4.4 Climatic Hotspots of Drought Variability and Impact

However, to also identify climatic hotspots where global climate may potentially drive changes in drought characteristics, the temporal patterns for each level of aggregation were correlated with those of climate teleconnection indices using a

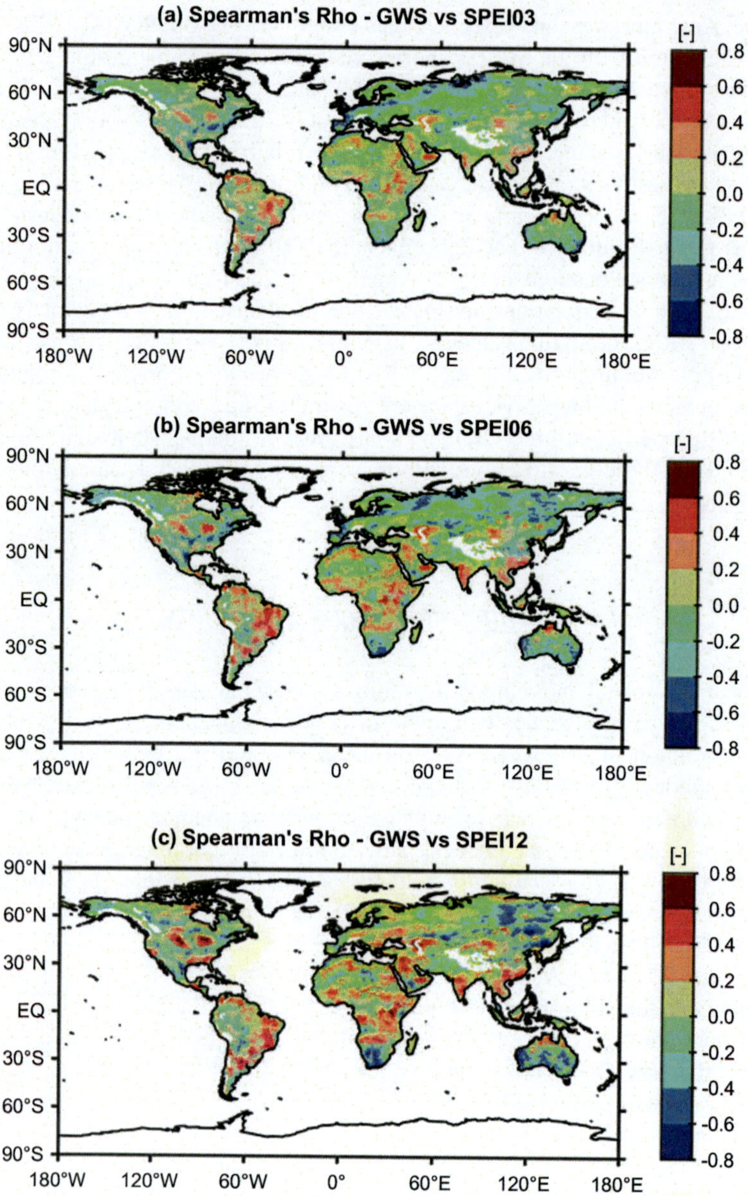

Fig. 11.8 Climate-groundwater interaction based on the Spearman's Rho correlation coefficient between groundwater and standardized precipitation evapotranspiration index. The correlations are based on three levels of integration periods or accumulation scales **a** SPEI-3 versus GWS, **b** SPEI-6 versus GWS, and **c** SPEI-12 versus GWS. Areas with no data were masked during the GWS estimation process to minimize uncertainties from regions (e.g., Patagonia, Alaska, Himalayas, Swiss Alps, etc.) with potentially strong variations in residual ice/snow cover. *Source* Ndehedehe et al. (2023)

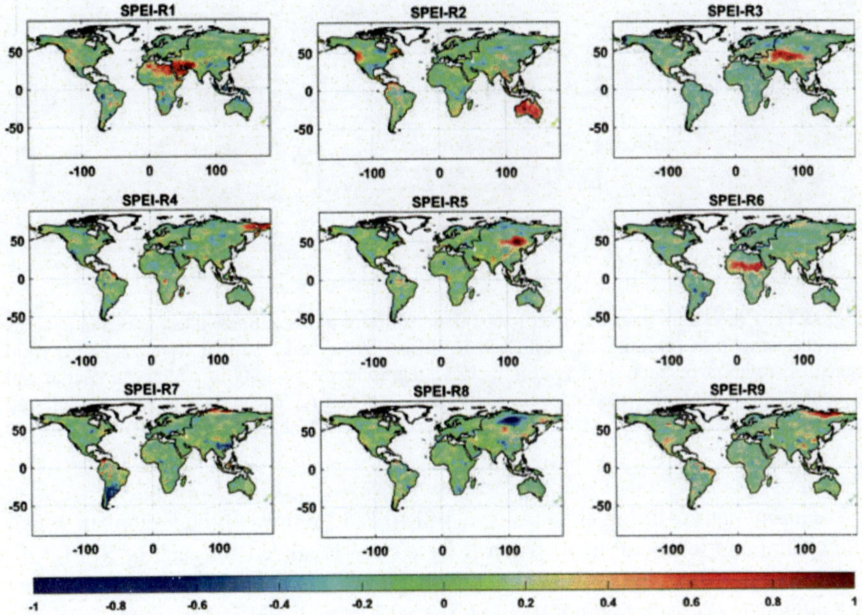

Fig. 11.9 Illustration of spatial patterns of localized SPEI-12 data for nine sub-regions (i.e., SPEI-R1 to SPEI-R9) using the Independent Component Analysis. These spatial patterns are jointly interpreted with the temporal evolutions indicated in Fig. 11.10. *Source* Ndehedehe et al. (2023)

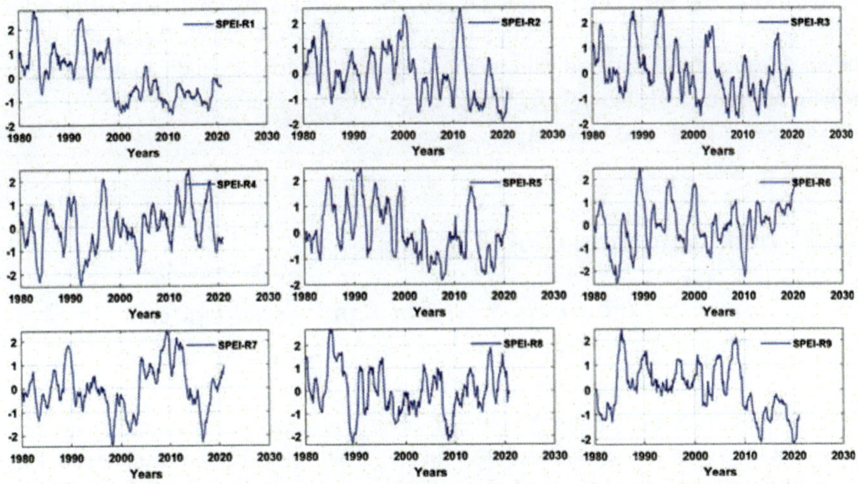

Fig. 11.10 Illustration of temporal evolutions of localized SPEI-12 data associated with the spatial patterns in Fig. 11.9 for nine sub-regions (i.e., SPEI-R1 to SPEI-R9). The spatial and temporal (standardized) patterns are jointly interpreted. *Source* Ndehedehe et al. (2023)

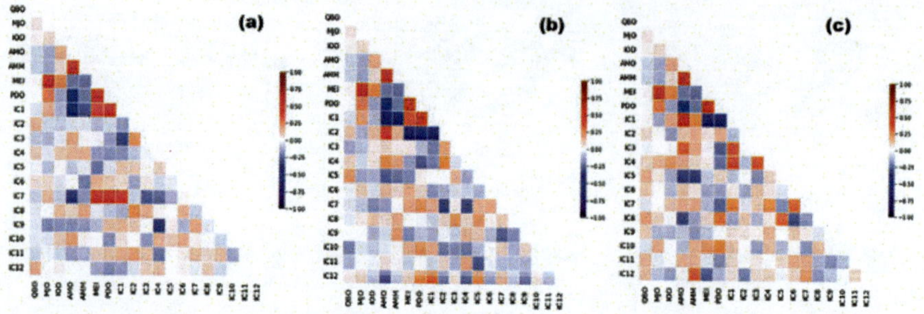

Fig. 11.11 Correlation matrix based on the relationships of global climate teleconnection patterns with the temporal evolutions of **a** SPEI-3, **b** SPEI-6 and **c** SPEI-12 that were generated from statistical rotation in Sect. 11.3.7. 1C1 to IC12 represent the regions in which the spatial and temporal SPEI patterns have been localized for SPEI-3, 6, and 12. The climate modes are indicated as QBO, MJO, IOD, AMO, AMM, MEI, and PDO. *Source* Ndehedehe et al. (2023)

correlation matrix. In Fig. 11.11a, b, c, 1C1 to IC12 represent the regions in which the spatial and temporal SPEI patterns have been localized for each of SPEI-3, 6, and 12 aggregation. The climate modes or global climate teleconnection patterns are indicated as QBO, MJO, IOD, AMO, AMM, MEI, and PDO (Sect. 11.3.7). The temporal variations of SPEI-3 in two regions (IC1 and IC7) show moderate positive and negative correlations ($r = 0.50$) with PDO and AMO, respectively, and IC7 is also moderately associated with MEI (Fig. 11.11a). In SPEI-6, IC1 and IC2 show moderately strong positive and negative correlations (slightly greater than 0.60) with AMO, PDO, and MEI (Fig. 11.11b). Lastly, SPEI-12 shows moderate correlations with four of these teleconnection patterns (Fig. 11.11c). Overall, there are several hotspots showing correlation values ranging from 0.30 to 0.45, indicating that these teleconnections contribute to significant proportion of variability in climatic conditions globally (Fig. 11.11a, b, c).

11.5 Implications on Global Hydrology

11.5.1 Twenty-First Century Global Land Areas Under Drought will Decrease but Their Durations will not

The findings from this study show that global land areas under drought will significantly decrease but their durations will not. There is evidence to insinuate that drought duration will vary under future emission scenarios. Regions like Australia, West Africa, South and North America, East Africa and much of Middle East and Asia will experience a rise in drought duration while North Africa will experience decrease. Results also indicate significant decline in the proportion of areas that will experience various drought intensities (moderate, severe, and extreme drought) in the future and for each category, drought affected areas will not reach 30% on average. But prolonged drought in the future, especially when it is severe or extreme, will have implications hydrological store (surface water, soil moisture, and groundwater) and changes in water availability. Given the devastating impacts of droughts on humans and the economy (Schroth et al. 2016; Rippey 2015), our understanding of future outlook in terms of drought characteristics and impacts underpins management efforts in water resources planning.

11.5.2 Can Global Climate Models be Trusted?

Global Climate Models (GCMs) are widely utilized tools for projecting and evaluating future climate conditions worldwide. The latest generation of these models, known as Coupled Model Intercomparison Project 6 (CMIP6), are utilized to generate climate projections under a range of scenarios, such as the Shared Socioeconomic Pathways (SSPs). In mountainous regions of China, the projected effects of these emissions scenarios include an increase in temperature and a decline in streamflow, which may have significant impacts on hydrology and water resources (e.g., Yang et al. 2022). Recent studies (Guo et al. 2022b; Cook et al. 2020) based on the CMIP6 models have reported on emerging trends, such as increasing drought conditions and drying conditions, across several regions including western North America, Central America, Europe and the Mediterranean, the Amazon, southern Africa, China, Southeast Asia, and Australia, particularly under the SSP5-8.5 emission scenarios.

Although some studies indicate that the proportion of areas experiencing drought may decline, rising temperature trends are predicted to exacerbate the duration of drought events in the latter half of the twenty-first century. Research conducted by Cook et al. (2020) shows that under the SSP585 emission scenarios, regions such as western/Central America, Europe and the Mediterranean, the Amazon, southern Africa, China, Southeast Asia, and Australia will be particularly affected by drought events. While the global extent of land areas affected by drought may diminish, the duration and intensity of these events may increase in most of these regions. Several studies, such as Ayugi et al. (2022), suggest that frequent droughts with prolonged duration and intensity are expected in East Africa under the emission scenarios in CMIP6 models. The GCMs appear to be consistent in projecting high warming conditions due to the increased greenhouse gases in the atmosphere, leading to a rise in evapotranspiration intensity and ultimately the duration of drought events.

11.5.3 Clime Model Prediction of Impacts on Water Availability

The impacts of climate change on the global hydrological cycle usually occur indirectly, though such impacts, e.g., prolonged droughts can sometimes have direct consequences, especially on surface water availability. When the climate warms, the aftermath can lead to decreased water resources because of the impacts of warming rising evapotranspiration. As the highest outgoing water flux, evapotranspiration can eventually exacerbate the impacts of droughts leading to substantial losses in surface water.

Several studies have used climate models to predict future climatic condition across different regions of the world. Under an extreme emission scenario in mountainous region of China, an increased temperature and declining streamflow have been projected for mid and end of century (Yang et al. 2022). The SSP5-8.5 emission scenario following the work undertaken by Guo et al. (2022b) also supports warming conditions in the second-largest river basin of China. Using the CMIP5 and CMIP6 model predictions, they found more warming conditions in the CMIP6 model predictions due to improved assumptions of human development based on fossil fuels/energy in extreme emission scenario used in the model. This rise in temperature is expected to result in decreased water resources. Regions that are typically dry and arid like the Middle East will experience increased temperature in the future. Climate model projections under the SSP8-8.5 emission scenario show that this rising temperature and falling precipitation will have negative impacts over the downstream countries of the Tigris-Euphrates basin toward the end of the twenty-first century (Dezfuli et al. 2022). Other regions where future warming based on the SSP5-8.5 emission scenarios are anticipated include South Asia, Mediterranean region, southern Africa, and Australia (e.g., Ayugi et al. 2022).

These increasing warming conditions are projected to have been attributed to the increased concentrations of greenhouse gases in the atmosphere (particularly under the SSP5-8.5 emission scenario). Warming in different regions of the world due to high temperatures will lead to more evapotranspiration. When this is combined with other climatic factors and surging human population (more water demand), several regions around the globe are expected to experience or be vulnerable to the drought events. Such events are usually associated with reduced water resources, which can ultimately have negative consequences not only for human beings, agriculture, but also for ecosystems and the services they support (Usman et al. 2021). In the glaciered regions of the world, rising temperatures can result in short-term increase of water resources (due to glacier melting). However, in a longer term perspective, danger of glacier disappearance is always present, which can also threaten water supply. Due to severe shortage of water resources, Gross Domestic Product (GDP), according to World Bank is also projected to decrease by up to 6% only in the mid of twenty-first century and this number could be significantly higher toward the end of the century. Emerging economies could be more devastated with these scenarios because of their low capacity for adaptation and limited access to resources.

11.5.4 *Will Groundwater be Affected by Future Changes in Drought Characteristics?*

The spatial correlations between SPEI for all accumulation scales and GWS are positive and negative across the globe. The strong positive correlations observed at the 12-month aggregation scale globally suggest that climate change could have a significant impact on groundwater. In some areas, the varying strength of negative correlations at short or longer SPEI aggregation scales may be due to non-linear hydrological systems or considerable lags between climate and groundwater. However, in general, the correlation between SPEI and groundwater increases as the accumulation of SPEI scales increases, such as at 3, 6, and 12 months (Fig. 11.8a, b, c). This means that short-term (SPEI-3) changes in rainfall and temperature are likely to have less impact on groundwater variations than prolonged rainfall deficits (e.g., SPEI-12). In arid areas like the Middle East and Australia, however, even short-to-moderate-term changes in rainfall can have a strong impact on changes in groundwater, especially with rising temperatures and evapotranspiration from the landscapes. In some regions, non-linear hydrological processes and delayed response of groundwater to rainfall changes due to aridity, aquifer properties, and over-exploitation can lead to strong opposite relationships, as observed in this study. It is worth noting that regions (such as Australia, Southern Africa, and Asia) with significantly negative and relatively higher correlations between SPEI and groundwater (Fig. 11.8c) are those where groundwater use is substantially higher and has management plans in place.

The observed negative correlations in some regions of the world could be attributed to either extensive groundwater use or the composite influence of geologic characteristics and aridity, leading to a delayed response. The linear relationship observed between climate data (SPEI) and groundwater suggests that changes in climate, rainfall intensity, and evapotranspiration will account for much of the variability in groundwater. Extended periods of rainfall are expected to increase recharge amplitudes, while prolonged drought will negatively impact this possibility. The crucial question now is whether future changes in drought characteristics will affect groundwater. Based on emission scenarios using SSP585, global drought extent or impact is projected to reduce in the near future (Sect. 11.5.1). However, drought duration will not decrease, at least for most locations. As drought is expected to increase in some hotspots, what does this mean for groundwater, particularly with strong positive correlations in those regions?

In the Anthropocene, groundwater depletion is a key problem in some regions around the world. Groundwater in about 50% of the world's largest aquifers are already in decline (Solander et al. 2017). The substantial loss in groundwater experienced in these aquifers jeopardize global food and water security and the sustainability of ecosystems, affecting human adaptation to climate change (Jasechko and Perrone 2021; Ndehedehe et al. 2021). Groundwater depletion is caused by long-term water extraction and mostly due to expansion of irrigated agriculture into semi-arid areas and increased urbanization (Bierkens and Wada 2019). However, prolonged extreme drought have also been identified as an important trigger that coalesce with human water extraction to trigger groundwater loss (Ojha et al. 2020). Increased water extraction will disrupt recharge, contributing to groundwater loss in time and space during prolonged drought events. The latter increases dependence on groundwater and decreased surface water flow and connection with aquifers. Recharge is sensitive to and affected by changes in rainfall intensity, flow, and temperature (Taylor et al. 2013). These factors regulate water balance and induce changes in the amount or timing of groundwater recharge. Since the advent of GRACE satellite both climatic and human footprints on groundwater depletion have been studied and the evidence on the composite influence of human water withdrawal and drought on groundwater loss is overwhelmingly critical and underpins effort to limit and increase resilience to impacts associated with groundwater loss. These impacts are deleterious to mankind and the environment and include, land subsidence, contribution to sea level change, groundwater salinization, loss of groundwater-dependent ecosystems, among others (Hugonnet et al. 2021; Bierkens and Wada 2019).

11.5.5 Global Climate Teleconnection Patterns are Important Predictors of Drought Impact

The spatial and temporal patterns of variability in SPEI were observed globally, with major hotspots located in the Middle East, North Africa, the West African Sahel, Australia, and some areas in Asia (for SPEI-3, 6, and 12). Significant teleconnection patterns were identified in some of these climatic hotspots, including the Atlantic Multi-decadal Oscillation (AMO), Pacific Decadal Oscillation (PDO), and Multivariate El-Niño Southern Oscillation Index (MEI). Although the lag response of these relationships was not assessed, the strong correlation values (positive or negative) of 0.5 and 0.6 suggest a significant contribution to the evolutionary patterns of climate extremes. Additionally, several hotspots showing correlation values between 0.30 and 0.45 indicate that climate teleconnection around the world, either directly or indirectly, drives changes in the characteristics of drought and flood events globally. For instance, the persistence of ENSO may lead to prolonged drought duration in some areas and flood events in other locations. These climate teleconnections can impact rainfall, leading to large variations in groundwater.

Some studies (e.g., Vissa 2019) have revealed that changes (i.e., decline and increase) in groundwater in northern India are linked to El-Niño and La Niña phases, respectively. The significant associations of SPEI with these teleconnections (PDO, ENSO, AMO) in arid and semi-arid domains that heavily rely on groundwater, such as the Middle East and North Africa, suggest that large variations in groundwater might be induced by these climate modes. The results from this study component confirm that AMO, PDO, and MEI are popular climate variability indices associated with changes in climatic conditions, emphasizing the critical role played by the Atlantic and Pacific oceans in large-scale climatic patterns that are expected to impact hydro-climatic conditions, predominantly characterized by multi-decadal and interannual oscillations.

11.5.6 How can we Maintain Balance, Justice, and Stability?

Anthropogenic activities are driving alterations in the Earth system, which along with climate change, are regulating the net contribution to sea-level rise (Ndehedehe 2023; Hugonnet et al. 2021; WCRP Group 2018). In the Anthropocene, the collective effort to shift the trajectory of the Earth system from damaging conditions to a stable and habitable state has been advocated (Steffen et al. 2018). This is because the resilience of the Earth system is being challenged. As earlier mentioned in Chap. 1 of this book, it's crucial to change this trajectory due to the interconnection between the biosphere, the atmosphere, the climate system, and their implications for human well-being and prosperity (Folke et al. 2021). Maintaining balance and pursuing a safe and just planet within the United Nations' Sustainable Development Goals (SDGs) is critical. The newly launched SDGSAT-1 satellite mission will accurately collect data on various human activities and provide data for various SDG indicators to support applications in climate-related hazards and natural disasters, among several others (Guo et al. 2002a).

To address the biodiversity decline problem, Obura et al. (2022) detailed six actions to achieve Earth system justice, including recommendations to reduce and reverse direct and indirect drivers causing decline. The assessment of Earth system planetary boundaries is receiving attention from a perspective that acknowledges human well-being as a function of Earth system stability and resilience (Rockström et al. 2023). The Earth Commission (https://earthcommission.org/) has began this assessment underpinned by several working groups as well as ongoing collaborations with relevant research institutions and advocacy groups, among others (Rockström et al. 2021). Reports by the Earth Commission, as well as other assessments by the Intergovernmental Panel on Climate Change and other sources (see, e.g., McKay et al. 2022; de Graaf et al. 2019; IMBIE 2018), document evidence suggesting that some of the thresholds for a healthy and stable planet and people, such as those for climate, water, aerosol, and biosphere, have already been transgressed (https://earthcommission.org/publications/). For example, in intensively irrigated areas of the world, the thresholds for groundwater have been transgressed, and environmental flow limits are expected to be reached for up to 79% watersheds by 2050 (de Graaf et al. 2019). Achieving resilience and stability of the Earth system, where humans are substantially less vulnerable to risk, is critical to maintaining justice within the Earth system and this will require that these thresholds and climate tipping points (McKay et al. 2022) are not exceeded further.

References

Adeyeri O, Zhou W, Laux P, Ndehedehe C, Wang X, Usman M, Akinsanola A (2023). Multivariate drought monitoring, propagation, and projection using bias-corrected general circulation models. Earth's Future 11(4):e2022EF003303. https://doi.org/10.1029/2022EF003303.

AghaKouchak A (2015) A multivariate approach for persistence-based drought prediction: Application to the 2010–2011 East Africa drought. J Hydrol 526:127–135. https://doi.org/10.1016/j.jhydrol.2014.09.063

AghaKouchak A, Mirchi A, Madani K, Baldassarre DB et al (2021) Anthropogenic drought: Definition, challenges and opportunities. Rev Geophys e2019RG000683. https://doi.org/10.1029/2019RG000683.

Agutu N, Awange J, Zerihun A, Ndehedehe C, Kuhn M, Fukuda Y (2017) Assessing multi-satellite remote sensing, reanalysis, and land surface models' products in characterizing agricultural drought in East Africa. Remote Sens Environ 194:287–302. https://doi.org/10.1016/j.rse.2017.03.041

Agutu N, Awange J, Ndehedehe C, Kirimi F, Kuhn M (2019) GRACE-derived groundwater changes over Greater Horn of Africa: temporal variability and the potential for irrigated agriculture. Sci Total Environ 693:133467. https://doi.org/10.1016/j.scitotenv.2019.07.273

Ahluwalia O, Singh PC, Bhatia R (2021) A review on drought stress in plants: Implications, mitigation and the role of plant growth promoting rhizobacteria. Resour Environ Sustain 5:100032. https://doi.org/10.1016/j.resenv.2021.100032

Ayugi B, Shilenje Z, Babaousmail H et al (2022) Projected changes in meteorological drought over East Africa inferred from bias-adjusted CMIP6 models. Nat Hazards. https://doi.org/10.1007/s11069-022-05341-8

Bierkens MFP, Wada Y (2019) Non-renewable groundwater use and groundwater depletion: a review. Environ Res Lett 14(6):063002. https://doi.org/10.1088/1748-9326/ab1a5f

Cardoso JF, Souloumiac A (1993) Blind beamforming for non-gaussian signals. IEE Proceed 140(6):362–370

Cenacchi N (2014) Drought risk reduction in agriculture: A review of adaptive strategies in East Africa and the Indo-Gangetic plain of South Asia. In: International food policy research institute (IFPRI), discussion Paper 1372. Retrieved from: http://ebrary.ifpri.org/cdm/ref/collection/p15738coll2/id/128277 23 Sept 2017

Chen J, Famigliett JS, Scanlon BR, Rodell M (2016) Groundwater storage changes: present status from grace observations. Surv Geophys 37(2):397–417. https://doi.org/10.1007/s10712-015-9332-4

Chiang F, Mazdiyasni O, AghaKouchak A (2020) Evidence of anthropogenic impacts on global drought frequency, duration, and intensity. Nat Commun 12:2754. https://doi.org/10.1038/s41467-021-22314-w

Cook BI, Mankin JS, Marvel K, Williams AP, Smerdon JE, Anchukaitis KJ (2020) Twenty-first century drought projections in the cmip6 forcing scenarios. Earth's Future 8(6):e2019EF001461. https://doi.org/10.1029/2019EF001461.

Cucchi M, Weedon GP, Amici A, Bellouin N, Lange S, Müller SH, Hersbach H, Buontempo C (2020) WFDE5: bias-adjusted ERA5 reanalysis data for impact studies. Earth Syst Sci Data 12(3):2097–2120. https://doi.org/10.5194/essd-12-2097-2020

Dai A (2013) Increasing drought under global warming in observations and models. Nat Clim Change 3:52–58. https://doi.org/10.1038/nclimate1633

de Graaf IE, Gleeson T, van Beek LR, Sutanudjaja EH, Bierkens MF (2019) Environmental flow limits to global groundwater pumping. Nature 574(7776):90–94. https://doi.org/10.1038/s41586-019-1594-4

Dezfuli A, Razavi S, Zaitchik BF (2022) Compound effects of climate change on future transboundary water issues in the middle east. Earth's Future 10(4):e2022EF002683

Diaz V, Corzo G, Lanen HAV, Solomatine DP (2019) Spatiotemporal drought analysis at country scale through the application of the STAND Toolbox. Spatiotemp Anal Extreme Hydrol Events 77–93. https://doi.org/10.1016/B978-0-12-811689-0.00004-5

FAO (2016) Climate change and food security: risks and responses. In: Food and agricultural organisation of the United Nations. Retrieved from http://www.fao.org/3/a-i5188e.pdf 20 Sept 2016

Farahmand A, AghaKouchak A (2015) A generalized framework for deriving nonparametric standardized drought indicators. Adv Water Resour 76:140–145. https://doi.org/10.1016/j.advwatres.2014.11.012

Ferreira V, Montecino H, Ndehedehe C, Heck B, Gong Z, Westerhaus M, de Freitas S (2018) Space-based observations of crustal deflections for drought characterization in Brazil. Scie Total Environ 644:256–273. https://doi.org/10.1016/j.scitotenv.2018.06.277

Folke C, Polasky S, Rockström J et al (2021) Our future in the anthropocene biosphere. Ambio 50:834–869. https://doi.org/10.1007/s13280-021-01544-8

Furrer M, Mostofi H, Spinler S (2022) A study on the impact of extreme weather events on the ceramic manufacturing in egypt. Resour Environ Sustain 7:100049. https://doi.org/10.1016/j.resenv.2022.100049

Giannini A, Saravanan R, Chang P (2003) Oceanic forcing of Sahel rainfall on interannual to decadal time scales. Science 302(5647):1027–1030. https://doi.org/10.1126/science.1089357

Gringorten II (1963) A plotting rule for extreme probability paper. J Geophys Rese 68(3):813–814. https://doi.org/10.1029/JZ068i003p00813

Guo H, Dou C, Chen H, Liu J, Fu B, Li X, Zou Z, Liang D (2022) SDGSAT-1: the world's first scientific satellite for sustainable development goals. Sci Bull. https://doi.org/10.1016/j.scib.2022.12.014

Guo Y, Yu X, Xu Y-P, Wang G, Xie J, Gu H (2022) A comparative assessment of CMIP5 and CMIP6 in hydrological responses of the Yellow River Basin. China. Hydrol Res 53(6):867–891. https://doi.org/10.2166/nh.2022.001

Hagenlocher M, Meza I, Anderson CC, Min A, Renaud FG, Walz Y, Siebert S, Sebesvari Z (2019) Drought vulnerability and risk assessments: state of the art, persistent gaps, and research agenda. Environ Res Lett 14(8):083002. https://doi.org/10.1088/1748-9326/ab225d

Hao Z, AghaKouchak A (2013) Multivariate standardized drought index: A parametric multi-index model. Adv Water Resour 57:12–18. https://doi.org/10.1016/j.advwatres.2013.03.009

Hao Z, AghaKouchak A (2014) A nonparametric multivariate multi-index drought monitoring framework. J Hydrometeorol 15(1):89–101. https://doi.org/10.1175/JHM-D-12-0160.1

Hayes M, Svoboda M, Wall N, Widhalm M (2011) The lincoln declaration on drought indices: Universal meteorological drought index recommended. In: Drought mitigation center faculty publications. Accessed from http://digitalcommons.unl.edu/droughtfacpub/14 9th June 2020

Hugonnet R, McNabb R, Berthier E et al (2021) Accelerated global glacier mass loss in the early twenty-first century. Nature 592:726–731. https://doi.org/10.1038/s41586-021-03436-z

IMBIE (2018) Mass balance of the antarctic ice sheet from 1992 to 2017. Nature 558:219–222. https://doi.org/10.1038/s41586-018-0179-y

Iturbide M, Casanueva A, Bedia J, Herrera S, Milovac J, Gutiérrez JM (2022) On the need of bias adjustment for more plausible climate change projections of extreme heat. Atmos Sci Lett 23(2):e1072. https://doi.org/10.1002/asl.1072

Jasechko S, Perrone D (2021) Global groundwater wells at risk of running dry. Science 372(6540):418–421. https://doi.org/10.1126/science.abc2755

Kiem AS, Johnson F, Westra S, van Dijk A et al (2016) Natural hazards in Australia: droughts. Clim Change 139(1):37–54. https://doi.org/10.1007/s10584-016-1798-7

Madani K, Shafiee-Jood M (2020) Socio-hydrology: A new understanding to unite or a new science to divide? Water 12(7). https://doi.org/10.3390/w12071941.

McKay DIA, Staal A, Abrams JF, et al (2022) Exceeding 1.5^{0}c global warming could trigger multiple climate tipping points. Science 377(6611):eabn7950. 10.1126/science.abn7950

McKee TB, Doeskin NJ, Kieist J (1993) The relationship of drought frequency and duration to time scales. In: Conference on applied climatology. American Meteorological Society, Boston, pp 179–184. Retrieved from www.ccc.atmos.colostate.edu/relationshipofdroughtfrequency.pdf. Accessed 27 June, 2014

Mishra AK, Singh VP (2010) A review of drought concepts. J Hydrol 391:202–216. https://doi.org/10.1016/j.jhydrol.2010.07.012

Naumann G, Alfieri L, Wyser K et al (2018) Global changes in drought conditions under different levels of warming. Geophys Res Lett

Ndehedehe C (2022a) Drought events. Springer International Publishing, Cham, pp 249–280. https://doi.org/10.1007/978-3-030-99577-5_11

Ndehedehe C (2022b) Satellite geodetic missions. Springer International Publishing, Cham, pp 53–70. https://doi.org/10.1007/978-3-030-99577-5_4.

Ndehedehe CE (2023) Understanding the earth system in the anthropocene. Resour Environ Sustain 12:100113. https://doi.org/10.1016/j.resenv.2023.100113

Ndehedehe CE, Anyah RO, Alsdorf D, Agutu NO, Ferreira VG (2019) Modelling the impacts of global multi-scale climatic drivers on hydro-climatic extremes (1901–2014) over the Congo basin. Sci Total Environ 651:1569–1587. https://doi.org/10.1016/j.scitotenv.2018.09.203

Ndehedehe CE, Agutu N, Ferreira VG, Getirana A (2020) Evolutionary drought patterns over the Sahel and their teleconnections with low frequency climate oscillations. Atmos Res 233:104700. https://doi.org/10.1016/j.atmosres.2019.104700

Ndehedehe CE, Ferreira VG, Agutu NO, Onojeghuo AO, Okwuashi O, Kassahun HT, Dewan A (2021) What if the rains do not come? J Hydrol 595:126040. https://doi.org/10.1016/j.jhydrol.2021.126040

Ndehedehe CE, Ferreira VG, Adeyeri OE, Correa FM, Usman M, Oussou FE, Kalu I, Okwuashi O, Onojeghuo AO, Getirana A, Dewan A (2023) Global assessment of drought characteristics in the anthropocene. Resour Environ Sustain 100105. https://doi.org/10.1016/j.resenv.2022.100105

Obura DO, DeClerck F, Verburg Peter H et al (2022) Achieving a nature- and people-positive future. One Earth. https://doi.org/10.1016/j.oneear.2022.11.013

Ojha C, Werth S, Shirzaei M (2020) Recovery of aquifer-systems in Southwest US following 2012–2015 drought: Evidence from InSAR, GRACE and groundwater level data. J Hydrol 587. https://doi.org/10.1016/j.jhydrol.2020.124943

O'Neill BC, Tebaldi C, van Vuuren DP, Eyring V, Friedlingstein P, Hurtt G, Knutti R, Kriegler E, Lamarque J-F, Lowe J, Meehl GA, Moss R, Riahi K, Sanderson BM (2016) The scenario model intercomparison project (ScenarioMIP) for CMIP6. Geosci Model Develop 9(9):3461–3482. https://doi.org/10.5194/gmd-9-3461-2016

Rippey BR (2015) The U.S. drought of 2012. Weather Clim Extremes 10(Part A):57–64. https://doi.org/10.1016/j.wace.2015.10.004.

Rockström J, Gupta J, Lenton TM, Qin D et al (2021) Identifying a safe and just corridor for people and the planet. Earth's Future 9(4):e2020EF001866. https://doi.org/10.1029/2020EF001866

Rockström J, Gupta J, Qin D et al (2023) Safe and just Earth system boundaries. Nature 619:102–111. https://doi.org/10.1038/s41586-023-06083-8

Samantaray AK, Ramadas M, Panda RK (2022) Changes in drought characteristics based on rainfall pattern drought index and the CMIP6 multi-model ensemble. Agric Water Manag 266:107568. https://doi.org/10.1016/j.agwat.2022.107568

Save H, Bettadpur S, Tapley BD (2016) High-resolution CSR GRACE RL05 mascons. J Geophys Res: Solid Earth 121(10):7547–7569. https://doi.org/10.1002/2016JB013007

Schroth G, Läderach P, Martinez-Valle AI, Bunn C, Jassogn L (2016) Vulnerability to climate change of cocoa in West Africa: Patterns, opportunities and limits to adaptation. Sci Total Environ 556:231–241. https://doi.org/10.1016/j.scitotenv.2016.03.024

Sheffield J, Wood EF (2008) Global trends and variability in soil moisture and drought characteristics, 1950–2000, from observation-driven simulations of the terrestrial hydrologic cycle. J Clim 21(3):432–458. https://doi.org/10.1175/2007JCLI1822.1

Solander K, Reager J, Wada Y et al (2017) Grace satellite observations reveal the severity of recent water over-consumption in the united states. Sci Rep 7:8723. https://doi.org/10.1038/s41598-017-07450-y

Spinoni J, Naumann G, Carrao H, Barbosa P, Vogt J (2014) World drought frequency, duration, and severity for 1951–2010. Int J Climatol 34(8):2792–2804. https://doi.org/10.1002/joc.3875

Steffen W, Rockström J, Richardson K et al (2018) Trajectories of the earth system in the anthropocene. Proceed Nat Acad Sci 115(33):8252–8259. https://doi.org/10.1073/pnas.1810141115

Tapley B, Bettadpur S, Watkins M, Reigber C (2004) The gravity recovery and climate experiment: Mission overview and early results. Geophys Res Lett 31:1–4. https://doi.org/10.1029/2004GL019920

Taylor R, Scanlon B, Döll P et al (2013) Ground water and climate change. Nat Clim Change 3:322–329. https://doi.org/10.1038/nclimate1744

Thomas B, Famiglietti J (2019) Identifying climate-induced groundwater depletion in GRACE. Sci Rep 9:4124. https://doi.org/10.1038/s41598-019-40155-y

Thomas B, Famiglietti J (2019) Identifying climate-induced groundwater depletion in GRACE. Sci Rep 9:4124. https://doi.org/10.1038/s41598-019-40155-y

Uddameri V, Singaraju S, Hernandez E (2019) Is Standardized Precipitation Index (SPI) a useful indicator to forecast groundwater droughts?-insights from a Karst aquifer. JAWRA J Am Water Resour Assoc 55(1):70–88. https://doi.org/10.1111/1752-1688.12698

Usman M, Ndehedehe CE, Farah H, Manzanas R (2021) Impacts of climate change on the streamflow of a large river basin in the Australian tropics using optimally selected climate model outputs. J Cleaner Prod 315:128091. https://doi.org/10.1016/j.jclepro.2021.128091

van der Molen M, Dolman A, Ciais P, Eglin T, Gobron N, Law B, Meir P, Peters W, Phillips O, Reichstein M, Chen T, Dekker S, Doubková M, Friedl M, Jung M, van den Hurk B, de Jeu R, Kruijt B, Ohta T, Rebel K, Plummer S, Seneviratne S, Sitch S, Teuling A, van der Werf G, Wang G (2011) Drought and ecosystem carbon cycling. Agric For Meteorol 151(7):765–773. https://doi.org/10.1016/j.agrformet.2011.01.018

Vicente-Serrano SM, Beguería S, López-Moreno JI (2010) A multiscalar drought index sensitive to global warming: The standardized precipitation evapotranspiration index. J Clim 23(7):1696–1718. https://doi.org/10.1175/2009JCLI2909.1

Vicente-Serrano SM, Beguería S, López-Moreno JI (2010) A multiscalar drought index sensitive to global warming: The standardized precipitation evapotranspiration index. J Clim 23(7):1696–1718. https://doi.org/10.1175/2009JCLI2909.1

Vissa NK, Anandh P, Behera M et al (2019) ENSO-induced groundwater changes in India derived from GRACE and GLDAS. J Earth Syst Sci Vol 128(115). https://doi.org/10.1007/s12040-019-1148-z.

WCRP Group, GSLB (2018) Global sea-level budget 1993-present. Earth Syst Sci Data 10(3):1551–1590. https://doi.org/10.5194/essd-10-1551-2018

Weedon GP, Balsamo G, Bellouin N, Gomes S, Best MJ, Viterbo P (2014) The WFDEI meteorological forcing data set: WATCH Forcing Data methodology applied to ERA-Interim reanalysis data. Water Resour Res 50(9):7505–7514. https://doi.org/10.1002/2014WR015638

White WB, Gershunov A, Annis JL, McKeon G, Syktus J (2004) Forecasting Australian drought using Southern Hemisphere modes of sea-surface temperature variability. Int J Climatol 24(15):1911–1927. https://doi.org/10.1002/joc.1091

Xu L, Chen N, Zhang X (2019) Global drought trends under 1.5 and 2 °c warming. Int J Climatol 39(4):2375–2385. https://doi.org/10.1002/joc.5958.

Yang M, Li Z, Anjum MN, Kayastha R, Kayastha RB, Rai M, Zhang X, Xu C (2022) Projection of streamflow changes under CMIP6 scenarios in the Urumqi river head watershed. Tianshan Mountain. China. Front Earth Sci 10. https://doi.org/10.3389/feart.2022.857854

Zhou L, Tian Y, Myneni RB, Ciais P et al (2014) Widespread decline of congo rainforest greenness in the past decade. Nature 509(7498):86–90. https://doi.org/10.1038/nature13265

Chapter 12
Floods and Wildfire Events

12.1 Introductory Remarks

The impacts of climate change on the environment are intensifying and alarming. Some of these impacts including those of floods and wildfire events are increasing and it is expected that during the coming decades, their intensity, frequency, and impacts will also increase significantly (Parjanne et al. 2021; Sayers et al. 2021; Viedma et al. 2018). For instance, the 2022 floods in Nigeria (Fig. 12.1) is one of the most devastating flood events in Africa in the last decade. This flood displaced 1.3 million people, causing more than 600 deaths and over 200,000 houses were destroyed.[1] As opposed to drought and wildfire events, the increasing humanitarian crisis and displacement caused by flood events across the world is arguably alarming. According to UNICEF, about 2.5 million people in Nigeria needed humanitarian support due to 2022 extreme flood. Devastating perilous floods appear to be very common in the Anthropocene, especially in the last 20 years. For example, the epic flood of 2017 caused by Hurricane Harvey in the United States led to one of the highest downpours observed in a tropical system.[2] This flood resulted in more than 25 trillion gallons of rainfall within four days, displacing more than 25,000 people and triggering power outages and other unprecedented physical damages in Texas (https://edition.cnn.com/specials/us/hurricane-harvey).

Furthermore, more than 400 people died in Sierra Leone due to mudslides and torrential flooding in the capital city of Sierra Leone (Ndehedehe 2017). In other coastal West African regions, extreme floods displaced about 800,000 and 600,000 people in 2007 and 2009, respectively and hundreds of death have been reported in each of these events (Ndehedehe 2017). In Asia and across several regions of the world, terrific and frequent devastating floods continue to cause damages. This includes

[1] https://www.unicef.org/press-releases/more-15-million-children-risk-devastating-floods-hit-nigeria.

[2] http://mashable.com/2017/08/29/harveyhouston-ood-by-the-numbers-worst-ood/.

C. Ndehedehe, *Hydro-Climatic Extremes in the Anthropocene*, Springer Climate,
https://doi.org/10.1007/978-3-031-37727-3_12

Fig. 12.1 A section of the 2022 floods along the river Niger floodplains, which ravaged several local communities in Nigeria, leading to more than 600 death and about 1.4 million people being displaced. *Source* NASA Earth Observatory. https://earthobservatory.nasa.gov/images/150463/flooding-inundates-southern-nigeria

Bangladesh, one of the world's notorious hotspots of flood events. Since the second half of 2022, for instance, Monsoon induced torrential rain has affected about 7.2 million people in Bangladesh (https://www.ifrc.org/emergency/bangladesh-floods). These floods have not only resulted in large proportion of human death and displacement but have caused severe damages to properties, public infrastructure, and facilities, as well as destroy croplands.

Furthermore, floods in extremely humid and considerably large hydrological systems such as the Amazon and Congo basins as well as coastal West Africa are

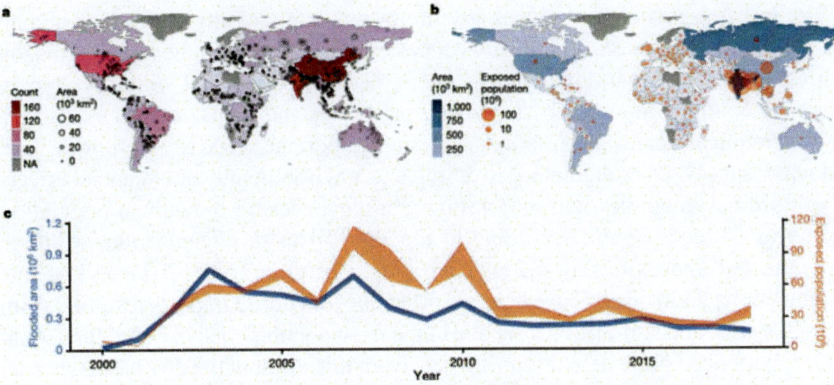

Fig. 12.2 Global flood database summarizing **a** country-specific cases of flood episodes in the global flood database (color scale), including the circular depiction of each flood event as well as their centroid positions **b** total exposed area and population per country from 2000–2015 **c** annual global population estimation. This Figure was originally adapted from Tellman et al. (2021) for use in the work of Ndehedehe (2022b) where it was first referenced

common. As captured by the Operational Land Imager-2 sensor onboard the US-owned Landsat 9 satellite, the floods in June 2022 in Nigeria escalated rapidly, ravaging considerably large communities along the river Niger floodplain and southern states in very early October 2022 (Fig. 12.1). The June 2021 extreme flood in the Amazon Basin triggered an unprecedented water level (above 29 m) lasting for 91 days in Brazil, surpassing the well-known historical flood of 2012 in terms of magnitude and impact (Espinoza et al. 2022). Other online reports[3] argue that the flood height in 2021 in the Amazon exceeded historical record since measurements began in 1903, further strengthening predictions for climate change linked to global warming. The impacts of floods on humans since 2000 (Fig. 12.2) and its characteristics (magnitude and severity) across several regions of the world, including Europe (where flash flood is frequent) and North America continue to change with increasing human population being vulnerable due to climate change. Even though much of the floods in the Anthropocene are the results of climate change (e.g., Tellman et al. 2021; Ndehedehe et al. 2021), in several emerging economies, the magnitude and impacts of these events can be facilitated by human activities, including development of residential buildings along floodplains and large watercourse (e.g., Dewan et al. 2017). The human footprint on climate extremes such as floods, heatwaves, and storms were identified earlier in the 2000s. About 71% of more than 500 occurrences of extreme climatic conditions identified were likely to be more severe due to human-induced climate change.[4] Put together, this reinforces the notion that climate and human activities are key drivers of extreme events like flood in the Anthropocene.

[3] https://news.mongabay.com/2021/07/lessons-from-the-2021-amazon-flood-commentary/.

[4] https://www.carbonbrief.org/mapped-how-climate-change-affects-extreme-weather-around-the-world/.

The growing intensity of global wildfire events is increasingly affecting human and environmental assets, and several ecosystem services. From the humid Amazon basin to dry and semi-arid domains in Australia, the frequency of wildfire events in the world continue to increase, destroying properties and crops, as well as disrupting communications, transportation, affecting important services like power, gas, and water supply. Overall, the negative impacts, environmental and financial effects of wildfire have complex and long lasting consequences on the human population and terrestrial ecosystem. According to the World Health Organization, wildfires have affected approximately 6.2 million people between 1998–2017 and of this amount, about 2400 deaths have been reported due to wildfire impacts resulting from injuries, burns, and suffocation.[5] As one of the most popular hotspot of wildfire, Australia experienced more deleterious impacts from this event in the 2019–2020 period because of its scale and intensity, which destroyed large hectares of land and affected billions of animals. The World Wide Fund for Nature a science-based organization estimated in their report (also summarized here[6]) that around *12.6 million hectares containing primarily forest and woodland* were destroyed and about 3 billion native vertebrates were affected. The latter includes 143 million mammals, 2.46 billion reptiles, 180 million birds, and 51 million frogs.

The changes in ecosystem functioning as well as the rate of anthropogenic and human activities on them constitute disparities in the natural range of fire sizes and resultant intensities, timings and frequencies, otherwise known as the fire regime. It is common knowledge that when terrestrial ecosystem has been significantly altered by human activities, it becomes more prone to wildfires. For instance, activities such as fire suppressions, timber harvesting, agricultural or urban encroachment, invasive species, and ignition patterns have been highlighted as environmental catalysts to wildfire events (Krawchuk et al. 2009). Unsurprisingly, policy strategies targeted at minimizing wildfires often emphasize on fuel reduction (Fernandes 2013). However, even where strategies recognize interacting environmental, cultural, and economic dimensions of wildfire, only a few address the complex land-use issues including urban sprawl and changes in land cover. The potential of escalating wildfire activity with climate change reiterates the need for a new path forward. In this chapter, changes in the characteristics of floods and wildfire events and how they have impacted human developments and progress are broadly discussed. The techniques and systems for improved assessments of these extreme events are also detailed.

[5] https://www.who.int/health-topics/wildfires/tab=tab_1.

[6] https://www.nbcnews.com/news/world/australian-wildfires-declared-among-worst-wildlife-disasters-modern-history-n1235071.

12.2 Global Flood Events

The rise in severe and extreme flood events is triggering more deleterious impacts on infrastructure, land use, human population, losses, and socio-economic problems (e.g., migration) (Ndehedehe 2022b). More people are now being affected by floods compared to any other environmental hazard that impacts on human progress and sustainable development. From the analyses of more than 900 large flood events (2000–2018) across the globe, the work undertaken by Tellman et al. (2021) showed that up to 290 million people were directly affected by floods. They reported an increase of 58–86 million people being affected by flood events between 2000 and 2015. Future climate projection suggests an increasing proportion of global population is expected to be exposed to floods (Tellman et al. 2021). The global flood database captures the rise in human population affected by floods in the last 20 years and also indicates a strong relationship between affected flooded areas and human population exposed to flood impacts (Fig. 12.2). This database contains maps showing the spatio-temporal distribution of 913 flood events that occurred between 2000–2018 (Tellman et al. 2021).

From the freely available large repositories of geospatial datasets that can support the assessment of water-related disasters in the Anthropocene, to the use of popular social media platforms like Twitter for flood tracking, reporting and early warning (Lindersson et al. 2020; de Bruijn et al. 2019), the need for empirical and easily accessible earth observation data to improve understanding on the changing characteristics of flood is important for risk management solutions. The awareness of society to the impacts of flood events was captured in the number of flood-related Tweets across the globe, US, Tanzania, and Indonesia (Fig. 12.3). The social media approach to flood reporting (de Bruijn et al. 2019) complements conventional methods of flood detection adding flavor to the knowledge and tracking of where these events are occurring on a near and/or real-time basis. The global flood monitor is a platform following the work of de Bruijn et al. (2019) that continuously broadcast new flood events in real-time across the world, providing information on hotpot of flood events (https://www.globalfloodmonitor.org/). However, research opportunities exist in flood monitoring and risk assessment that leverages freely available rich repositories of large earth observation data (Lindersson et al. 2020). For instance, the global flood database by Tellman et al. (2021), which collected and successfully mapped major 913 flood events using about 12,719 image scenes from the Terra and Aqua MODIS sensors, is an example of such repository.

12.2.1 Flood Risk and Impacts in the Anthropocene

Flooding is no doubt one of the most devastating natural hazards that have caused death and severe damage of properties, infrastructures, and the environment (Fig. 12.2). Floods have also led to substantial socio-economic loss and an estimated 2.5 billion

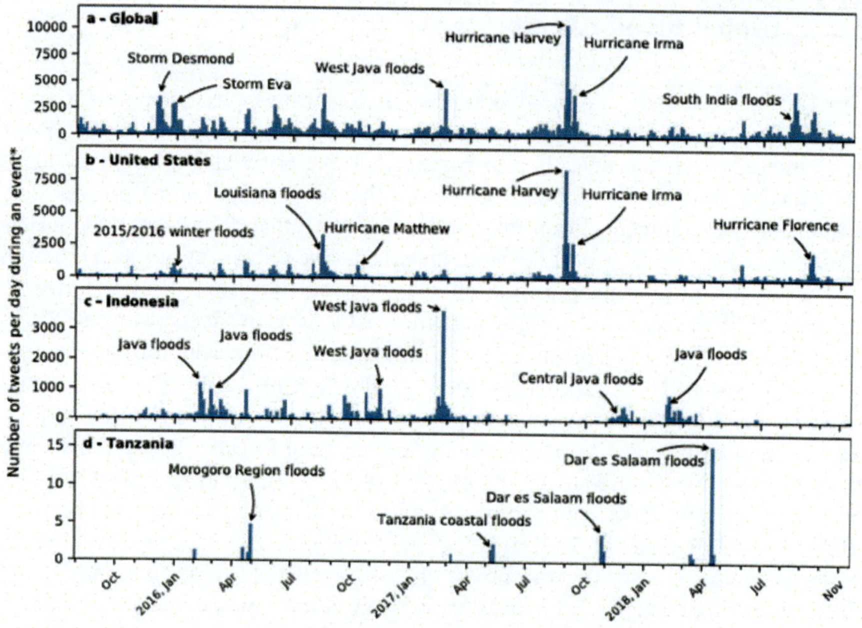

Fig. 12.3 Relevant social media interactions from Twitter associated with flood events across the **a** globe, **b** United States, **c** Indonesia, and **d** Tanzania. *Source* de Bruijn et al. (2019)

dollars of socio-economic loss per year from 1960 to 2017 (Petit-Boix et al. 2017). Another account indicates that flood damages resulted in an estimated 651 billion dollars loss during the 2000–2019 period (CRED 2020). In Thailand lost, damages worth 59 billion dollars due to the extreme 2011 flood events were reported in (Hall et al. 2014). In other regions of the world (India, US, Bangladesh, Nepal), flood events have left behind more economically-damaging devastating impacts including large displacements of human settlements, death, power outages, among other physical damages (Ndehedehe 2022b).

Intensifying global climate change and anthropogenic influences are responsible for the increasing rate of flood incidents, especially due to surge in the frequency and intensity of precipitation patterns (Pham et al. 2021). To manage these incessant flood events and their consequences on the ecosystem, it becomes necessary to adopt flood risk assessment mitigation strategies. Flood risk assessment for future extremes is a key aspect of flood forecasting which borders on protecting the environment from dangerous climatic conditions. Risk assessments are primarily focused on indicators of how prone a given basin is to flooding, taking into account the precipitation rates and soil moisture conditions of the region. Flood risks integrate the possibility that an event could occur and the corresponding consequence associated with that event. Many regions are prone to volatile flood hazards which can be a factor of growing human population there. This shows that even if flood hazards reduce, flood risks can become intensified as a result of increased vulnerability or exposure to multiple climatic and anthropogenic factors.

12.2.2 Assessing Flood Risk

In the past, several techniques, methods, and approaches have been proposed for efficient flood risk assessment. Some of these have attempted the use of statistical methods, multi-criteria analysis, analysis based on remote sensing and geographic information system techniques, and scenario-based inundation analysis to develop robust and efficient models that can tackle real-life scenarios of flooding (Lyu et al. 2019). To further strengthen the reliability of the developed system models, some studies have integrated the use of long-term historic records for risk assessment exercises. Using this approach, Nott (2006) who argued that the use of short historic datasets for risk assessment of natural hazards fails to identify and fully capture the natural variability of the hazard was able to provide useful resources, that helped planners understand the trend and potential impacts of natural hazards using long-term historic records. Some thoughts around techniques in flood risk assessments are detailed below.

- Other popular approaches which leverage statistical methods and novel machine learning frameworks have contributed immensely to designing feasible models that critically address flood risks. For example, Black and Burns (2002) attempted to improve the characterization of catchment controls on flood magnitudes over Scotland by developing statistical models that allow improved estimations of flood risks, extents, and depth over the study region. A key issue discussed in the work of Black and Burns (2002) is that flood risk is expected to increase over a decadal time scale. These impacts were tied to the clustering of maximum recorded floods over some large rivers in the region and increase in peaks over threshold frequencies. Their study also confirmed the identification and use of historic flood records as being increasingly evident for efficient flood risk assessment strategies, as comprehensive and long-term historic records which include exceptional floods tend to provide some level of protection against the dangers of basing assessments on relatively short-term datasets.
- The application of the multi-criteria evaluation approach for flood risk management offers the benefit of elucidating flood risk zones through the integration of spatio-multi-temporal historic satellite observations, with socio-economic datasets, infrastructure, and updates from land-use vulnerabilities. This argument is based on the fact that most flood risk occurrence is a combination of both natural and anthropogenic influences, thereby warranting the need for knowledge regarding the spatial extent of flooding areas. Therefore, the use of multi dataset

as drivers becomes a potential source for more efficient and reliable flood management strategies and mitigation programmes (Danumah 2016).

- Some studies argue that the lack of supporting information to strengthen intervention is a major cause of inefficient flood risk preparedness and response, resulting in more devastating flood impacts (e.g., Schipper and Pelling 2006). Chau et al. (2013) suggested that the high recurrent flooding, including fatalities and ecosystem disruption could have been limited if there was sufficient, up to date, and comprehensive data. The need to map hazardous zones, develop response strategies that are best suited for affected regions, and mitigation strategies that minimize the influence of possible flood events like flood early warning systems are critical to developing future flood risk assessments. This led to increased application of geographic information system (GIS) and remote sensing datasets in flood disaster monitoring and management (Fig. 12.4). Remote sensing data, including optical and microwave sensors, possess the capabilities of delineating flood-prone regions (Fig. 12.4). They have also been useful in developing flood hazard maps by providing primary datasets to assess critical drivers of the flood such as precipitation intensity, land use/cover, among others. Today, satellite earth resources have bridged data gaps, enabling data deficient regions to monitor floodplain productivity and landscapes that are frequently flooded (e.g., Ndehedehe 2022b; Tellman et al. 2021).
- To minimize the devastating effects and reduce the loss from the occurrence of flood events, Kabenge et al. (2017) created spatially distributed flood hazard maps for a data-scarce area of western Uganda using remote sensing methods. The flood hazard index developed in their study was based on remote sensing and geospatial methods, and showed a clear percentage of the high risk and low risk region susceptible to flooding. Apart from the production of a spatially distributed composite remote sensing based flood hazard index and a flood hazard map that proved efficient for coordinating emergency responses in cases of flood events, the primary strength of developing flood mitigation strategy based on remote sensing and geospatial techniques (Ennouri et al. 2021) is on its applicability at watershed scales and in data-scarce regions, which can be prone to flooding (Fig. 12.4). The reporting of flood events to support mitigation and urban development and transition have been underpinned by remote sensing data. Change detection (based on image differentiation, post-classification, etc.), single and multi-band inundation metrics using multi-temporal satellite images have been employed to assess flood extents and impacts for flood risk assessment (e.g., Ndehedehe et al. 2021; Ndehedehe 2022b; Ennouri et al. 2021; Wolski et al. 2017).

Scenario-based inundation analysis is a quantitative technique that applies scenario analysis to forecast flood risk prior to its occurrence. Applying this method over the UK, Pant et al. (2018) presented an integrated framework for quantifying wider spatial consequences of infrastructure failures due to flooding by modeling flood inundation impacts on critical infrastructure systems. This led to the development of an informed and reliable flood resilience planning model. Using this scenario-based inundation analysis, they were able to identify, which infrastructure assets lie within

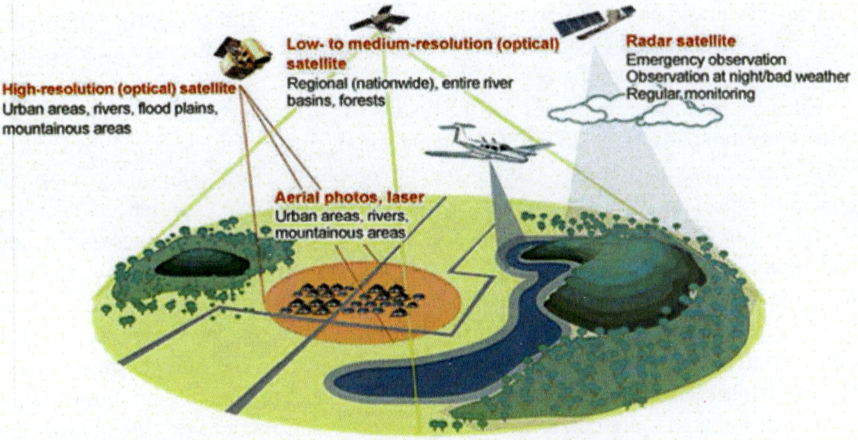

Fig. 12.4 Remote sensing capabilities and techniques to support flood risk assessment and management. *Source* https://www.kkc.co.jp/english/service/env_energy/k_green_solar.html

the flooded zones and the number of residents affected by floods, as well as the infrastructures outside the flooded areas. Critical information such as this proved useful in deciding where to prioritize flood protection/mitigation measures, which could be around flooded assets that generate widespread network interruptions. Ultimately, this framework aids urban planners in narrowing down and strategizing spatial flood management planning for enhanced resilience to flooding or inundations.

12.2.3 Changes in the Magnitude of Flood Events

The intensification of the hydrologic cycle due to climate change impacts has led to growing concerns that flood events will occur more frequently and increase in magnitude (Milly et al. 2002). Flood events are natural hazards that are both complex and pervasive with significant threat to our environment and terrestrial ecosystem. The complicated nature of flood events, which includes multiple flood initiation patterns, suggests that flood events are less likely to respond to changes in climate oscillations in an orderly and uniform manner. This claim is backed up by two recent studies. Do et al. (2017) investigated the presence of trends in yearly maximum daily stream flow for 9213 stations across the globe. They reported substantial decreasing trends in more stations than several other stations with substantial increasing trends. The conclusion of Hodgkins et al. (2017) that the fluctuations in the occurrence rate of enormous flood events were more affected by multi-decadal variability than by long-term trends was based on trend analysis of over 1200 gauges of diverse catchments across North America and Europe. These two studies clearly confirm the existing uncertainty in global flood patterns, especially when averaged over global scales or considerably larger regions.

Changes to flood magnitudes are the results of several factors, including climate teleconnection patterns (e.g., ENSO) and anthropogenic activities. Given the volatility of these flood magnitudes, there is need to emphasize frequent flood assessment analysis, especially for catchments and watersheds that have experienced land-use changes. Tan and Gan (2015) conducted this exercise on watersheds across Canada and found evidence of non-stationarity in annual maximum stream flow datasets. Given that the impact of non-stationarities is not limited to natural climate change or anthropogenic impacts alone, the analysis helped to investigate other factors that could contribute to the non-stationary characteristics of the annual stream flow. Results from their analysis showed that, for the Canadian annual maximum stream flow, sharp changes are detected more frequently than monotonic trends; and this can be attributed to the recent regulation of several rivers in the region over the course of the twentieth century. The impacts of land-use changes and anthropogenic influences on large scale drainage basins contribute to the ease in tracking temporal trends in annual maximum stream flows. This, however, is not the case with pristine river basins with stable land-use conditions, as climate change impacts resulting from anthropogenic influences tend to contribute to monotonic trends, and possibly, sharp changes in annual maximum stream flow series.

Changes in flood regimes can be heightened as a result of increased global warming and greenhouse gas emissions that amplify the water cycle and intensify river floods. Rivers usually have peak flows in specific periods of the year with the exception of floods caused by fitful weather events such as hurricanes. This was evidenced in the study by Rood et al. (2016), that investigated historic trends in the durations, magnitude, and timing of the annual peak flows of rivers that drain the Rocky Mountains around the North American hydrographic apex. Reports from the study indicated global warming would intensify the water cycle, thereby leading to extreme flooding. Alternatively, it is possible that the increased warming conditions could lower and minimize flooding events. This logic matches the seasonal patterns in which weather conditions of some tropical regions such as North America and Sub-Saharan Africa produce river floods as a result of the changes in spring season that are wet and cool, but exhibit fewer floods and low precipitation during the warmer summer. The study found stronger evidence of flood moderation rather than flood intensification over their study region and also identified the influenced of teleconnection patterns from the Pacific across Canada.

Finally, it is pertinent to point out that the rate, frequency, and magnitude of flood dynamics over a large portion of the earth surface are significantly influenced by several factors. These include

- land use of the region, catchment, or watershed in question,
- climate change impacts,

- global climate teleconnection patterns,
- human activities and anthropogenic influences,
- sea-level rise due to melting of the polar caps.

The erratic nature of these impacts makes it very difficult to accurately forecast increase (++), decrease (−−), or moderation (+−) in long term flood magnitudes for any region, as well as limits the use of traditional statistical and prediction algorithms for flood management systems. However, with the advent of machine learning and deep learning methods, the task of sorting the severe non-linearity of climatic variables looks promising, and the past decade has witnessed increased accuracy of emerging studies that applied these techniques especially in flood management and water resource controls.

12.3 Flood Prediction

Globally, large human populations or the most populated cities of the world across several continents are primarily located along the coast. The vulnerability of these coastal regions to extreme events like flooding has resulted in intensified efforts to improve its modeling and prediction. The non-linearity of climate change processes particularly with human footprints and anthropogenic influences redefining these processes makes accurate prediction even more difficult. One reason to account for this could be that changes in the characteristics of hydrological extremes (droughts and floods) drive substantial variations in hydrological processes on both local and global scales. These changes imply that the hydrological response of some catchments to climate change can vary substantially due to its characteristics (e.g., population, climatology, topography, land use, etc.) and location.

12.3.1 Climate Models

The seasonal discharge and extreme flows on the Pearl River in China under different Regional Climate Model scenarios provide some insights to the unique response of catchments to climate variability and change. Under five different climate models, Yan et al. (2015) showed variability in the reduction of average low flow over the Pearl River basin. The RCP4.5 climate mode used in the experiment depicted a varying degree of reduction across the basin at a quotient ranging from 6% to 48%. The resulting problem from this is the likelihood of varying river discharge aggravating water stress and leading to temporary water shortages. The uncertainties associated with these climate models limit their application in the estimation of future flows. Additionally, their faithfulness (climate models) is being challenged due to these uncertainties and inability to accurately replicate changes in the characteristics (e.g., intensity, magnitude, etc.) of rainfall. The emergence of machine learning techniques is now facilitating the characterization of non-linear climate-based scenarios and offers better opportunities for improved estimations of extreme climate events like floods. The use of machine learning in flood prediction is therefore discussed further in the next section.

12.3.2 Statistical and Machine Learning Methods

In the past, flood mapping and its characterization were usually either qualitative (e.g., Analytic Hierarchy Process) or quantitative (e.g., statistical and machine learning-based approaches methods) in nature (Adnan et al. 2023). A general problem with statistical models (e.g., weights of evidence, multivariate approaches) is their inability to capture existing nonlinear relationships between flood episodes and their key drivers (Adnan et al. 2023). The earlier work of Gudiyangada Nachappa et al. (2020), which compared the suitability of statistical models (analytical hierarchical process and analytical network process) and machine in flood susceptibility mapping confirmed this. They demonstrated the predictive superiority of machine learning models (Random Forest and support vector machine) over statistical models in flood vulnerability mapping. Machine learning-based techniques are increasingly being used in flood forecasting because of their capability in big data processing to improve knowledge on nonlinear relationships between floods and their causes (e.g., Adnan et al. 2023; Gudiyangada Nachappa et al. 2020). Some of the new and gradually emerging machine learning techniques have shown significant improvement in the assessment of flood risk, flood vulnerability, and even flood forecasting (Furquim et al. 2016). For instance, the Ensemble Empirical Mode Decomposition (EEMD) has been implemented to assess the impacts associated with coastal disasters along the coast of Taiwan, showing that sea level is expected to rise on the average by 5.8cm between 2020 and 2039 (Hsu et al. 2017). However, the plenitude of most flood forecasting studies in the twenty-first century now use machine learning algorithms. Some of these algorithms include the artificial neural network (ANN), support vector machine (SVM), and several variants of the ANN that integrates logistic regression, radial basis function (RBF), and metaheuristic optimization algorithms have been developed for flood disaster modeling, flood risk assessment or its susceptibility.

The integration of these techniques with robust database systems of flood patterns over well-documented water sheds have further improved the efficiency and applicability of these models in real-life scenarios. For instance, Khoirunisa et al. (2021) highlighted the possibility and applicability of integrating machine learning techniques with database systems for improved flood studies. They demonstrated the use of a GIS-based ANN model to assess the spatial distribution of flood events over Keelung City. A validation result showing correlation coefficient of 0.814 represents the accuracy and predictability of the integrated system. With the potential of integrating machine learning algorithms and database systems, the possibility to combine several important raster layers derived from remote sensing, in situ observations, and hydrological models into ANNs for improved flood prediction models is achievable (Fig. 12.5). However, as with the use of hybrid formations of machine learning to improve classification of hyperspectral and PolSAR data (Okwuashi and Ndehedehe 2020; Okwuashi et al. 2021), the accuracy of machine-learning-based flood forecast can be improved using an ensemble of several machine learning models (Gudiyangada Nachappa et al. 2020). The illustration in Fig. 12.6 is an example in southwestern Bangladesh where the capability of individual machine learning and hybridized machine learning models in flood susceptibility mapping was assessed. Even though flood vulnerable areas have been captured by all models, the presence of uncertainties in these models is revealed through differences in flood magnitude. As was highlighted by Adnan et al. (2023), optimized machine models resulting from several ensemble models have better predictive accuracy with reduced classification errors, thus potentially useful in supporting risk-based assessments and early warning systems.

12.4 Wildfires

A wildfire is a fire that spreads rapidly and can also occur in wetlands and woodland areas. Wildfire destroys human life and wildlife and affects infrastructures (e.g., buildings), ecology (e.g., vegetation communities), and economy, causing damage that disrupts several natural and ecological processes on a large scale (Ndehedehe 2022c; Yeung and Li 2018). For instance, this may include increasing the probability of flood events and triggering a rise in global warming. However, wildfire is an essential ecological process contributing to the health of global forest ecosystem. This process is ecologically disturbing and, controlled by three key requirements;

- environmental conditions that stimulate burning,
- vegetative resources to burn, and
- ignitions.

The intricacies of combustion are theoretically straightforward, but comprehending the relative effect of abiotic and biotic controls on observed, conventional fire regimes is of interest in ecological research. Also, the role of natural events, anthropogenic influence and human drivers that are changing the landscapes to more or less flammable are also factors to be considered in the assessments. As a result of the interactions, influences and feedbacks among fire, atmospheric and terrestrial systems on climate dynamics, human health and policy adaptation, global and interdisciplinary interest in fire research has seen an unprecedented rise over the past decade (Krawchuk et al. 2009).

Climate change has led to changes in the duration of wildfire through early onset of fire seasons and late termination (Jolly et al. 2015). It has equally been associated with

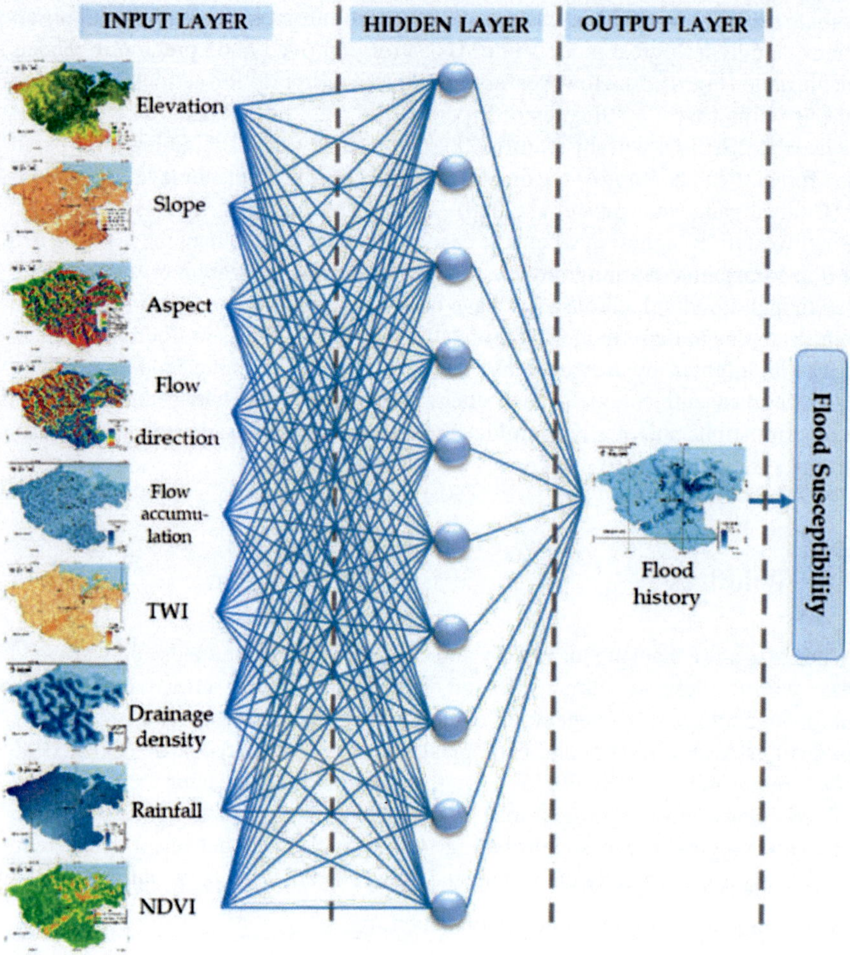

Fig. 12.5 A machine learning model to assess flood susceptibility. *Source* Khoirunisa et al. (2021)

more extreme wildfire events in terms of spatial extent, severity, intensity, and loss of lives and properties (Lannom 2014). Over the United States, An et al. (2015) projected an increase in wildfire risk over a significant percentage of all the states under both moderate and precipitous greenhouse gas (GHG) accumulation scenarios. The study listed the southern central states as the region with the highest risk (Fig. 12.7). With the accumulation of the GHG gases, the authors estimated a rise in the magnitude of projected climate change effects on wildfire activities in the long run (2031–2050) than is being experienced in the short run (2011–2030).

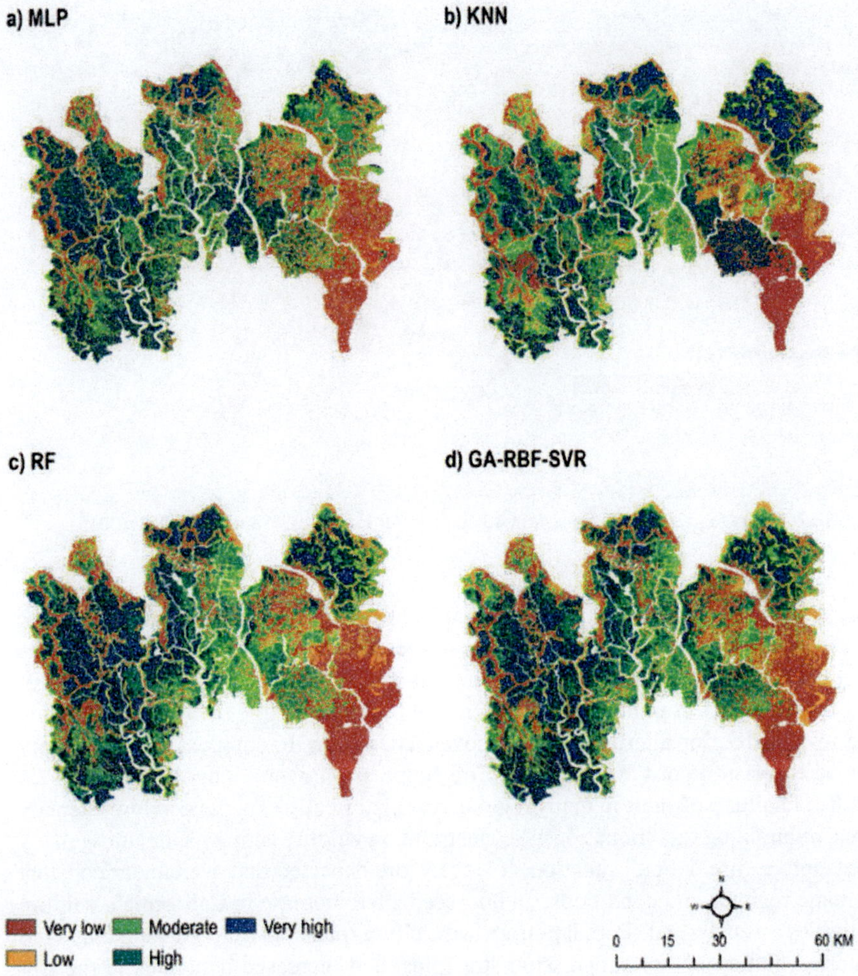

Fig. 12.6 Performance of four machine learning flood models in flood vulnerability mapping. These machine learning models are the **a** multi-Layer Perceptron-MLP, **b** K-Nearest Neighbor-KNN, **c** Random Forest-RF, and **d** Gaussian approximation and Support Vector Regression-GA-SVR. *Source* Adnan et al. (2023)

Determining the comparative role of climate change, biomes, and previous fire management practices in influencing these extreme fire episodes is critical for efficient wildfire policies and mitigation strategies. Other factors such as climatology, vegetation cover, and diverse human settlement patterns are also responsible for playing major roles in the complexities surrounding wildfire spread and increased intensity. Some spatially heterogonous regions have documented the potential impacts of climate change on their regions and signaled the complexities of separating the impacts of climate change and non-climatic processes in wildfire assessments. For

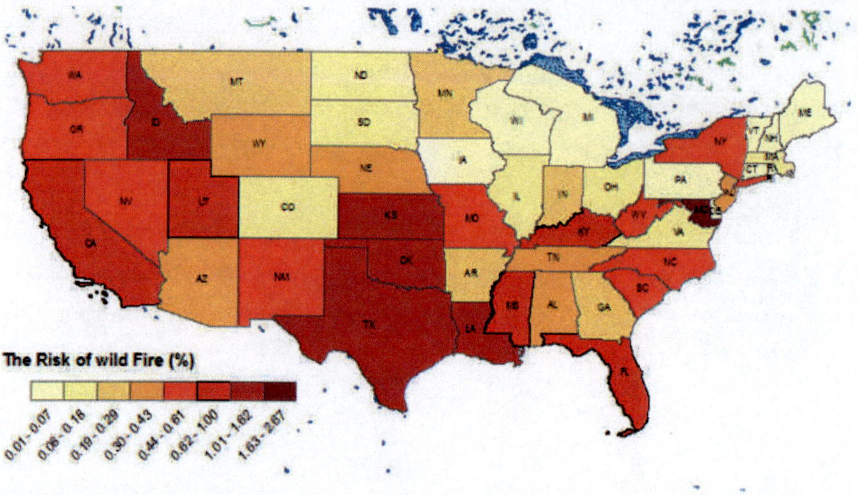

Fig. 12.7 Average annual wildfire risk in 48 continental US states. *Source* An et al. (2015)

instance, Williams et al. (2019) noted that California is a particularly difficult place to disentangle the drivers of wildfire activities due to the divergent nature of the ecosystem. As a result, apart from the impacts of climate change, human and anthropogenic activities dominating the wildfire regimes over a large fraction of the region are responsible for altering the land cover, attempting to suppress fires, and supplying the vast majority of ignitions. Fire suppression over a long time allowed for artificial buildup of fuels in many regions that historically experienced low intensity fires, minimizing fuel limitation as a constraint on wildfire activity and putting many areas into a "fire deficit" situation. It is therefore expected that even under constant climatic regimes and conditions; an unprecedented increase in California's wildfire activities is envisaged. Population growth, city expansions into surrounding wild lands, evolution of fire suppression strategies, and increased dynamics in the frequency of human-ignited wildfires will also aid in this projected increase in wildfire activities over California. The impacts of these non-climatic factors in fire-climate interactions make it difficult to isolate and identify the influence of climate change on wild fire activities over California and other regions with similar ecological settings.

12.4.1 Long-Term Changes in Global Fire Occurrence

Climate change as well as landscape changes and socio-economic factors have been identified as important drivers in global fire occurrence (Viedma et al. 2018). With increasing human factors causing ignitions and driving changes in landscape that enhances flammability, there is now focus on assessing the major roles that humans play in fire occurrence. Even though, climatic drivers are well established factors that cause large burned areas across the world (e.g., Turco et al. 2019; Viedma et al. 2018), the occurrence and impact (extent of burning) of burned areas across the world have changed substantially in the last 20 years. The recent review of Zubkova et al. (2023) on the uncertainties associated with the use of satellite earth observation data to quantify global trends in burned areas identified important drivers of decreasing burned areas, including agricultural expansion. It is noteworthy to point out that the effect of climate on global wildfire events (Fig. 12.8) is highly seasonal and changes across vegetation types. Historically, the global long-term (1982—2018) Burned Area database from the work of Otón et al. (2021) shows the outcome of a binary machine-learning-based classification. Between 1982 and 2018, an estimated 165.26 Mkm^2 global areas were burned areas with an annual average of 4.59 Mkm^2 (Otón et al. 2021). The illustration in Fig. 12.8 shows the global changes in the spatial distribution of burned area for two epochs (the 1982—2000 and 2001—2018 periods) based on the AVHRR3 data. The occurrence of fire events is predominant in tropical rainforests in Africa and South America, Eastern India, Central Asia, and wet-dry tropics in northern Australia (Fig. 12.8). Apparently, much of these burned areas are located in West and central Africa (Fig. 12.8) and the most extensively burned continent (i.e., Africa) in the world with approximately 66% of global burned areas reported (Otón et al. 2021). Africa is indeed a globally recognized hotspot of burned areas as a recent report (Cardil et al. 2023) suggests the continent accounts for about 70% of burned areas and the plenitude of these fires occurred in savannas as well as other biomes, including the tropical rainforest in central Africa.

For most of the tropical regions, fires are more frequent during summer and large burned portions are advanced by the combined drying effects of atmospheric aridity and precipitation dearth, especially in forested ecosystems where the availability of fuel is not a limiting factor. During Fall, most destructive fires occur in coastal shrub lands and are mostly driven by extreme offshore downslope wind events whose intensity are regulated by climatic conditions of the respective region. The magnitude of these wind events vary from place to place. Strong offshore winds with low relative humidity has the potential of quickly drying fuels and distributing large wildfires before the beginning of the winter precipitation season in most tropical regions (Billmire et al. 2014). Recently, the spread and magnitude of burned area across the globe have been linked to climate teleconnection patterns (Cardil et al. 2023). These teleconnections are large scale oceanic phenomena and their roles in modulating weather conditions as well as changes in terrestrial hydrology, drought, and moisture availability globally have been heavily established in the literature (e.g., Ndehedehe et al. 2023; Kalu et al. 2022; Vissa 2019; Ndehedehe and Ferreira 2020; Kiem et al. 2016; Phillips et al. 2012; Boening et al. 2012). However, in several regions of the world and across numerous biomes, significant relationships exist between climate teleconnection patterns and burned area. For instance, the recent work of Cardil et al. (2023) showed that these teleconnections drive changes in about 52.9% (more than half) of global burned area, with the Tropical North Atlantic mode being the most critical from the suite of climate teleconnection patterns such as El-Niño-Southern Oscillation, Arctic Oscillation, Indian Ocean Dipole, Pacific Decadal Oscillation,

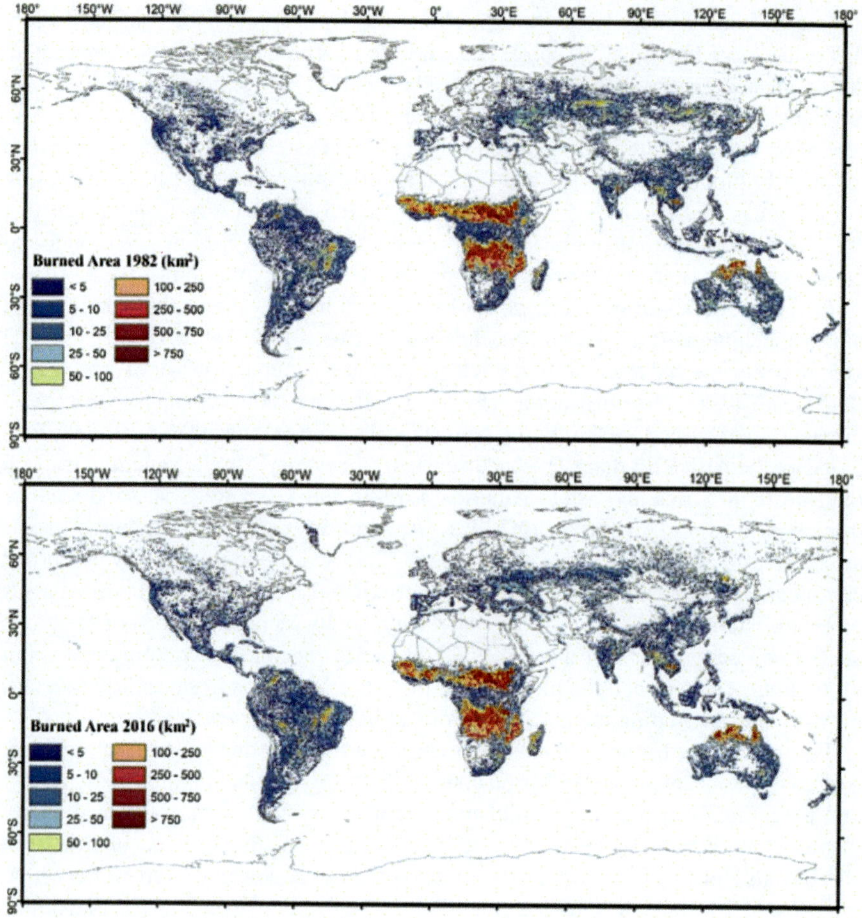

Fig. 12.8 Long-term changes in global burned areas based on the advanced very high resolution radiometer satellite data. *Source* Otón et al. (2021)

Atlantic Multi-decadal Oscillation, and Pacific North American, etc. Be it in Europe, Asia, Africa, or Australia, the influence of these global climate teleconnections as important drivers of fire episodes or burned areas have been established (Cardil et al. 2023).

As a result of increasing GHG's, several climate models have projected substantial climate change scenarios by the end of this century, an overall rise in global land surface temperature, and a drying trend in the tropics (Liu et al. 2010). From this projection, it is likely that cases of wildfires events are expected to heighten in the coming years. This is expected because climate change could ignite more intense fires that could spread faster in fire prone regions like California (Fried et al. 2004), Australia (Fisher et al. 2021), and South America (Bustillo Sánchez et al. 2021).

The effect of climate change on boreal forests could lead to a 50% increase in wildfire occurrence by the end of this century (Flannigan et al. 2009). Therefore, it is important to study and understand future wildfire trends under projected climate change to clearly assess its potential impacts both on the environment, humans, and properties. The key to the development of necessary and critical measures to mitigate the impacts of wildfire events lie in a proper understanding of the wildfire patterns and occurrence, especially in relation to both climatic and non-climatic stimulants.

12.4.2 Improved Techniques and Systems for Enhanced Wildfire Assessment

The extent of economic, environmental, and human damage caused by wildfires has warranted the need for enhanced and improved techniques and systems for its assessment. In Australia, for example, extreme wildfires at the urban interfaces of major cities have destroyed millions of hectares of land (Ashe et al. 2009). The wildfires of 2019/2020 destroyed an unprecedented 5.7 million hectares of forested regions across Australia and burnt an estimated 2500 houses (Filkov et al. 2020). These losses re-emphasize the need for an enhanced fire prediction systems for early warning of this natural hazard. These systems are important, especially given the expected rise in the frequency of fire events in local communities (Fig. 12.9). The work undertaken by Viedma et al. (2018) in a large rural community in west-central Spain reported 23,435 fire episodes, which burned a total of 456,253 ha between 1979 and 2008 with much of these fire events characterized as small fires (Fig. 12.9). The other important reason is the changing role of several predictors (e.g., biophysical and human-related factors) of wildfires in rural communities where these tend to frequently occur. For instance, topography, changes in land-use/land cover were major predictors of fire events while climate and socio-economic variables, and forest interfaces played a minor role, but the impact of these drivers, in terms of their predictive powers changed over time, thus constraining the ability to predict future fires (Viedma et al. 2018).

As opposed to the prevalence of increasing frequency of small fire events in the rural communities of west-central Spain (Fig. 12.9), the unprecedented widespread large fire events in the Amazon rainforest in the last few years (Fig. 12.10) suggests even the most humid regions of the world are vulnerable to fire threats. In the last few years (that is, since 2012), both the intensity and frequency of these fires increased in the Amazon forest (Fig. 12.10). Figure 12.10a shows the temporal patterns of a satellite-based active fire detections by fire types for the southern Amazon, which are indicated in a range of color types and compared to average fire detections during the 2012–2019 period. The temporal variation of the cumulative fire event based on satellite record from 2012 to 2020 indicates an increased fire event in August of each year, leading up to a strong rise in daily fire counts in September (Fig. 12.10b). But the 2020 fire events show one of the largest fire counts in the region compared

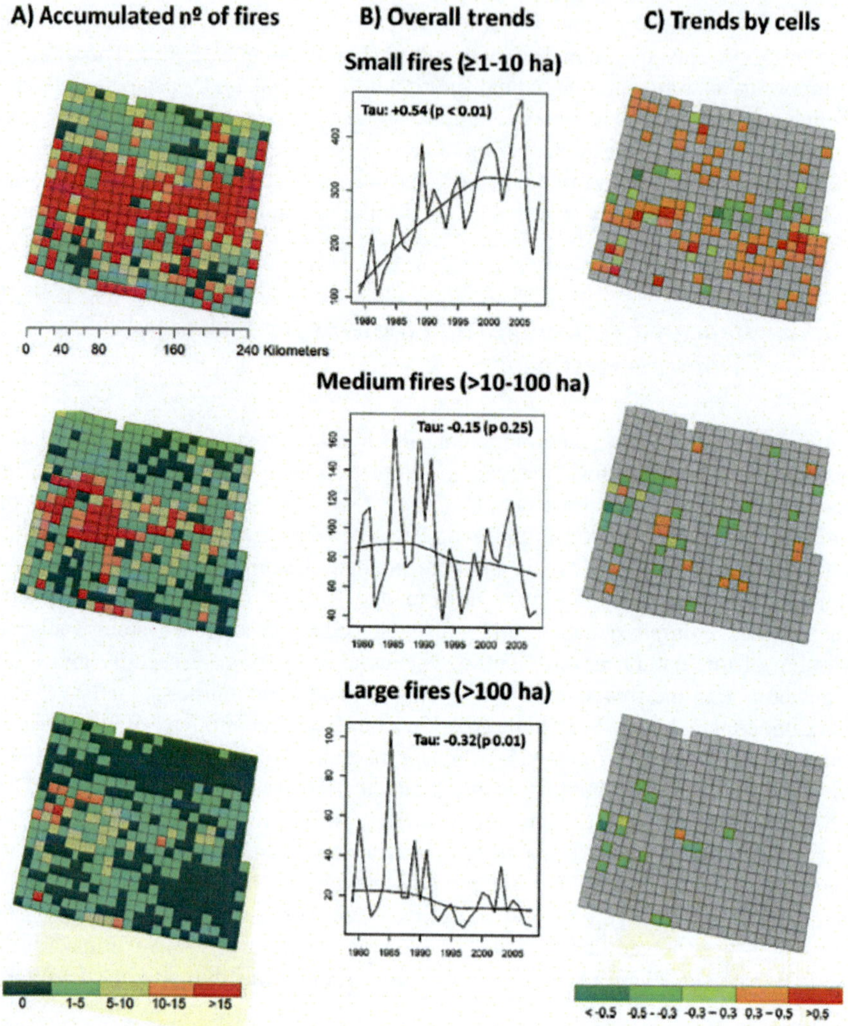

Fig. 12.9 Wild fire trends in west-central area of Spain. The data is based on **a** accumulated number of fire events **b**, temporal trends of fire events **c** trends based on cell during the 1979–2008 period. *Source* Viedma et al. (2018)

to the amplitude of detected fire counts in previous years (Fig. 12.10b). Given the complexity of fire events and how difficult they are to predict and extinguish, the use of big satellite data that leverage data mining algorithms such as support vector machine and neural network is emerging as innovative techniques and strategies to improve the prediction of fire events (Sayad et al. 2019). This section details important aspects of these Artificial Intelligence methods in fire related studies and assessments. This includes how the growing repository of satellite Earth observations

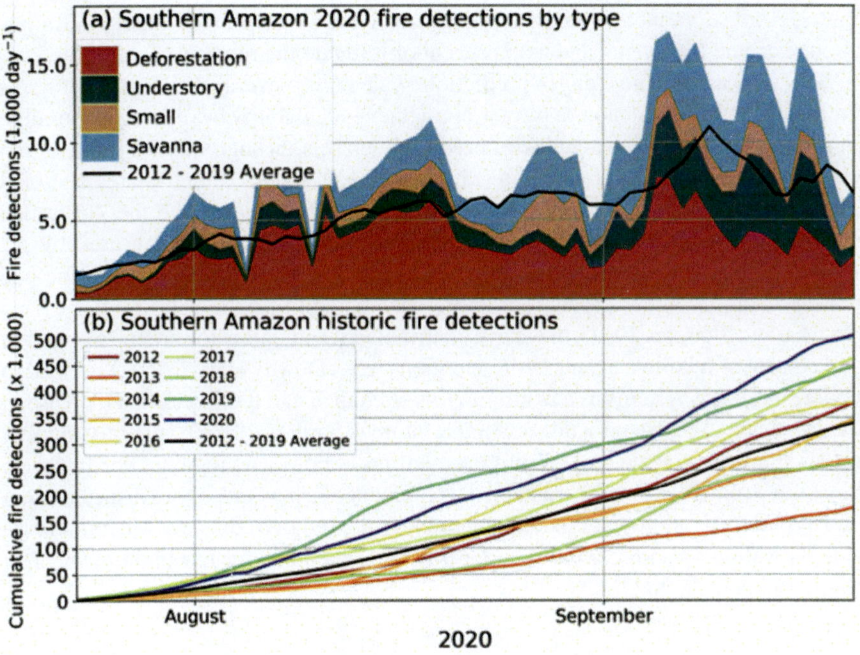

Fig. 12.10 Historical fire trends in the southern Amazon rainforest since 2012. The 2020 fire event by **a** type (i.e., based on the cause) is compared to **b** temporal patterns of past fire detections since 2012. *Source* https://globalfiredata.org/pages/2020/09/22/amazon-fire-activity-in-2020-surpasses-2019/

(e.g., Moderate Resolution Imaging Spectroradiometer—MODIS, Landsat, Sentinel-1/2 and indicators like Normalized Difference Vegetation Index-NDVI, Enhanced Vegetation Index-EVI, etc.) underpin the monitoring, assessment, and prediction of wildfire.

12.4.2.1 Wildfire Prediction Models

Wildfire prediction models have been developed to improve wildfire assessments and mitigation strategies by predicting fire behavior (Shah et al. 2022). These models rely on key parameters such as fire spread rate, direction, heat generation, and causes, to provide more accurate information for assessing the natural hazard. Various types of wildfire prediction models have been developed, including functional fire models that use the rate of fire spread, fire intensity, and available heat per square foot to predict fire spread. These models are calibrated through fire containment procedures outlined in Sullivan (2009). Response variables that possess specific characteristics, such as scalability to different wildfire danger conditions and possibility for validation, are used as fire danger indicators in these models (McRae and Sharples 2011). By using these response variables, wildfire prediction models can provide timely warnings about extreme fire events and help prevent or mitigate the impacts of wildfires.

The operational fire danger prediction systems of several developed countries including the USA, Canada, and Australia achieved scalability by relating all possible ranges of the response variables to predictor datasets through quasi-empirical and empirical fire behavior procedures (Shah et al. 2022). The United States National Fire Danger Rating System, the Canadian Fire Weather Index, and the Australian McArthur Fire Danger Rating Systems are current fire systems being used by the USA, Canada, and Australia respectively. Despite their limitations which include non-verification, under prediction of fire hazards and missing extreme fire events (McRae and Sharples 2011), they have contributed significantly toward the mitigation and control of wildfires events in these regions (e.g., Dowdy et al. 2010). The need to update these fire systems as a result of climate change and the forecasted increase in the frequency of extreme wildfires in the future (Sharples 2016) has led to ongoing research aimed at replacing the fire drought indices of some regions. For instance, there is a strong focus on the trial and development of a new index for Australia using more contemporary science, and this includes, the design of the new Australian Fire Danger Rating System currently under trial and scheduled for launch in the nearest future (Shah et al. 2022).

12.4.2.2 Satellite Remote Sensing of Wildfire Events

Remote sensing observations for wildfire assessments have contributed to redefining conventional wildfire assessment programs. Their mode of operation makes it possible to have better estimation of spatial and temporal variations of fire conditions which is critical for fire prevention. The vast coverage capabilities and steady revisit times of satellite remote sensing techniques which make it easy to quantify where and when vegetation is dry enough to facilitate propagation or fire ignition, is essential for strategic planning. Also, this information is vital in the location of containment lines and equally aid in firefighting strategies. Big remote sensing data on the state of the crops based on NDVI have been combined with fire indicator "Thermal Anomalies" derived from "MODIS" data to optimize fire prediction accuracy using Artificial Intelligence techniques (Sayad et al. 2019).

Fire assessments and monitoring can be facilitated with remote sensing data because of data availability, broad range of spectral channels that aids the fire detection, and the repetitive coverage of satellite missions (daily, weekly, etc.). Additionally, as with the use of remote sensing indicators to detect the presence of water on landscapes, important wildfire variables like vegetation (using NDVI) can be easily extracted from satellite data (Ndehedehe 2022c). NDVI can capture the condition of crop health through changes in vegetation greenness and other fire related variables like Leaf Area Index (captures canopy light conditions and Thermal Anomalies (TA) and Land Surface Temperature are indicators and variables based on remote sensing (Sayad et al. 2019). Thus, facilitating the use of satellite remote sensing in mapping, predicting, and assessing wildfire events.

The use of satellite earth observations such as Landsat, MODIS, and Sentinel-2 is helping to provide opportunities to monitor and assess impacts of wildfires on the environment. These satellite data are available and handy to help assess vegetation response to wildfire events under a changing climate. This assessment is critical to improving our understanding of vegetation regrowth post wildfire recovery (Ndehedehe 2022c). Remote sensing indicators have been developed to detect changes in biophysical variables (vegetation and water), including areas affected by wildfire. The latter can be estimated by quantifying burn severity. To this end, the normalized burn ratio (NBR) is used and it is given as,

$$NBR = b_{NIR} - b_{SWIR}/b_{NIR} + b_{SWIR} \tag{12.1}$$

where NIR is the near-infrared band and $SWIR$ is the short wave infrared band of any optical satellite system (e.g., landsat, Sentinel-2, etc.). The burn severity is estimated by computing NBR of the period before the fire and post-fire. The use of this indicator for burn severity assessment is based on the sensitivity of NIR to vegetation greenness and $SWIR$ to bare soil and charcoal (Ndehedehe 2022c; Yeung and Li 2018). Additionally, the difference between pre- and post-fire images from NBR can also be useful in burn severity assessment. This difference is called the Differenced Normalized Burn Ratio (i.e., $dNBR{=}NBR_{prefire} - NBR_{postfire}$). In the classification of NBR values (range from -1 to $+1$), higher values represent green vegetation while lower values represent areas affected by wildfire (i.e., burned areas), or bare soil, water or cloud. But in the dNBR (typically ranging from -2 to $+2$) classification, negative values indicate vegetation regrowth while values near zero (0) represent unburned areas (Yeung and Li 2018). Values greater than 0.1 in the dNBR indicate burned areas.

Compared to the MODIS and Landsat data, Sentinel-2 data provide a better opportunity to quantify burn area on a much higher spatial resolution. Novel remote sensing indicators like the Burned Area Index for Sentinel-2 (BAIS2) are facilitating the assessment of wildfires using the Sentinel-2 multi-spectral data (Filipponi 2018). The false color composite of the Macquarie Marshes pre-wildfire and during the wildfire, which started in June 2019 show some visible differences (Fig. 12.11a, b). The BAIS2 is a multi-band indicator that uses five multi-spectral Sentinel-2 bands (20 m spatial resolution) to map and identify areas affected by fire. The spatial locations with red pixels are burned areas showing that much of southern areas of the marshes were affected by fire while the yellow pixels are unaffected or mildly affected locations (Fig. 12.11c).

Furthermore, remotely sensed observations provide key updates in the execution of preseason planning when fire agencies and land management organizations formulate their seasonal view toward fire ignitions, regions at risk, and projected burn intensities. It is important for fire managers to know the moisture content of vulnerable and risk regions to know if the burn will be successful because if a region is too wet, the fire will not progress, whereas a dry region can lead to over burn or even uncontrolled burning. Nowadays, the contribution of remote sensing techniques to wildfire assessments comprises a broad range of sensors, which includes, ground

Fig. 12.11 False color composites from Sentinel-2 for the period **a** before and **b** during wildfire event and **c** spatial patterns for the 2019 fire events in the Macquarie Marshes in New South Wales, Australia. The burned area assessment is based on Sentinel-2 data using the Burned Area Index for Sentinel-2. Red pixels are burned areas. *Source* Ndehedehe (2022c)

spectroradiometers, unmanned aerial vehicles, airplanes and satellite mission, some of which are specifically designed to target fire activities. Also, the availability of thermal bands and band combination methods also provide more suitable approaches for wildfire assessments over various spatial and temporal scales. Over the last decade, the range of interpretation methods for remotely sensed wildfire assessments has expanded

- toward synergy between optical and thermal datasets,
- between optical and radar sets,
- between optical passive and Light detection and ranging (Lidar) datasets,
- the extended application of radiative transfer models,
- change detection analysis, and
- machine learning algorithms.

As a final note on the utility of remote sensing in wildfire assessments, challenges and caveats exist. The gaps and limitations in our understanding of global fire events are mostly caused by the quality of existing data and recently, the uncertainties associated with using satellite remote sensing data to assess the extent of burned areas have recently been reported Zubkova et al. (2023). Some of the thoughts they shared regarding these challenges and caveats include the coarse resolution of earth observation products used in wild fire mapping that causes bias in the detection of larger fires. Given that several wildfires now occur on a local scales and with the decline in global fire events associated with farming operations and changes to

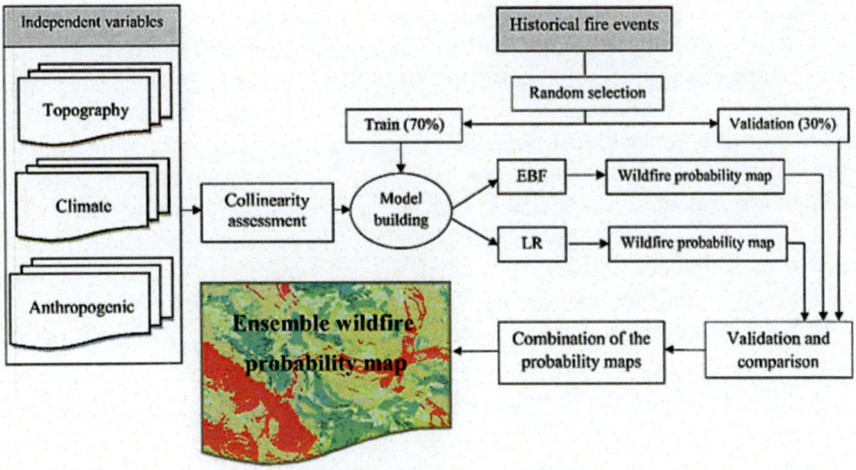

Fig. 12.12 Predictor selection role in wildfire assessment using machine learning. *Source* Jaafari et al. (2019)

the environment and socio-economic factors (Otón et al. 2021; Turco et al. 2019; Viedma et al. 2018), such limitations of remotely sensed data should be important considerations in their usage to quantify impacts on different spatial and temporal scales. This is particularly important to facilitate improved characterization or the accuracy of trends and trajectories in burnt areas globally and locally, which currently, is still debatable (Zubkova et al. 2023).

12.4.2.3 Machine Learning

Machine learning algorithms are robust and powerful predictive tools which have been frequently used for the prediction of extreme climatic episodes including wildfire and its danger possibilities (Rodrigues and de la Riva 2014). Using independent variables which can equally be referred to as, the predictor datasets, they undergo a vigorous training phase where they attempt to learn the relationship between the climate predictors and the predictand (e.g., wildfire). The non-linearity of climate variables often overwhelm traditional machine learning techniques, which resulting in poor predictions and reduced accuracies. However, deep-learning and meta-learning approaches have emerged, providing a sustainable framework for disintegrating and learning these climate variables regardless of their non-linearity. Machine learning techniques provide predictive evaluations ranging from location-based fire ignitions, fire durations, and fire intensities per area (Fig. 12.12). Their testing phase enables the validation of the predicted models to ascertain the reliability of the designed machine learning model, especially for real-life scenarios.

The performance of the algorithm is usually considered to be good if a greater percentage of the fires from the evaluation datasets (training) fall in higher predicted probabilities (testing) (Ngoc Thach et al. 2018). To get better prediction results from machine learning methods, it is important to filter and fine tune the predictors so as not to include dormant climatic variables which do not play any role on the wildfire spread (Fig. 12.12). For example, when using an artificial neural network to model wildfire assessment, it is pertinent to include variables that have strong relationship with the fire intensity and spread, such as soil moisture content, latent heat transfer, normalized difference vegetation index (NDVI), and wind. The inclusion of variables without any significant relationship/correlation with the predictand only tends to overwhelm the machine learning process and may lead to poor predictions (Kalu et al. 2022).

12.4.3 Cloud-Computing for Wildfire Mapping

One of the highly notable cloud computing platforms for geospatial data processing and analyses is the Google Earth Engine (GEE). The GEE is a computational platform that provides an interface for application programming that is freely and publicly available and aids easy access to several geospatial data and remotely sensed observations across the globe (Gorelick et al. 2017). This computational platform (see, https://developers.google.com/earth-engine) enables data integration, harmonization, and development of large-scale cloud computing resources for the advancement of Earth science and environmental monitoring applications. These applications occur on a broad range of spatial and temporal scales, thus making it a very unique and highly sophisticated computational platform for planetary-scale geospatial analysis (Ndehedehe 2022a). The goal of this section is not to discuss GEE as it has been widely discussed in the literature (Ndehedehe 2022a; Tamiminia et al. 2020; Kumar and Mutanga 2018; Gorelick et al. 2017) but to explore the applications of the platform in wildfire mapping and assessment (e.g., Tavakkoli Piralilou et al. 2022; Ndehedehe 2022a). For instance, several reputable international agencies like The National Aeronautics and Space Administration, European Space Agency have retrieved large satellite data (e.g., Sentinel-2) to track the spread of fires, estimate burn severity, and assess the impact of fires on the landscape using the GEE platform. The capability to assess wildfire has further been underpinned by machine learning models embedded in the GEE platform (Fig. 12.11). For instance, using the GEE, the study undertaken by Tavakkoli Piralilou et al. (2022) evaluated the effects of several multi-resolution satellite data (Landsat 8, Shuttle Radar Topographic Mission, Sentinel-2, and ALOS) on the prediction of wildfire susceptibility underpinned by machine learning models (e.g., random forest and support vector machine).The illustration in Fig. 12.11 shows the methodology or process of how this was evaluated in the forested areas of the Guilan Province, Iran using the GEE platform. As shown in Fig. 12.11, QGIS software, which is an open access geospatial information system software was used to implement the Dempster–Shafer theory for data fusion

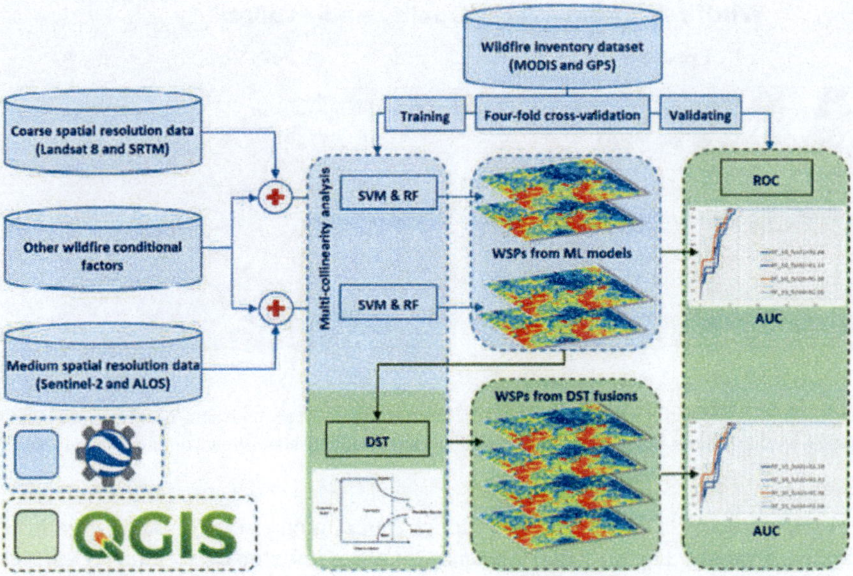

Fig. 12.13 Integration of machine learning models (support vector machines and random forest) with geospatial data for wildfire susceptibility prediction using the Google Earth Engine. *Source* Tavakkoli Piralilou et al. (2022)

as well as wildfire susceptibility prediction optimization, and assessing accuracy (Tavakkoli Piralilou et al. 2022) (Fig. 12.13).

Furthermore, there are several computational and online resources to help with mapping fire events using the GEE platform. For instance, using the Normalized Burn Ratio, the United Nations repository provides a step-by-step procedure generating a burn severity map to help with detecting areas affected by wildfires (https://un-spider.org/advisory-support/recommended-practices/recommended-practice-burn-severity/burn-severity-earth-engine).

12.4.4 *Impacts of Wildfire on Freshwater (Groundwater and Surface Water)*

Wildfires are a common occurrence that pose a significant threat to the environment, and managing them is critical. Preventing and controlling the damage caused by wildfires is considered one of the most important challenges in environmental hydrology (e.g., Gray et al. 2018). The frequency and intensity of wildfires have increased due to climate change, lack of forest management practices, and global warming (Girona-García et al. 2021; Shakesby 2006). The impacts of wildfire on debris flows, flash floods, and erosion are essential to environmental hydrology (Girona-García et al.

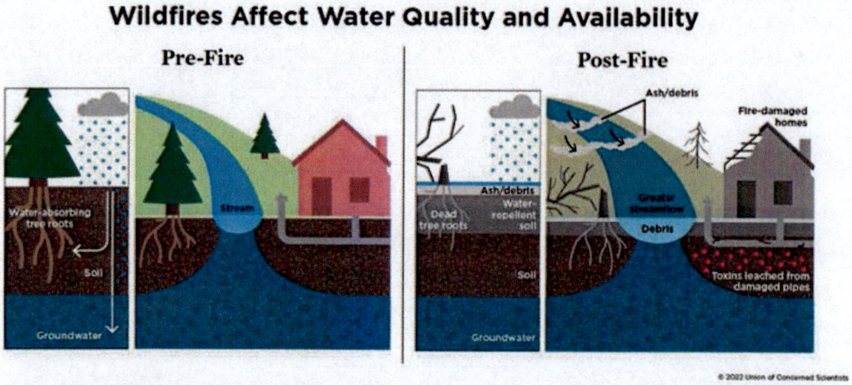

Fig. 12.14 Varying actions of pre-fire and post-fire processes on groundwater resources. *Source* https://www.ucsusa.org/resources/wildfire-water-western-united-states#read-online-content

2021; Ebel 2020; Shakesby 2006). Wildfires affect groundwater systems through complex processes. During wildfire events, ash mixes with topsoil, creating a water-repellent layer, making the soil unreceptive and impervious (Fig. 12.14). This unreceptiveness leads to an increase in runoff rates from wildfire-affected areas, accelerating erosion and causing a significant amount of soil loss. Some studies suggest that wildfires can enhance, induce, or even destroy soil water repellency (e.g., Fernández et al. 2019; Larson-Nash et al. 2018), which controls the interaction of soils and water. Additionally, wildfires disrupt soil-dwelling organisms such as termites and ants, which are essential contributors to efficient ecosystem servicing through the creation of macropores that control infiltration (Certini et al. 2021). Other studies (Robichaud et al. 2016) however, have shown increased infiltration rates caused by wildfires.

Groundwater depletion can occur due to various management activities, such as vegetation management and water extraction for agricultural and domestic purposes. These activities could affect groundwater discharge, and groundwater dependent ecosystems such as springs, streams, and wetlands. For instance, vegetation management can cause a temporary rise in groundwater and the water table through reduced evapotranspiration. Changes in the water budget of a groundwater system (difference of evapotranspiration plus groundwater pumping and groundwater recharge, which is equivalent to groundwater discharge) in a river basin can potentially impact groundwater availability, but prescribed burning and vegetation management have different effects on groundwater[7]. Mechanical thinning and prescribed burning can increase groundwater levels due to reduced evapotranspiration, but changes in soil hydrologic function resulting from prescribed burning, such as high soil burn severity or water-repellent soils, can lead to increased runoff and decreased infiltration, thereby reducing groundwater recharge.

[7] https://www.chrome-extension://efaidnbmnnnibpcajpcglclefindmkaj/https://www.fs.usda.gov/Internet/FSE_DOCUMENTS/stelprdb5398909.pdf

Studies on the impact of wildfires on groundwater quality have produced mixed results, making it difficult to establish factual claims on their interactions and correlations. Giambastiani et al. (2018) reported a positive feedback of forest fire on the quality and quantity of groundwater resources over their study region. They found a reduction in salinity, along with an increase in infiltration and freshwater lens thickness, attributed to reduced canopy interception and evapotranspiration caused by vegetation absence after the fire incident. However, Mansilha et al. (2020) reported an increase in fluoride, nitrogen, sulfate, iron, manganese, and chromium and variability in pH levels of groundwater samples collected over burned areas in Peri-urban. Their study also reported an increase in carcinogenic post-fire concentration of polycyclic aromatic hydrocarbons, especially after extreme winter and spring rain events, but noted that the levels did not exceed stipulated guidelines for safe drinking water. Despite these findings, a critical evaluation of past studies has not shown any evidence of wildfires causing severe contamination of groundwater quality, rendering it unsafe for human consumption or unfit as an aquatic habitat. Giambastiani et al. (2018) concluded that wildfires above a thick vadose zone have little or no impact on groundwater quality.

In contrast, the impact of wildfire on surface water quality is more evident than its impact on groundwater quality and quantity (Mansilha et al. 2020). Wildfires can have significant effects on water quality and quantity, including increased erosion, sedimentation, and nutrient runoff into streams and rivers, which can negatively impact aquatic ecosystems and human water supplies (Gomez Isaza et al. 2022). The severity and duration of these impacts can vary depending on factors such as wildfire size, intensity, and proximity to water sources, as well as post-fire weather conditions. Therefore, understanding the impacts of wildfires on water resources is crucial for effective management and mitigation of the effects of wildfires on ecosystems and human societies and local communities.

Wildfires have caused changes in several water quality parameters of interest to water systems (see, e.g., Nunes et al. 2017; Tecle and Neary 2015). The burning of forest catchments may lead to a long-lasting legacy of deteriorating surface water quality, whose persistence and magnitude can be measured from a few months to several years after the wildfire incident. Post-fire surface water quality concerns are complicated, and they vary substantially from region to region depending on the intensity, duration, and severity of the wildfire. Other factors that affect post-fire water quality include the concentration and amount of precipitation during post-fire rain episodes, the geomorphological and geological nature of the terrain, the vegetation cover, and soil characteristics of the affected region (Tecle and Neary 2015). The impact of wildfire on surface water could also be due to indirect actions from smoke and aerial deposition of ashes (Poursanidis and Chrysoulakis 2017). As humans increasingly rely on groundwater, particularly for drinking purposes, high quality groundwater is maintained because it is stored in underground aquifers. This means that the risk of contamination is significantly reduced.

References

Adnan MSG, Siam ZS, Kabir I, Kabir Z, Ahmed MR, Hassan QK, Rahman RM, Dewan A (2023) A novel framework for addressing uncertainties in machine learning-based geospatial approaches for flood prediction. J Environ Manag 326:116813. https://doi.org/10.1016/j.jenvman.2022.116813

An H, Gan J, Cho SJ (2015) Assessing climate change impacts on wildfire risk in the United States. Forests 6(9):3197–3211. https://doi.org/10.3390/f6093197

Ashe B, McAneney KJ, Pitman AJ (2009) Total cost of fire in australia. J Risk Res 12(2):121–136. https://doi.org/10.1080/13669870802648528

Billmire M, French N, Loboda T, Owen R, Tyner M (2014) Santa ana winds and predictors of wildfire progression in southern california. Int J Wildland Fire 23(18):1119–1129. https://doi.org/10.1071/wf13046

Black AR, Burns JC (2002) Re-assessing the flood risk in Scotland. Sci Total Environ 294(1):169–184. https://doi.org/10.1016/S0048-9697(02)00062-1

Boening C, Willis JK, Landerer FW, Nerem RS, Fasullo J (2012) The 2011 La Niña: So strong, the oceans fell. Geophys Res Lett 39(19):L19602. https://doi.org/10.1029/2012GL053055

Bustillo Sánchez M, Tonini M, Mapelli A, Fiorucci P (2021) Spatial assessment of wildfires susceptibility in Santa Cruz (Bolivia) using random forest. Geosciences 11(5). https://doi.org/10.3390/geosciences11050224.

Cardil A, Rodrigues M, Tapia M et al (2023) Climate teleconnections modulate global burned area. Nat Commun 14(427). https://doi.org/10.1038/s41467-023-36052-8.

Certini G, Moya D, Lucas-Borja ME, Mastrolonardo G (2021) The impact of fire on soil-dwelling biota: A review. For Ecol Manag 488:118989. https://doi.org/10.1016/j.foreco.2021.118989

Chau VN, Holland J, Cassells S, Tuohy M (2013) Using gis to map impacts upon agriculture from extreme floods in vietnam. Appl Geogr 41:65–74. https://doi.org/10.1016/j.apgeog.2013.03.014

CRED (2020) Human cost of disasters. An overview of the last 20 years: 2000–2019. Scientific reports. Accessed from (https://reliefweb.int/report/world/human-cost-disasters-overview-last-20-years2000-2019) on 1st August, 2021

Danumah JH, O.-S. S. B et al (2016) Flood risk assessment and mapping in abidjan district using multi-criteria analysis (AHP) model and geoinformation techniques, (cote d'ivoire). Geoenviron Disasters 3(10):1–13. https://doi.org/10.1186/s40677-016-0044-y

de Bruijn J, de Moel H, Jongman B et al (2019) A global database of historic and real-time flood events based on social media. Sci Data 6(311). https://doi.org/10.1038/s41597-019-0326-9.

Dewan A, Corner R, Saleem A, Rahman MM, Haider MR, Rahman MM, Sarker MH (2017) Assessing channel changes of the Ganges-Padma River system in Bangladesh using Landsat and hydrological data. Geomorphology 276:257–279. https://doi.org/10.1016/j.geomorph.2016.10.017

Do HX, Westra S, Leonard M (2017) A global-scale investigation of trends in annual maximum streamflow. J Hydrol 552:28–43. https://doi.org/10.1016/j.jhydrol.2017.06.015

Dowdy AJ, Mills GA, Finkele K, de Groot W (2010) Index sensitivity analysis applied to the Canadian forest fire weather index and the mcarthur forest fire danger index. Meteorol Appl 17(3):298–312. https://doi.org/10.1002/met.170

Ebel BA (2020) Temporal evolution of measured and simulated infiltration following wildfire in the Colorado Front Range, USA: shifting thresholds of runoff generation and hydrologic hazards. J Hydrol 585:124765. https://doi.org/10.1016/j.jhydrol.2020.124765

Ennouri K, Smaoui S, Triki M (2021) Detection of urban and environmental changes via remote sensing. Circular Econ Sustain 1:1423–1437. https://doi.org/10.1007/s43615-021-00035-y

Espinoza J-C, Marengo JA, Schongart J, Jimenez JC (2022) The new historical flood of 2021 in the Amazon River compared to major floods of the 21st century: Atmospheric features in the context of the intensification of floods. Weather Clim Extremes 35:100406

Fernandes PM (2013) Fire-smart management of forest landscapes in the mediterranean basin under global change. Landscape Urban Plan 110:175–182. https://doi.org/10.1016/j.landurbplan.2012.10.014

Fernández C, Fontúrbel T, Vega JA (2019) Wildfire burned soil organic horizon contribution to runoff and infiltration in a pinus pinaster forest soil. J For Res 24(2):86–92. https://doi.org/10.1080/13416979.2019.1572091

Filipponi F (2018) Bais2: Burned area index for sentinel-2. Proceedings 2(7). https://doi.org/10.3390/ecrs-2-05177

Filkov AI, Ngo T, Matthews S, Telfer S, Penman TD (2020) Impact of Australia's catastrophic 2019/20 bushfire season on communities and environment. Retrospective analysis and current trends. J Saf Sci Resil 1(1):44–56. https://doi.org/10.1016/j.jnlssr.2020.06.009

Fisher R, Heckbert S, Garnett S (2021) Reframing wildfire simulations for understanding complex human–landscape interactions in cross-cultural contexts: A case study from Northern Australia. Fire 4(3). https://doi.org/10.3390/fire4030046

Flannigan M, Stocks B, Turetsky M, Wotton M (2009) Impacts of climate change on fire activity and fire management in the circumboreal forest. Glob Change Biol 15(3):549–560. https://doi.org/10.1111/j.1365-2486.2008.01660.x

Fried J, Torn M, Mills E (2004) The impact of climate change on wildfire severity: A regional forecast for northern California. Clim Change 64:169–191. https://doi.org/10.1023/B:CLIM.0000024667.89579.ed

Furquim G, Pessin G, Faiçal B et al (2016) Improving the accuracy of a flood forecasting model by means of machine learning and chaos theory. Neural Comput Appl 27:1129–1141. https://doi.org/10.1007/s00521-015-1930-z

Giambastiani BMS, Greggio N, Nobili G, Dinelli E, Antonellini M (2018) Forest fire effects on groundwater in a coastal aquifer (Ravenna, Italy). Hydrol Proces 32(15):2377–2389. https://doi.org/10.1002/hyp.13165

Girona-García A, Vieira DC, Silva J, Fernández C, Robichaud PR, Keizer JJ (2021) Effectiveness of post-fire soil erosion mitigation treatments: A systematic review and meta-analysis. Earth-Sci Rev 217:103611. https://doi.org/10.1016/j.earscirev.2021.103611

Gomez Isaza DF, Cramp RL, Franklin CE (2022) Fire and rain: A systematic review of the impacts of wildfire and associated runoff on aquatic fauna. Glob Change Biol 28(8):2578–2595. https://doi.org/10.1111/gcb.16088

Gorelick N, Hancher M, Dixon M, Ilyushchenko S, Thau D, Moore R (2017) Google earth engine: Planetary-scale geospatial analysis for everyone. Remote Sens Environ 202:18–27. https://doi.org/10.1016/j.rse.2017.06.031

Gray ME, Zachmann LJ, Dickson BG (2018) A weekly, continually updated dataset of the probability of large wildfires across western US forests and woodlands. Earth Syst Sci Data 10(3):1715–1727. https://doi.org/10.5194/essd-10-1715-2018

Gudiyangada Nachappa T, Tavakkoli Piralilou S, Gholamnia K, Ghorbanzadeh O, Rahmati O, Blaschke T (2020) Flood susceptibility mapping with machine learning, multi-criteria decision analysis and ensemble using Dempster Shafer Theory. J Hydrol 590:125275. https://doi.org/10.1016/j.jhydrol.2020.125275

Hall JW, Grey D, Garrick D, Fung F, Brown C, Dadson SJ, Sadoff CW (2014) Coping with the curse of freshwater variability. Science 346(6208):429–430. https://doi.org/10.1126/science.1257890

Hodgkins GA, Whitfield PH, Burn DH, Hannaford J, Renard B, Stahl K, Fleig AK, Madsen H, Mediero L, Korhonen J, Murphy C, Wilson D (2017) Climate-driven variability in the occurrence of major floods across North America and Europe. J Hydrol 552:704–717. https://doi.org/10.1016/j.jhydrol.2017.07.027

Hsu T.-W, Shih D.-S, Li C.-Y, Lan Y.-J, Lin Y.-C (2017) A study on coastal flooding and risk assessment under climate change in the mid-western coast of Taiwan. Water 9(6). https://doi.org/10.3390/w9060390.

Jaafari A, Mafi-Gholami D, Thai Pham B, Tien Bui D (2019) Wildfire probability mapping: Bivariate vs. multivariate statistics. Remote Sens 11(6). https://doi.org/10.3390/rs11060618.

Jolly W, Cochrane M, Freeborn P et al (2015) Climate-induced variations in global wildfire danger from 1979 to 2013. Nat Commun 6:7537. https://doi.org/10.1038/ncomms8537

Kabenge M, Elaru J, Wang H et al (2017) Characterizing flood hazard risk in data-scarce areas, using a remote sensing and GIS-based flood hazard index. Nat Hazards 89:1369–1387. https://doi.org/10.1007/s11069-017-3024-y

Kalu I, Ndehedehe CE, Okwuashi O, Eyoh AE, Ferreira VG (2022) An assimilated deep learning approach to identify the influence of global climate on hydrological fluxes. J Hydrol 614:128498. https://doi.org/10.1016/j.jhydrol.2022.128498

Khoirunisa N, Ku C.-Y, Liu C.-Y (2021) A GIS-based artificial neural network model for flood susceptibility assessment. Int J Environ Res Pub Health 18(3). https://doi.org/10.3390/ijerph18031072.

Kiem AS, Johnson F, Westra S, van Dijk A, Evans JP, O'Donnell A, Rouillard A, Barr C, Tyler J, Thyer M, Jakob D, Woldemeskel F, Sivakumar B, Mehrotra R (2016) Natural hazards in Australia: droughts. Clim Change 139(1):37–54. https://doi.org/10.1007/s10584-016-1798-7

Krawchuk MA, Moritz MA, Parisien M-A, Van Dorn J, Hayhoe K (2009) Global pyrogeography: the current and future distribution of wildfire. PLOS ONE 4(4):1–12. https://doi.org/10.1371/journal.pone.0005102

Kumar L, Mutanga O (2018) Google Earth Engine applications since inception: Usage, trends, and potential. Remote Sens 10(10). https://doi.org/10.3390/rs10101509.

Lannom Karen O, Tinkham WT et al (2014) Defining extreme wildland fires using geospatial and ancillary metrics. Int J Wildland Fire 23(3):322–337

Larson-Nash SS, Robichaud PR, Pierson FB, Moffet CA, Williams CJ, Spaeth KE, Brown RE, Lewis SA (2018) Recovery of small-scale infiltration and erosion after wildfires. J Hydrol Hydromech 66(3):261–270. https://doi.org/10.1515/johh-2017-0056

Lindersson S, Brandimarte L, Mård J, Di Baldassarre G (2020) A review of freely accessible global datasets for the study of floods, droughts and their interactions with human societies. WIREs Water 7(3):e1424. https://doi.org/10.1002/wat2.1424

Liu Y, Stanturf J, Goodrick S (2010) Trends in global wildfire potential in a changing climate. For Ecol Manag 259(4):685–697. https://doi.org/10.1016/j.foreco.2009.09.002

Lyu H-M, Shen S-L, Zhou A, Yang J (2019) Perspectives for flood risk assessment and management for mega-city metro system. Tunn Undergr Space Technol 84:31–44. https://doi.org/10.1016/j.tust.2018.10.019

Mansilha C, Melo A, Martins ZE, Ferreira IMPLVO, Pereira, AM, Espinha Marques J (2020) Wildfire effects on groundwater quality from springs connected to small public supply systems in a peri-urban forest area (Braga Region, NW Portugal). Water 12(4). https://doi.org/10.3390/w12041146.

McRae R, Sharples J (2011) A conceptual framework for assessing the risk posed by extreme bushfires. Aust J Emerg Manag 26(2):47–53

Milly P, Wetherald R, Dunne K et al (2002) Increasing risk of great floods in a changing climate. Nature 415(6871):514–517. https://doi.org/10.1038/415514a

Ndehedehe CE (2017) Remote sensing of West Africa's water resources using multi-satellites and models. PhD Thesis, Curtin University, Bentley, Perth, Western Australia. Retrieved from: http://hdl.handle.net/20.500.11937/59637 on 12th January 2018

Ndehedehe C (2022a) Cloud-based geospatial analysis. Springer International Publishing, Cham, pp 73–95

Ndehedehe C (2022b) Floodplain productivity in wet-dry tropics. Springer International Publishing, Cham, pp 527–576. https://doi.org/10.1007/978-3-030-99577-5_18.

Ndehedehe C (2022c) Remote sensing of surface vegetation. Springer International Publishing, Cham, pp 131–176

Ndehedehe CE, Ferreira VG (2020) Assessing land water storage dynamics over Southern America. J Hydrol 580:124339. https://doi.org/10.1016/j.jhydrol.2019.124339

Ndehedehe CE, Onojeghuo AO, Stewart-Koster B, Bunn SE, Ferreira VG (2021) Upstream flows drive the productivity of floodplain ecosystems in tropical queensland. Ecol Indic. https://doi.org/10.1016/j.ecolind.2021.107546

Ndehedehe CE, Ferreira VG, Adeyeri OE, Correa FM, Usman M, Oussou FE, Kalu I, Okwuashi O, Onojeghuo AO, Getirana A, Dewan A (2023) Global assessment of drought characteristics in the Anthropocene. Res Environ Sustain 12:100105. https://doi.org/10.1016/j.resenv.2022.100105

Ngoc Thach N, Bao-Toan Ngo D, Xuan-Canh P, Hong-Thi N, Hang Thi B, Nhat-Duc H, Dieu TB (2018) Spatial pattern assessment of tropical forest fire danger at Thuan Chau area (Vietnam) using GIS-based advanced machine learning algorithms: A comparative study. Ecol Inform 46:74–85. https://doi.org/10.1016/j.ecoinf.2018.05.009

Nott J (2006) Extreme events: a physical reconstruction and risk assessment. Cambridge University Press

Nunes B, Silva V, Campos I, Pereira JL, Pereira P, Keizer JJ, Gonçalves F, Abrantes N (2017) Off-site impacts of wildfires on aquatic systems–biomarker responses of the mosquitofish Gambusia holbrooki. Sci Total Environ 581-582:305–313. https://doi.org/10.1016/j.scitotenv.2016.12.129

Okwuashi O, Ndehedehe CE (2020) Deep support vector machine for hyperspectral image classification. Pattern Recognit 103:107298. https://doi.org/10.1016/j.patcog.2020.107298

Okwuashi O, Ndehedehe CE, Olayinka DN, Eyoh A, Attai H (2021) Deep support vector machine for PolSAR image classification. Int J Remote Sens 42(17):6502–6540. https://doi.org/10.1080/01431161.2021.1939910

Otón G, Lizundia-Loiola J, Pettinari ML, Chuvieco E (2021) Development of a consistent global long-term burned area product (1982–2018) based on AVHRR-LTDR data. Int J Appl Earth Obs Geoinform 103:102473. https://doi.org/10.1016/j.jag.2021.102473

Pant R, Thacker S, Hall J, Alderson D, Barr S (2018) Critical infrastructure impact assessment due to flood exposure. J Flood Risk Manag 11(1):22–33. https://doi.org/10.1111/jfr3.12288

Parjanne A, Rytkönen A-M, Veijalainen N (2021) Framework for climate proofing of flood risk management strategies in Finland. Water Secur 14:100096. https://doi.org/10.1016/j.wasec.2021.100096

Petit-Boix A, Sevigné-Itoiz E, Rojas-Gutierrez LA, Barbassa AP, Josa A, Rieradevall J, Gabarrell X (2017) Floods and consequential life cycle assessment: Integrating flood damage into the environmental assessment of stormwater best management practices. J Cleaner Prod 162:601–608. https://doi.org/10.1016/j.jclepro.2017.06.047

Pham BT, Luu C, Phong TV, Nguyen HD, Le HV, Tran TQ, Ta HT, Prakash I (2021) Flood risk assessment using hybrid artificial intelligence models integrated with multi-criteria decision analysis in Quang Nam Province. Vietnam. J Hydrol 592:125815. https://doi.org/10.1016/j.jhydrol.2020.125815

Phillips T, Nerem RS, Fox-Kemper B, Famiglietti JS, Rajagopalan B (2012) The influence of ENSO on global terrestrial water storage using GRACE. Geophys Res Lett 39(L16705):2012. https://doi.org/10.1029/2012GL052495

Poursanidis D, Chrysoulakis N (2017) Remote sensing, natural hazards and the contribution of ESA sentinels missions. Remote Sens Appl: Soc Environ 6:25–38. https://doi.org/10.1016/j.rsase.2017.02.001

Robichaud PR, Wagenbrenner JW, Pierson FB, Spaeth KE, Ashmun LE, Moffet CA (2016) Infiltration and interrill erosion rates after a wildfire in western Montana, USA. CATENA 142:77–88. https://doi.org/10.1016/j.catena.2016.01.027

Rodrigues M, de la Riva J (2014) An insight into machine-learning algorithms to model human-caused wildfire occurrence. Environ Model Soft 57:192–201. https://doi.org/10.1016/j.envsoft.2014.03.003

Rood SB, Foster SG, Hillman EJ, Luek A, Zanewich KP (2016) Flood moderation: Declining peak flows along some rocky mountain rivers and the underlying mechanism. J Hydrol 536:174–182. https://doi.org/10.1016/j.jhydrol.2016.02.043

Sayad YO, Mousannif H, Al Moatassime H (2019) Predictive modeling of wildfires: A new dataset and machine learning approach. Fire Saf J 104:130–146. https://doi.org/10.1016/j.firesaf.2019.01.006

Sayers P, Gersonius B, den Heijer F, Klerk WJ, Fröhle P, Jordan P, Ciocan UR, Rijke J, Vonk B, Ashley R (2021) Towards adaptive asset management in flood risk management: A policy framework. Water Secur 12:100085. https://doi.org/10.1016/j.wasec.2021.100085

Schipper L, Pelling M (2006) Disaster risk, climate change and international development: scope for, and challenges to, integration. Disasters 30(1):19–38. https://doi.org/10.1111/j.1467-9523.2006.00304.x

Shah SU, Yebra M, Van Dijk AIJM, Cary GJ (2022) A new fire danger index developed by random forest analysis of remote sensing derived fire sizes. Fire 5(5):152. https://doi.org/10.3390/fire5050152

Shakesby RA, Doerr S (2006) Wildfire as a hydrological and geomorphological agent. Earth-Sci Rev 74(3):269–307. https://doi.org/10.1016/j.earscirev.2005.10.006

Sharples JJ, C. G. F.-H. P et al (2016) Natural hazards in australia: extreme bushfire. Clim Change 139:85–99. https://doi.org/10.1007/s10584-016-1811-1

Sullivan AL (2009) Wildland surface fire spread modelling, 1990–2007. 2: Empirical and quasi-empirical models. Int J Wildland Fire 18:369–386

Tamiminia H, Salehi B, Mahdianpari M, Quackenbush L, Adeli S, Brisco B (2020) Google Earth Engine for geo-big data applications: A meta-analysis and systematic review. ISPRS J Photogram Remote Sens 164:152–170. https://doi.org/10.1016/j.isprsjprs.2020.04.001

Tan X, Gan TY (2015) Nonstationary analysis of annual maximum streamflow of canada. J Clim 28(5):1788–1805

Tavakkoli Piralilou S, Einali G, Ghorbanzadeh O, Nachappa TG, Gholamnia K, Blaschke T, Ghamisi P (2022) A google earth engine approach for wildfire susceptibility prediction fusion with remote sensing data of different spatial resolutions. Remote Sens 14(3). https://doi.org/10.3390/rs14030672.

Tecle A, Neary D (2015) Water quality impacts of forest fires. J Pollut Eff Control 3(3). https://doi.org/10.4172/2375-4397.1000140.

Tellman B, Sullivan J, Kuhn C, Kettner A, Doyle C, Brakenridge G, Erickson T, Slayback D (2021) Satellite imaging reveals increased proportion of population exposed to floods. Nature 596:80–86. https://doi.org/10.1038/s41586-021-03695-w

Turco M, Jerez S, Augusto S, Tarín-Carrasco P, Ratola N, Jiménez-Guerrero P, Trigo RM (2019) Climate drivers of the 2017 devastating fires in portugal. Sci Rep 9:13886. https://doi.org/10.1038/s41598-019-50281-2

Viedma O, Urbieta I, Moreno J (2018) Wildfires and the role of their drivers are changing over time in a large rural area of west-central Spain. Sci Rep 8:17797. https://doi.org/10.1038/s41598-018-36134-4

Vissa NK, A. P. B.-M et al (2019) ENSO-induced groundwater changes in India derived from GRACE and GLDAS. J Earth Syst Sci Vo 128(115). https://doi.org/10.1007/s12040-019-1148-z.

Williams AP, Abatzoglou JT, Gershunov A, Guzman-Morales J, Bishop DA, Balch JK, Lettenmaier DP (2019) Observed impacts of anthropogenic climate change on wildfire in California. Earth's Fut 7(8):892–910. https://doi.org/10.1029/2019EF001210

Wolski P, Murray-Hudson M, Thito K, Cassidy L (2017) Keeping it simple: Monitoring flood extent in large data-poor wetlands using MODIS SWIR data. Int J Appl Earth Obs Geoinf 57:224–234. https://doi.org/10.1016/j.jag.2017.01.005

Yan D, Werners SE, Ludwig F, Huang HQ (2015) Hydrological response to climate change: The Pearl River, China under different RCP scenarios. J Hydrol: Reg Stud 4:228–245. https://doi.org/10.1016/j.ejrh.2015.06.006

Yeung CA, Li R (2018) Comparison of vegetation regeneration after wildfire between mediterranean and tundra ecosystems by using landsat images. Ann GIS 24(2):99–112. https://doi.org/10.1080/19475683.2018.1424740

Zubkova M, Humber ML, Giglio L (2023) Is global burned area declining due to cropland expansion? how much do we know based on remotely sensed data? Int J Remote Sens 44(4):1132–1150. https://doi.org/10.1080/01431161.2023.2174389

Index

A
Accelerometers, 228
Activation function, 367
Active systems, 186
Adaptive Neuro-Fuzzy Inference System, 362
Advanced Microwave Scanning Radiometer, 205
Advanced Microwave Scanning Radiometer2, 205
Advanced Scatterometer, 205
Aerosol Optical Depth, 113
Africa, 11
Agenda for Sustainable Development, 213
Airbone, 185
Akosombo, 8
ALOS-PALSAR, 189, 198
Alos PalsarRTC-DEM, 121
Altimetry, 100
Amazon rainforest, 208
AMSR2, 205
AMSR-E, 205
Analytical hierarchy method, 194
Antarctic ice, 92
Antarctic ice sheet, 10
Antarctica, 3
Anthropocene, 5, 6, 409
Anthropocene era, 46
Anthropogenic drought, 388
Anthropogenic footprints, 7
Anthropogenic forcings, 7, 385
Aquitard, 200
Aral Sea, 9
Areas under drought, 395
Artificial drainage, 167
Artificial Neural Network, 108, 335, 361, 378
Artificial recharge, 199
Asia, 400
Atlantic Meridional Mode, 396
Atlantic Multi-Decadal Oscillation, 241, 285, 341, 396
Australia, 400
Australian Water Resources Assessment, 49
Automated machine learning, 335
Autoregressive exogenous, 335
Autoregressive integrated moving average, 106

B
Backcasting, 107
Back propagation, 368
Back Propagation Neural Network, 251, 353, 362
Backward reconstruction, 107
Bayesian data fusion, 95
Bayesian regularisation, 362
Biogeochemical cycles, 12
Blue water, 17, 59
Boosted regression trees, 107
Brazil, 9
Building resilience, 74
Burned Area Index for Sentinel-2, 437

C
Canadian Space Agency, 101
Carbon emissions, 389
Center for Space Research, 242
Centre National D'Etudes Spatiales, 101, 112
Central African rainforest, 15
CHAllenging Minisatellite, 89
CHAMP, 89
Chloride data, 162

C. Ndehedehe, *Hydro-Climatic Extremes in the Anthropocene*, Springer Climate,
https://doi.org/10.1007/978-3-031-37727-3

Chloride mass balance, 162
Climate adaptation, 169
Climate change, 4, 427
Climate oscillations, 32
Climate projections, 12, 111
Climate risk, 70
Climate teleconnections, 14
Cloud condensation nuclei, 119
CMIP6, 12, 394, 406
Coastal aquifers, 291
Coincidence map, 193, 194
Constituents constants, 366
Consumptive water, 17
Conventional Least Squares, 379
Convoluted neural network, 106
Convolutional Long Short-Term Memory, 108
Convolution kernel, 341
Convolution Neural Networks (CNN), 106, 335, 341
Coupled Model Intercomparison Project 6, 405
Coupled Model Intercomparison Project Phase, 388
Coupled Model Intercomparison Project Phase 6, 12, 225, 396
Critical value, 375
Cross-validation, 371
Crustal displacements, 208
Cubesats, 83

D
Dams, 10
Datum, 365
Deep neural network, 335
Deforestation, 15
Deutsches Zentrum fur Luft and Raumfahrt, 228
Dewatering, 5
Differenced Normalized Burn Ratio, 437
Diffuse percolation, 162
Direct recharge, 161
Discharge, 164
Diurnal, 361
Downscaling, 108
Drought assessments, 386
Drought duration, 387, 399
Drought events, 387
Drought extent, 399
Drought impact, 386
Drought metric, 399
Drought occurrence, 401
Drought onset, 387
Drought risk, 385

E
Earth Commission, 7, 409
Earth movements, 188
Earth resources satellites, 46
Earth system, 6, 7, 16, 409
East Africa, 42, 400
Economical water scarcity, 67
El Nino, 112
El Niño–Southern Oscillation, 3, 14, 32, 385
Empirical risk minimization, 327
Ensemble-based Kalman filter, 335
Ensemble Kalman filtering, 305
Ensemble Neural Network, 362
Environmental Science Services Administration, 81
ENVISAT, 86, 101
Epoch parameter, 372
Error threshold value, 372
ESSA, 81
European Drought Observatory, 49
European Flood Alert System, 49
European Remote Sensing Scatterometer, 205
European Space Agency, 47, 82, 192, 333
Eutrophication, 17
Extreme events, 190
Extreme gradient boosting, 335

F
Feed forward back propagation, 362
Feed-forward neural network, 367
Floods, 415, 426
Food scarcity, 39
Food security, 69, 224
Form number, 376
Forward prediction, 364
Fractured aquifers, 193
Freshwater, 224
Frobenius, 342

G
Gamma distribution function, 394
Gaussian process regression, 353
Generalized linear model, 335
Generalized Neural Network, 362
Generalized Regression Neural Network, 362

General regression neural networks, 335
Geocenter, 89
Geodetic pertubations, 188
Geophysical phenomena, 208
Geopotential, 89
Geopotential spherical harmonic Stokes coefficient, 395
Geospatial methods, 194
Glacial-isostatic adjustment, 230
Glacier, 3
Glacier mass, 3, 6
Glacier mass balance, 92
Global climate, 391
Global Climate Models (GCMs), 388, 391, 405
Global Climate Observation System, 111
Global Forest Watch, 15
Global groundwater, 395
Global hydrological models, 105
Global Land Analysis and Discovery, 15
Global Land Data Assimilation Model, 340
Global Navigation Satellite System, 207
Global population, 11
Global Positioning System (GPS), 89, 208
Global reservoirs and dams, 8
Global sea level, 7
Global warming, 225
Global water cycle, 17, 223
Global water resources, 64
Goddard Space Flight Center, 239
GRACE, 95, 230, 395
GRACE-FO, 107, 330
GRACE Follow On, 200
GRACE Groundwater Drought Index, 150
Gravimetry, 3
Gravity field and steady-state Ocean Circulation Explorer (GOCE), 89
Gravity models, 107
Gravity Recovery and Climate Experiment (GRACE), 82, 95, 195, 228, 230, 325, 395
Gray water, 17
Great Artesian Basin, 295
Greater Horn of Africa, 40
Greenland, 3
Greenland Ice Sheet, 10
Green water, 17
Gringorten plotting position, 394
Groundwater, 94, 95, 136, 407
Groundwater aquifers, 5, 96
Groundwater changes, 391
Groundwater dependent ecosystems, 94, 147
Groundwater depletion, 5, 13, 17, 407
Groundwater discharge, 164, 290, 291
Groundwater drought, 135, 387
Groundwater drought index, 150
Groundwater hydrology, 196
Groundwater level drought, 153
Groundwater level index, 153
Groundwater over-exploitation, 61
Groundwater pumping, 208
Groundwater recharge, 17, 160
Groundwater storage anomalies, 195

H

Harmonic constituents, 365, 376
Hidden layer, 367
Hidden neurons, 367, 372
High tides, 361
Human actions, 7
Human-managed reservoirs, 7
Human population, 11
Hurricane Harvey, 415
Hydrologic connectivity, 75
HydroSat, 122, 123
HydroSat database, 123
Hydrosphere, 81
Hyperspectral, 181

I

ICESat-2, 8
Ice sheet, 3
Independent component analysis, 249, 396
Indian Ocean Dipole, 14, 242, 285, 298, 341, 396
Indian Space Research Organisation, 187
Indirect recharge, 161
Input neuron, 367
Integrated machine learning, 337
Interconnection weights, 368
Interferograms, 86, 199
Interferometer, 118
Interferometric radiometer, 203
Interferometric Synthetic Aperture Radar (InSAR), 86, 97, 120
Intergovernmental Panel on Climate Change, 7, 394, 409
International Panel Climate Change, 62
International Space Station, 82
International Water Management Institute, 68
IPCC-AR4, 390
Isostatic corrections, 92

J

Jarque-Bera test, 363

K

KaRIn instrument, 104
Kling–Gupta efficiency, 121
Kurtosis, 370

L

Lagrange multiplier, 366
Lake Chad, 4, 34, 35, 68
Lake Urmia, 36
Lake Volta, 8
Land reclamation , 11
Landsat, 81
Landsat 7 ETM, 193
Landsat 8, 46, 193
Landsat 9, 46
Landscape transformations, 11
Land subsidence, 199
Land surface models, 93
Land use land cover, 13
Land water storage, 90
La Niña, 32, 33, 112
Lasso regularization, 343
Least squares, 361
LiDAR, 191
Linear statistical models, 335
Local agitations, 42
Localized recharge, 162
Low tides, 361

M

Machine learning, 108, 326
Machine learning downscaling, 88
Macro-pore flow, 162
Maiden Julian Oscillation, 396
Man-made lakes, 17
Marginal probability, 394
Mass change and Geosciences International Constellation (MAGIC), 90
Microwave satellites, 203
Middle East, 42, 400
Millennium Drought, 31, 390
Mitigation strategies, 210
Moderate Resolution Imaging Spectroradiometer, 114
MODIS16A2, 239
Monsoon rainfall, 91
Multi-angle imaging spectroradiometer, 114
Multi-influencing factors, 194
Multilayer perceptrons, 341
Multiple Linear Regression, 362
Multi-satellitE Retrievals for Global Precipitation Measurements, 239, 296
Multispectral, 185
Multivariate drought index, 394
Multivariate ENSO Index, 341
Murray–Darling basin, 35, 330

N

NASA-ISRO, 185
NASA-ISRO SAR, 48
National Aeronautics and Space Administration (NASA), 84, 228
National Center for Environmental Prediction, 282
National Water Agency, 9
Nigeria, 415
Ninos 1+2, 341
NISAR, 185
NOAA, 396
Nodal factor, 365
Non-linear autoregressive, 335
Non-linear regressions, 106
Non-parametric, 394
Nonparametric model, 362
Non-renewable groundwater, 17
North Africa, 400
North America, 400
North Atlantic Multi-decadal Oscillation, 285
North Atlantic Oscillation, 241, 341
North Atlantic Oscillation Index, 341
North Crimean Canal, 41
Numerical modeling, 163
Numerical Terradynamic Simulation Group, 239

O

Observational networks, 76
Observational record, 76
Ocean-atmosphere interactions, 390
Ocean bottom pressure, 92
Optical, 181
Orientations of fractures, 193
OSTM/Jason-2, 112
Output neurons, 367, 372
Overfitting, 333
Over-parameterization, 362

P
Pacific Decadal Oscillation, 3, 14, 241, 285, 299, 341, 396
Paleoclimate, 170
Palmer Drought Severity Index, 226
PALSAR, 191, 193
Partial least regression, 108
Passive remote sensing, 185
Passive systems, 185
Petabytes, 83
Physical water scarcity, 67
Polarimetric radar-imaging, 188
PolSAR, 186
Population, 11
Poroeslastic, 98
Predictands, 326, 351
Prediction, 326
Preferential flow, 162
Pumping, 147

Q
Quasi-Biennial Oscillation, 396

R
Radar altimetry, 92, 117
Radar remote sensing, 186
Radarsat-2, 199
Radial Basis Function, 362
Random Forest, 107, 335
Random seed number, 370, 372
Recharge, 4, 159
Recharge flux, 159
Recharge mechanisms, 161
Regression, 426
Regularized Neural Network, 362
RELU, 343
Remote sensing, 193
Remote sensing hydrology, 82
Renewable freshwater resource, 59
Reservoirs, 8
Retrieval algorithm, 202
River discharge, 123
Root Mean Squared Error, 372

S
Sahelian drought, 35
Salinization, 168
Satellite altimetry, 82
Satellite geodetic programmes, 82
Satellite geodetic system, 85
Satellite laser ranging, 89
Satellites, 181
Satellite systems, 81
Scanning Multichannel Microwave Radiometer, 205
SDGSAT-1, 47, 213
SDGSAT-1 satellite, 409
SDGSAT-1 sensors, 215
Sea level, 3, 10
Sea level heights, 362
Sea-level rise, 3, 10
Sea surface temperature, 341
Semidiurnal, 361, 364
Sentinel-1, 199
Shallow learning Feed Forward Neural Networks, 108
Shared Socioeconomic Pathways (SSPs), 388, 405
Sigmoid function, 367
Singular spectrum analysis, 333
SMOS-INRA-CESBIO, 203
Sobel filters, 193
Socioeconomic Pathways, 12
Soil moisture, 201
Soil Moisture Active Passive (SMAP), 203, 205
Soil Moisture and Ocean Salinity (SMOS), 203, 205
Space-based sensors, 81
SpaceX, 102
Special Sensor Microwave Imager, 205, 296
Sputnik, 81
SSP585 emission, 394
Standard Neural Network, 361
Standardised Evapotranspiration Index (SPEI), 391
Standardised runoff index, 394
Standardised soil moisture index, 394
Standardized Precipitation Index (SPI), 392
Standardized water-level index, 152
Statistical machine learning, 108
Stormwater harvesting, 68
Streamflows, 161
Subsidence, 208
Support vector machines, 332, 335
Support vector regressors, 342
Surface water, 33, 99, 196
Surface Water and Ocean Topography (SWOT), 48, 85, 101, 104
Surface water area variation, 123
Surface water extent, 123
Surface Water Ocean Topography, 101
Sustainable development goals, 213
SWOT mission, 118

Synthetic Aperture Radar (SAR), 85, 199

T

Taylor series, 366
Teleconnection, 341
Television and Infrared Observation Satellite (TIROS), 81
Terraclimate, 240
Terrestrial water balance, 118
Terrestrial water storage, 75, 92, 117, 195, 196, 228, 295, 325
Terrestrial water storage anomalies, 105
Thermal stress, 12, 13
Three Gorges Dam, 8
Threshold coefficients, 368
Tidal constituent, 365
Tidal form factor, 376
Tide, 361
Tide gauge, 364
Tide prediction, 361
TOPEX/Poseidon, 112
Total drainable water storage, 230
Total Solar Irradiance, 112
Training error, 370, 372
Transboundary water conflict, 43
Tritium data, 162
TRMM Microwave Imager, 296
Tropical Rainfall Measuring Mission, 117

U

UNICEF, 415
United Nations, 11
Unite Nations Climate Conference, 62
United Nation 2030 Agenda for Sustainable Development, 47
United Nations' Sustainable Development Goals, 409

V

Vegetation, 201
Vegetation optical depth, 203, 205
Vertical crustal displacements, 208
Visible and Infrared Scanner, 296

W

WATCH Forcing Data, 394
Water budget, 195
Water budget equation, 195
Water crises, 60
Water cycle, 62, 92
Water deficit, 70
Water demand, 72
Water diplomacy, 42, 71
Water extraction, 14
Water governance, 37
Water impoundment, 17
Water infrastructures, 70
Water level, 123
Water reservoirs, 18
Water resource management, 208
Water scarcity, 67
Water security, 72
Water storage anomaly, 123
Water table, 147
Water table fluctuation, 160
WaterGAP, 48
Wavelet artificial neural network, 106
Weight coefficient, 368
Weighted residuals, 366
West Africa, 209
West Antarctica, 92
WGHM, 105
Wildfire, 427
World Climate Conference-3, 62
World Economic Forum, 68
World meteorological organization, 393
World Wide Fund, 418